W0231914

CELL–CELL INTERACTIONS IN THE RELEASE OF INFLAMMATORY MEDIATORS

Eicosanoids, Cytokines, and Adhesion

ADVANCES IN EXPERIMENTAL MEDICINE AND BIOLOGY

Editorial Board:

NATHAN BACK, *State University of New York at Buffalo*

IRUN R. COHEN, *The Weizmann Institute of Science*

DAVID KRITCHEVSKY, *Wistar Institute*

ABEL LAJTHA, *N.S. Kline Institute for Psychiatric Research*

RODOLFO PAOLETTI, *University of Milan*

Recent Volumes in this Series

A Continuation Order Plan is available for this series. A continuation order will bring delivery of each new volume immediately upon publication. Volumes are billed only upon actual shipment. For further information please contact the publisher.

CELL–CELL INTERACTIONS IN THE RELEASE OF INFLAMMATORY MEDIATORS

Eicosanoids, Cytokines, and Adhesion

Edited by

Patrick Y-K Wong

New York Medical College
Valhalla, New York

and

Charles N. Serhan

Harvard Medical School
Boston, Massachusetts

PLENUM PRESS • NEW YORK AND LONDON

Library of Congress Cataloging-in-Publication Data

Cell-cell interactions in the release of inflammatory mediators :
  eicosanoids, cytokines, and adhesion / edited by Patrick Y-K Wong
  and Charles N. Serhan.
        p.   cm. -- (Advances in experimental medicine and biology ;
  v.314)
     "Based on the proceedings of the Federation of American Societies
  for Experimental Biology Symposium on Cell-Cell Interactions in the
  Release of Inflammatory Mediators, held April 1-6, 1990, in
  Washington, D.C."--T.p. verso.
     Includes bibliographical references and index.

     1. Inflammation--Mediators--Congresses.   2. Cytokines--Congresses.
  3. Eicosanoic acid--Derivatives--Congresses.   4. Cell adhesion-
  -Congresses.   I. Wong, Patrick Y.-K.   II. Serhan, Charles N.
  III. Federation of American Societies for Experimental Biology
  Symposium on Cell-Cell Interactions in the Release of Inflammatory
  Mediators (1990 : Washington, D.C.)
     [DNLM: 1. Cell Adhesion--immunology--congresses.   2. Cytokines-
  -immunology--congresses.   3. Eicosanoids--immunology--congresses.
  4. Immunity, Cellular--physiology--congresses.   5. Inflammation-
  -immunology--congresses.   QW 700 C3925 1990]
  RB131.C43 1991
  616'.0473--dc20
  DNLM/DLC
  for Library of Congress   91-45567
                                                                  CIP

Based on the proceedings of the Federation of American Societies for
Experimental Biology Symposium on Cell-Cell Interactions in the Release
of Inflammatory Mediators, held April 1-6, 1990, in Washington, D.C.

ISBN-13: 978-1-4684-6026-1   e-ISBN-13: 978-1-4684-6024-7
DOI: 10.1007/978-1-4684-6024-7

© 1991 Plenum Press, New York
Softcover reprint of the hardcover 1st edition 1991

A Division of Plenum Publishing Corporation
233 Spring Street, New York, N.Y. 10013

All rights reserved

No part of this book may be reproduced, stored in a retrieval system, or transmitted
in any form or by any means, electronic, mechanical, photocopying, microfilming,
recording, or otherwise, without written permission from the Publisher

PREFACE

This volume constitutes, in part, the proceedings of the symposium on "Cell-Cell Interaction and Release of Inflammatory Mediators" organized by Drs. Patrick Y-K Wong and Charles N. Serhan and presented at the FASEB meeting in Washington, D.C. in April, 1990. It contains chapters by the symposium speakers as well as contributions from investigators in this field.

Readers will find exciting advances in this volume, which contains chapters dedicated to state-of-the-art knowledge in the field of Cell-Cell Interaction and the functions of released mediators in inflammatory diseases. This book includes "cutting edge" investigations on transcellular eicosanoid biosynthesis, cytokines, PAF, and adhesion as well as interactions of inflammatory cells with endothelium, lung, and kidney. Also, the control and regulation of renal function by lipid mediators generated during cell-cell interactions between renal mesangial cells and leukocytes has generated insight into the cell biology and regulatory role of these mediators in the kidney. Moreover, the relationship between these areas is discussed in sequelae of both asthmatic and renal diseases.

We hope that some of the enthusiasm and excitement present in this research are also evident here and that this volume will serve as a reference for researchers, teachers, and students to survey this rapidly growing field.

The Editors

CONTENTS

LEUKOCYTE ADHESION: MOLECULAR BASIS AND RELEVANCE IN INFLAMMATION

Manuel Patarroyo, Lennart Lindbom and Claes
Lundberg

Depts. of Immunology and Physiology
Karolinska Institutet, Stockholm; and
Inflammation Res., Pharmacia, Uppsala, Sweden

INTRODUCTION

Leukocytes interact with one another, with other cell
types such as vascular endothelial cells, and with
extracellular matrices to traffic to extravascular tissues,
and to generate immune and inflammatory responses. Although
some of these interactions are mediated by soluble molecules
such as cytokines, others require firm leukocyte-cell or
leukocyte-matrix stickiness, a process refered to as
adhesion. This adhesiveness is transient and usually
subsequent to cell activation. It has a molecular basis and
a profound biological relevance in host defense and tissue
injury. The present article will summarize our studies on
the biology and molecular basis of leukocyte adhesion, and
its central role in leukocyte functions and inflammatory
responses. Comprehensive reviews have been recently
published by us[1-4] and by other scientists[5-7].

PHORBOL ESTER-LYMPHOCYTE AGGREGATION: AN ADHESION-SPECIFIC ASSAY

In early studies, phorbol ester was found to induce, at
nanomolar concentrations, aggregation of human blood
lymphocytes. The phenomenon occured within minutes, and was
preceded by induction of morphological changes, such as
uropod-like structures and membrane ruffles[8,9] (Fig.1).
Further analysis demonstrated that the cell aggregation was
a metabolic process which required participation of
microfilaments, preformed cell surface proteins and
extracellular divalent cations, mainly Mg^{++} but also Ca^{++}.
It was then concluded that phorbol ester was able to induce
a cell adhesive (binding) phenotype in lymphoid cells, and
that cell surface moieties, refered to as cell adhesion
(binding) molecules (CAMs), mediated this antigen-

Cell-Cell Interactions in the Release of Inflammatory Mediatiors
Edit ed by P. Y-K Wong and C.N. Serhan, Plenum Press, New York, 1991

1

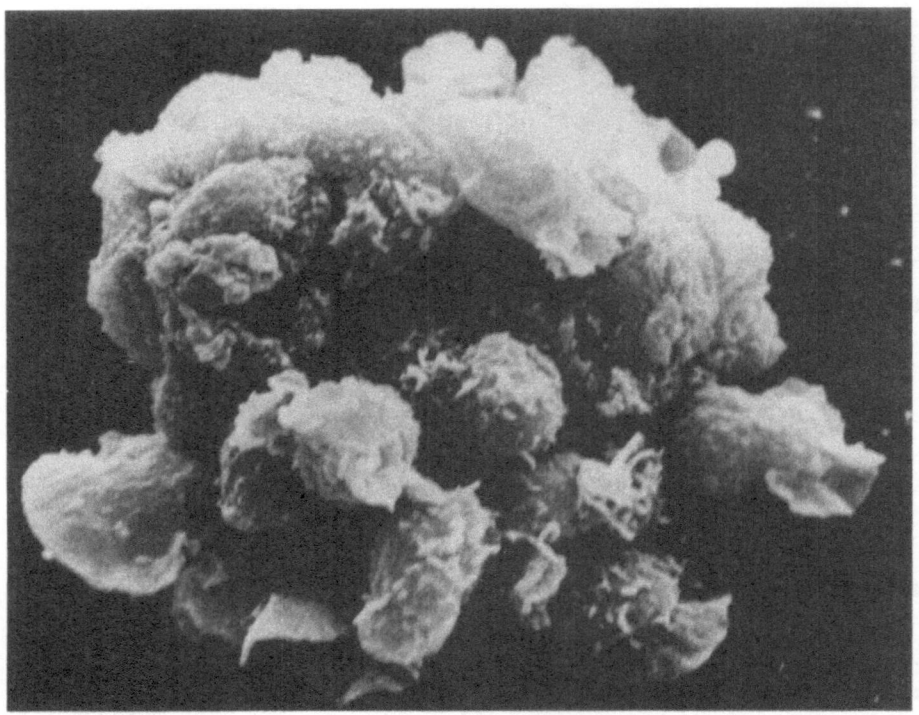

Fig. 1. Scanning electron micrograph of aggregated blood
lymphocytes after brief treatment with phorbol
ester. Note the uropod-like structures and
membrane ruffles of the aggregated cells.

independent cell clustering[1,8,9]. In parallel studies other
groups demonstrated that phorbol ester rapidly enters cells
and that once in the cytoplasm this compound behaves as
diacylglycerol, an endogenous intracellular messenger, and
permanently activates protein kinase C (PKC)[1]. Thus, phorbol
ester induces cellular responses, such as adhesion, by
bypassing cell surface molecules which, under physiological
conditions, mediate stimulus-recognition and activation
events. These characteristics made the phorbol ester-induced
lymphocyte aggregation an adhesion-specific assay, most
suitable for identifying the CAMs with the aid of blocking
antibodies.

Leu-CAMs (CD11/CD18 MOLECULES)

Mouse monoclonal antibody (mAb) 60.3 was the first mAb in
our hands, after testing more than 100, which was able to
block the phorbol ester-induced lymphocyte aggregation[10]
(Fig.2). The antibody had been recently reported by Beatty
el al.[11] and defined a novel cell surface specificity common
to human leukocytes, presently known as CD18. IgG and Fab
fragments were similarly effective but did not inhibit, as
expected, other simultaneous membrane processes, such as the
morphological changes and lateral redistribution of membrane
proteins[10]. Since the antibody also reacted strongly with

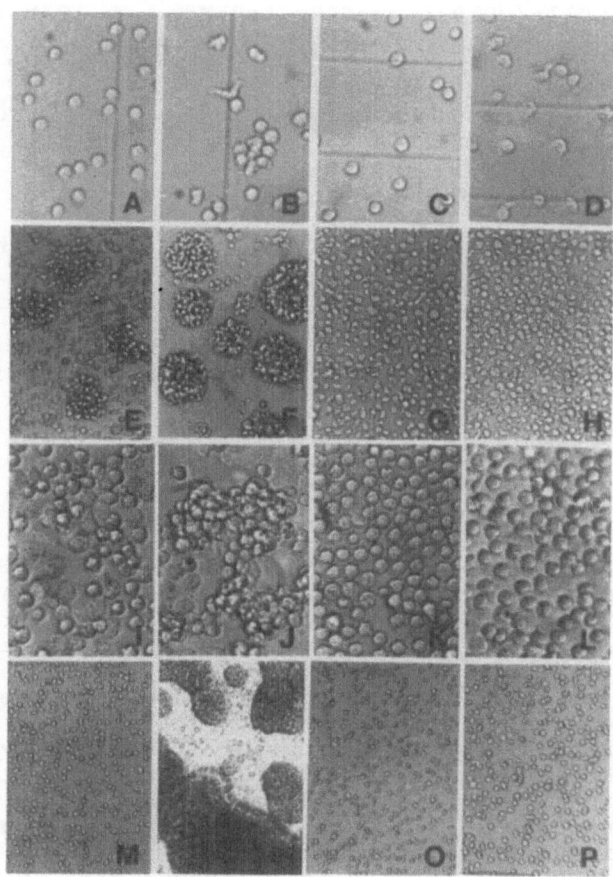

Fig. 2. Aggregation of differnt types of leukocytes
induced by phorbol ester treatment and effect of
Fab fragments of mAb 60.3 to CD18 on the
intercellular adhesion. A-D: blood T cells; E-H:
EBV-immortalized normal B cells; I-L: monocytes;
M-P: PMN. B, D, F, H, J, L, N and P were treated
with 60 nM phorbol ester for 20 min. C, D, G, H,
K, L, O and P were incubated with 20 µg/ml of
Fab fragments of mAb 60.3. Cells were rotated at
100 rpm at 37°C.

Table 1. The integrin family

Subfamilies	Members	Ligands
Leu-CAMs (β2)	CD11a/CD18 (Leu-CAMa,LFA-1)	ICAM-1(CD54),ICAM-2
	CD11b/CD18 (Leu-CAMb,Mac-1)	ICAM-1,iC3b,FX,FB,?
	CD11c/CD18 (Leu-CAMc,Leu-M5)	iC3b,?
VLA proteins (β1)	CD49-/CD29 (VLA-1)	LM,CO
	CD49b/CD29 (VLA-2)	LM,CO
	CD49-/CD29 (VLA-3)	FN,LM,CO
	CD49d/CD29 (VLA-4)	FN,VCAM-1
	CD49-/CD29 (VLA-5)	FN
	CD49f/CD29 (VLA-6)	LM
Cytoadhesins (β3)	CD41/CD61 (gpIIb/IIIa)	FB,FN,vWF
	CD51/CD61 (VNR)	VN,FN,vWF,TSP

LM, laminin; CO, collagen; FN, fibronectin; FB, fibrinogen; FX, factor X; VN, vitronectin; vWF, von Willebrand factor; TSP, thrombospondin.

polymorphonuclear leukocytes (PMN), its effect was tested on the phorbol ester-induced aggregation of these cells. Antibody 60.3 was equally inhibitory but did not affect the simultaneous superoxide generation and lysozyme release[12] (Fig.2). Interestingly, the phorbol-ester induced cell adhesion to plastic was also blocked, as well as the cell aggregation induced by the ionophore A23187 and the chemotactic tripeptide formyl-methionyl-leucyl-phenylalanine (fMLP)[12]. In collaboration with Dr. Carl Gahmberg at the Univ. of Helsinki, the antigen recognized by mAb 60.3 was further characterized. From unfractionated blood leukocytes the antibody immunoprecipitated four cell surface glycopolypeptides (GP) refered to as GP90, GP160, GP155 and GP130, and presently known as CD18, CD11a, CD11b and CD11c, respectively. Dissociation of the protein complex followed by immunoprecipitation indicated that antibody 60.3 reacted with GP90 (CD18), and that the larger glycopolypeptides were non-covalently associated. Due to the reactivity of the antibody with leukocytes exclusively, these structures were designated leukocytic-CAMs (Leu-CAMs). Comparative studies with mAbs produced by other research groups indicated that the Leu-CAMs corresponded to TA-1/LFA-1 (CD11a/CD18), Mac1/OKM1/Mo1 (CD11b/CD18) and p150,95/LeuM5 (CD11c/CD18), three cell surface protein heterodimers with distinct α-subunits and a common β-subunit[1,5,6] (Table 1). The adhesive function of these molecules had not been unambiguosly demonstrated. In additional studies, the aggregation of Epstein Barr virus (EBV)-immortalized normal B cells and of monocytes (Fig.2), and the adhesion of T lymphocytes to the

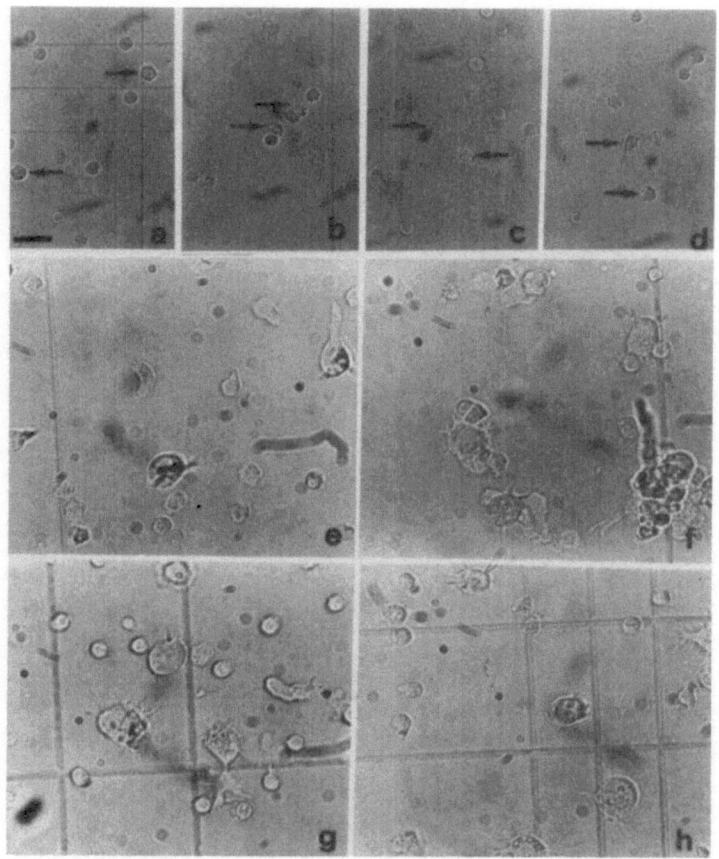

Fig. 3. Adhesion of blood T cells to monocytes and EBV-
immortalized normal B cells induced by phorbol
ester treatment, and inhibition of the
intercellular adhesion by Fab fragments of mAb
60.3 to CD18. a-d: T cells and monocytes
(indicated by arrows); e-h: T cells and EBV-
immortalized normal B cells (large cells).
b,d,f,h: cells were treated with 60 nM phorbol
ester for 20 min. c,d,g,h: cells were incubated
with 20 µg/ml of Fab fragments of mAb 60.3.

former cells[13,14] (Fig.3), as well as of each type of
leukocyte to vascular endothelial cells[15] (Fig.4) in
presence of phorbol ester was analysed. Fab fragments of
antibody 60.3 were always inhibitory.

CD11a/CD18 is expressed by most leukocytes, but
strongly by lymphocytes and monocytes. Myelomonocytic cells
and natural killer cells also express CD11b/CD18 and
CD11c/CD18. Interestingly, the former cells contain, in
addition to the cell surface CD11/CD18 molecules,
intracellular pools of CD11b/CD18 and CD11c/CD18[1,5,6].

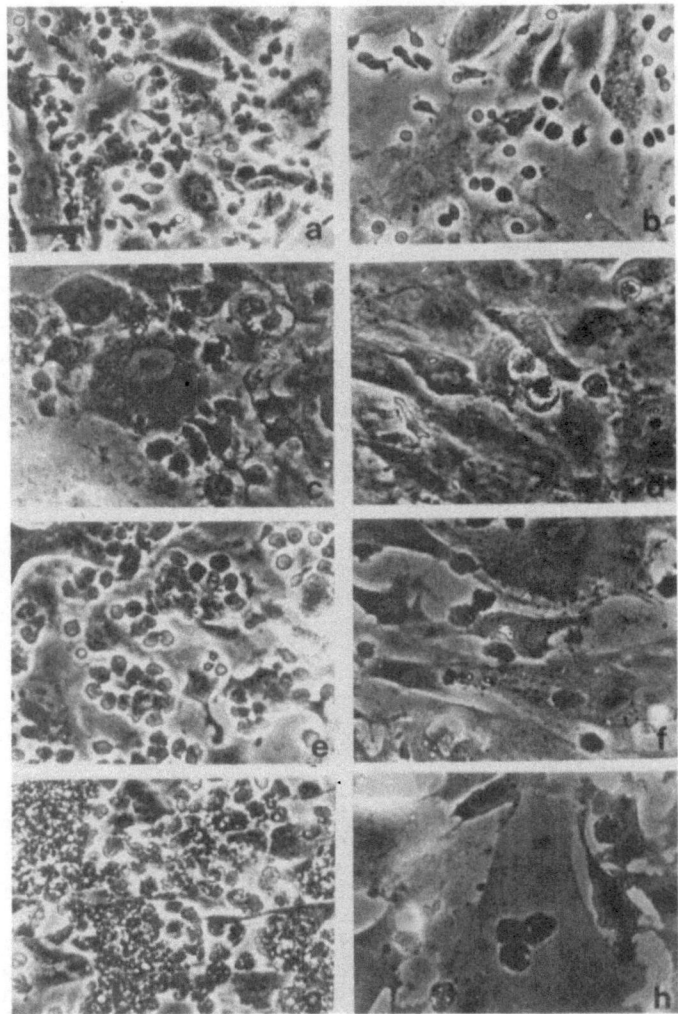

Fig. 4. Adhesion of different types of leukocytes to
vascular endothelial cells induced by phorbol
ester treatment and effect of Fab fragments of
mAb 60.3 to CD18 on the intercellular binding.
a,b: T cells; c,d: EBV-immortalized normal B
cells; e,f: monocytes; g,h: PMN. b,d,f,h: cells
were incubated with 20 μg/ml of Fab fragments.
All cells were treated with 60 nM phorbol ester
for 20 min at 37°C.

Table 2. Other adhesion molecules

Receptor	Ligand
CD2 (Leu-5,OKT11,LFA-2)	CD58 (LFA-3)
CD56 (Leu-19,N-CAM)	CD56
LAM-1 (Leu-8,MEL-14)	?
Syalyl-Lewis X determinant	ELAM-1
CD15 (lacto-N-fucopentaose III)	CD62 (GMP-140)
CD44 (H-CAM,Pgp-1)	Hyaluronan,CO

N-CAM, neural-CAM; H-CAM, homing associated-CAM; CO, collagen; LAM, lymphocyte associated molecule; ELAM, endothelial leukocyte adhesion molecule; GMP, granule membrane protein.

Comparative structural analysis has demonstrated that the Leu-CAMs belong to a large family of adhesive protein heterodimers known as integrins (Table 1)[4-6]. This family also includes the VLA proteins and the cytoadhesins. Most of these molecules are strongly expressed by nonhemopoietic cells, but are also expressed by leukocytes. Among the VLA proteins, VLA-4 is widely found on lymphocytes and monocytes. This molecule is a receptor for vascular cell adhesion molecule 1 (VCAM-1), an adhesive ligand expressed by endothelial cells following stimulation with IL-1. VLA proteins and cytoadhesins are major receptors for extracellular matrix components such as fibronectin, collagen, laminin and vitronectin. Expression of most of the integrins by white cell indicates interaction of the leukocytes with matrices.

ICAM-1 (CD54)

In additional studies, a mAb to the human lymphoblast antigen LB-2 was found to block the phorbol ester-induced aggregation of lymphoblasts, and hence to identify another adhesion molecule[16]. CD54, the antigen, had been originally reported by Clark et al. in human (BB-2/LB-2)[17,18], and by Takei in mouse (MALA-2)[19]. Comparative studies indicated that this adhesion molecule corresponded to intercellular adhesion molecule 1 (ICAM-1), an adhesion molecule which had been recently identified by Dr. Tim Springer's group at Harvard University in Boston by using the phorbol ester-adhesion assay[6]. CD54 is a single glycoprotein weakly expressed by blood monocytes, NK cells and other lymphocytes. In frozen sections CD54 is detected in germinal center B cells, dendritic cells, macrophages and vascular endothelium, including high endothelial venules (HEV). Its expression is selectively induced or increased within hours by inflammatory mediators such as interferon γ (IFNγ), interleukine 1β (IL-1β) and tumor necrosis factor α (TNFα) in lymphoid, epithelial and endothelial cells, and by IL-2 in T lymphocytes and NK cells[1,4]. ICAM-1 is an adhesive ligand for Leu-CAMa (CD11a/CD18), and also appears to bind to Leu-CAMb (CD11b/CD18) (Table 1)[4,6]. Together with Dr. Fumio Takei at Terry Fox Laboratory in Vancouver, mouse ICAM-1 was similarly identified[20]. More recently, Springer's group found ICAM-2, another adhesive ligand for CD11a/CD18, by using a functional cDNA cloning assay[6]. Further studies together with Dr. Carl Gahmberg have indicated that ICAM-2 is a cell surface glycoprotein with an apparent molecular weight of 55.000 expressed by endothelial cells and lymphoblasts[21].

OTHER ADHESION MOLECULES

The complexity of leukocyte adhesion is illustrated by the existance of additional adhesion molecules, shown in Table 2. CD2, which is mainly expressed by T cells, mediates adhesion to a large variaty of cells, by recognizing CD58 on the latter cells[1,4,6]. CD56, an isoform of neural cell adhesion molecule (N-CAM), is mainly expressed by NK cells, and binds to CD56-positive cells, in a homophilic fashion[4]. Three related molecules known as selectins or LEC-CAMs, recognize carbohydrate ligands and mediate adhesion between leukocytes and endothelial cells[4,6]: Lymphocyte associated molecule-1 (LAM-1, MEL-14) is expressed by most leukocytes and participates in lymphocyte homing into lymphoid tissues. Endothelial leukocyte adhesion molecule 1 (ELAM-1) is expressed by endothelial cells within a few hours after stimulation with inflammatory cytokines, while CD62 expression is induced on both endothelial cells and platelets a few minutes after stimulation. Both ELAM-1 and CD62 mediate adhesion between endothelial and myelomonocytic cells. CD44, a molecule structurally unrelated to the previous ones, is expressed by most leukocytes, functions as a receptor for hyaluronan and collagen and may also participate in lymphocyte homing[4]. Leukocytes can use more than one adhesion pathway simultaneously.

Table 3. Adhesion-dependent functions of leukocytes

-Leukocyte-mediated cytotoxicity (CTL, NK, LAK, ADCC).
-Ag-induced lymphocyte proliferation.
-Helper activity for Ig production.
-Ag-induced lymphokine production.
-Leukocyte binding to vascular endothelium, leukocyte-dependent plasma leakage and leukocyte extravasation.
-PMN-mediated endothelial cell injury.
-Leukocyte spreading, migration and chemotaxis.
-Lymphocyte homing, localization and recirculation.
-Phagocytosis of large particles.
-Substrate-induced oxidative burst and deganulation.

MECHANISMS AND REGULATION OF LEUKOCYTE ADHESION

Adhesion mediated by the Leu-CAMs seems to be the result of a conformational change of these molecules at the cell surface. Although translocation of intracellular pools of CD11b/CD18 and CD11c/CD18 to the plasma membrane in myelomonocytic cells is associated to the adhesion process, it is not necessary for adhesion to occur[4,5,22]. While chemoattractants such as fMLP, C5a and IL-8 stimulate Leu-CAMb-dependent PMN adhesion, specific antigen induces a Leu-CAMa-dependent adhesion of lymphocytes[4-7]. Interestingly, H_2O_2 also induces a Leu-CAMb-dependent adhesion of myelomonocytic cells[23], and the rapid adhesive response of both PMN and endothelial cells induced by LTB_4 is mediated by CD11/CD18 molecules[24].

Another regulatory level of adhesion is induction of cell surface expression, by translocation or de novo synthesis, of the adhesion molecules. While CD62 expression is induced within minutes, expression of ELAM-1, ICAM-1 and VCAM-1 is induced by inflammatory cytokines after hours of stimulation[4,6]. Use of a particular adhesion pathway, with characteristic kinetics and leukocyte type involved, appears to be determined by the original stimulus and its target cell.

ADHESION-DEPENDENT FUNCTIONS OF LEUKOCYTES

Most leukocyte functions require adhesion (Table 3). These complex phenomena have been studied by us and others both in vitro and in vivo by using blocking or stimulatory mAbs to the adhesion molecules[1-7]. Cytotoxic leukocytes have to adhere to their targets (virus-infected cells, allogeneic cells, or tumor cells) to kill them. In collaboration with Dr. Tuomo Timonen at the University of Helsinki, NK cells have been found to use the three Leu-CAMs, CD2, ICAM-1 and receptors for RGD, a motif of some matrix proteins, to bind and lyse tumor cells[25]. Monocyte-macrophages and allogeneic cells need to adhere to T cells to present antigen and to induce proliferation of the lymphocytes. Induction of IFNγ and IL-2 production by T lymphocytes also requires adhesion molecules[1,26], and T cells interact physically with B cells to induce maturation of the latter cells and antibody production. Moreover, stimulation of CD11/CD18, CD44 and CD58 adhesion molecules in monocytes results in production of IL-1β and TNFα[4]. CD11b/CD18 also participates in phagocytosis. Large particles have to adhere to the phagocyte surface before they are ingested. In addition, H_2O_2 production and degranulation in PMN induced by physiological stimuli are dependent on Leu-CAMb-mediated adhesion[4,22,27,28], and leukocytes adhere to substrates to obtain the traction that allows migration and chemotaxis. Nonlytic injury (detachment) of endothelial cells by activated PMN also requires CD11/CD18.

The fundamental role of CD11/CD18 molecules in vivo was directly demonstrated by Arfors et al.[29] by using mAb 60.3 to CD18. PMN accumulation in the rabbit, as measured by myeloperoxidase, following i.d. injections of purified

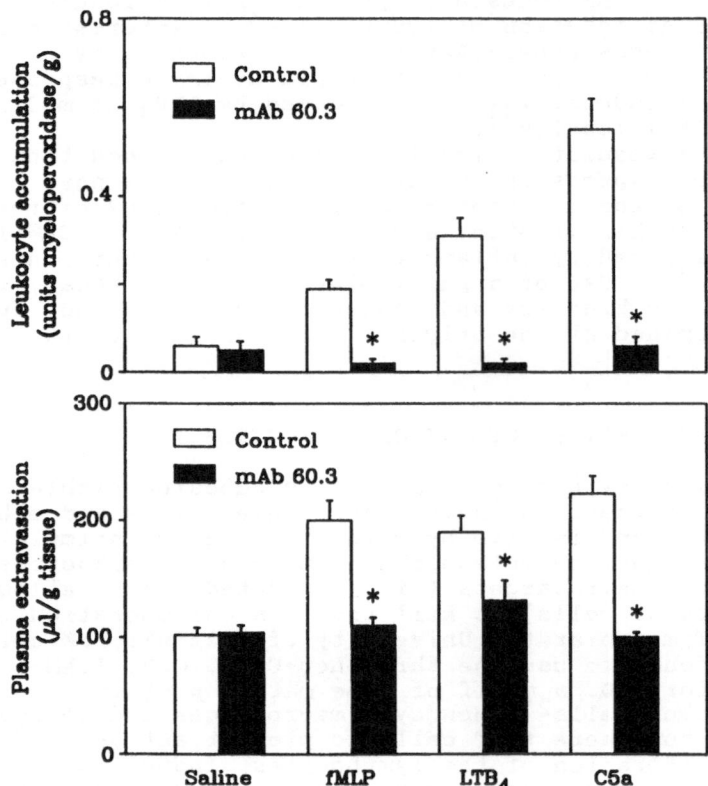

Fig. 5. Leukocyte accumulation and plasma extravasation
following intradermal injections of inflammatory
mediators in the dorsal skin region of the
rabbit. The animals were pretreated with either
saline (control) or mAb 60.3 (2 mg/kg)
intravenously 10 minutes prior to the i.d.
injections. Values represent mean ±SEM of six
animals.

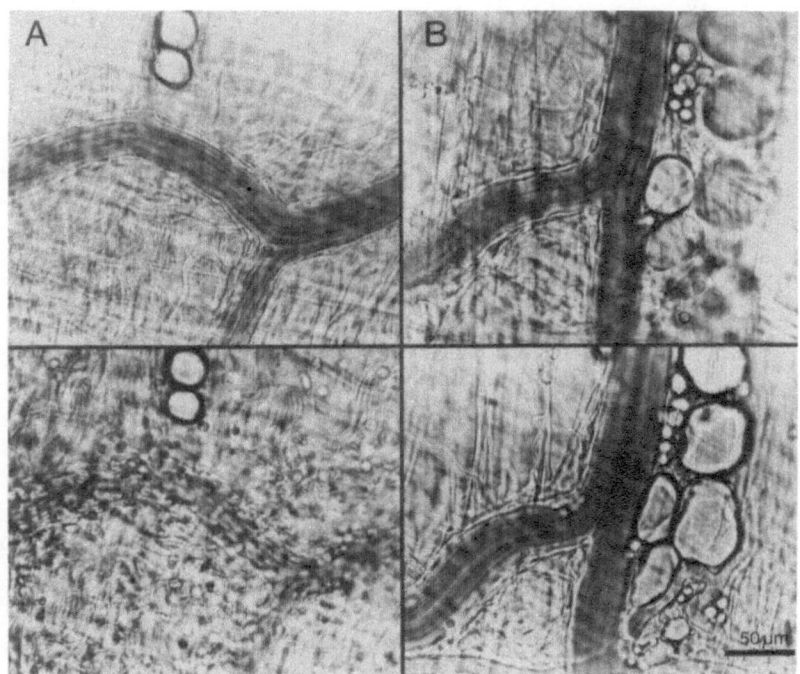

Fig. 6. <u>In vivo</u> micrographs of venules in the rabbit
        tenuissimus muscle before (upper panels) and 15
        min after (lower panels) topical application of
        chemotactic stimulus. A: animal pretreated
        intravenously with saline. B: animal pretreated
        with mAb to CD18 (5 mg/Kg). Note profound
        adherence of leukocytes in the saline-treated
        animal (lower left) and lack of adherent cells
        in the antibody-treated animal (lower right).

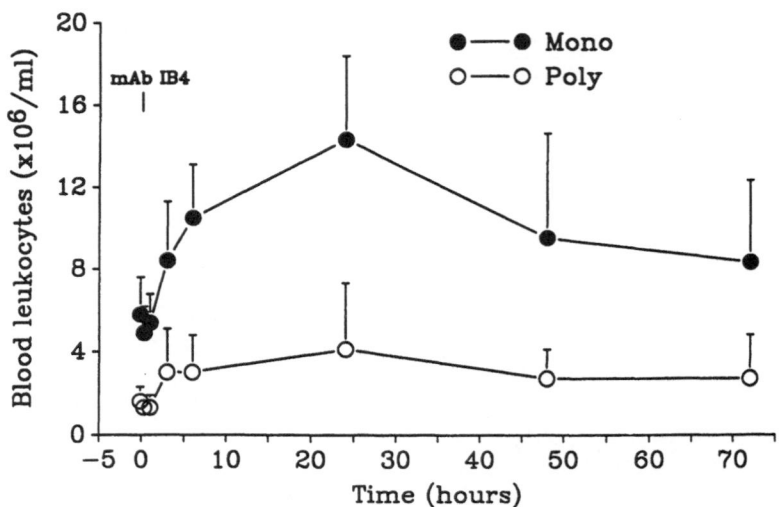

Fig. 7. Number of circulating mononuclear and polymorphonuclear leukocytes in peripheral blood during a 72 h-period after intravenous treatment with a mAb to CD18 (5 mg/kg). Values represent mean±SD of six animals.

inflammatory substances, i.e. fMLP, LTB$_4$ and C5a was totally abolished by pretreatment of the animal with the antibody (Fig.5). Moreover, PMN-dependent plasma leakage induced by these mediators was also inhibited (Fig.5). This _in vivo_ function of CD11/CD18 has later been confirmed in similar models[4,7,30,31]. Intravital microscopy of the rabbit tenuissimus muscle superfused with inflammatory mediators has clearly demonstrated that the effect of the anti-CD18 mAb relies on inhibition of leukocyte adhesion to post-capillary venular endothelium (Fig.6)[29,30]. Lack of the adhesive function precludes further steps in the chemotactic response of the stimulated leukocytes, and thus prevents these cells from reaching the interstitial space. Rolling of leukocytes along the venular endothelium was unaffected by anti-CD18 mAb treatment, indicating that this physical interaction between leukocytes and endothelial cells has a different molecular basis. Interestingly, CD11/CD18 molecules do not participate in the transient neutropenia associated with PMN accumulation in the lung, caused by systemically injected fMLP[31]. We have observed a major increase in the number of peripheral blood lymphocytes 24 hours after i.v. injection of anti-CD18 mAbs in rabbits (Fig.7)[30]. This redistribution of lymphocytes is probably due to interference with lymphocyte homing. We have further found that tissue swelling in response to a delayed-type hypersensitivity reaction (DTH), which is mainly T-lymphocyte mediated, is reduced in rabbits pretreated with anti-CD18 mAbs (Fig.8)[30].

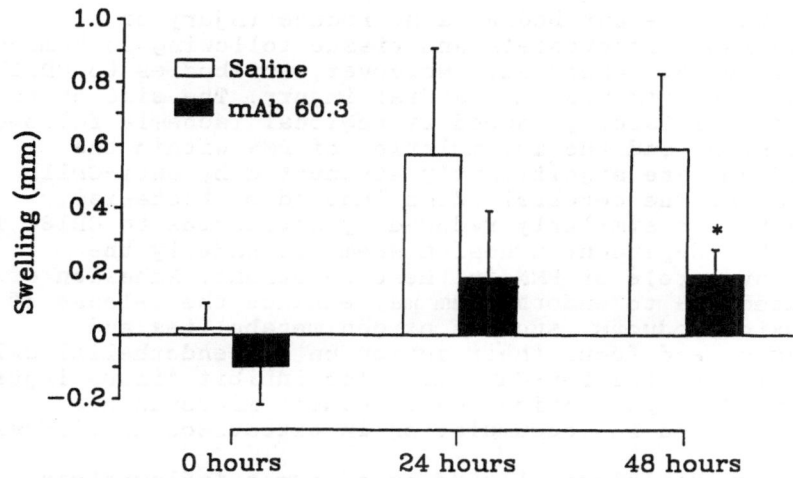

Fig. 8. Swelling of ear tissue after local challenge
with tuberculin PPD in _Mycobacterium
tuberculosis_-immunized animals systematically
treated with saline or mAb 60.3 to CD18.
Swelling was measured prior to and at 24 and 48
hr after challenge, and calculated as the paired
difference between thickness of PPD-injected ear
and saline-injected contalateral ear. Values
represent mean±SD of four animals in each
group.

LEUKOCYTE ADHESION IN PATHOLOGICAL INFLAMMATORY REACTIONS

The effect of adhesion blocking mAbs on noxious
inflammatory reactions has been tested in several animal
models by various groups (Table 4)[7]. Anti-CD18 antibodies
reduce organ injury and improve survival from hemorrhagic
shock and resuscitation. In animals treated with these
antibodies, the gut and liver injuries caused by the
generalized ischemia-reperfusion are either absent or much

Table 4. Effect of mAbs to leukocyte adhesion molecules
on noxious immune and inflammatory reactions

-Prevention of death and reduction of tissue (gut and
 liver) injury following hemorrhagic shock and
 resuscitation in rabbits and primates (CD18 mAb).
-Reduction of injury to endothelium, microvessels and
 tissue following ischemia-reperfusion of cat intestine
 and rabbit ear (CD18 mAb).
-Reduction of cerebral edema induced by bacterial
 meningitis in rabbits (CD18 mAb).
-Reduction of infarct size following ischemia-reperfusion
 of canine myocardium (CD11b mAb).
-Delay of renal allograft rejection in primates (CD54
 mAb).

---

attenuated. The antibodies also reduce injury of endothelium, microvessels and tissue following ischemia-reperfusion of rabbit ear. Moreover, antibodies to CD11b decrease experimental myocardial injury. The size of the myocardial infarct produced by regional ischemia followed by reperfusion, and the accumulation of PMN within the myocardium, are significantly attenuated by anti-CD11b antibodies. The cerebral edema induced by bacterial meningitis is similarly reduced by antibodies to CD18. Thus, CD11b/CD18-dependent adhesion seems to underly the detrimental role of PMN in these reactions. Adhesion of activated PMN to endothelium may enhance the release of PMN cytotoxic products, such as oxygen metabolites and proteases, and focus their action on the endothelial cells. Antibodies to the Leu-CAMs may also inhibit tissue injury by blocking PMN aggregation, which causes microvascular occlusion, and PMN accumulation in extravascular tissues.

Immunohistological studies of human inflammatory disorders such as thyroiditis, dermatitis, nephritis, encephalitis and asthma have recently demonstrated increased ICAM-1 expression in the inflamed tissues. In these histological studies, the distribution of ICAM-1 correlates topographically with areas of tissue damage and infiltrating leukocytes[4].

CONCLUDING REMARKS

Adhesion to cells and matrices is fundamental in the physiology of T and B lymphocytes, NK cells, monocyte-macrophages and PMN, and contributes to the generation of inflammatory responses. Leukocyte adhesion participates in the release of cytokines, oxygen metabolites and granule components and, in turn, some of these inflammatory mediators induce leukocyte adhesion. Participation of adhesion in several leukocyte responses induced by physiological stimuli can be partially explained by the signalling function of the adhesion molecules[4,32]. Further studies will analyse the participation of adhesion in the production of lipid inflammatory mediators, such as eicosanoids and platelet-activating factor. It is likely that adhesion of PMN to platelets and/or vascular endothelium contributes to the transcellular and cooperative synthesis of leukotrienes and lipoxins[33-35].

ACKNOWLEDGEMENTS

This work was supported by the Swedish Cancer Society, the Medical Research Council and the Karolinska Institute.

REFERENCES

1. M. Patarroyo and W. Makgoba, Leukocyte adhesion to cells: Molecular basis, physiological relevance and abnormalities, Scand J Immunol 30:129 (1989).
2. M. Patarroyo and W. Makgoba, Leukocyte adhesion to cells in immmune and inflammatory responses, The Lancet ii 1139 (1989).

3. M. Patarroyo, J. Prieto, J. Rincon, T. Timonen, C. Lundberg, L. Lindbom, B. Åsjö and C. Gahmberg, Leukocyte-cell adhesion: A molecular process fundamental in leukocyte physiology, _Immunol Rev_ 114:67 (1990).

4. M. Patarroyo, Leukocyte adhesion in host defense and tissue injury, _Clin Immunol Immunopath_, 60:333 (1991).

5. A. Arnaout, Structure and function of the leukocyte adhesion molecules CD11/CD18, _Blood_ 75:1037 (1990).

6. T. Springer, Adhesion receptors of the immune system, _Nature_ 346:425 (1990).

7. T. M. Carlos and J. Harlan, Membrane proteins involved in phagocyte adherance to endothelium, _Immunol Rev_ 114:5 (1990).

8. M Patarroyo, G. Yogeeswaran, P. Biberfeld, E. Klein and G. Klein, Morphological changes, cell aggregation and cell membrane alterations caused by phorbol 12,13-dibutyrate in human blood lymphocytes, _Int J Cancer_ 30:707 (1982).

9. M. Patarroyo, Effects of Epstein Barr virus infection and phorbol ester treatment on the lytic and binding interactions between human lymphoid cells, _PhD Thesis_, Karolinska Institutet, Stockholm (1982).

10. M. Patarroyo, P. Beatty, J. Fabre and C. Gahmberg, Identification of a cell surface protein complex mediating phorbol ester-induced adhesion (binding) among human mononuclear leukocytes, _Scand J Immunol_ 22:171 (1985).

11. P. Beatty, J. Ledbetter, P. Martin, T. Price and J. Hansen, Definition of a common leukocyte cell surface antigen (Lp95-150) associated with diverse cell-mediated immune functons, _J Immunol_ 131:2913 (1983).

12. M. Patarroyo, P. Beatty, C. Serhan and C. Gahmberg, Identification of a cell surface glycoprotein mediating adhesion in granulocytes, _Scand J Immunol_ 22:619 (1985).

13. M. Patarroyo, P. Beatty, K. Nisson and C. Gahmberg, Identification of a cell-surface glycoprotein mediating cell-adhesion in EBV-immortalized normal B cells, _Int J Cancer_ 38:539 (1986).

14. M. Patarroyo, J. Prieto, P. Beatty, E. Clark and C. Gahmberg, Adhesion-mediating molecules of human monocytes, _Cell Immunol_ 113:278 (1988).

15. J. Prieto, P. Beatty, E. Clark and M. Patarroyo, Molecules mediating adhesion of T and B cells, monocytes and granulocytes to vascular endothelial cells, _Immunol_ 63:631 (1988).

16. M. Patarroyo, E. Clark, J. Prieto, C. Kantor and C. Gahmberg, Identification of a novel adhesion molecule in human leukocytes by monoclonal antibody LB-2, _FEBS Lett_ 210:127 (1987).

17. E. Clark and T. Yokochi, Human B cell and B cell blast-associated surface molecules defined with monoclonal antibodies, _in_: "Leukocyte typing", A. Bernard et al. eds., Springer-Verlag, Berlin, p. 339 (1984).

18. E. Clark, J. Ledbetter, R. Holly, P. Dinndorf and G. Shu, Polypeptides on human B cell lymphocytes

associated with cell activation, *Hum Immunol* 16:100 (1986).

19. F. Takei, Inhibition of mixed lymphocyte response by a rat monoclonal antibody to a novel murine lymphocyte activation antigen (MALA-2), *J Immunol* 134:1403 (1985).

20. J. Prieto, F. Takei, R. Gendelman, B. Christenson, P. Biberfeld and M. Patarroyo, MALA-2, mouse homologue of human adhesion molecule ICAM-1 (CD54), *Eur J Immunol* 19:1551 (1989).

21. P. Nortamo, R. Salcedo, T. Timonen, M. Patarroyo and C. Gahmberg, A monoclonal antibody to the human leukocyte adhesion molecule ICAM-2. Cellular distribution and characterization of the antigen, *J Immunol*, 146:2530 (1991).

22. B. Schleiffenbaun, R. Moser, M. Patarroyo and J. Fehr, The cell surface glycoprotein Mac-1 (CD11b/CD18) mediates neutrophil adhesion and modulates degranulation independently of its quantitative cell surface expression, *J Immunol* 142:4100 (1989).

23. G. Skoglund, I. Cotgreave, J. Rincon, M. Patarroyo and M. Ingelman-Sundberg, H_2O_2 activates CD11b/CD18-dependent cell adhesion, *Biochem Biophys Res Comm* 157:443 (1988).

24. P. Lindström, R. Lerner, J. Palmblad and M. Patarroyo, Rapid adhesive responses of endothelial cells and of neutrophils induced by LTB_4 are mediated by leukocytic adhesion protein CD18, *Scand J Immunol* 31:737 (1990).

25. T. Timonen, C. Gahmberg and M. Patarroyo, Participation of CD11a-c/CD18, CD2 and RGD-binding receptors in endogenous and interleukine-2-stimulated NK activity of CD3-negative large granular lymphocytes, *Int J Cancer* 46:1035 (1990).

26. C. Lindqvist, M. Patarroyo, P. Beatty and H. Wigzell, A monoclonal antibody inhibiting leukocyte adhesion blocks induction of IL-2 production but not IL-2 receptor expression, *Immunol* 60:579 (1987).

27. S. Shappell, C. Toman, D. Anderson, A. Taylor, M. Entman and W. Smith, Mac-1 (CD11b/CD18) mediates adherence-dependent hydrogen peroxide production by human and canine neutrophils, *J Immunol* 144:2702 (1990).

28. J. Richter, J. Ng-Sikorski, I. Olsson and T. Andersson, Tumor necrosis factor-induced degranulation in adherent human neutrophils is dependent on CD11b/CD18-integrin-triggered oscillations of cytosolic free Ca^{++}, *Proc Natl Acad Sci USA* 87:9472 (1990).

29. K. Arfors, C. Lundberg, L. Lindbom, K. Lundberg, P. Beatty and J. Harlan, A monoclonal antibody to the membrane glycoprotein complex CD18 inhibits polymorphonuclear leukocyte accumulation and plasma leakage *in vivo*, *Blood* 69:338 (1987).

30. L. Lindbom, C. Lundberg, J. Prieto, J. Raud, P. Nortamo, C. Gahmberg and M. Patarroyo, Rabbit leukocyte adhesion molecules CD11/CD18 and their participation in acute and delayed inflammatory

responses and leukocyte distribution *in vivo*, <u>Clin Immunol Immunopath</u> 57:105 (1990).

31. C. Lundberg and S. Wright, Relation of the CD11/CD18 family of leukocyte antigens to the transient neutropenia caused by chemoattractants, <u>Blood</u> 76:1240 (1990).

32. J. Ng-Sikorski, R. Andersson, M. Patarroyo and T. Andersson, Calcium signalling capacity of the CD11b/CD18 integrin on human neutrophils, <u>Exp Cell Res</u> 195:504 (1991).

33. A. Fradin, J. Zirrolli, J. Maclouf, L. Vausbinder, P. Henson and R. Murphy, Platelet-activating factor and leukotriene biosynthesis in whole blood. A model for the study of transcellular arachidonate metabolism, <u>J Immunol</u> 143:3680 (1989).

34. F. Grimminger, B. Kreusler, U. Schneider, G. Becker and W. Seeger, Influence of microvascular adherence on neutrophil leukotriene generation. Evidence for cooperative eicosanoid synthesis, <u>J Immunol</u> 144:1866 (1990).

35. S. Fiore and C. Serhan, Formation of lipoxins and leukotrienes during receptor-mediated interactions of human platelets and recombinant human granulocyte/macrophage colony-stimulating factor-primed neutrophils, <u>J Exp Med</u> 172:1451 (1990).

PRODUCTS OF INFLAMMATORY CELLS SYNERGISTICALLY ENHANCE SUPEROXIDE
PRODUCTION BY PHAGOCYTIC LEUKOCYTES

John A. Badwey[1,2], Jiabing Ding[1], Paul G. Heyworth[3] and
John M. Robinson[4]

[1] Department of Cell Physiology, Boston Biomedical Research
Institute; [2]Department of Biological Chemistry and Molecular
Pharmacology, Harvard Medical School, Boston, MA; [3]Department
of Molecular and Experimental Medicine, Research Institute of
Scripps Clinic, La Jolla, CA; and [4]Department of Cell
Biology, Neurobiology and Anatomy, Program in Molecular,
Cellular and Developmental Biology and the Ohio State
Biochemistry Program, The Ohio State University, Columbus, OH

Superoxide (O_2^-) is a major component of the oxygen-dependent
antimicrobial and cytocidal arsenal of neutrophils [1,2]. The oxidase
system that generates this substance is dormant and disassembled in
unstimulated cells and consists of both membrane-bound and soluble
("cytosolic factors") components[3,4]. The known membrane-components are a
low-potential, heterodimeric b-cytochrome[5,6] and a ras-related GTP-
binding protein[7]. The most thoroughly characterized cytosolic factors
are proteins with molecular masses of 47 (p47) and 67kDa[8-11]. Upon
stimulation of neutrophils, there is a translocation of the soluble
components to the plasmalemma where the oxidase is assembled[12,13] (Figure
1). This assembly requires the presence of the b-cytochrome[12,14] and is
associated with and/or organized by cytoskeletal proteins [15]. The intact
system produces O_2^- according to the following stoichiometry[2]:

$$NAPDH + 2O_2 \rightarrow 2O_2^- + NADP^+ + 2H^+.$$

Cell-Cell Interactions in the Release of Inflammatory Mediators
Edit ed by P. Y-K Wong and C.N. Serhan, Plenum Press, New York, 1991

There are at least three stimulatory pathways in neutrophils which can trigger activation of the oxidase system[16,17]. The most topical and thoroughly studied pathway involves an intense phosphorylation of p47 catalyzed by protein kinase C (PKC)[18,19]. While p47 has been characterized[20], cloned and its sequence predicted[10,21], its exact function remains obscure. The sequence of p47 contains a highly cationic carboxy terminus with six serine residues in configurations highly favorable for phosphorylation by PKC (i.e., with arginine residues on the N-terminal side or both sides of the serine moiety[22])[10,21]. Several phosphorylated isoforms of p47 are observed in vivo[20,23].

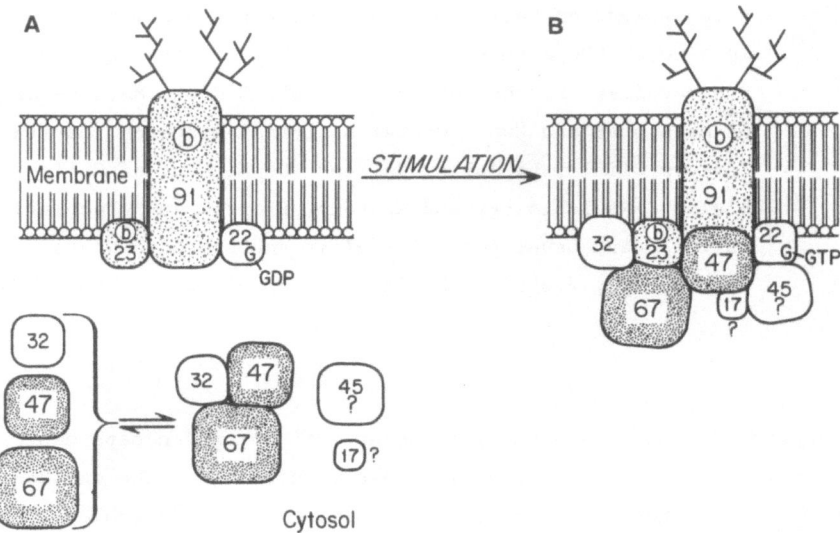

Figure 1.   General view of the NADPH-oxidase System. In unstimulated cells (A) the oxidase is dormant and dissociated, with components residing in both the cytosol and plasmalemma. Upon stimulation (B) there is a translocation of the soluble components to the plasmalemma where the active oxidase complex is assembled.

A diverse array of products of inflammatory and other cells can synergistically enhance superoxide production by neutrophils. The products include cytokines [e.g., granulocyte-macrophage colony-stimulating factor (GM-CSF), gamma interferon][24,25], vasoconstrictive

20

peptides (i.e., endothelins)[26], chemotactic factors[27] and
biologically active lipids [e.g., 5(S)-hydroxy-6-trans-8,11,14-cis-
eicosatetraenoic acid (5-HETE), platelet activating factor][28,29]. In
this chapter, we will briefly review studies from our laboratories on the
synergistic stimulation of neutrophils by activators of PKC and agents
capable of elevating intracellular Ca^{2+}. A model which can explain the
latter mentioned type of synergy will be presented and the physiological
consequences of this phenomenon will be discussed. Detailed reviews are
available elsewhere on oxygen radicals[1], activation of the NADPH-oxidase
system[2,30] and PKC[31].

SYNERGISTIC STIMULATION OF NEUTROPHILS

    Neutrophils stimulated with a variety of agents exhibit a rapid
activation of phospholipases C and D[32-35]. Phospholipase C catalyzes the
hydrolysis of phosphatidylinositol 4,5-bisphosphate to produce two second
messengers, sn-1,2-diglyceride (DG) and inositol 1,4,5-
trisphosphate[Ins(1,4,5)P_3]. The former activates PKC[36]. The latter
stimulates the release of Ca^{2+} from intracellular storage
sites [37,38]. Phospholipase D catalyzes the hydrolysis of certain
phospholipids to provide phosphatidate, which is rapidly dephosphorylated
to DG[37].

    The advantages of having two second messengers produced
simultaneously can be dissected pharmacologically. Stimulation of PKC
and elevation of intracellular Ca^{2+} can be achieved separately in cells
by treatment with activators of the kinase or Ca^{2+} ionophores,
respectively. Xenobiotic activators of PKC include a diverse array of
potent tumor-promoters (e.g., phorbol esters, mezerein) which bind to the
DG-site on the kinase[36,39]. Mezerein is a particularly attractive
reagent for such studies because, unlike the phorbol esters, its
structure does not embody a diglyceride like moiety[39]. Neutrophils
treated with optimal amounts of mezerein or phorbol esters generate large
quantities of O_2^- [40,41], whereas cells treated with ionophore A23187 do
not[41]. However, treatment of neutrophils with <u>suboptimal</u> amounts of both
mezerein and ionophore A23187 results in a dramatic synergistic
stimulation of O_2^- release[41]. Other combinations of tumor-promoters and
calcium ionophores have been effective in producing a similar response[42-45].

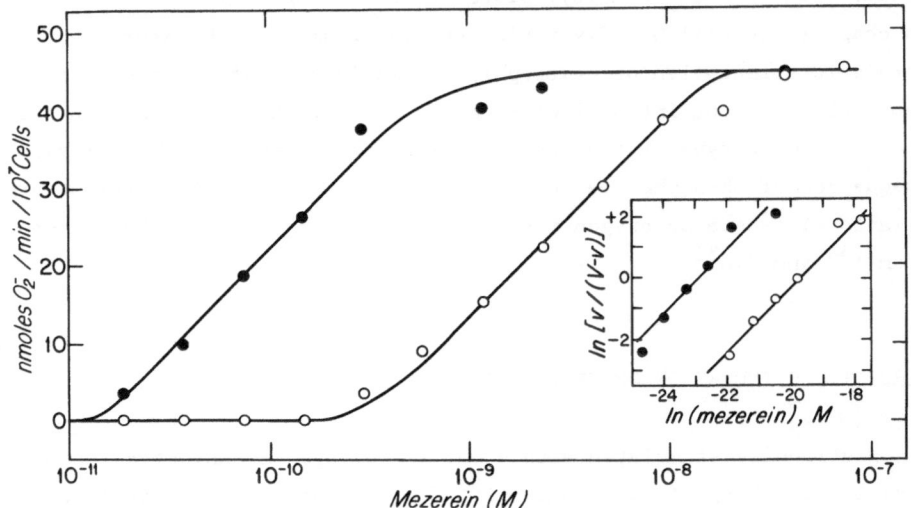

Figure 2.   Effect of the calcium ionophore A23187 on the dose-response
curve for mezerein. Superoxide release was measured after
stimulation of neutrophils with different concentrations of
mezerein in the absence (O) or presence (●) of ionophore
A23187 (0.025μM). The inset shows the Hill plots of these
data (for details, see reference 41).

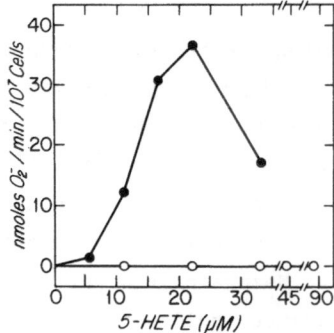

Figure 3.   Effect of 5-HETE on superoxide release by neutrophils.
Neutrophils were treated with different concentrations of 5-
HETE in the presence (●) and absence (O) of mezerein
(0.15nM). Mezerein alone at this concentration was
ineffective in stimulating the release of O_2^- (for details,
see 41).

22

The effect of ionophore A23187 on the dose-response curve for mezerein is presented in Figure 2. Addition of the ionophore to neutrophils dramatically reduces the amount of mezerein required to stimulate half-maximal O_2^- production, but does not affect the rate of this response at optimal amounts of mezerein[41]. Tumor promoters and diglyceride activate PKC by lowering the requirement for Ca^{2+} [36]. The situation described in Figure 2 may be considered the obverse of that, i.e., increased Ca^{2+} lowers the requirement for mezerein. Thus, elevation of intracellular Ca^{2+} markedly increases the responsiveness of cells to activators of PKC. This situation is very different from that observed during the synergistic stimulation of serotonin release from platelets[46]. In that study, the ionophore stimulated the maximal response. Stimulation of platelets requires activation of both PKC and a calmodulin stimulated protein kinase[46]. As we shall see, synergistic stimulation of neutrophils can be explained by activation of PKC alone.

A ROLE FOR 5-HETE IN SUPEROXIDE GENERATION

A similar synergistic stimulation of neutrophils is also observed with 5-HETE and low amounts of tumor-promoters[28,41]. Neutrophils produce this derivative of arachidonic acid when treated with agents that elevate intracellular Ca^{2+} [47,48]. 5-HETE alone is completely ineffective in stimulating O_2^- release at concentrations of 5 to $90\mu M$ (Figure 3). However, when 5-HETE is combined with a suboptimal amount of mezerein (0.15nM), the cells release large quantities of O_2^- [36.8 ± 6.5 (SD,n=6) nmoles O_2^-/min/10^7 cells]. This "priming" effect of 5-HETE is dose-dependent and the majority of the response occurs between 10 to $25\mu M$[41] (Figure 3). The level of endogenous 5-HETE produced during stimulation of neutrophils is ca. $16\mu M$[49].

Synergy with ionophore A23187 exhibits a partial requirement for external Ca^{2+} (ca. 50%), whereas that with 5-HETE is nearly totally dependent upon this cation being present in the external medium[41]. The partial requirement for external Ca^{2+} with A23187 is likely due to the ability of this substance to mobilize Ca^{2+} from both the external medium and from the intracellular storage sites. The near absolute requirement for external Ca^{2+} with 5-HETE may suggest that this lipoxygenase product does not function simply as an ionophore, but in terms of O_2^- release is active only at the plasmalemma, perhaps by opening a channel or gate for Ca^{2+}. Alternately, 5-HETE may be completely metabolized at the plasmalemma and thus incapable of reaching the intracellular Ca^{2+} depots.

Neutrophils stimulated with a variety of physiological agents exhibit an elevation in intracellular Ca^{2+} that occurs in two phases: an initial rapid increase which is transient, followed by a second, more gradual increase which is prolonged[34,50,51]. The initial phase is independent of external Ca^{2+} and is due to the $Ins(1,4,5)P_3$-mediated release of Ca^{2+} from the intracellular stores[34,37]. The second phase is dependent upon an influx of Ca^{2+} across the plasmalemma[51]. Agents which elevate intracellular Ca^{2+} promote the production of 5-HETE in neutrophils by activating phospholipase A_2 [which catalyzes the release of arachidonate from membrane phospholipids[52]] and the 5-lipoxygenase[53]. 5-HETE accumulates in cells for ca. 5 min. after stimulation, then declines rapidly as a result of its incorporation into phospholipids and triglyceride[47,48]. Thus, 5-HETE produced as a result of the initial increase in cellular Ca^{2+} may itself serve as a second-messenger for the recruitment of Ca^{2+} from the external medium. 5-HETE at 3.0μM has been shown to increase $^{45}Ca^{2+}$ uptake by rabbit neutrophils ca. 75%[54]. Whether the unesterified or an esterified derivative of 5-HETE is the active entity remains to be determined. A role for lipoxygenase derived products as second messengers which modulate potassium channels in neurons has been described[55]. It will be interesting to see whether a general role emerges for different HETEs in regulating ion channels. A number of cytokines prime neutrophils to produce increased amounts of both lipoxygenase products and O_2^- upon stimulation[24,56]. Causal links between these phenomena should be sought.

PROTEIN KINASE C AND SYNERGY

Neutrophils stimulated with tumor-promoters exhibit a translocation of PKC from a soluble to particulate fraction that is not reversed by the presence of Ca^{2+} chelators (i.e., "membrane inserted" or "chelator resistant" PKC)[57,58]. Association of PKC with the membrane fulfills the phospholipid requirement of this enzyme. This translocation precedes the activation of the oxidase system and exhibits a dose-response curve with phorbol esters similar to that for the stimulation of O_2^- production[58]. In contrast, synergistically stimulated neutrophils do not exhibit "chelator-resistant" translocation of PKC, even though the cells release large quantities of O_2^- [41]. Two types of associations of PKC with phospholipid vesicles have been demonstrated in vitro[59,60]. The initial complex is reversible and Ca^{2+}-dependent. This loosely associated form of the enzyme can subsequently undergo insertion into the bilayer to form a complex that is not reversed by chelation of Ca^{2+} [59,60]. Could the loosely bound, membrane associated form of PKC be active during synergy? Activity for this form of PKC had previously not been demonstrated in cells and has only recently been demonstrated in a cell-free system[61].

To obtain evidence that this form of PKC was indeed involved during synergy, we investigated the proteins phosphorylated under these circumstances along with the effects of antagonists of PKC on these phenomena. Compounds that inhibit PKC by binding to the ATP [1-(5-isoquinolinylsulfonyl)-2-methylpiperazine (H-7)[62]] or the DG/tumor-promoter binding sites (sphingosine[63]) on the kinase inhibit O_2^- release from cells stimulated either synergistically or with optimal amounts of tumor-promoters[41]. In contrast, N-(2-guanidinoethyl)-5-isoquinoline sulfonamide (HA1004), an antagonist of the cAMP and cGMP-dependent kinases[62], does not inhibit O_2^- release under either of these circumstances[41]. It must be noted, however, that these inhibitors affect protein kinases other than PKC[62,64] and it is not certain if neutrophils are equally permeable to H-7 and HA1004.

The most prominent alterations in the phosphoprotein pattern of neutrophils stimulated with activators of PKC are an intense incorporation of ^{32}P into two proteins with molecular masses of ca. 47 (p47) and 49kDa[19,65,66]. As noted above, p47 is a subunit of the NADPH-oxidase system[8-10,21] and is a substrate for PKC[67,68]. The nature of the 49kDa protein (p49) is unknown. Inhibitors of PKC block the phosphorylation of these proteins in cells stimulated with optimal amounts of activators of PKC[16,69]. Phosphorylation of p47 and p49 during synergistic stimulation of neutrophils with mezerein and 5-HETE is presented in Figure 4. Treatment of the cells with 5-HETE alone has no effect on these proteins (lane b), whereas treatment with a suboptimal amount of mezerein alone results in a small increase in the phosphorylation of p47 (lane c). The combination of these agents, however, results in a marked phosphorylation of p47, and perhaps p49, but to a small extent (lane d)[70]. H-7 (200μM) dramatically inhibits the phosphorylation of p47 under these conditions, whereas HA1004 does not[70].

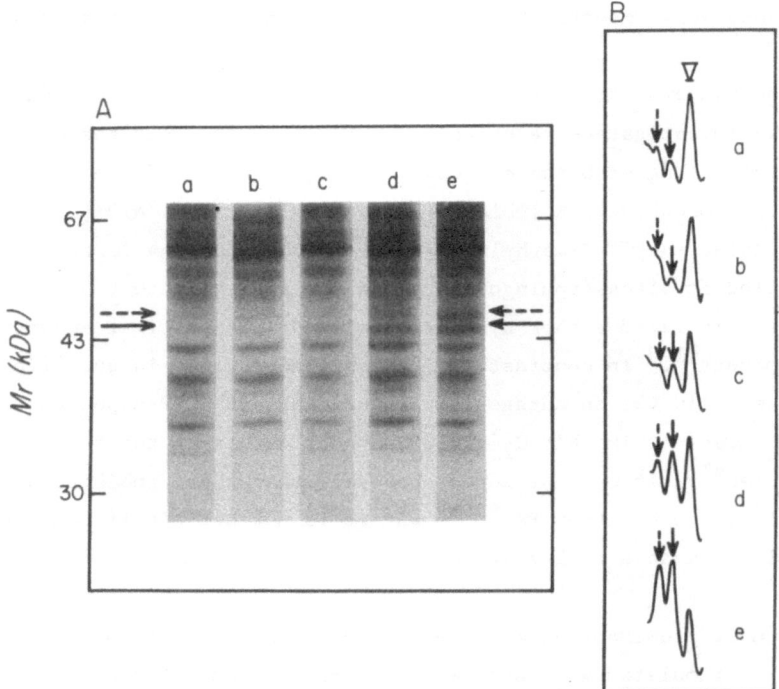

Figure 4.   Phosphorylation of p47 during synergistic stimulation of
             neutrophils with low amounts of mezerein and 5-HETE. Part A
             compares portions of autoradiographs from neutrophils
             treated for 3 min with: (a) Me$_2$SO (0.25% v/v); (b) 5-HETE
             (24μM); (c) a suboptimal amount of mezerein (0.30nM); (d) a
             suboptimal amount of mezerein (0.30nM) plus 5-HETE (24μM);
             and (e) an optimal amount of mezerein (38nM). The 47- and
             49-kDa proteins are indicated by the unbroken and broken
             arrows, respectively. Part B presents the densitometric
             scans of these bands, along with peak V which is shown for
             the purpose of orientation (for details, see reference 70).

Thus, we demonstrate that a substrate of PKC (i.e., p47) is phosphorylated under conditions where PKC does not undergo "chelator resistant" translocation into the plasmalemma and that this phosphorylation is blocked by H-7. A translocation of PKC to the membrane can be observed in neutrophils stimulated synergistically with 5-HETE if Ca^{2+} chelators are omitted from the extraction buffer[71]. Thus, several lines of evidence are consistent with the reversible, membrane associated form of PKC being active during synergy. It must be noted, however, that kinases other than PKC may conceivably phosphorylate p47. A recent study has shown that synergistic stimulation of platelets results in the activation of several novel and yet uncharacterized protein kinases[72].

MOLECULAR BASIS OF SYNERGY AND PHYSIOLOGICAL RELEVANCE

A number of kinetic studies on purified PKC have demonstrated that the catalytically active form of this kinase consists of a quaternary complex of enzyme protein, phospholipid, Ca^{2+} and the activator (e.g., DG, tumor promoter)[73-75]. With regard to synergy, a striking interdependence exists among the cofactors; i.e., changing the concentration of one can markedly alter the affinity of the enzyme for the other two[75]. The concentration of enzyme protein and phospholipid (membrane) may be considered constant in stimulated neutrophils. Therefore, increased cellular Ca^{2+} mediated by 5-HETE or ionophore A23187 may compensate, at least partially, for a suboptimal amount of mezerein and allow formation of an active PKC complex. This synergy in activating PKC is reflected in the phosphorylation of p47 (Figure 4) and ultimately in the generation of O_2^- (Figures 2 & 3).

A qualitative explanation for the interdependence of the cofactors of PKC has been presented on the basis of model studies[76-78]. As noted above, increased levels of Ca^{2+} can lead to a reversible binding of PKC to the membrane. This interaction with membrane sensitizes or permits the enzyme to respond with increased activity towards suboptimal amounts of tumor-promoters or DG[76-78]. One can speculate that the basis for this increased responsiveness may involve conformational changes and/or proximity effects (e.g., protein, Ca^{2+} and phospholipid are now physically associated and the activator will preferentially partition into the lipid phase of the system). With regard to this model, it may be noted that the concentrations of Ca^{2+} that are achieved in neutrophils by the Ins(1,4,5)P_3 signal can result in a significant fraction of PKC becoming associated with the plasmalemma[79]. A Ca^{2+} independent mechanism of PKC translocation to the membrane may also exist with certain agonists that prime neutrophils[29]. Thus, association of PKC with the plasmalemma, rather than the increase in Ca^{2+} itself, may be the crucial step for synergy. What are the physiological advantages of synergy? Three possibilities are particularly noteworthy:

1.   Synergy provides an amplification mechanism which can allow cells to respond massively to small changes in the levels of Ca^{2+} and/or DG.

2.   Under the conditions of synergy, any enzyme which can affect the cellular levels of Ca^{2+} [e.g., Ins(1,4,5)P_3-phosphatase] or DG [e.g., diglyceride kinase] may exert a major influence over the generation of O_2^-. Thus, several control sites are available for modulating the cellular response.

3.   The membrane inserted ("chelator resistant") form of PKC is chronically stimulated (i.e., active in the absence of Ca^{2+} and DG), whereas the reversible, membrane associated form is not and requires the continuous presence of DG and Ca^{2+} for activity[59-61]. The latter complex is functional during synergy and can allow for transient stimulation of neutrophils.

Thus, elevation of both DG and Ca^{2+} during cell stimulation and the interdependence of these cofactors in forming the activated complex of PKC provide enormous versatility in the modulation of O_2^-. This is not surprising considering the antimicrobial and cytocidal mechanisms associated with this process and their possible pathological consequences.

Acknowledgements

    The authors are grateful to Ms. Lucy Hassey for secretarial help in preparing this paper.

    Studies from these laboratories were supported by grants AI-28342, AI-23323 and AI-24321 from the National Institutes of Health, and the Council of Tobacco Research 2065. J.M.R. was also supported by a Bremer Foundation Grant and a Seed Grant from Ohio State University.

    Figures 2, 3 & 4 were reprinted with the permission of the J. Biol. Chem.

References

1. J. A. Badwey, and M. L. Karnovsky, Active oxygen species and the functions of phagocytic leukocytes, <u>Annu. Rev. Biochem.</u> 49:695 (1980).

2. J. T. Curnutte and B. M. Babior, Chronic granulomatous disease, <u>Adv. Human Genet.</u> 16:229 (1987).

3. R. A. Heyneman and R. E. Vercauteren, Activation of a NADPH-oxidase from horse polymorphonuclear leukocytes in a cell-free system, <u>J. Leukocyte Biol.</u> 36:751 (1984).

4. Y. Bromberg and E. Pick, Unsaturated fatty acids stimulate NADPH-dependent superoxide production by cell free system derived from macrophages, <u>Cell Immunol.</u> 88:213 (1984).

5. A. W. Segal and O. T. Jones, Novel cytochrome b system in phagocytic vacuoles from human granules, <u>Nature</u> 276:515 (1978).

6. C. A. Parkos, R. A. Allen, C. G. Cochrane, and A. J. Jesaitis, Purified cytochrome b from human granulocyte plasma membrane is comprised of two polypeptides with relative molecular weights of 91,000 and 22,000, <u>J. Clin. Invest.</u> 80:73 (1987).

7. M. T. Quinn, C. A. Parkos, L. Walker, S. H. Orkin, M. C. Dinauer, and A. J. Jesaitis, Association of a Ras-related protein with cytochrome b of human neutrophils, <u>Nature</u> 342:198 (1989).

8. B. D. Volpp, W. M. Nauseef, and R. A. Clark, Two cytosolic neutrophil oxidase components absent in autosomal chronic granulomatous disease, <u>Science</u> 242:1295 (1988).

9. H. Nunoi, D. Rotrosen, J. I. Gallin, and H. L. Malech, Two forms of autosomal chronic granulomatous disease are deficient in distinct neutrophil cytosol factors, <u>Science</u> 242:1298 (1988).

10. B. D. Volpp, W. M. Nauseef, J. E. Donelson, D. R. Moser, and R. A. Clark, Cloning of the cDNA and functional expression of the 47-kilodalton cytosolic component of human neutrophil respiratory burst oxidase, <u>Proc. Natl. Acad. Sci. USA</u> 86:7195 (1989).

11. T. L. Leto, K. J. Lomax, B. D. Volpp, H. Nunoi, J. M. G. Sechler, W. M. Nauseef, R. A. Clark, J. I. Gallin, and H. L. Malech, Cloning of a 67-kD neutrophil oxidase factor with similarity to a noncatalytic region of p60^{c-src}, <u>Science</u> 248:727 (1990).

12. P. G. Heyworth, C. F. Shrimpton, and A. W. Segal, Localization of the 47kDa phosphoprotein involved in the respiratory burst oxidase of phagocytic cells, <u>Biochem. J.</u> 260:243 (1989).

13. R. A. Clark, B. D. Volpp, K. G. Leidal, and W. M. Nauseef, Two cytosolic components of the human respiratory burst oxidase translocate to the plasma membrane during cell activation, <u>J. Clin. Invest.</u> 85:714 (1990).

14. D. Rotrosen, M. E. Kleinberg, H. Nunoi, T. Leto, J. I. Gallin, and H. L. Malech, Evidence for a functional cytoplasmic domain of phagocyte oxidase cytochrome b_{558}, <u>J. Biol. Chem.</u> 265:8745 (1990).

15. M. T. Quinn, C. A. Parkos, and A. J. Jesaitis, The lateral organization of components of the membrane skeleton and superoxide generation in the plasma membrane of stimulated human neutrophils, <u>Biochim. Biophys. Acta</u> 987:83 (1989).

16. J. A. Badwey, J. M. Robinson, P. G. Heyworth, and J. T. Curnutte, 1,2-Dioctanoylglycerol can stimulate neutrophils by different mechanisms. Evidence for a pathway that does not involve phosphorylation of p47, <u>J. Biol. Chem.</u> 264:20676 (1989).

17. J. M. Robinson, P. G. Heyworth, and J. A. Badwey, Utility of staurosporine in uncovering differences in the signal transduction pathways for superoxide production in neutrophils, <u>Biochim. Biophys. Acta</u> 1052:299 (1990).

18. A. W. Segal, P. G. Heyworth, S. Cockroft, and M. M. Barrowman, Stimulated neutrophils from patients with autosomal recessive chronic granulomatous disease fail to phosphorylate a Mr-44,000 protein, <u>Nature</u> 316:547 (1985).

19. P. G. Heyworth, and A. W. Segal, Further evidence for involvement of a phosphoprotein in the respiratory burst oxidase of human neutrophils, <u>Biochem. J.</u> 239:723 (1986).

20. N. Okamura, J. T. Curnutte, R. L. Roberts, and B. M. Babior, Relationship of protein phosphorylation to the activation of the respiratory burst in human neutrophils: defects in the phosphorylation of a group of closely related 48-kDa proteins in two forms of chronic granulomatous disease, <u>J. Biol. Chem.</u> 263:6777 (1988).

21. K. J. Lomax, T. L. Leto, H. Nunoi, J. I. Gallin, and H. L. Malech, Recombinant 47 kilodalton cytosol factor restores NADPH oxidase in chronic granulomatous disease, <u>Science</u> 246:987 (1989).

22. C. House, R. E. Wettenhall, and B. E. Kemp, The influence of basic residues on the substrate specificity of protein kinase C, <u>J. Biol. Chem.</u> 262:772 (1987).

23. D. Rotrosen and T. L. Leto, Phosphorylation of neutrophil 47kDa cytosolic oxidase factor. Translocation to membrane is associated with distinct phosphorylation events, <u>J. Biol. Chem.</u> 265:19910 (1990).

24. R. H. Weisbart, D. W. Golde, S. C. Clark, G. G. Wong, and J. C. Gasson, Human granulocyte-macrophage colony stimulating factor is a neutrophil activator, <u>Nature</u> 314:361 (1985).

25. G. Berton, L. Zeni, M. A. Cassatella, and F. Rossi, Gamma interferon is able to enhance the oxidative metabolism of human neutrophils, <u>Biochem. Biophys. Res. Commun.</u> 138:1276 (1986).

26. K. Ishida, K. Takeshige, and S. Minakami, Endothelin-1 enhances superoxide generation of human neutrophils stimulated by the chemotactic peptide N-formyl-methionyl-leucyl-phenylalanine, <u>Biochem. Biophys. Res. Commun.</u> 173:496 (1990).

27. L. C. McPhail, C. C. Clayton, and R. Snyderman, The NADPH-oxidase of human polymorphonuclear leukocytes. Evidence for regulation by multiple signals. <u>J. Biol. Chem.</u> 259:5768 (1984).

28. J. T. O'Flaherty, J. D. Schmitt, and R. L. Wykle, Interactions of arachidonate metabolism on protein kinase C in mediating neutrophil function. <u>Biochem. Biophys. Res. Commun.</u> 127:916 (1985).

29. J. T. O'Flaherty, J. F. Redman, D. P. Jacobson, and A. G. Rossi, Stimulation and priming of protein kinase C translocation by a Ca^{2+} transient-independent mechanism. Studies in human neutrophils challenged with platelet activating factor and other receptor agonists, <u>J. Biol. Chem.</u> 265:2169 (1990).

30. P. G. Heyworth and J. A. Badwey, Protein phosphorylation associated with the stimulation of neutrophils. Modulation of superoxide production by protein kinase C and calcium, <u>J. Bioenerg. Biomembr.</u> 22:1 (1990).

31. U. Kikkawa, A. Kishimoto, and Y. Nishizuka, The protein kinase C family: heterogeneity and its implications, <u>Annu. Rev. Biochem.</u> 58:31 (1989).

32. M. Volpi, R. Yassin, P. H. Naccache, and R. I. Sha'afi, Chemotactic factor causes rapid decrease in phosphatidylinositol 4,5-bisphosphate and phosphatidylinositol 4-monophosphate in rabbit neutrophils, <u>Biochem. Biophys. Res. Commun.</u> 112:957 (1983).

33. C. N. Serhan, M. J. Broekman, H. M. Korchak, J. E. Smolen, A. J. Marcus, and G. Weissman, Changes in phosphatidylinositol and phosphatidic acid in stimulated neutrophils. Relationship to calcium mobilization, aggregation and superoxide radical generation, <u>Biochim. Biophys. Acta</u> 762:420 (1983).

34. H. Ohta, F. Okajima, and M. Ui, Inhibition by islet-activating protein of a chemotactic peptide-induced early breakdown of inositol phospholipids and Ca^{2+} mobilization in guinea pig neutrophils. J. Biol. Chem. 260:1571 (1985).

35. J-K. Pai, M. I. Siegel, R. W. Egan, and M. M. Billah, Activation of phospholipase D by chemotactic peptide in HL-60 granulocytes, Biochem. Biophys. Res. Commun. 150:355 (1988).

36. M. Castagna, Y. Takai, K. Kaibuchi, K. Sano, U. Kikkawa, and Y. Nishizuka, Direct activation of calcium activated, phopholipid-dependent protein kinase by tumor-promoting phorbol esters. J. Biol. Chem. 257:7847 (1982).

37. H. Streb, R. F. Irvine, M. J. Berridge, and I. Schulz, Release of Ca^{2+} from a nonmitochondrial intracellular store in pancreatic acinar cells by inositol-1,4,5-trisphosphate, Nature 306:67 (1983).

38. M. Prentki, C. B. Wollheim, and P. D. Lew, Ca^{2+} homeostasis in permeabilized human neutrophils. Characterization of Ca^{2+}-sequestering pools and the action of inositol-1,4,5-trisphosphate. J. Biol. Chem. 25:1377 (1984).

39. A. Couturier, S. Bazgar, and M. Castagna, Further characterization of tumor-promoter-mediated activation of protein kinase C, Biochem. Biophys. Res. Commun. 121:448 (1984).

40. J. M. Robinson, J. A. Badwey, M. L. Karnovsky, and M. J. Karnovsky, Release of superoxide and change in morphology by neutrophils in response to phorbol esters. Antagonism by inhibitors of calcium binding proteins. J. Cell Biol. 101:1052 (1985).

41. J. A. Badwey, J. M. Robinson, W. Horn, R. J. Soberman, M. J. Karnovsky, and M. L. Karnovsky, Synergistic stimulation of neutrophils. Possible involvement of 5-hydroxy-6,8,11,14-eicosatetraenoate in superoxide release. J. Biol. Chem. 263:2779 (1988).

42. J. M. Robinson, J. A. Badwey, M. L. Karnovsky, and M. J. Karnovsky, Superoxide release by neutrophils: synergistic effects of a phorbol ester and a calcium ionophore. Biochem. Biophys. Res. Commun. 122:734 (1984).

43. F. Di Virgilio, D. P. Lew, and T. Pozzan, Protein kinase C activation of physiological processes in human neutrophils at vanishingly small cytosolic Ca^{2+} levels, Nature 310:691 (1984).

44. A. Penfield and M. M. Dale, Synergism between A23187 and 1-oleoyl-2-acetyl-glycerol in superoxide production by human neutrophils, Biochem. Biophys. Res. Commun. 125:332 (1984).

45. T. H. Finkel, M. J. Pabst, H. Suzuki, L. A. Guthrie, J. R. Forehand, W. A. Phillips, and R. B. Johnston Jr., Priming of neutrophils and macrophages for enhanced release of superoxide anion by the calcium ionophore A23187. Implications for regulation of the respiratory burst, J. Biol. Chem. 262:12589 (1987).

46. K. Kaibuchi, Y. Takai, M. Sawamura, M. Hoshijima, T. Fujikura, and Y. Nishizuka, Synergistic functions of protein phosphorylation and calcium mobilization in platelet activation, J. Biol. Chem. 258:6701 (1983).

47. W. F. Stenson and C. W. Parker, Metabolism of arachidonic acid in ionophore-stimulated neutrophils. Esterification of a hydroxylated metabolite into phospholipids, J. Clin. Invest. 64:1457 (1979).

48. R. W. Bonser, M. I. Siegel, S. M. Chung, R. T. McConnell, and P. Cuatrecasas, Esterification of an endogenously synthesized lipoxygenase product into granulocyte cellular lipids, Biochemistry 20:5297 (1981).

49. C. J. Meade, G. A. Turner, and P. E. Bateman, The role of polyphosphoinositides and their metabolic products in A23187-induced release of arachidonic acid from rabbit polymorphonuclear leukocytes, Biochem. J. 238:425 (1986).

50. H. M. Korchak, L. E. Rutherford, and G. Weissman, Stimulus response coupling in the human neutrophil. Kinetic analysis of changes in calcium permeability. J. Biol. Chem. 259:4070 (1984).

51. D. Pittet, D. P. Lew, G. W. Mayr, A. Monod, and W. Schlegel, Chemotactic receptor promotion of Ca^{2+} influx across the plasma membrane of HL-60 cells. A role for cytosolic free calcium elevations and inositol (1,3,4,5)-tetrakisphophate production. J. Biol. Chem. 264:7251 (1989).

52. F. Alonso, P. M. Henson, and C. C. Leslie, A cytosolic phospholipase in human neutrophils that hydrolyzes arachidonyl-containing phosphatidylcholine, Biochim. Biophys. Acta 878:273 (1986).

53. B. Samuelsson and C. D. Funk, Enzymes involved in the biosynthesis of Leukotriene B_4, J. Biol. Chem. 264:19469 (1989).

54. P. H. Naccache, R. I. Sha'afi, P. Borgeat, and E. J. Goetzl, Mono- and dihydroxyeicosatetraenoic acids alter calcium homeostasis in rabbit neutrophils. J. Clin. Invest. 67:1584 (1981).

55. D. Piomelli, A. Volterra, N. Dale, S. A. Siegelbaum, E. R. Kandel, J. H. Schwartz, and F. Belardetti, Lipoxygenase metabolites of arachidonic acid as second messengers for presynaptic inhibition of Aplysia sensory cells, Nature 328:38 (1987).

56. J. F. DiPersio, P. Billing, R. Williams, and J. C. Gasson, Human granulocyte-macrophage colony stimulating factor and other cytokines prime human neutrophils for enhanced arachidonic acid release and leukotriene B_4 synthesis, J. Immunol. 140:4315 (1988).

57. J. Nishihira and J. T. O'Flaherty, Phorbol myristate acetate receptors in human polymorphonuclear neutrophils, J. Immunol. 135:3439 (1985).

58. M. Wolfson, L. C. McPhail, V. N. Nasrallah, and R. Snyderman, Phorbol myristate acetate mediates redistribution of protein kinase C in human neutrophils: potential role in the activation of the respiratory burst enzyme, J. Immunol. 135:2057 (1985).

59. M. D. Bazzi and G. L. Nelsestuen, Properties of membrane-inserted protein kinase C, Biochemistry 27:7589 (1988).

60. M. D. Bazzi and G. L. Nelsestuen, Properties of the protein kinase C-phorbol ester interaction, Biochemistry 28:3577 (1989).

61. M. D. Bazzi and G. L. Nelsestuen, Differences in the effects of phorbol esters and diacylglycerols on protein kinase C, Biochemistry 28:9317 (1989).

62. H. Hidaka, M. Inagaki, S. Kawamoto, and Y. Sasaki, Isoquinoline sulfonamides, novel and potent inhibitors of cyclic nucleotide dependent protein kinase and protein kinase C, J. Biol. Chem. 23:5036 (1984).

63. E. Wilson, M. C. Olcott, B. M. Bell, A. H. Merrill, and J. D. Lambeth, Inhibition of the oxidative burst in human neutrophils by sphingoid long-chain bases. Role of protein kinase C in activation of the burst. J. Biol. Chem. 261:12616 (1986).

64. A. B. Jefferson and H. Schulman, Sphingosine inhibits calmodulin-dependent enzymes, J. Biol. Chem. 263:15241 (1988).

65. C. Schneider, M. Zanetti, and D. Romeo, Surface reactive stimuli selectively increase protein phosphorylation in human neutrophils, FEBS Lett. 127:4 (1981).

66. J. A. Badwey, P. G. Heyworth, and M. L. Karnovsky, Phosphorylation of both 47 and 49kDa proteins accompanies superoxide release by neutrophils, Biochem. Biophys. Res. Commun. 158:1029 (1989).

67. J. R. White, C.-H.-Huang, J. M. Hill, P. H. Naccache, E. L. Becker, and R. I. Sha'afi, Effect of phorbol 12-myristate 13-acetate and its analogue 4-α-phorbol-12,13-didecanoate on protein phosphorylation and lysosomal enzyme release in rabbit neutrophils. J. Biol. Chem. 259:8605 (1984).

68. I. M. Kramer, R. L. Verhoeven, R. L. van der Bend, R. S. Weening, and D. Roos, Purified protein kinase C phosphorylates a 47-kDa protein in control neutrophil cytoplasts but not in cytoplasts from patients with the autosomal form of chronic granulomatous disease, J. Biol. Chem. 263:6777 (1988).

69. J. A. Badwey, W. Horn, P. G. Heyworth, J. M. Robinson, and M. L. Karnovsky, Paradoxical effects of retinal in neutrophil stimulation, J. Biol. Chem. 264:14947 (1989).

70. P. G. Heyworth, M. L. Karnovsky, and J. A. Badwey, Protein phosphorylation associated with synergistic stimulation of neutrophils, J. Biol. Chem. 264:14935 (1989).

71. J. T. O'Flaherty and J. Nishihira, 5-Hydroxyicosatetraenoate promote Ca^{2+} and protein kinase C mobilization in neutrophils, Biochem. Biophys. Res. Commun. 148:575 (1987).

72. J. E. Ferrel Jr. and G. S. Martin, Thrombin stimulates the activities of multiple previously unidentified protein kinases in platelets, J. Biol. Chem. 264:20723 (1989).

73. U. Kikkawa, Y. Takai, Y. Tanaka, R. Miyake, and Y. Nishizuka, Protein kinase C as a possible receptor protein of tumor-promoting phorbol esters, J. Biol. Chem. 258:11442 (1983).

74. Y. A. Hannun and R. M. Bell, Phorbol ester binding and activation of protein kinase C on triton X-100 mixed micelles containing phosphatidylserine, J. Biol. Chem. 261:9341 (1986).

75. Y. A. Hannun, C. R. Loomis, and R. M. Bell, Protein kinase C activation in mixed micelles. Mechanistic implications of phospholipid, diacylglycerol, and calcium interdependencies, J. Biol. Chem. 261:7184 (1986).

76. M. Wolf, P. Cuatrecasas, and N. Sahyoun, Interaction of protein kinase C with membranes is regulated by Ca^{2+}, phorbol esters, and ATP, J. Biol. Chem. 260:15718 (1985).

77. M. Wolf, H. LeVine III, W. S. May Jr., P. Cuatrecasas, and N. Sahyoun, A model for intracellular translocation of protein kinase C involving synergism between Ca^{2+} and phorbol esters, Nature 317:546 (1985).

78. W. S. May Jr., N. Sahyoun, M. Wolf, and P. Cuatrecasas, Role of intracellular Ca^{2+} mobilization in the regulation of protein kinase C-mediated membrane processes, Nature 317:549 (1985).

79. W. A. Phillips, T. Fujiki, M. W. Rossi, H. M. Korchak, and R. B. Johnston, Influence of calcium on the subcellular distribution of protein kinase C in human neutrophils. Extraction conditions determine partitioning of histone-phosphorylating activity and immunoreactivity between cytosol and particulate fractions, J. Biol. Chem. 264:8361 (1989).

GRANULOCYTE-MACROPHAGE COLONY-STIMULATING FACTOR AND THE

NEUTROPHIL: MECHANISMS OF ACTION

Julian Gomez-Cambronero and Ramadan I. Sha'afi

Department of Physiology. University of Connecticut
Health Center. Farmington, Connecticut 06030, U.S.A.

1. INTRODUCTION

The granulocyte-macrophage colony-stimulating factor (GM-CSF) is a hemopoietic growth factor that regulates the production and functional activity of granulocytes and macrophages. GM-CSF is part of the complex cytokine/interleukine network. This complex network is composed of mediator molecules that participate in the immune system (see Table I). GM-CSF is also a member of the colony-stimulating factors (CSF's) family. It has been known for a long time that CSF's are humoral factors necessary for the control and induction of hemopoiesis (1,2). Its name was derived from the observation that when GM-CSF is added to a semisolid culture of human bone marrow cells it stimulates the formation of pure colonies of both macrophages and neutrophils (3). GM-CSF plays a crucial role in the commitment of bone marrow stem cells, directing the proliferation and differentiation into specific precursors of the different lineages, and enhancing the survival of both progenitors and mature cells (4,5). This cytokine is also able to potentiate the activation of mature phagocytes (neutrophils, eosinophils and macrophages) and in certain cases, it can activate directly mature phagocytes (6). Thus, GM-CSF plays an important role in host defense (7).

The possibility that GM-CSF (as well as other cytokines) could have therapeutic potentials, prompted investigators to clone the gene in order to produce it in large quantities. Recombinant human GM-CSF (rhGM-CSF) can now be obtained from either bacteria or from mammalian cells (8-11). As a result of the commercial availability of rhGM-CSF, a large body of literature has accumulated in the past few years. This review will focus primarily on the priming and activating effects of GM-CSF

Cell-Cell Interactions in the Release of Inflammatory Mediators
Edited . by P. Y-K Wong and C.N. Serhan, Plenum Press, New York, 1991

Table I
GM-CSF AS A PART OF THE COMPLEX
CYTOKINE/INTERLEUKINE NETWORK

Activating Cytokine	Cell Type	Effect Produced
Il-1, TNF-α	Fibroblasts	Release of GM-CSF
	Vascular	Release of GM-CSF
	Endothelial Cells	
	Mesanchymall Cells	Release of CSFs
GM-CSF alone	PMNs	Release of cyto-kines (IL-1)
	Nonhemopoietic tumor cells monocytes,macrophages	Clonal growth
GM-CSF + other cytokines (IL-1,IL-6,TNF-α, M-CSF, G-CSF.) (*)	Stem cell in B. marrow	Proliferation, granulocytic diff. and maturation
None	Nonhemopoietic tumor cells	Constitutive CSFs production

(*) GM-CSF requires synergistic interaction with other CSFs and accesory cells to fully stimulate myeloid progenitors.

on mature cells (specifically the neutrophil) rather than on its proliferative action.

2. CELLULAR ORIGIN OF GM-CSF

Early *in vitro* studies demonstrated that the main sources of this cytokine were the T-lymphocytes and activated macrophages sensibilized against the specific antigen in the inflammatory process (12). Both cell types are known to interact cooperatively, through the release of several interleukins, to elaborate GM-CSF and other CSF's (4,13). Resting peripheral blood mononuclear cells populations, and purified T-lymphocyte and monocyte populations do not produce transcripts (14,15) for GM-CSF or for other CSF's (14,16). However, following stimulation with PMA, γ-IFN and TNF-α, or lipopolysaccaride, the expression of GM-CSF and other CSF's mRNAs increases dramatically

(14-16). Endogenous IL-1 regulates the production of GM-CSF in human thymic non-lymphoid cells (17). In T cells (TH1 clones), GM-CSF is released by the combined action of IL-2 and PGEs, suggesting a role of cAMP in the process (18).

The human T-lymphoblastoid cell line known as Mo, infected with the human T-cell leukemia virus (HTLV)-II, was first used to purify the human GM-CSF (12). Endothelial cells and fibroblasts stimulated by IL-1 or TNF-α (19-21) and cloned bone marrow stromal cell lines stimulated with IL-1 (22) are also sources of GM-CSF *in vitro*. Fibroblasts constitutively transcribe GM-CSF, and the transcription is enhanced by TNF-α and PMA (19). In keranocytes, the addition of IL-1 has been shown to lead to a dose-dependent increase of GM-CSF production (23). The keranocyte derived T-cell growth factor (KTGF) reported in some studies (24) was later shown to be identical to GM-CSF (25).

Several studies have also shown that primary human leukemia blasts cells express the gene for GM-CSF (26), and in some tumor cell-host systems, GM-CSF-like factors are produced constitutively by the tumor cells. These factors may play a role in the development of tumor metastasis (27).

Although there is a great deal of information about the production of GM-CSF *in vitro*, little is known about the process of *in vivo* production of GM-CSF and the other CSFs. GM-CSF is not present in measurable quantities in normal serum. Moreover, its presence in bone marrow cultures has been difficult to detect because either the rate of utilization equals that of its production or because once produced, it is sequestered in the extracellular matrix (28,29).

3. THE GM-CSF MOLECULE AND ITS GENE

GM-CSF is a single subunit protein of 14 kDa molecular weight in its nonglycosylated form and 22 kDa in its N-glycosylated form (8). Glycosylation is not required for activity, but it might affect the pharmacokinetics of administered GM-CSF to patients (30). In 1985, the complete sequence was determined in two different laboratories by N-terminal amino acid sequence analysis (11) and by translation of the nucleotide sequences of cDNA clones (10,11). GM-CSF is composed of a single chain of 127 amino acids, containing 4 cysteines in the C-terminal half part of the molecule. The sequence is different from any other known CSF, and there is very limited conservation between sequences of different species (4). Although there is no X-ray crystallographic data as yet, structure-function studies have provided some light on the secondary or tertiary structure organization. These studies have made use of fragments and chemi-

cally-synthesized analogs (31) of mutant cDNAs, encoding truncated forms of the proteins (32) or hybrid cDNA molecules that
contain coding regions for human GM-CSF and murine-GM-CSF (33).
They predict the existence of two α-helices in the N-terminal
end (residues 13-27 and 31-46), that could assume an antiparal-
lel structure because they are free of disulfide bonds (32). A
similar configuration is also found in the N-terminus of IL-2
(34). Complete formation of the two α-helices is essential to
full activity of the molecule. It is not clear whether the
first helix is required for the binding of GM-CSF to its receptor or to provide a correct folding of the whole molecule (31).
Other regions (residues 38-48 and 95-111) have been proposed to
be implicated in the binding of the molecule to its receptors
(33).

The human GM-CSF gene exists as a single copy of 2.5 kilo-
bases containing four exons and three introns (35,36). There is
only about 60% homology with the murine gene (11). The gene is
mapped in humans on the long arm of chromosome 5 (37), close to
the IL-3 gene (multi-CSF) (38), and on chromosome 11 in mice
about 230 kb apart for the IL-3 gene (39). A region spanning
two cytokine-specific sequences (CK-1 and CK-2) in the GM-CSF
promoter region binds two nuclear proteins (nuclear factors NF-
GMa and NF-GMb) that are responsible for regulating GM-CSF
transcription (40). The expression of the GM-CSF gene is very
regulated in humans. Only activated cells express GM-CSF and
other CSF genes (5).

4. THE GM-CSF RECEPTOR

Description

The presence of specific receptors for GM-CSF has been
demonstrated in bone marrow stem cells, myeloid and erythroid
progenitors, promyelocytic cell lines such as HL-60, human
acute myeloid leukemic cells, in mature phagocytes (eosinophils, macrophages and neutrophils) and in human placenta (41-
43). The number of receptors varies with the maturity state of
the cells along the promyelomonocytic lineage (41). In humans,
the most mature myelomonocytic cells (macrophages, eosinophils
and neutrophils) express the highest numbers of a single class
of high affinity receptor: Kd=37 pM, with 293-1000 sites/cell
(41). Some nonhematopoietic tumor cells also have GM-CSF receptors (44,45).

Crosslinking studies indicated that the molecular weight
of the GM-CSF receptor was 85 kDa (41). More recently, studies
of expression cloning (46) have demonstrated the existence of a
GM-CSF receptor precursor (which is the unglycosylated form of
the receptor) with 400 amino acids (Mr 45 kDa). Besides the

extracellular domain, it has a single transmembrane domain, and a short (54 amino acids) intracytoplasmatic tail. It has significant sequence homology with receptors for IL-6, erytropoietin and the β-chain of IL-2 (46). Others (45,47) have suggested that the cloned receptor is actually one component (named α-chain) of the receptor. The α-chain has been purified and characterized as having an Mr of 80 kDa and a Kd in the subnanomolar range. The other component (β chain) has a Mr of 135 kDa and a Kd in the subpicomolar range. The distribution of the β-chain is restricted to hematopoietic cells, while the α-chain is also expressed in other cells, including those from some solid tumors (45).

The GM-CSF receptor does not bear a tyrosine kinase catalytic domain (46). This fact sets it apart from other growth factor receptors (48), including M-CSFR, (49,50), but makes it similar to the case of IL-2R and IL-3R, other receptors too small to encode the intracellular domain (51).

Regulation

As mentioned above, expression of receptors depends on the maturity state of the cells. Experimentally, it has been possible to observe the increase associated with differentiation. In HL-60 cells, dimethylsulfoxide and retinoic acid increase the number of high affinity receptors for GM-CSF, coincident with their neutrophilic differentiation (41). It is not known, however, whether this represents *de novo* synthesis of high affinity binding sites, a decrease in receptor turnover, or a recruitment of high affinity sites from a large pool of low affinity sites (41). With the availability of the cloned GM-CSF and cDNA probes, it should not be difficult to study the regulation of the receptor expression in differentiating cells.

Studies of receptor regulation associated with activation of the cells are somewhat more advanced. It has been suggested that internalization and degradation occurs naturally and immediately upon binding of the GM-CSF to its receptor (4), and is not a rate-limiting step in the signal transduction process (52). In neutrophils, the GM-CSF receptor is downregulated by GM-CSF itself, by active phorbol esters and fMLP and by other granulocyte-activating agents (41,53-56). The neutrophil receptor is not competed for by IL-3 (54). In contrast, the GM-CSF receptor of the myeloblast is resistant to downregulatory effects by GM-CSF itself, PMA or A23187, although IL-3 completely inhibited binding (54). The two different patterns of regulation and binding have prompted Cannistra*et al.* (54) to name those affected by PMA (neutrophil-like) as Class I receptors and those unaffected (myeloblast-like) as Class II receptors. There is a coexpression of Class I and II in some cases of acute myeloid leukemia.

The down regulation of GM-CSF receptor by PMA represents another case of phorbol esters interferring with signal transduction (57). In neutrophils, both the binding of GM-CSF and its associated actions are inhibited in PMA-treated cells (55). It is not clear if the action of PMA is mediated through protein kinase C activation or through a direct effect on the plasma membrane. If the PMA effect is mediated through protein kinase C activation, then one can speculate that the PKC-phosphorylated GM-CSF-receptors cannot bind GM-CSF or that they are internalized and therefore become unavailable for binding. There are several serine/threonine site candidates for a putative phosphorylation (46). In the case of the EGF receptor, it has been demonstrated clearly that activation of PKC by PMA inhibits the binding of EGF to its receptors, and this inhibition is associated with the phosphorylation of specific amino acids on the EGF receptors (58,59).

As for internalization after phosphorylation, May *et al.* (60) have shown that PMA causes a rapid and reversible internalization of transferrin-receptor in HL-60 cells that is temporarily associated with reversible increased phosphorylation. Another possible mechanism for the observed down regulation of the GM-CSF-receptors by activation of PKC would involve a PKC-phosphorylated protease, which could be translocated to the membrane. This protease would then cleave the receptors for GM-CSF, thus inhibiting the binding of GM-CSF. Such a mechanism has been described for the colony-stimulating factor-1 receptor (CSF-1R) (61).

5. CELLULAR FUNCTIONS ACTIVATED BY GM-CSF

As indicated above, there are high affinity receptors for GM-CSF in bone marrow stem cells, as well as in mature phagocyte cells. Binding of GM-CSF to these receptors initiates several responses both *in vitro* and *in vivo*. GM-CSF potentiates greatly the effects of major chemoattractants and other agents such as those generated in immunological reactions. Therefore, GM-CSF can be considered a "priming" agent. For example, many of the responses stimulated by fMet-Leu-Phe, such as the oxidative burst, degranulation and chemotaxis are greatly enhanced in GM-CSF-treated human neutrophils. In all cases, this cytokine plays a role of amplifying the inflammatory response to chemotactic factors and physiological stimuli. Some of these effects are summarized in Table II.

GM-CSF as a Hemopoietic Growth Factor

GM-CSF is a potent inductor of DNA synthesis*in vitro* (42). It stimulates the proliferation and differentiation of

Table II
BIOLOGICAL ACTIONS OF GM-CSF

In Vitro		In Vivo
In Bone Marrow	**In Phagocytes**	
• Enhancement of Cell Survival and Life Span		• Increases the stem cell number • Relocates the site of hemopoiesis to the spleen • Elevates the number of circulating granulocytes (in patients with AIDS and in those undergoing chemotherapy) • Activates granulocytes
• Stimulation of cell proliferation	• Activation of functional activity	
• Induction of cellular differentiation	• Potentiation of the inflammatory responses by increasing:	

• Commitment of biopotential precursors to enter granulocytic or monocytic lineage	• chemotaxis
	• intracellular cGMP levels
	• antibody-dependent cell mediated toxicity
	• phagocytosis
	• degranulation
	• expression of membrane antigens release of : $O_{\bar{2}}$
	  PAF,AA,LTs
	  IL-1,IFNs	

precursors of the myelomonocytic lineage. The order of decreasing responsiveness among progenitor cells is: macrophage > neutrophil > neutrophil-macrophage > eosinophil > megakaryocyte (1,4). The survival of GM-colony-forming units (CFU-GM) requires the continuous presence of GM-CSF and other CSF's at concentrations in the pM range, and their removal leads to cessation of proliferation and death (62). Cooperation between cytokines for cell proliferation and survival is probably necessary. A synergistic increase in the number of granulocytic colonies is observed when GM-CSF's at suboptimal doses, and IL-6 at effective doses, are both present in a CFU-GM clonogenic assay (63). Maximal production of eosinophils *in vitro* requires the presence, initially, of IL-3 and GM-CSF, followed by IL-5 (64). Combinations of IL-3 and GM-CSF enhance proliferation of early progenitor cells in mice bone marrow (65). Moreover, intravenous administration of GM-CSF and IL-3 to monkeys produced a strong leukocytosis (66).

Clinical Applications

The effectiveness of GM-CSF as a granulopoietic hormone has been confirmed *in vivo* (67-71). Its ability to increase the number of progenitor and mature cells, as well as to prolong the survival of both, has made this compound extremely useful in treating some human diseases in which hemopoiesis is compromised. In the presence of erythropoietin, human GM-CSF interacts with early progenitors and mature cells, as well as to prolong the survival of both, has made this compound extremely useful in treating some human diseases in which hemopoiesis is compromised. In the presence of erythropoietin, human GM-CSF interacts with early progenitors and supports the proliferation of erythroprogenitors.

Results from formal Phase I trials in humans, demonstrated a significant increase of the white blood cell counts, especially when GM-CSF was administered as a continuous infusion rather than as a bolus injection (72,73). In patients with myelodysplasia or aplastic anemia, administration of GM-CSF produced a dose-dependent rise of leukocyte counts (74-77), a result clouded by the fact that no increase in circulating neutrophils was observed in patients with severe aplastic anemia. GM-CSF, nevertheless, could still be useful in patients with minimal residual myelopoiemia (76). GM-CSF has been used also in Phase I/II studies to reduce the myelosupression produced by chemotherapy treatments. GM-CSF is normally well tolerated and is biologically active in those leukopenic patients (78, 79). Also, GM-CSF is being used to increase hemopoietic reconstitution following high dose chemotherapy or radiotherapy needed for bone marrow transplantation (80-83). In AIDS patients, GM-CSF administration has been used to reverse the neutropenia, reduced neutrophil function and lymphopenia (75). In addition, GM-CSF induces macrophage tumorcidal activity (84), although this effect has not been tested in vivo. Overall, the results of these studies have been very encouraging. In some cases, better results are obtained when GM-CSF is used in conjunction with other CSF's (such as G-CSF).

Recent findings suggest that GM-CSF may have a more general regulatory function on nonhemopoietic cell types than previously anticipated, and this could affect the usefulness of this cytokine in some specific diseases. For example, it has been shown that certain tumor cells have receptors for GM-CSF. These cells include myeloblasts from patients with acute myeloid leukemia (42), small cell carcinoma lines (SCCL), SV-40 transformed African green monkey kidney cell line COS (44), non-smal cell lung cancer, stomach cancer, colon cancer (85), osteosarcoma cells (85,86) a breast carcinoma cell line (86), and choriocarcinoma cells (45). A possible role of GM-CSF in gene activation in the evolution of solid tumor and in the pathogenesis of myelofibrosis has yet to be defined (86). On the other hand, when myeloid cell lines are transfected with retrovirally activated GM-CSF gene, the autocrine stimulation of cells by aberrant synthesis of GM-CSF leads to malignant transformation (87). GM-CSF and G-CSF stimulate the growth in culture of blasts from patients with acute myeloblastic leukemia (87). Moreover, the development of some murine and human leukemias *in vivo* is associated with the constitutive activation by genetic rearrangement of genes of GM-CSF and other cytokines, especially IL-3.

Taken together, these findings stress the fact that it may be necessary to monitor for adverse side effects of GM-CSF therapy in patients whose malignant cells may be directly stimulated by the cytokine. Also, GM-CSF primes phagocytes so the release of oxygen radicals, lipid mediators and enzymes is potentiated. These effects should be considered in the administration of GM-CSF, specially at high doses.

Oxidative Metabolism

In 1985, Gasson and her associates reported that the fMet-Leu-Phe-induced superoxide anion release was enhanced by more than fourfold in GM-CSF-treated cells (89). The enhancement was dose-dependent and dependent on the length of incubation with GM-CSF. However, when added alone to the cells, GM-CSF has little effect, if any, on the respiratory burst. This initial observation has been confirmed in several laboratories (90-96). Moreover, stimulated superoxide release was also found to be enhanced in granulocytes harvested from patients receiving GM-CSF infusion (97). GM-CSF primes neutrophils for enhanced superoxide release when cells are also stimulated with C5a, LTB_4 (90), PAF (95), but not with PMA (90,95) or Zymosan (95). The magnitude of potentiation is greatest in the case of fMet-Leu-Phe (98). This potentiation is further enhanced by PKC inhibitors H-7 and H-8, thus ruling out a possible role for PKC in GM-CSF-primed oxidative burst (52). GM-CSF has no effect on monocyte superoxide response (99). Both glycosylated and nonglycosylated forms of GM-CSF have similar effects on priming (62). GM-CSF does not potentiate superoxide production by cytoplasts stimulated with fMet-Leu-Phe. This lack of stimulation by GM-CSF is most likely due to the loss of the granules, the nucleus and other intracellular compartments, and suggests the importance of those cellular components to the priming action of GM-CSF (95,98).

GM-CSF and Lipid Mediators

In vitro, GM-CSF primes neutrophils for enhanced arachidonic acid (AA) release to a number of stimuli such as fMet-Leu-Phe, platelet-activating factor (PAF), leukotriene B_4 (LTB_4), and the calcium ionophore A23187 (100). Indomethacin and cycloheximide have no effect on this enhancement (100). GM-CSF at doses between 1-2 pM, enhances the A23187-induced generation of LTC_4 in eosinophils (101). Preincubation of neutrophils with 200 pM GM-CSF for one hour enhances synthesis of LTB_4, its all-trans isomers and omega-oxidation products, and 5-hydroxyeicosatetraenoic acid in response to both the calcium ionophore A23187 and fMet-Leu-Phe (102). Similar results are found when the stimulus is C5a (103) or when thrombin and FMLP are added to PMN co-incubated with platelets (104). All these data suggest that GM-CSF potentiates the activation of phospholipase A_2.

GM-CSF pretreatment causes an augmentation of phosphatidic acid (PA) release (Yasui, K. and Sha'afi, R.I., unpublished data) and diacylglycerol (DAG) production (105, Yasui, K. and Sha'afi, R.I., unpublished data) in human neutrophils stimulated with fMet-Leu-Phe. The pathway which appears to be potentiated by GM-CSF is phospholipase D and not phospholipase C specific phosphoinositides (Kozo, Y. and Sha'afi, R.I., unpublished data). This conclusion is based on the fact that none of the stimulated hydrolysis of the various phosphoinositides are affected by GM-CSF-treatment (106).

GM-CSF also primes neutrophils for enhanced platelet-activating factor (PAF) release (107-111). In GM-CSF-primed neutrophils the chemotactic peptide fMet-Leu-Phe triggers PAF synthesis by activating phospholipase A_2(PLA2) (109). This conclusion was based on the use of different PLA_2 inhibitors including an amino acid peptide derived from a conserved region of the calpactin superfamily (109).

In separate studies, the level of lyso-PAF: acetyl-CoA acetyltransferase activity in five cancer patients with normal hemopoiesis was increased by GM-CSF administration (108). It has been suggested that PAF production from neutrophils may explain some of the toxicity observed during treatment with high doses of GM-CSF (108).

Chemotaxis

A neutrophil-migration inhibitory factor derived from T-lymphocytes (NIF-T) that inhibits a neutrophil's ability to migrate under agarose (112), was later found to be identical to GM-CSF (12). Weisbart et al. (113) have shown that a brief (five to ten minutes) exposure to physiological concentrations of GM-CSF (10-100 pM) enhances fMet-Leu-Phe-induced neutrophil chemotaxis, concomitant with an increase in the number of fMet-Leu-Phe receptors. GM-CSF stimulates the migration of human endothelial cells (114). Using the Boyden chamber assay, Kownatzki et al. (115) found that GM-CSF (at concentrations ranging from 0.1 to 10,000 U/ml) had neither chemokinetic nor chemotactic activity on the neutrophil but it inhibited the chemotactic migration towards fMet-Leu-Phe and C5a. Inhibition of chemotaxis and of random migration in granulocytes by GM-CSF was also reported by others (91,92,99). The results of these studies were difficult to compare because of the use of dissimilar concentrations and times of incubation with GM-CSF, implicating that a generalized conclusion about the effect of GM-CSF on chemotaxis could not be made. In addition, neutrophil migration to a sterile inflammatory site is defective during GM-CSF infusion after autologous bone marrow transplantation (116) or prior to chemotherapy in patients with malignant diseases (117).

Based on all the published and unpublished (Yassui, K. and Sha'afi, R.I., unpublished results) data, it is reasonable to conclude that the effect of GM-CSF on chemotaxis depends on its concentration and the length of incubation prior to the addition of the stimulus. Low concentrations of GM-CSF and a short period of incubation enhances chemotaxis. On the other hand, high concentrations of GM-CSF or a longer period of incubation inhibits chemotaxis.

Degranulation

Cells preincubated with GM-CSF and subsequently stimulated with fMLP, PAF or A23187, demonstrated enhanced degranulation of lysosomal enzymes. They include release of lysozyme (93,118), lactoferrin, myeloperoxidase (MPO) (119), vitamin B_{12}-binding protein (a marker of specific granules) (120), elastase (91), proteinase (92) and N-acetylglucosaminidase (55). At least in one case, it is indicated that the process of primed degranulation might be dependent on G-protein activation and calcium (119). The effect of GM-CSF on lysozyme release from Candida is abrogated by anti-Mo1 antibody (118). GM-CSF, when added alone, is also able to originate vitamin B_{12}-binding protein release (120,121). The effect is quantitatively similar to that elicited by cytochalasin B or fMLP (120). In another study (121), it was found that human neutrophils exposed to GM-CSF resulted in a time- and concentration-dependent (3-100 U/ml) extracellular release of vitamin B_{12}-binding protien but only when cytochalasin B was present in the assay. The effect was reduced with EGTA (121).

In vitro, incubation of whole blood with GM-CSF produces a rise in plasma lactoferrin (LF) and transcobalamine (TC). Results in purified neutrophils were, however, variable (122). Consistent with this, there were found elevated levels of lactoferrin and transcobalamine in plasma of patients administered GM-CSF. These patients' neutrophils showed secondary granule depletion and were hypofunctional (122).

Phagocytosis and Cellular-Mediated Cytotoxicity

GM-CSF and G-CSF potentiate neutrophil antibody-dependent cellular-mediated cytotoxicity (ADCC) and phagocytosis of tumor targets (melanoma and neuroblastoma) (123), antibody-coated staphylococci (124) or of serum opsonized yeast (93). Also GM-CSF enhanced eosinophil cytotoxicity towards Schistosome mansoni larvae (101). *In vivo* phagocytosis of Crytococcus neoformans is similar before and during GM-CSF infusion of patients undergoing chemotherapy treatment (116).

GM-CSF as a Direct Cell Activator

In addition to priming neutrophils to the subsequent stimulation by a second agonist, GM-CSF on its own is able to activate certain cell responses and to produce several biochemical
changes. Some of these actions occur in adherent cells and
others occur in suspended cells. It is interesting to note that
while GM-CSF is unable to stimulate the oxidative burst when
added alone to suspended cells (see above), it is able to activate this system in adherent cells (125-127). Also, MPO release is not seen in cells in suspension (121), but it can be
demonstrated in adherent neutrophils (119). Although the basis
for these differences is not clear, it will be worthwhile to
elucidate, since its understanding may provide important clues
concerning the mechanism of GM-CSF action.

Binding of GM-CSF to its receptor upregulates certain surface proteins, such as receptors and adhesion molecules. For
example, constitutive expression of FcγRIII is enhanced by GM-
CSF (128). Moreover, GM-CSF upregulates G-CSF, M-CSF, γ-IFN
fMet-Leu-Phe and complement receptors. GM-CSF affects other
surface proteins such as granulocytic functional antigens 1 and
2 and Mo1 (CD11b) antigen (52,93,129). The effect of GM-CSF is
rapid and not mediated by protein synthesis nor by protein kinase C (52). It has also been observed that the continuous infusion of GM-CSF into patients rapidly upregulates the expression of granulocyte CD11b (129).

A logical consequence of the aforementioned GM-CSF-induced
increase in surface expression of certain proteins is the increase of neutrophil adhesiveness. CD11b (Mo1), selectively
expressed after neutrophils are exposed to GM-CSF, is a member
of a family of surface glycoproteins that are essential for
adhesion-dependent granulocyte functions. Brief exposure of
granulocytes to GM-CSF *in vitro* increases the adherence to
plastic surfaces (93,125,126) and to nylon fibers (94). In the
latter case, the optimal effect is obtained with 25-50 ng/ml of
GM-CSF for 30 min (94). Cell-cell increased adhesion has also
been demonstrated (130). On the other hand, GM-CSF has no effect on monocyte adhesiveness, even when they were used at
unphysiological concentrations. GM-CSF increases adherence to
the substratum by inducing the development of polarized filopo-
dia (125) and by increasing the number of intracellular vesicles (125,126). Neutrophils adherent to plastic surfaces release superoxide anions and hydrogen peroxide, unlike those in
suspension, when exposed to GM-CSF (125,126) or to TNF's (127).

In vivo, GM-CSF treatment causes a transient neutropenia
within 15 minutes of its infusion (129,131). This is believed
to be due to the migration of neutrophils and monocytes to the
pulmonary vasculature (131) and is associated with a marked

increase in the surface antigen CD11b (129) and cellular adhesion molecule (CAM) (131). The influence of this phenomenon on *in vivo* granulocyte aggregation may be clinically relevant (129). GM-CSF enhances the attachment of neutrophils to monolayers of human umbilical vein endothelial cells (93,132). This is due to its action on neutrophils and not on the endothelial cells (132). This observation was not reproduced by other investigators (93).

Kapp *et al.* (125) have shown that GM-CSF alone, at concentrations ranging from 1 to 10 U/ml enhances the chemoluminescense response in neutrophils. The response is rapid and increases with increasing times of incubation, reaching a maximum at 60-90 minutes. There after the cells are completely deactivated to further restimulation with GM-CSF or with TNF, but they respond to other activators of the oxidative burst (125). In addition to its direct action, GM-CSF potentiates the fMet-Leu-Phe-induced chemiluminescence (133).

6. BIOCHEMICAL CHANGES INITIATED BY THE ADDITION OF GM-CSF TO NEUTROPHILS

Na$^+$/H$^+$ Exchanger

The plasma membrane of a wide variety of cells, including neutrophils, contain a 1:1 tightly coupled antiport that exchanges extracellular Na$^+$ internal H$^+$ (134). In neutrophils this antiporter can be activated by various agents (135). It is generally found that this Na/H$^+$ exchanger system becomes deactivated in GM-CSF-treated neutrophils (136-138). While the stimulated increase in cell alkalinization is inhibited in GM-CSF-treated cells, the rapid acidification produced by fMet-Leu-Phe is magnified in these cells (136,138). Simons and her associates (136) conclude that this increased stimulated acidification in GM-CSF-primed cells does not result from disordered proton excretion, but instead, from increased release of intracellular free acid which is only partially coupled to glucose catabolism or to the generation of superoxide anion. The basis for the observed deactivation of Na$^+$/H$^+$ antiporter by GM-CSF is not known. One possibility is that the addition of GM-CSF activates this system directly (137,139). The data concerning the effect of GM-CSF on the Na$^+$/H$^+$ antiporter are contradictory. The mechanism(s) by which GM-CSF could affect the Na$^+$/H$^+$ antiport machinery are, at the present, unknown. Recently, it has been shown in hamster fibroblasts that growth factors induce direct phosphorylation of the antiporter (140).

Tyrosine Phosphorylation

The effect of GM-CSF on tyrosine phosphorylation has been

studied in different systems, including human neutrophils
(137,141), murine hematopoietic progenitor cell lines (51),
human myeloid leukemia HL-60 cells (137) and a multipotential
hematopoietic cell line (143). In HL-60 cells, prominent
phosphorylation of a protein of Mr 75 kDa (p75) occurred on
serine and tyrosine residues. Tyrosine phosphorylation of p75
is also demonstrated to occur following treatment with other
cytokines (G-CSF, IFN-γ and TNF-α) (142). The phosphorylation,
both in tyrosine and in other residues, is mediated by kinases
other than protein kinase C (51,143).

Gomez-Cambronero et al., (137) have shown that incubation
of intact neutrophils with GM-CSF triggers rapid phosphorylation on tyrosine residues of several protein substrates. The
majority of those proteins (especially one with an Mr of 41
KDa) are also phosphorylated by other neutrophil agonists such
as PAF, PMA, fMet-Leu-Phe, TNF-α and ionophore A23187
(55,137,141,144) Interestingly, in fibroblasts stimulated by a
variety of mitogens, the presence of a protein of 42 kDa has
been found to be intensely and transiently phosphorylated on
tyrosine residues (145). This protein has been postulated to be
the microtubule-associated protein kinase (MAP kinase). In
other systems, MAP kinase phosphorylates MAP-2 and S6 kinase on
serine and threonine residues (146,147). A protein of about 42
kDa has also been purified from neutrophils and shows a kinase
activity (148). The MAP kinase found in fibroblasts and the
pp41 in neutrophils (137) share some features. They are both
found in the cytosol and have a similar timecourse of tyrosine
phosphorylation (Gomez-Cambronero, J. and Sha'afi, R.I., unpublished data). The tyrosine phosphorylation of p42 may be associated with the process of secretion. Both secretion and p42
phosphorylation require extracellular calcium (149).

Protein kinases specific of tyrosine (PTK) were initially
detected in oncogenic products of viral tumors (pp60^{v-src}) and in
some growth factor receptors (150). Until recently, protein
phosphorylation on tyrosine residues was only related with cellular growth and transformation (150). During the past few
years, however, a still growing number of laboratories are re-
porting the presence of tyrosine kinase activities in non-proliferative tissues and cells. Tyrosine-specific kinases, such
as the products of the proto-oncogenes c-src(151), c-hck
(152), c-fes/fps (153) and c-fgr (154), are known to be present
in white blood cells and some are expressed at a high level.
In the course of degranulation, c-fgr is translocated from
granules to the plasma membrane (154). Other tyrosine kinase
activities have been observed in neutrophils as well (155).
Recently, attempts have been made to purify kinase activities
specific of tyrosine in the neutrophil from membrane and cytosol preparations (156). Using Western blots, it has been
demonstrated (55,137,141,144,157-162) that tyrosine-specific

protein phosphorylation occurs when human or rabbit neutrophils are stimulated with a variety of agonists, including fMet-Leu-Phe, PMA and A23187.

The importance of this type of phosphorylation in the GM-CSF-induced effects are as yet unknown. It has been suggested that PTKs might play an important role in transducing signals to the cell interior (163), or perhaps in activating some functional responses, such as superoxide production (160). Several possibilities for a physiological role of tyrosine phosphorylation in the neutrophil activation are depicted in Table III. Activation of a tyrosine kinase system is an early response to the binding of GM-CSF to its receptor (137). In this respect, the situation is analogous to that of other cytokines and growth factors, such as G-CSF, IL-3, TNF—α, γ-IFN and M-CSF (51,142,143,164). But unlike some other growth factors, the GM-CSF receptor does not bear an intrinsic tyrosine kinase activity (46). GM-CSF binding could trigger the activation of kinases, possibly closely associated to the receptor.

Gene Expression

Gene expression activation could explain some of the slow effects triggered by GM-CSF such as proliferation and differentiation. Priming effects are usually more rapid, but, even in one hour of incubation, certain genes are known to be activated. Using Northern blot analysis, it has been demonstrated that c-*fos* is constitutively expressed in neutrophils, a finding surprising enough (165). Augmentated expression of c-*fos* can be demonstrated after incubation with a variety of stimulants (165). Induction of c-*fos* mRNA is impaired by treating neutrophils with toxin of Bordetella pertussis (166). The existence of early-response genes (TIS7,8,10 and 11) in both proliferative and terminally-differentiated myeloid cells has been reported.

Table III

Possible Role of Tyrosine Phosphorylation in Neutrophils

Superoxide Production

Adhesion

Gene Expression

Protein synthesis

Regulation of Signal Transduction by phosphorylating certain key substrates:

    -PI Kinase
    -Phospholipases: PLC, PLD, PLA$_2$
    -Subunits of GTP-binding proteins

The significance of this finding is still unclear (167). On the other hand, the activation of certain proto-oncogene kinases is a common feature in activation by growth factors. M-CSF in human placental trophoblasts (169) as well as IL-3 and GM-CSF in murine myeloid cell lines (164) activate *raf* kinase activity.

In a variety of cells, GM-CSF increases IL-1, G-CSF, M-CSF and TNF (170,171) mRNa transcripts. The levels of these cytokines (except those of IFN-α, according to Ref. 171) in culture medium rose markedly by GM-CSF treatment as measured by radioimmunoassays. This indicates that protein synthesis is initiated by GM-CSF. Newly synthesized cytokines would play a role in recruiting more cells and in enhancing their performance at the site of infection.

De Novo Synthesis of Proteins

It has been shown that GM-CSF stimulates uridine incorporation into RNA (109). The addition of GM-CSF to [^{32}S]methionine-labeled neutrophils increased the incorporation of the label into TCA-precipitable material (172). Two classes of newly labeled polypeptides were detected in 2-D gels. One class changed very little upon GM-CSF treatment, whereas the relative rate of labelling of a second group increased several times (172). In separate studies using murine neutrophils (173), it was found that, out of a total of 180 proteins, the amount of label contained in 11 proteins was significantly higher in the presence of GM-CSF, while three proteins, apparently of cytoplasmatic origin, contained less label than control cells. GM-CSF, as well as TNF-α, stimulates the synthesis and secretion of a 23 kDa protein not yet characterized (174).

In addition to the rapid action of GM-CSF on surface molecules mentioned earlier, there are long-term effects involving protein synthesis. After 8 hours of treatment with GM-CSF, neutrophils synthesize more MHC class I protein, CR1, CD11b and CR3 β-chain (CD18) compared with untreated control cells (175). These proteins are constitutively expressed in neutrophils (176) but are overproduced in response to GM-CSF.

7. SIGNAL TRANSDUCTION PATHWAYS

G-Proteins

Many of the known growth factors (EGF, PDGF, ILGF) exert their actions by binding to specific receptors that bear tyrosine kinase activities (48). The need for other intermediate elements such as GTP-binding regulatory proteins (G-proteins) for the transducing of the signal is not apparent and, therefore, may not exist. On the other hand, agonists whose receptors do not contain a tyrosine kinase domain require that their receptors interact with some type of transducing mechanism. However, M-CSF which clearly belongs to the first group, might activate some intracellular enzymes through a process dependent also on G-proteins (177,178).

As stated earlier, the GM-CSF receptor does not contain a tyrosine kinase domain (46), and therefore, its receptor could be coupled to one or more G-proteins. Numerous studies have been carried out to investigate the role of known G-proteins focusing on: a) the measurement of GTPase activity, b) the measurement of G-protein subunits expression in membranes, and c) the treatment of whole cells with several bacterial toxins that are known to affect some G-proteins.

GTPase basal activity is significantly higher in GM-CSF-treated cells. Pertussis toxin reduces the increase in the basal and stimulated activity in GM-CSF-treated cells. Significant effects of GM-CSF on the GTPase activity can be seen only when the cytokine is added to the intact cells prior to the preparation of the membrane. Although not likely, the method used to prepare membranes could account for the lack of effect of GM-CSF when added directly to isolated membranes (98). In a totally different approach, the same authors (Gomez-Cambronero, J. and Sha'afi, R.I., unpublished data) have found, using immunoblots of isolated membrane preparation treated with anti-$G_i\alpha_2$ antibodies, that the amount of $G_i\alpha_2$ is higher is membranes prepared from GM-CSF-treated cells as compared to control.

Neither botulinum D toxin nor cholera toxin inhibits the fMet-Leu-Phe-induced superoxide generation in controls or GM-CSF-treated neutrophils (95). On the the other hand, pertussis toxin inhibits some biological actions produced by GM-CSF. These include: Na^+ and PO_4^{3-}-fluxes, (137), the induction of c-fos mRNA (166), intracellular alkalinization (137), membrane-associated GTPase activity (137) and, to a lesser degree, the levels of tyrosine phosphorylation of some proteins (141). Pertussis treatment also inhibits some of the functions enhanced by preincubation of neutrophils with GM-CSF, such as the enhancement of A23187-induced leukotriene synthesis (166) and PAF-induced calcium rise (95,166).

Preincubation of human neutrophils with the human hormone granulocyte-macrophage colony stimulating factor (GM-CSF) inhibits the specific binding of leukotriene B_4, [^3H]LTB$_4$, but not the non-metabolizable bioactive platelet-activating factor, [^3H]C-PAF, to intact cells (179). This inhibition requires that the GM-CSF interacts with intact cells, and it cannot be reproduced if GM-CSF is added to membrane preparation (179).

The action of GM-CSF is both dose and time dependent, and is not prevented by pertussis toxin or the protein kinase C inhibitor staurosporine. Moreover, the rise in calcium produced by LTB$_4$ but not PAF is also inhibited in human neutrophils pretreated with GM-CSF (179). Recently, it has been shown that GM-CSF directly induces neutrophils platelet-activating factor synthesis, and it stimulates neutrophils to synthesize and secrete leukotriene B$_4$ (180).

Based on all the available data, it is reasonable to state that a direct coupling of the GM-CSF receptor to any of the known G-proteins cannot be clearly established. The observation that certain actions of GM-CSF are reduced in pertussis toxin cells may reflect only an indirect dependency of some GM-CSF actions on the pertussis toxin sensitive G-proteins, and not that its receptors are coupled to any of these G-proteins.

Phospholipases

The effect of GM-CSF on various phospholipases such as phospholipases C, D and A$_2$ have been investigated in several laboratories (87,100,111,181, Yasui, K. and Sha'afi, R.I., unpublished results). The fMet-Leu-Phe-induced increases in release of various water soluble inosited such as inositol 1-phosphate (IP), inositol 1,4-bisphosphate (IP$_2$) and inositol 1-1,4,5-trisphosphate (IP$_3$) was not potentiated in GM-CSF-treated cells (182). This strongly suggests that GM-CSF does not potentiate the activation of phospholipase C. On the other hand, the fMet-Leu-Phe-induced release of arachidonic acid is greatly enhanced by GM-CSF suggesting that it potentiates the activity of phospholipase A$_2$ Finally, the fMet-Leu-Phe-induced release of PA and the production of DAG, through phospholipase D activation, are greatly increased by GM-CSF (Yasui, K. and Sha'afi, R.I., unpublished results). It is proposed that the GM-CSF potentiation of the FMLP-stimulated generation of diacylglycerol (DAG) comes through the phospholipase D pathway. All these data indicate that GM-CSF potentiates the activity of phospholipase A$_2$ and phospholipase D, but not the activity of phospholipase C. In all cases the potentiation by GM-CSF on the fMet-Leu-Phe-induced activation of the sensitive phospholipases is inhibited by pertussis toxin treatment. This strongly suggests that, at least, part of the GM-CSF action is distal to the receptor G-protein interaction.

Calcium and Protein Kinases C

In accordance with the results just mentioned, several laboratories have reported findings on the levels of calcium, a second messenger whose levels rise as a consequence of PLC ac-

tivation in the neutrophil (135). GM-CSF promotes no changes
in resting levels of calcium ions, although it enhances the
free calcium rise (94,121,138,183,184) when cells are subsequently stimulated with fMet-Leu-Phe or PAF. On the other hand,
calcium ions are more effective transduction signals in GM-CSF
treated cells (185).

It is unlikely that the calcium and phosphatidylserine
activated protein kinase C system is involved in the priming
action of GM-CSF. This view is supported by two experimental
observations. First, GM-CSF does not translocate this enzyme
system to the membrane. Second, H-7 has no effect on the superoxide release produced by fMLP in GM-CSF-treated neutrophils
(98). H-7 and H-8 enhance fMet-Leu-Phe-stimulated oxidative
burst in GM-CSF-treated cells and inhibits internalization of
the GM-CSF receptor (52).

In conclusion, GM-CSF could activate discrete pathways
unrelated to PKC but related to other kinases, such as tyrosine-specific kinases, MAP kinases or other serine/threonine
protein kinases like c-raf proto-oncogene product (169), as
stated earlier. In the case of the other CSF, (M-CSF) in which
studies are more advanced, signal transducing molecules may
include PI-3-kinase, PLCγ, C-Raf-1 protein and GTPase activating protein (GAP) (168).

Cyclases

The addition of GM-CSF to human neutrophils has been shown
to increase the intracellular level of cyclic GMP and to decrease the level of cyclic AMP (182). This is mediated, at
least in part, by affecting the activities of the two enzymes
(guanylate cyclase and adenylate cyclase) responsible for the
generation of the two cyclic nucleotides (182). The role of
these two second messengers in GM-CSF-induced actions is not
clear.

8. REGULATION OF OTHER AGONISTS RECEPTORS

GM-CSF binding to its receptors alters the expression of
other receptors and regulates the binding of their respective
ligands. GM-CSF up-regulates the receptors of γ-IFN (186), cer
tain complement components [C3bi (94), CR1 (CD35), CR3(or Mac-
1)(175)], M-CSF, and G-CSF. The upregulation of complement receptors is rapid and not dependent on protein synthesis. It is
due instead to mobilization of receptors from intracellular
compartments. There is mounting evidence that GM-CSF upregulates the fMet-Leu-Phe receptors (113,183,187) [not observed by
Yuo et al. 94)] and that it might represent the way by which
GM-CSF enhances other agonists' effects. During short times of

incubation, there is an increase in both the number and the affinity. In longer times, there is a shift to overexpression of only low affinity receptors. However, McColl et al., (188) have claimed that GM-CSF and TNFα alter the actions of certain agonists, including fMet-Leu-Phe, independently of a modulation in the expression of their cell surface receptors. In conclusion, since not all the actions produced by fMet-Leu-Phe are potentiated by GM-CSF, the significance of fMet-Leu-Phe receptors upregulation in GM-CSF-induced priming is still not completely clear.

Contrary to the above mentioned cases, GM-CSF, as stated earlier, also down-regulates leukotriene B_4 receptors. While this action is not affected by pertussis toxin or inhibitors of protein kinase C, it can be prevented by the tyrosine phosphorylation inhibitor erbstatin (179,189). The leukotriene B_4-induced calcium rise is inhibited in GM-CSF-treated cells (179). This inhibition can be overcome by increasing the concentration of leukotriene B_4 (197). This may explain why other leukotriene B_4-induced actions, such as superoxide production, which require a high dosage of leukotriene B_4 are not inhibited and, in fact, are somewhat potentiated in GM-CSF-treated cells.

GM-CSF also down-regulates IL-3 receptors in eosinophils (18,33,190) and IL-2 receptors in monocytes (191). Human GM-CSF partially inhibits the binding of human IL-3 to its receptors in neutrophils and vice versa (192), suggesting that there is a common receptor for human IL-3 and for GM-CSF, although there is no homology between both receptors.

REFERENCES

1. D. Metcalf, The granulocyte-macrophage colony-stimulating factors, Science 229: 16-22 (1985).
2. E.R. Stanley, Hematopoietic growth factors. In: Hematopoiesis: Methods in hematology. Edited by D.W. Golde, pp. 319-332. Churchill Livingstone, New York (1984).
3. D.W. Golde and Gasson, J.C., Hormones that stimulate the growth of blood cells, Scientific American, July 1988, 62-70 (1988).
4. N.A. Nicola, Hemopoietic cell growth factors and their receptors, Annu. Rev. Biochem, 58: 45-77 (1989).
5. S.C. Clark, and R. Kamen, Human hematopoietic colony-stimulating factors, Science 236: 1229-1237 (1987).
6. D.W. Golde, and J.C. Gasson, Cytokines: Myeloid growth factors, In: Inflammation: Basic Principles and Clinical Correlates, ed. J.I. Gallin, I.M. Goldstein, and R. Snyderman, Raven Press, Ltd., New York, pp. 253-261 (1988).
7. K.-I. Arai, F. Lee, A. Miyajima, S. Miyatake, N. Arai, and T. Yokota, Cytokines: Coordinators of immune and inflammatory responses, Annu. Rev. Biochem., 59: 783-836 (1990).

8. M.A. Cantrell, D. Anderson, D.P. Cerretti, V. Price, K. McK-ereghan, R.J. Tushinski, D.Y. Mochizuki, A. Larsen, K. Grab-stein, S. Gillis, and D. Cosman, Cloning, sequence, and ex-pression of a human granulocyte-macrophage colony-stimulat-ing factor. Proc. Natl. Acad. Sci. U.S.A., 82: 6250-6254 (1985).

9. A.W. Burgess, C.G., Begley, G.R. Johnson, A.F. Lopez, D.J. Williamson, J.J. Mermod, R.J. Simpson, A. Schmitz, and J.F. DeLamarter, Purification and properties of bacterially syn-thesized human granulocyte-macrophage colony-stimulating fac-tor, Blood, 69: 43-51 (1987).

10. F. Lee, T. Yokota, T. Otsuka, L. Gemmell, N. Larson, J. Luh, K.I. Arai, and D. Rennick, Isolation of cDNA for a human granulocyte-macrophage colony-stimulating factor by a functional expression in mammalian cells.Proc. Natl. Acad. Sci. U.S.A., 82: 4360-4364 (1985).

11. G.G. Wong, J.S. Witek, P.A. Temple, K.M. Wilkens, A.C. Leary, D.P. Lusenburg, S.S. Jones, E.L. Brown, R.M. Kay, E.C. Orr, C. Shoemaker, D.W. Golde, R.J. Kaufman, R.M. Hewick, E.A. Wang, and S.C. Clark, Human GM-CSF: molecu-lar cloning of the complementary DNA and purification of the natural and recombinant proteins.Science, 228: 810-815 (1985).

12. J.C. Gasson, R.H. Weisbart, S.E. Kaufman, S.C. Clark, R.M. Hewick, G.G. Wong, and D.W. Golde, Purified human granulo-cyte-macrophage colony-stimulaing factor: Direct action on neutrophils, Science, 226: 1339-1342 (1984).

13. E.P. Cronkite, Analytical review of structure and regula-tion of hemopoiesis, Blood Cells 14: 313-328 (1988).

14. W. Oster, A. Lindermann, R. Mertelsmann and F. Herrmann, Regulation of gene expression of M-, G-, GM-, and multi-CSF in normal and malignant hematopoietic cells,Blood Cells , 14: 443-462 (1988).

15. C.A. Sieff, C.M. Niemeyer, S.J. Mentzer and D.V. Faller, Interleukin-1, tumor necrosis factor, and the production of colony-stimulating factors by cultured mesenchymal cells, Blood, 72: 1316-1323 (1988).

16. T.J. Ernst, A.R. Ritchie, G.D. Demetri and J.D. Griffin, Regulation of granulocyte- and monocyte-colony stimulating factor mRNA levels in human monocytes is mediated primarily at a post-transcriptional level,J. Biol. Chem. , 264: 5700-5703 (1989).

17. D. Ridgway, M.S. Borzy, and G.C. Bagby, Granulocyte macro-phage colony-stimulating activity production by cultured human thymic nonlymphoid cells is regulated by endogenous interleukin-1, Blood, 72: 1230-1236 (1988).

18. H. Quill, A. Gaur, and R.P. Phipps, Prostaglandin E2-de-pendent induction of granulocyte-macrophage colony-stimu-lating factor secretion by cloned murine helper T cells,J. Immunol., 142: 813-818 (1989).

19. H.P. Koeffler, J. Gasson, and A. Tobler, Transcriptional

and posttranscriptional modulation of myeloid colony-stimulating factor expression by tumor necrosis factor and other agents, <u>Molec. & Cell. Biol.</u>, 8: 3432-3438 (1988)

20. K.M. Zsebo, V.N. Yuschenkoff, S. Schiffer, D. Chang, E. McCall, C.A. Dinarello, M.A., Brown, B. Altrock, and G.C. Bagby Jr., Vascular endothelial cells and granulopoiesis: interleukin-1 stimulates release of G-CSF and GM-CSF, <u>Blood</u>, 71: 99-103 (1988).

21 J.R. Zucali, C.A. Dianrello, D.J. Oblon, M.A. Gross, L. Anderson and R.S. Weiner, Interleukin 1 stimulates fibro-blasts to produce granulocyte-macrophage colony-stimulating activity and prostaglandin E2, <u>J. Clin. Invest.</u>, 77: 1857-1863 (1986).

22. W.E. Fibbe, J. Van Damme, A. Billiau, H.M. Goselink, P.J. Voogt, G. VanEeden, P. Ralph, B.W. Altrock and F.Falkenburg, Interleukin 1 induces human marrow stromal cells in longterm culture to produce granulocyte colony-stimulating factor and Macrophage Colony-Stimulating Factor, <u>Blood</u>, 71: 430-435 (1988).

23. T.S. Kupper, F. Lee, N. Birchall, S. Clark, and S. Dower, Interleukin 1 binds to specific receptors on human kera-tinocytes and induces granulocyte macrophage colony-stimu-lating factor mRNA and protein. A potential autocrine role for inter-leukin 1 in epidermis, <u>J. Clin. Invest.</u>, 82: 1787-1792 (1988).

24. T.S. Kupper, D. Coleman, J. McGuire, D. Goldminz and M. Horowitz, Keranocyte derived T cell growth factor: a T cell growth factor functionally distinct from interleukin-2, <u>Proc. Natl. Acad. Sci. U.S.A</u>, 83: 4451-4455 (1986).

25. T.S. Kupper, F. Lee, D. Coleman, J. Chodakewitz, P. Flood and M. Horowitz, Keratinocyte derived T-cell growth factor (KTGF) is identical to granulocyte macrophage colony stimu-lating factor (GM-CSF), <u>J. Investigative Dermatol</u>, 91: 185-188 (1988).

26. S. Okamura, S. Hayashi, Y. Asano, T. Shibuya, T. Otsuka and Y. Niho, Expression of the granulocyte-macrophage colony-stimulating factor gene in leukemic blast cells from pa-tients with acute non-lymphocytic leukemia, <u>Biomedicine & Pharmacotherapy</u>, 42: 65-67 (1988).

27. Y. Tsuchiya, M. Igarashi, R. Suzuki and K. Kumagai, Produc-tion of colony-stimulating factor by tumor cells and the factor-mediated induction of suppressor cells, <u>J. Immunol.</u>, 141: 699-708 (1988).

28. M.Y. Gordon, G.P. Riley, S.M. Watt and M.F. Greaves, Com-part-mentalization of a haematopoietic growth factor (GM-CSF) by glycosaminoglycans in the bone marrow microenvironment, <u>Nature</u>, 326: 403-405 (1987).

29. R. Roberts, J. Gallagher, E. Spooncer, T.D. Allen, F. Bloomfield and T.M. Dexter, Heparan sulphate bound growth factors: a mechanism for stromal cell mediated haemopoi-esis, <u>Nature</u>, 332: 376-378 (1988).

30. P. Mayer, C. Lam, H. Obenaus, E. Liehl and J. Besemer, Re-

combinant human GM-CSF induces leukocytosis and activates peripheral blood polymorphonuclear neutrophils in nonhuman primates, <u>Blood</u>, 70: 206-213 (1987).

31. I. Clark-Lewis, A.F. Lopez, L.B. To, M.A. Vadas, J.W. Schrader, L.E. Hood, and S.B.H. Kent, Structure-function studies of human granulocyte-macrophage colony-stimulating factor, <u>J. Immunol.</u>, 141: 881-889 (1988).

32. N.M. Gough, D. Grail, D.P. Gearing and D. Metcalf, Mutagenesis of murine granulocyte-macrophage colony-stimulating factor reveals critical residues near the N terminus, <u>Eur. J.Biochem.</u> 169: 353-358 (1987).

33. K. Kaushansky, S.G. Shoemaker, S. Alfaro, and C. Brown, Hematopoietic activity of granulocyte-macrophage colony-stimulating factor is dependent upon two distinct regions of the molecule: Functional analysis based upon the activities of interspecies hybrid growth factors,<u>Proc. Natl. Acad. Sci. U.S.A.</u>, 86: 1213-1217 (1989).

34. F.E. Cohen, P.A. Kosen, I.D. Kuntz, L.B. Epstein, T.L. Ciardelli and K.A. Smith, K.A., Structure-activity studies of interleukin-2, <u>Science</u>, 234: 349-352 (1986).

35. S. Miyatake, T. Otsuka, T. Yokota, F. Lee, and K. Arain, Structure of the chromosomal gene for granulocyte-macrophage colony-stimulating factor: comparison of the mouse and human genes, <u>EMBO J</u>. 4: 2561-2568 (1985).

36. E. Stanley, D. Metcalf, P. Sobieszczuk, N.M. Gough, A.R. Dunn, The structure and expression of the murine gene encoding gran-ulocyte-macrophage colony-stimulating factor: evidence for utilisation of alternative promoters,<u>EMBO J.</u> 4: 2569-2573 (1985).

37. M.M. LeBeau, N.D. Epstein, S.J. O'Brien, A.W. Nienhuis, Y.-C. Yang, S.C. Clark and J.D. Rowley, The interleukin 3 gene is located on human chromosome 5 and is deleted in myeloid leukemias with a deletion of 5q,<u>Proc. Natl. Acad. Sci. U.S.A.</u>, 84: 5913-5917.

38. K. Huebner, M. Isobe, C.M., Croce, D.W. Golde, S.E. Kaufman and J.C. Gasson, The human gene encoding GM-CSF is at 5q21-q32, the chromosome region deleted in the 5q anomaly,<u>Science</u>, 230: 1282-1285 (1985).

39. D.P. Barlow, M. Bucan, H. Lehrach, B.L. Hogan, and N.M. Gough, Close genetic and physical linkage between the murine haemopoietic growth factor genes GM-CSF and Multi-CSF, (IL3) <u>EMBO J.</u>, 6: 617-623 (1987).

40. M.F. Shannon, J.R. Gamble and M.A. Vadas, Nuclear proteins interacting with the promoter region of the human granulocyte-macrophage colony-stimulating factor gene,<u>Proc. Natl. Acad. Sci. U.S.A.</u>, 85: 674-678 (1988).

41. J. DiPersio, P. Billing, S. Kaufman, P. Eghtesady, R.E. Williams and J.C. Gasson, Characterization of the human granulocyte-macrophage colony-stimulating factor receptor, <u>J. Biol. Chem.</u>, 263: 1834-1841 (1988).

42. L.M. Budel, I.P. Touw, R. Delwel, S.C. Clark and B. Lowen-

berg, Interleukin-3 and granulocyte-monocyte colony-stimulating factor receptors on human acute myelocytic leukemia
cells and relationship to the proliferative response,
Blood, 74: 565-571 (1989).

43. H. Uzumaki, T. Okabe, N. Sasaki, K. Hagiwara, F. Takaku, M.
Tobita, K. Yasukawa, S. Ito and Y. Umezawa, Identification
and characterization of receptors for granulocyte colony-
stim-ulating factor on human placenta and trophoblastic
cells, Proc. Natl. Acad. Sci. U.S.A, 86: 9323-9326 (1989).

44. G.C. Baldwin, J.C. Gasson, S.E. Kaufman, S.G. Quan, R.E.
Williams, B.R. Avalos, A.F. Gazdar, D.W. Golde and J.F. Di-
Persio, Nonhematopoietic tumor cells express functional GM-
CSF receptors, Blood, 73: 1033-1037 (1989).

45. S. Chiba, K. Shibuya, Y.-F. Piao, A. Tojo, N. Sasaki, S.
Matsuki, K. Miyagawa, K., Miyazono and F. Takaku, Identification and cellular distribution of distinct proteins
forming human GM-CSF receptor, Cell Regulation, 1: 327-335
(1990).

46. D.P. Gearing, J.A. King, N.M. Gough and N.A. Nicola, Expression cloning of a receptor for human granulocyte-macrophage colony-stimulating factor, EMBO J., 8: 3667-3676
(1989).

47. S. Chiba, K. Shibuya, K. Miyuazono, A. Tojo, Y. Oka, K.
Miyagawa and F. Takaku, Affinity purification of human
granulocyte macrophage colony-stimulating factor receptor
α-chain, Demonstration of binding by photoaffinity
labeling, J. Biol. Chem. 265: 19777-19781 (1990).

48. Y. Yarden, Growth factor receptor tyrosine kinases, Ann.
Rev. Biochem. 57: 443-478 (1988).

49. C.J. Sherr, M.F. Roussel and Rettenmier, Colony-stimulating
factor-1 receptor (c-fms) J. Cell Biochem. 38:179-187
(1988)

50. G. Carpenter and S. Cohen, Epidermal Growth Factor, J.
Biol. Chem. 265: 7709-7712 (1990).

51. A.O. Morla, J, Schreurs, A. Miyajima, and J.Y.J. Wang, Hematopoietic growth factors activate the tyrosine phosphorylation of distinct sets of proteins in interleukin-3-de-
pendent murine cell lines, Molec. & Cell. Biol., 8: 2214
(1988).

52. A. Khwaja, P.J. Roberts, H.M. Jones, K. Yong, M.S. Jaswon
and D.C. Linch, Isoquinolinesulfonamide protein kinase
inhibitors H7 and H8 enhance the effects of granulocyte-
macrophage colony-stimulaing factor (GM-CSF) on neutrophil
function and inhibit GM-CSF receptor internalization, Blood
76: 996-1003 (1990).

53. F. Walker, N.A. Nicola, D. Melcalf and A.W. Burgess, Hier-
archial down-modulation of hemopoietic growth factor receptors, Cell, 43: 269-276 (1985).

54. S.A. Cannistra, M. Koenigsmann, J. DiCarlo, P. Groshek and
J.D. Griffin, Differentiation-associated expression of two
functionally distinct classes of granulocyte-macrophage

colony-stimulaing factor receptors by human myeloid cells, J. Biol. Chem., 265: 12656-12663 (1990).

55. J. Gomez-Cambronero, C.-K. Huang, M. Yamazaki, E. Wang, T.F.P. Molski, E.L. Becker and R.I. Sha'afi, Phorbol ester inhibits granulocyte-macrophage colony-stimulating factor binding and tyrosine phosphorylation. Am. J. Physiol (Cell Physiol). In press.(1992).

56. N.A. Nicola, M.A., Vadas, and A.F. Lopez, Down-modulation of receptors for granulocyte colony-stimulating factor on human neutrophils by granulocyte-activating agents, J. Cell. Physiol., 128: 501-509 (1986).

57. R.I. Sha'afi and T.F.P. Molski, Inhibition of stimulated cell responses by phorbol esters and other activators of protein kinase C: Sites of action. Membrane Biochem. 7: 143-152 (1985).

58. T. Hunter, N. Ling and J.A. Cooper, Protein kinase C phosphorylation of the EGF receptor at a threonine residue close to the cytoplasmatic face of the plasma membrane, Nature, 311: 480-483 (1985).

59. M.I. Wahl, S. Nishibe and G. Carpenter, Cancer cells, Growth factor signaling pathways: phosphoinositide metabolism and phosphorylation of phospholipase C, Cancer Cells, 1: 101-107 (1989).

60. W.S. May, S. Jacobs and P. Cuatrecasas, Association of phorbol ester-induced hyperphosphorylation and reversible regulation of transferrin membrane receptors in HL60 cells, Proc. Natl. Acad. Sci. U.S.A. 81: 2016-2020 (1984).

61. J.R. Downing, M.F. Roussel and C.J. Sherr, Ligand and protein kinase C downmodulate the colony-stimulating factor 1 receptor by independent mechanisms, Mol. Cell. Biol. 9: 2890-2896 (1989).

62. E.M. Bonnem and G. Morstyn, Granulocyte macrophage colony stimulating factor (GM-CSF) current status and future development, Seminars in Oncology, 15: 46-51 (1988).

63. D. Caracciolo, S.C. Clark and G. Rovera, Human interleukin-6 supports granulocytic differentiation of hematopoietic progenitor cells and acts synergistically with GM-CSF, Blood, 73: 666-670 (1989).

64. D.J. Warren, and M.A. Moore, Synergism among interleukin 1, interleukin 3, and interleukin 5 in the production of eosinophils from primitive hemopoietic stem cells, J. Immunol.140:, 94-99 (1988).

65. H.E. Broxmeyer, D.E. Williams, G. Hangoc, S. Cooper, S. Gillis, R.K. Shadduck and D.C. Bicknell, Synergistic myelopoietic actions in vivo after administrations of purified natural murine colony-stimulating factor 1, recombinant murine interleukin 3, and recombinant murine granulocyte/macrophage colony-stimulating factor, Proc. Natl. Acad. Sci. U.S.A. 84: 3871-3875 (1987).

66. R.E. Donahue, M. Seehra, M. Metzger, D. Lefbvre, B. Rock, S. Carbone, D.G. Nathan, M. Garnick, P.K. Sehgal, and D.

Laston, Human IL-3 and GM-CSF act syndergistically in stimulating hematopoiesis in primates,<u>Science</u> 241: 1820-1823 (1988).

67. W.P. Peters, The effect of recombinant human colony-stimulating factors on hematopoietic reconstiution following autologous bone marrow transplantation,<u>Seminars in Hematology</u>, 26: 18-23 (1989).

68. G. Morstyn, G.J. Lieschke, W. Sheridan, J. Layton, J. Cebon and R.M. Fox, Clinical experience with recombinant human granulocyte colony-stimulating factor and granulocyte macrophage colony-stimulating factor,<u>Seminars in Hemotology,</u> 26: 9-13 (1989).

69. W.P. Steward, N. Thatcher and S.B. Kaye, Clinical applications of myeloid colony stimulating factors,<u>Cancer Treatment Re-views</u>, 17: 77-87 (1990).

70. J.L. Abrilove, Introduction and overview of hematopoietic growth factors, <u>Seminars in Hematology</u> 26: 1-4 (1989).

71. J.A. Glaspy and D.W. Golde, Clinical applications of the myeloid growth factors, <u>Seminars in Hematology</u>, 26: 14-17 (1989).

72. F. Herrmann, G. Schulz, A. Lindemann, W. Meyenburg, W. Oster, D. Krumwieh and R. Mertelsmann, Hematopoietic responses in patients with advanced malignancy treated with recombinant human granulocyte-macrophage colony-stimulating factor, <u>J. Clin. Oncology,</u> 7: 159-167 (1989).

73. W.P. Steward, J.H. Scarffe, R. Austin, E. Bonnem, N. Thatcher, G. Morgenstern and D. Crowther, Recombinant human granulocyte macrophage colony stimulating factor (rhGM-CSF) given as daily short infusion. A phase I dose-toxicity study, <u>British J. Cancer</u>, 59: 142-145 (1989).

74. J.H. Antin, B.R. Smith, W. Holmes, and D.S. Rosenthal, Phase I/II study of recombinant human granulocyte-macrophage colony-stimulating factor in aplastic anemia and myelodysplastic syndrome, <u>Blood</u>, 72: 705-713 (1988).

75. R.E. Champlin, S.D. Nimer, P. Ireland, D.H. Oette and D.W. Golde, Treatment of refractory aplastic anemia with recombinant human granulocyte-macrophage colony-stimulating factor, <u>Blood</u>, . 73: 694-699 (1989).

76. C. Nissen, A. Tichelli, A. Grathwohl, B. Speck, A. Milne, E.C. Gordon-Smith and J. Schaedelin, Failure of recombinant human granulocyte-macrophage colony-stimulating factor therapy in aplastic anemia patients with very severe neutropenia, <u>Blood</u>, 72: 2045-2047 (1988).

77. S. Vadhan-Raj, M. Keating, A. LeMaistre, W.M. Hittelman, K. McCredie, J.M. Trujillo, H.E. Broxmeyer, H.E., C. Henney and J.U. Gutterman, Effects of recombinant human granulocyte-macrophage colony-stimulating factor in patients with myelodysplastic syndromes,<u>New England J. Med.</u>, 317: 1545-1452 (1987).

78. K.S. Antman, J.D. Griffin, A. Elias, M.A. Socinski, L. Ryan, S.A. Cannistra, D. Oette, M. Whitley, E. Frei and

L.E. Schnipper, Effect of recombinant human granulocyte-macrophage colony-stimulating factor on chemotherapy-induced myelosuppression, <u>New England J. Med.</u>, 319: 593-598 (1988).

79. J.H. Edmonson, H.J. Long, J.A. Jeffries, J.C. Buckner, G. Culen-Otero, T.R. Fitch, Amelioration of chemotherapy-induced thrombocytopenia by GM-CSF: apparent dose and schedule dependency, <u>J. Natl. Cancer Inst.</u>, 81: 1510-1512 (1989).

80. B.R. Blazar, J.H. Kersey, P.B. McGlave, D.A. Vallera, L.C. Lasky, R.J. Haake, B. Bostrom, D.R. Weisdorf, C. Epstein and N.K. Ramsay, In vivo administration of recombinant human granulocyte-macrophage colony-stimulating factor in acute lymphoblastic leukemia patients receiving purged autografts, <u>Blood</u>, 73: 849-857 (1989).

81. S.J. Brandt, W.P. Peters, S.K. Atwater, J. Kurtzberg, M.J. Borowitz, R.B. Jones, E.J. Shpall, R.C. Bast, C.J. Gilbert and D.H. Oette, Effect of recombinant human granulocyte-macrophage colony-stimulating factor on hematopoietic reconstituion after high-dose chemotherapy and autologous bone marrow transplantion, <u>New England J. Med</u>, 318: 869-876 (1988).

82. J. Nemunaitis, J.W. Singer, C.D. Buckner, R. Hill, R. Storb, E.D. Thomas, and F.R. Appelbaum, Use of recombinant human granulocyte-macrophage colony-stimulating factor in autologous marrow transplantation for lymphoid malignancies, <u>Blood</u>, 72: 834-836 (1988).

83. M.A. Socinski, S.A. Cannistra, A. Elias, K.H. Antman, L. Schnipper and J.D. Griffin, Granulocyte-macrophage colony-stimulating factor expands the circulating haemopoietic progenitor cell compartment in man, <u>Lancet</u>, 1: 1194-1198 (1988).

84. K.H. Grabstein, D.L. Urdal, R.J. Tushinski, D.Y. Mochizuki, V.L. Price, M.A. Cantrell, S. Gillis and P.J. Conlon, Induction of macrophage tumoricidal activity by granulocyte-macrophage colony-stimulating factor, <u>Science</u>: 232: 506-508 (1986).

85. K. Miyagawa, S. Chiba, S., Shibuyak, Y.-F. Piao, S. Matsuki, J. Yokota, M. Terada, K. Miyazono and F. Takaku, F., Frequent expression of receptors for granulocyte-macrophage colony-stimulating factor of human nonhematopoietic tumor cell lines. <u>J. Cell Physiol.</u> 143: 483-487 (1990).

86. S. Dedhar, L. Gaboury, P. Galloway and C. Eaves, Human granulocyte-macrophage colony-stimulating factor is a growth factor active on a variety of cell types of nonhemopoietic origin, <u>Proc. Natl. Acad. Sci. U.S.A</u>, 85: 9253-9257 (1988).

87. C. Laker, C. Stocking, U. Bergholz, N. Hess, J.F. De Lamarter and W. Ostertag, Autocrine stimulation after transfer of the granulocyte-macrophage colony-stimulating factor gene and au-tonomous growth are distinct but interdependent

steps in the oncogenic pathway, Proc. Natl. Acad. Sci. U.S.A., 84: 8458-8462 (1987).

88. G.Y. Cheng, C.A. Kelleher, J. Miyauchi, C. Wang, G. Wong, S.C. Clark, E.A. McCulloch and M.D. Minden, Structure and expression of genes of GM-CSF and G-CSF in blast cells from patients with acute myeloblastic leukemia, Blood, 71: 204-208 (1988).

89. R.H. Weisbart, D.W. Golde, S.C. Clark, G.G. Wong and J.C. Gasson,1 Human granulocyte-macrophage colony-stimulaing factor is a neutrophil activator, Nature, 314: 361-363 (1985).

90. R.H. Weisbart, L. Kwan, D.W. Golde and J.C. Gasson, Human GM-CSF primes neutrophils for enhanced oxidative metabolism in response to the major physiological chemoattractants, Blood, 69: 18-21 (1987).

91. M. Klausmann, K.H. Pfluger, D. Krumwieh, F.R. Seiler and K. Havemann, Influence of recombinant human granulocyte-macrophage colony-stimulating factor on granulocyte functions, Behring Institute Mitteilungen 83: 265-269 (1988).

92. M. Klausmann, K.H. Pfluger, D. Krumwieh, F.R. Seiler and K. Havemann, Modulation of functions of granulocytes by recombinant human GM-CSF and possible complications of GM-CSF therapy, Leukemia, 2: 63S-72S (1988).

93. A.F. Lopez, D.J. Williamson, J.R. Gamble, C.G. Begley, J.M. Harlan, S.J. Klebanoff, A. Waltersdorph, G. Wong, S.C. Clark and M.A. Vadas, Recombinant human granulocyte-macrophage colony-stimulating factor stimulates in vitro mature human neutrophil and eosinophil function, surface receptor expression, and survival, J. Clin. Invest., 78: 1220-1228 (1986).

94. A. Yuo, S. Kitagawa, A. Ohsaka, M. Ohta, K. Miyazono, T. Okabe, A. Urabe, M. Saito and F. Takaku, Recombinant human granulocyte colony-stimulating factor as an activator of human granulocytes: Potentiation of responses triggered by receptor-mediated agonists and stimulation of C3bi receptor expression and adherence, Blood, 74: 2144-2149 (1989).

95. J.L. Mege, J. Gomez-Cambronero, T.F. Molski, E.L. Becker and R.I. Sha'afi, Effect of granulocyte-macrophage colony-stimulating factor on superoxide production in cytoplasts and intact human neutrophils: Role of protein kinase and G-proteins, J. Leukocyte Biol., 46: 161-168 (1989).

96. R. Sullivan, J.P. Fredette, J.L. Leavitt, A.S. Gadenne, J.D. Griffin and E.R. Simons, Effects of recombinant human granulocyte-macrophage colony-stimulating factor (GM-CSFrh) on transmembrane electrical potentials in granulocytes: Relationship between enhancement of ligand-mediated depolarization and augmentation of superoxide anion O_2^- production, J. Cell. Physiol., 139: 361-369 (1989).

97. R. Sullivan, J.P. Fredette, M. Socinski, A. Elias, K. Antman, L. Schnipper, and J.D. Griffin, Enhancement of superoxide anion release by granulocytes harvested from patients

receiving granulocyte-macrophage colony-stimulaing factor, Br. J. Haematol., 71: 475-479 (1989).

98. R.I. Sha'afi, J. Gomez-Cambronero, M. Yamazaki, M. Durstin, T.F.P. Molski and C.-K. Huang, Activation of human neutrophils by granulocyte-macrophage colony-stimulating factor: Role of guanine-nucleotide binding proteins, In:Biology of Cellular Transducing Signals eds. Vanderhoek, J.Y., Axelrod, J., Jelesma, C. and Moody, T.W. , Plenum Press, NY, pp. 153-162.

99. A. Kharazmi, H. Nielsen and K. Bendtzen, Modulation of human neutrophil and monocyte chemotaxis and superoxide responses by recombinant TNFα and GM-CSF, Immunobiol., 177: 363-370 (1988).

100. J.F. DiPersio, P. Billing, R. Williams and J.C. Gasson, Human granulocyte-macrophage colony-stimulating factor and other cytokines prime human neutrophils for enhanced archidonic acid release and leukotriene B_4 synthesis, J. Immunol., 140: 4315-4322 (1988).

101. D.S. Silberstein, W.F. Owen, J.C. Gasson, J.F. DiPersio, D.W. Golde, J.C. Bina, R. Soberman, K.F. Austen and J.R. David, Enhancement of human eosinophil cytotoxicity and leukotriene synthesis by biosynthetic (recombinant) granulocyte-macrophage colony-stimulating factor, J. Immunol., 137: 3290-3294 (1986).

102. J.F. DiPersio, P.H. Naccache, P. Borgeat, J.C. Gasson, M.H. Nguyen and S.R. McColl, Characterization of the priming effects of human granulocyte-macrophage colony-stimulating factor on human neutrophil leukotriene synthesis, Prostaglandins, 36: 673-691 (1988).

103. C.A. Dahinden, J. Zingg, F.E. Maly and A.L. de Weck, Leukotriene production in human neutrophils primed by recombinant human granulocyte-macrophage colony-stimulaing factor and stimulated with the complement component C5A and FMLP as second signals, J. Exper. Med., 167: 1281-1295 (1988).

104. S. Fiore and C.N. Serhan, Formation of lipoxins and leukotrienes during receptor-mediated interactions of human platelets and recombinant human granulocyte/macrophage colony-stimulating factor-primed neutrophils, J. Exp. Med. 172: 1451-1457 (1990).

105. S.R. Tyagi, E.F. Winton and J.D. Lambeth, Granulocyte-macrophage colony-stimulating factor primes human neutrophils for increased diacylglycerol generation in response to chemoattractant, FEBS Lett., 257: 188-190 (1989).

106. S.J. Corey and P.M. Rosoff, Granulocyte-macrophage colony-stimulating factor primes neutrophils by activating a pertussis toxin-sensitive G protein not associated with phosphatidyl-inositol turnover, J. Biol. Chem., 264: 14165-14171 (1989).

107. U. Wirthmueller, A.L. De Weck and C.A. Dahinden, Platelet-activating factor production in human neutrophils by se-

quential stimulation with granulocyte-macrophage colony-stimulating factor and the chemotactic factors C5A or formyl-methionyl-leucyl-phenylalanine, J. Immunol., 142: 3213-3218 (1989).

108. M. Aglietta, C. Monzeglio, F. Apr'a, C. Mossetti, A.C. Stern, G. Giribaldi and F. Bussolino, In vivo priming of human normal neutrophils by granulocyte-macrophage colony stimulating factor: Effect on the production of platelet activating factor, Br. J. Haematology, 75: 333-339 (1990).

109. U. Wirthmueller, A.L. de Weck and C.A. Dahinden, Studies on the mechanism of platelet-activating factor production in GM-CSF primed neutrophils: Involvement of protein synthesis and phospholipase A_2 activation, Biocem. Biophys. Res. Commun., 170: 556-562 (1990).

110. M. Yamazaki, J. Gomez-Cambronero, M. Durstin, T.F.P. Molski, E.L. Becker and R.I. Sha'afi, Phorbol 12-myristate 13-acetate inhibits binding of leukotriene B_4 and platelet-activating factor and the responses they induce in neutrophils: Site of action, Proc. Natl. Acad. Sci. U.S.A. 86: 5791-5794 (1989).

111. J. Gomez-Cambronero, M. Durstin, T.F.P. Molski, P.H. Naccache and R.I. Sha'afi, Calcium is necessary but not sufficient for the platelet-activating factor release in human neutrophils stimulated by physiological stimuli, J. Biol. Chem. 264: 21699-21704 (1989).

112. R.H. Weisbart, D.W. Golde, L. Spolter, P. Eggena and H. Rinderknecht, Neutrophil migration inhibition factor from T-lymphocytes (NIF-T): a new lymphokine. Clin. Immunol. Immunopath-ol., 14: 441-448 (1979).

113. R.H. Weisbart, D.W. Golde and J.C. Gasson, Biosynthetic human GM-CSF modulates the number and affinity of neutrophil fMet-Leu-Phe receptors, J. Immunol., 137: 3584-3587 (1986).

114. F. Busssolino, J.M. Wang, P. Defilipii, F. Turrini, F., Sanavio, C.J. Edgell, M. Aglietta, P. Arese and A. Mantonavi, Granulocyte and granulocyte-macrophage colony-stimulating factors induce human endothelial cells to migrate and proliferate, Nature, 337:471-473 (1989).

115. E. Kownatzki, E. Liehl, H. Aschauer and S. Uhrich, Inhibition of chemotactic migration of human neutrophilic granulocytes by recombinant human granulocyte-macrophage colony-stimulating factor, Immunopharm., 19: 139-143 (1990).

116. W.P. Peters, A. Stuart, M.L. Affronti, C.S. Kim and R.E. Coleman, Neutrophil migration is defective during recombinant human granulocyte-macrophage colony-stimulating factor infusion after autologous bone marrow transplantation in humans, Blood, 72: 1310-1315 (1988).

117. I.E. Addison, B. Johnson, S. Devereux, A.H. Goldstone and D.C. Linch, Granulocyte-macrophage colony-stimulating factor may inhibit neutrophils migration in vivo, Clin. & Exper. Immunol., 76: 149-153 (1989).

118. Y. Kletter, I. Bleiberg, D.W. Golde and I. Fabian, Antibody to Mol abrogates the increase in neutrophil phagocytosis and degranulation induced by granulocyte-macrophage colony-stimulating factor, <u>Eur. J. Haematology</u>, 43: 389-396 (1989)

119. J. Richter, T. Anderssson and I. Olsson, Effect of tumor necrosis factor and granulocyte-macrophage colony-stimulating factor on neutrophil degranulation,<u>J. Immunol.</u>, 142: 3199-3205 (1989).

120. S.E. Kaufman, J.F. DiPersio and J.C. Gasson, Effects of human GM-CSF on neutrophil degranulation in vitro,<u>Exper. Hematol.</u>, 17: 800-804 (1989).

121. R.J. Smith, J.S. Justen and L.M. Sam, Recombinant human granulocyte-macrophage colony-stimulating factor induces granule exocytosis from human polymorphonuclear neutrophils, <u>Inflammation</u>, 14: 83-92 (1990).

122. S. Devereux, J.B. Porter, K.P. Hoyes, R.D. Abeysinghe, R., Saib and D.C. Linch, Secretion of neutrophil secondary granules occurs during granulocyte-macrophage colony stimulating factor induced margination,<u>Br. J. Haematol.</u>, 74: 17-23 (1990).

123. B.H. Kushner and N.K. Cheung, GM-CSF enhances 3F8 monoclonal antibody-dependent cellular cytotoxicity against human melanoma and neuroblastoma,<u>Blood</u>, 73: 1926-1941 (1989).

124. J. Fleischmann, D.W. Golde, R.H. Weisbart and J.C. Gasson, Granuloctye-macrophage colony-stimulating factor enhances phagocytosis of bacteria by human neutrophils, <u>Blood</u>, 68: 708-711 (1986).

125. A. Kapp, G. Zeck-Kapp, M. Danner and T.A. Luger, Human granulocyte-macrophage colony-stimulating factor: An effective direct activator of human polymorphonuclear neutrophilic granulocytes. <u>J. Invest. Dermatol</u>. 91: 49-55 (1988).

126. C.F. Nathan, Respiratory burst in adherent human neutrophils: Triggering by colony-stimulating factors CSF-GM and CSF-G, <u>Blood</u>, 73: 301-306 (1989).

127. C.F. Nathan, Neutrophil activation on biological surfaces. Massive secretion of hydrogen peroxice in response to products of macrophages and lymphocytes.<u>J. Clin. Invest</u>. 80: 1550-1560 (1987).

128. A.M. Buckle, Y. Jayaram and N. Hogg, Colony-stimulating factors and interferonγ differentially affect cell surface molecules shared by monocytes and neutrophils,<u>Clin. & Exp. Immunol.</u>, 81: 339-345 (1990).

129. M.A. Socinski, S.A. Cannistra, R. Sullivan, A. Elias, K. Antman, L. Schnipper and J.D. Griffin, Granulocyte-macrophage colony-stimulating factor induces the expression of the CD11b surface adhesion molecule on human granulocytes in vivo, <u>Blood</u>, 72: 691-697 (1988).

130. M.A. Arnaout, E.A. Wang, S.C. Clark and C.A. Sieff, Human

recombinant granuloctye-macrophage colony-stimulating factor increases cell-to-cell adhesion and surface expression of adhesion-promoting surface glycoproteins on mature granulocytes. J. Clin. Invest. 78: 597-601 (1986).

131. S. Devereux, H.A. Bull, D. Campos-Costa, R. Saib and D.C. Linch, Granulocyte macrophage colony stimulating factor induced changes in cellular adhesion molecule expression and adhesion to endothelium: in vitro and in vivo studies in man, Br. J. Haematology, 71: 323-330 (1989).

132. J.R. Gamble, T.H. Rand, A.F. Lopez, I. Clark-Lewis and M.A. Vadas, Heterogeneity of recombinant granulocyte-macrophage colony-stimulating factor-mediated enhancement of neutrophil adherence to endothelium, Exper. Hematology, 18: 897-902 (1990).

133. S. Yamaga, S. Okamura, T. Otsuka and Y. Niho, Effect of granulocyte-macrophage colony-stimulating factor on chemiluminescence of human neutrophils, Intl. J. Cell Cloning 7: 50-58 (1989).

134. R.I. Sha'afi and T.F.P. Molski, Effects of neutrophil and platelet activators. In: Na$^+$/H$^+$ exchange. Ed. S. Grinstein, CRC Press Boca Raton, Florida (1988).

135. R.I. Sha'afi and T.F.P. Molski, Activation of the neutrophil. Progress in Allergy 42: 1-64 (1988).

136. R. Sullivan, J.D. Griffin, J. Wright, D.A. Melnick, J.L. Leavitt, J.P. Fredette, J.H. Horne, C.A. Lyman, K.G. Lazzari and E.R. Simons, Effects of recombinant human granulocyte-macrophage colony-stimulating factor on intracellular pH in mature granulocytes, Blood, 72: 1665-1673 (1988).

137. J. Gomez-Cambronero, M. Yamazaki, F. Metwally, T.F.P. Molski, V.A. Bonak, C.-K. Huang, E.L. Becker and R.I. Sha'afi, Granulocyte-macrophage colony-stimulating factor and human neutrophils: Role of guanine nucleotide regulatory proteins, Proc. Natl. Acad. Sci.U.S.A, 86: 3569-3573 (1989).

138. P.H. Naccache, N. Faucher, P. Borgeat, J.C. Gasson and J.F. DiPersio, Granulocyte-macrophage colony-stimulating factor modulates the excitation-response coupling sequence in human neutrophils, J. Immunol., 140:3541-3546 (1988).

139. F. Bussolino, J.M. Wang, F. Turrini, D. Alessi, D. Ghigo, C. Costamagna, G. Perscarmona, A. Mantovani and A. Bosia, Stimulation of the Na/H$^+$ exchanger in human endothelial cells activated by granulocyte- and granulocyte-macrophage colony-stimulating factor, J. Biol. Chem. 264: 18284-18287 (1989).

140. C. Sardet, L. Counillon, A. Franchi and J. Pouyssegur, Growth factors induce phosphorylation of the Na/H$^+$ antiporter, a glycoprotein of 110 kD, Science 247: 723-726, (1990).

141 J. Gomez-Cambronero, C.-K. Huang, V.A. Bonak, E. Wang, J.E. Casnellie, T. Shiraishi and R.I. Sha'afi, Tyrosine phos-

phorylation in human neutrophil,<u>Biochem. Biophys. Res. Commun</u>. 162: 1478-1485 (1989).

142. J.P.M. Evans, A.R. Mire-Sluis, A.V. Hoffbrand and R.G. Wick-remasinghe, Binding of G-CSF, GM-CSF, tumor necrosis factor-α, and γ-interferon to cell surface receptors on human myeloid leukemia cells triggers rapid tyrosine and serine phosphorylation of a 75-Kd protein,<u>Blood</u>, 75: 88-95 (1990).

143. P.H. Sorensen, A.L. Mui, S.C. Murthy and G. Krystal, Interleukin-3, GM-CSF, and TPA induce distinct phophorylation events in an interleukin 3-dependent multipotential cell line, <u>Blood</u>, 73: 406-418 (1989).

144. J. Gomez-Cambronero, E. Wang. G. Johnson, C.-K. Huang and R.I. Sha'afi, Platelet activating factor induces tyrosine phosphorylation in human neutrophils, <u>J. Biol. Chem.</u> 266: 6240-6245 (1991).

145. A.J. Rossomando, D.M. Payne, M.J. Weber and T.W. Sturgill, Evidence that pp42, a major tyrosine kinase target protein, is a mitogen-activated serine/threonine protein kinase, <u>Proc. Natl. Acad. Sci. U.S.A</u>, 86: 6940-6943 (1989).

146. E. Erikson, D. Stefanovic, J. Blenis, R.L. Erikson and J.L. Maller, Antibodies to Xenopus egg S6 kinase II recognize S6 kinase from progesterone- and insulin-stimulated Xenopus oocytes and from proliferating chicken embryo fibroblasts. <u>Mol. Cell. Biol.</u> 7: 3147-3155 (1987).

147. L.B. Ray and T.W. Sturgill, Rapid stimulation by insulin of a serine/threonine kinase in 3T3-L1 adipocytes that phosphorylates microtubule-associated protein 2 in vitro. <u>Proc. Natl. Acad. Sci. U.S.A</u>.84: 1502-1506 (1987).

148. K.F. Balazovich and E.L. McEwen, Purification and characterization of a soluble 42 kilodalton protein kinase from human neutrophils. <u>J. Cell. Biol.</u> III, 49a (abstract) (1990).

149. C.M. Ely, K.M. Oddie, J.S. Litz, A.J. Rossomando, S.B. Kanner, T.W. Sturgill and S.J. Parsons, A 42-kD tyrosine kinase substrate linked to chromaffin cell secretion exhibits an associated MAP kinase activity and is highly related to a 42 kD mitogen-stimulated protein in fibroblasts, <u>J. Cell Biol.</u>, 110: 731-742 (1990).

150. T. Hunter and J.A. Cooper, Protein tyrosine kinases <u>Ann. Rev. Biochem.</u> 54: 897-930 (1985).

151. C.E. Gee, J. Griffin, L. Sastre, L.J. Miller, T.A., Springer, H. Piwnica-Worms and T.M. Roberts, Differentiation of myeloid cells is accompanied by increased levels of pp60c-src. protein and kinase activity<u>Proc. Natl. Acad. Sci. U.S.A</u>. 83: 5131-5135 (1986).

152. S.F. Ziegler, C.B. Wilson and R.M. Perlmutter, Augmented ex-pression of a myeloid-specific protein tyrosine kinase gene (hck) after macrophage activation. <u>J Exp. Med.</u> 168: 1801-1810 (1988).

153. G. Yu, T.E. Smithgall and R.I. Glazer, K562 leukemia cells transfected with human c-fes gene acquire the ability to undergo myeloid differentiation. J. Biol. Chem. 264: 10276-10281 (1989).

154. J.S. Gutkind and K.C. Robins, Translocation of the FGR protein-tyrosine kinase as a consequence on neutrophil activation. Proc. Natl. Acad. Sci. U.S.A. 86: 8783-8787 (1989).

155. A.S. Kraft and R.L. Berkow, Tyrosine kinase and phosphotyrosine phosphatase activity in human promyelocytic leukemia cells and human polymorphonuclear leucocytes. Blood, 70: 356-362 (1987).

156. R.L. Berkow, R.W. Dodson and A.S. Kraft, Human neutrophils contain distinct cytosolic and particulate tyrosine kinase activities: Possible role in neutrophil activation. Biochem. Biophys. Acta 997: 292-301 (1989).

157. R.L. Berkow and R.W. Dodson, Tyrosine-specific protein phosphorylation during activation of human neutrophils, Blood, 75: 2445-2452 (1990).

158. C.-K. Huang, V. Bonak, G.R. Laramee and J.E. Casnellie, Protein tyrosine phosphorylation in rabbit peritoneal neutrophils, Biochem. J. 269: 431-436 (1990).

159. C.-K. Huang, G.R. Laramee and J.E. Casnellie, Chemotactic factor induced tyrosine phosphorylation of membrane associated proteins in rabbit peritoneal neutrophils. Biochem. Biophys. Res. Commun, 151: 794-801 (1988).

160. P.H. Naccache, C. Gilbert, A.C. Caon, M. Gaudry, C.-K. Huang, V.A. Bonak, K. Umezawa and S.R. McColl, Selective inhibition of human neutrophil functional responsiveness by Erbstatin, an inhibitor of tyrosine protein kinase, Blood, 76: 2098-2104 (1990).

161. P.E. Nasmith, G.B. Mills and S. Grinstein, Guanine nucleotides induce tyrosine phosphorylation and activation of the respiratory burst in neutrophils. Biochem. J. 257: 893-897 (1989).

162. T.C. Wright, M.J. Karnovsky and J.M. Robinson, Tyrosine phosphorylation of 43 Kd and 41 Kd proteins occurs during PMN activation, J. Cell. Biol. 107: 57a (1988).

163. M. Katan and P.J. Parker, Oncogenes and cell control Nature 32: 203 (1988).

164. J.R. Downing, C.W. Rettenmier and C.J. Sherr, Ligand-induced tyrosine kinase activity of the colony-stimulating factor 1 receptor in a murine macrophage cell line Mol. Cell. Biol. 8: 1795-1799 (1988).

165. F. Colotta, J.M. Wang, N. Polentaruttim and A. Mantovani, Expression of c-fos proto-oncogene in normal human peripheral blood granulocytes, J. Exp. Med. , 165: 1224-1229 (1987).

166. S.R. McColl, C. Kreis, J.F. DiPersio, P. Borgeat and P.H. Naccache, Involvement of guanine nucleotide binding pro-

teins in neutrophil activation and priming by GM-CSF, _Blood,_ 73: 588-591 (1989).

167. B.C. Varnum, R.W. Lim, D.A. Kujubu, S.J. Luner, S.E. Kaufman, J.S. Greenberger, J.C. Gasson and H.R. Herschman, Granulocyte-macrophage colony-stimulating factor and tetradecanoyl phorbol acetate induce a distinct, restricted subset of primary response TIS genes in both proliferating and terminally differentiated myeloid cells, _Molec. & Cell. Biol.,_ 9: 3580-3583 (1989).

168. G.G. Choudhury, V.L. Sylvia, A. Pfeifer, L.-M. Wang, E.Z. Smith and A.Y. Sakaguchi, Human colony stimulating factor-1 receptor activates the C-raf-1 proto-oncogene kinase, _Biochem. Biophys. Res. Commun._ 172: 154-159 (1990).

169. M.P. Carroll, I. Clark-Lewis U.R. Rapp and W.S. May, Interleukin-3 and granulocte-macrophage colony-stimulating factor mediate rapid phosphorylation and activation of cytosolic c-raf. _J. Biol. Chem._ 265: 19812-19817 (1990).

170A. Lindemann, D. Riedel, W. Oster, S.C. Meuer, D. Blohm, R.H. Mertelsmann, and F. Herrmann, Granulocyte/macrophage colony-stimulating factor induces interleukin 1 production by human polymorphonuclear neutrophils, _J. Immunol.,_ 140: 837-839 (1988).

171. A. Lindenmann, D. Riedel, W. Oster, H.W. Ziegler-Heitbrock, R. Mertelsmann and F. Herrmann, Granulocyte-macrophage colony-stimulating factor induces cytokine secretion by human polymorphonuclear leukocytes. _J. Clin. Invest.,_ 83: 1308-1312 (1989).

172. S.W. Edwards, C.S. Holden, J.M. Humphreys and C.A. Hart, Granulocyte-macrophage colony-stimulating factor (GM-CSF) primes the respiratory burst and stimulates protein bio-synthesis in human neutrophils, _FEBS Lett.,_ 256: 62-69 (1989).

173. I.J. Stanley and A.W. Burgess, Granulocyte macrophage-colony stimulating factor stimulates the synthesis of membrane and nuclear proteins in murine neutrophils, _J. Cell. Biochem.,_ 23: 241-258 (1983).

174. S.R. McColl, R. Paquin and A.D. Beaulieu, Selective synthesis and secretion of a 23 Kd protein by neutrophils following stimulation with granulocyte-macrophage colony-stimulating factor and tumor necrosis factor-alpha, _Biochem. Biophys. Res. Commun._ 172: 1209-1216 (1990).

175. E. Neuman, J.W. Huleatt and R.M. Jack, Granulocyte-macrophage colony-stimulaing factor increase synthesis and expression of CR1 and CR3 by human peripheral blood neutrophils, _J. Immunol.,_ 145: 3325-3332 (1990).

176 R.M. Jack and D.T. Fearon, Selective synthesis of mRNA and proteins by human peripheral blood neutrophils. _J. Immunol._ 140: 4286-4293 (1988).

177. Y. He, E. Hewlett, D. Temeles and P. Quesenberry, Inhibition of interleukin 3 and colony-stimulating factor 1-stimulated marrow cell proliferation by pertussis toxin, _Blood,_ 71: 1187-1195 (1988).

178 K. Imamura and D. Kufe, Colony-stimulating factor 1-induced Na$^+$ influx into human monocytes involves activation of a pertussis toxin-sensitive GTP-binding protein,J. Biol. Chem., 263: 14093-14098 (1988).

179. M. Yamazaki, T.F.P. Molski, T. Stevens, C.-K. Huang, E.L. Becker and R.I. Sha'afi, Binding of leukotriene Band platelet-activating factor to neutrophils: Effects of granulocyte-macrophage colony-stimulating factor, phorbol 12-myristate 13-acetate and fMet-Leu-Phe,Am. J. Physiol (Cell Physiol), In Press (1991).

180. J.F. DiPersio, S. Aggarival, D.W. Golde, GM-CSF directly induces neutrophil platelet activating factor (PAF)

181. R. Sullivan, J.D. Griffin, E.R. Simons, A.I. Schafer, T. Meshulam, J.P. Fredette, A.K. Maas, A.-S. Gadenne, J.L. Leavitt and D.A. Melnick, Effects of recombinant human granulocyte and macrophage colony-stimulating factors on signal transduction pathways in human granulocytes,J. Immunol., 139: 3422-3430 (1987)

183. R.G. Coffey, J.S. Davis, and J.Y. Djeu, Stimulation of guanylate cyclase activity and reduction of adenylate cyclase activity by granulocyte-macrophage colony-stimulating factor in human blood neutrophils,J. Immunol., 140: 2695-2701 (1988).

184. D. English, H.E. Broxmeyer, T.G. Gabig, L.P. Akard, D.E. Williams and R. Hoffman,Temporal adaptation of neutrophil oxidative responsiveness to n-formyl-methionyl-leucyl-pheny-lalanine; Acceleration by granulocyte-macrophage colony-stimulating factorJ. Immunol., 141: 2400-2406 (1988).

185. R. Sullivan, J.P. Fredette, J.D.., Griffin, J.L. Leavitt, E.R. Simons and D.A. Melnick, An elevation in the concentration of free cytosolic calcium is sufficient to activate the oxidative burst of granulocytes primed with recombinant human granulocyte-macrophage colony-stimulating factor. J.Biol. Chem. 264: 6302-6309 (1989).

186. S.H. Zuckerman and R.D. Schreiber, Up-regulation of gamma interferon receptors on the human monocytic cell line U937 by 1,25-dihydroxyvitamin D3 and granulocyte-macrophage colony-stimulating factor,J. Leukocyte Biol. 44: 187-191 (1988).

187. Y.H. Atkinson, A.F. Lopez, W.A. Marasco, C.M. Lucas, G.G., Wong, G.F. Burns and M.A. Vadas, Recombinant human granulocyte-macrophage colony-stimulating factor (rH GM-CSF) regulates fMet-Leu-Phe receptors on human neutrophils, Immunology, 64: 519-525 (1988).

188. S.R. McColl, D. Beauseigle, C. Gilbert and P.H. Naccache, Priming of the human neutrophil respiratory burst by granulocyte-macrophage colony-stimulating factor and tumor necrosis factor-α involves regulation at a post-cell surface receptor level, J. Immunol., 145: 3047-3053 (1990).

189. A.F. Lopez, J.M. Eglinton, D. Gillis, L.S. Park, S. Clark

and M.A. Vadas, Reciprocal inhibition of binding between interleukin 3 and granulocyte-macrophage colony-stimulating factor to human eosinophils, <u>Proc. Natl. Acad. Sci. U.S.A.</u>, 86: 7022-7026 (1989).

190. W.W. Hancock, M.E. Pleau and L. Kobzik, Recombinant granulocyte-macrophage colony-stimulating factor down-regulates expression of IL-2 receptor on human mononuclear phagocytes by induction of prostaglandin E., <u>J. Immunol.</u>, 140: 3021-3025 (1988).

191. L.S. Park, D. Friend, V. Price, D. Anderson, J. Singer, K.S. Prickett and D.L. Urdal, Heterogeneity in human interleukin-3 receptors. A subclass that binds human granulocyte-macrophage colony-stimulating factor, <u>J. Biol. Chem.</u>, 264: 5420-5427 (1989).

TRANSCELLULAR METABOLISM OF ARACHIDONIC ACID IN PLATELETS AND POLYMORPHONUCLEAR LEUKOCYTES ACTIVATED BY PHYSIOLOGICAL AGONISTS: ENHANCEMENT OF LEUKOTRIENE B4 SYNTHESIS

Rémi Palmantier and Pierre Borgeat

Unité de Recherche Inflammation, Immunologie et Rhumatologie, Centre de recherche du CHUL, 2705 boul. Laurier, Québec, G1V 4G2, Canada

INTRODUCTION

Polymorphonuclear leukocytes (PMNL) are known to play a major role in the inflammatory process in part through their ability to produce and respond to chemotactic factors. Leukotriene (LT) B_4, a metabolite of arachidonic acid derived from the 5-lipoxygenase pathway, is produced by phagocytes and has potent chemotactic and chemokinetic effects on these cells[1]. PMNL stimulated with the ionophore A23187 synthesize large amounts of LTB_4[2], while receptor-mediated activation of PMNL and monocytes-macrophages by agonists such as the chemotactic peptide N-formyl-Met-Leu-Phe (fMLP), the complement fragment C5a, platelet-activating factor (paf-acether) or by phagocytosis also leads to LTB_4 synthesis[3-7]. However LTB_4 synthesis induced by natural agonists is of lower magnitude as it is often not detectable by HPLC procedures[8]. Evidence has accumulated during the last decade, supporting that platelet/leukocyte interactions occur in several pathophysiological situations[9], and in particular that platelets might modulate inflammation. Indeed, activated platelets release arachidonic acid metabolites, paf-acether, platelet-derived growth factor (PDGF), platelet factor 4 (PF4), serotonin and adenine nucleotides which could affect PMNL functions such as migration, degranulation, adherence and production of superoxide anion[10-13]. Also, several aspects of the transcellular metabolism of eicosanoids between platelets and PMNL, which might contribute to regulatory mechanism of inflammation, have already been documented. For instance, arachidonic acid released from platelets is metabolized into LTB_4 by the PMNL when both cell types are stimulated with the ionophore A23187[14]. In addition, PMNL stimulated with the ionophore A23187 in presence of platelets transform platelet-derived 12S-hydroxy-5,8,10,14-(Z,Z,E,Z)-eicosatetraenoic acid (12-HETE) into 5S,12S-dihydroxy-6,8,10,14(E,Z,E,Z)-eicosatetra-enoic acid (5S,12S-DiHETE)[14,15] and subsequently into 5S,12S,20-TriHETE through ω-oxidation. Similarly, 12-HETE synthesized in platelets is metabolized to 12S,20-dihydroxy-5,8,10,14(Z,Z,E,Z)-eicosatetraenoic acid (12,20-DiHETE) by unstimulated neutrophils[16]. Furthermore, it was shown that LTA_4 released by

Cell-Cell Interactions in the Release of Inflammatory Mediatiors
Edited by P. Y-K Wong and C.N. Serhan, Plenum Press, New York, 1991

73

PMNL is metabolized by platelets into LTC$_4$[17] and lipoxins[18]. In previous studies we have also shown that addition of platelets to PMNL incubated with a high concentration of arachidonic acid (60μM) results in an increased synthesis of 5-lipoxygenase products; the amplification of 5-lipoxygenase product synthesis in this coincubation system was attributed to the synthesis of 12S-hydroperoxy-5,8,10,14(Z,Z,E,Z)eicosatetra-enoic acid (12-HpETE) by the platelet 12-lipoxygenase[19]. Amplification of LTB$_4$ synthesis in zymosan-stimulated PMNL was also observed in presence of thrombin-stimulated platelets[20]. It is noteworthy that in most studies (except for ref 20) enhancement of LTB$_4$ synthesis in platelet/PMNL coincubations was observed under conditions of cell activation by either the ionophore A23187 or exogenous substrate; more recently the studies of Fiore and Serhan[21] and studies from our laboratory[22,23] clearly demonstrated that platelet-PMNL interactions leading to increased formation of LTB$_4$ could also occur under conditions of receptor-mediated cell activation.

In the present study, we further investigated the synthesis of LTB$_4$ in platelet/PMNL suspensions treated with the hematopoietic growth factor granulocyte-macrophage colony-stimulating factor (GM-CSF), thrombin and various soluble PMNL agonists, and documented the mechanism of enhancement of LTB$_4$ synthesis by platelets.

MATERIALS AND METHODS

Reagents

    5,8,11,14(all cis)-eicosatetraenoic acid (arachidonic acid), adenosine 5'-diphosphate (ADP), fMLP, L-α-phosphatidylcholine,β-acetyl-γ-O-hexadecyl (paf-acether), L-α-lysophosphatidylcholine-γ-O-hexadecyl (lyso-paf-acether), prostaglandin B$_2$ (PGB$_2$), soybean lipoxydase (type I) and NaBH$_4$ were obtained from Sigma Chemical Company (Saint Louis, MO, USA). [^{14}C(U)]-arachidonic acid (1.0 Ci/mmol) was purchased from New England Nuclear (Boston, MA, USA). Thrombin was from Parke-Davis (Scarborough, Ontario). Hank's balanced salt solution (HBSS) and N-2-hydroxyethylpiperazine-N'-2-ethanesulfonic acid (HEPES) were purchased from GIBCO (Burlington, Ontario, Canada), Ficoll-paque was purchased from Pharmacia (Dorval, Quebec, Canada) and solvents were HPLC grade from Fisher and J.T. Baker (Montreal, Canada). Biosynthetic recombinant human granu-locyte-macrophage colony-stimulating factor (rhGM-CSF) was a generous gift from the Genetics Institute (Cambridge, MA, USA). rhGM-CSF was diluted in HBSS containing 2.5 % fetal calf serum under sterile conditions to a stock concentration of 100 nM, and stored at -20°C. Recombinant human C5a was a generous gift from Dr Henry Showell (Pfizer Pharmaceuticals, Groton, CT, USA); it was kept in HBSS at a stock concentration of 30 μM and stored at -20°C. Arachidonic acid was purified by silicic acid chromatography before use. Arachidonic acid and paf-acether were kept in hexane and chloroform, respectively, and aliquots of stock solutions were evaporated and redissolved in DMSO for addition to cell suspensions; the stock solution of fMLP was prepared in DMSO.

Cell preparation procedures

Venous blood was obtained from healthy donors and collected on citric acid-dextrose anticoagulant. Blood was centrifuged at 200 x g for 15 min at 20°C. The platelet-rich plasma (PRP) was removed and platelets were isolated as described previously[24]. Briefly, the PRP was acidified to pH 6.4 with 0.15 M citric acid and centrifuged at 900 x g for 10 min at 20°C in siliconized tubes. Supernatants were removed, and tube walls and platelet pellet surfaces were gently rinsed with Ca^{++}- and Mg^{++}-free HBSS containing 10 mM HEPES buffer pH 7.4 (HBSS-HEPES). Platelets were finally resuspended with the same buffer at 4 X 10^8 platelets/ml.

PMNL were prepared by sequential dextran sedimentation and centrifugation on Ficoll-paque cushions[25]. Contaminating erythrocytes were eliminated by hypotonic lysis. PMNL were then resuspended in Ca^{++}- and Mg^{++}-free HBSS-HEPES buffer at the cell density of 16 x 10^6 /ml, and the viability was evaluated by the trypan blue exclusion test. Cell viability was always greater than 95%.

Incubation procedures

Equal volumes of the platelet and PMNL suspensions were mixed resulting in a cell suspension containing 2 x 10^8 platelets /ml and 8 x 10^6 PMNL/ml; $CaCl_2$ and $MgCl_2$ were added to the final concentrations of 2 mM and 0.5 mM, respectively, and the cells were pre-incubated 5 min at 37°C. The cells were then incubated 10 min at 37°C in presence of various stimuli. In experiments where platelets and PMNL were incubated separately, the platelet and PMNL suspensions were diluted one-fold with HBSS-HEPES to obtain final cell concentrations as in the coincubation system. In experiments where rhGM-CSF was used, PMNL were preincubated with the cytokine at the concentration of 200 pM, 60 min at 37°C in HBSS-HEPES containing 2 mM $CaCl_2$ and 0.5 mM $MgCl_2$ at the cell density of 16 x 10^6 PMNL/ml; coincubation of rhGM-CSF-primed PMNL with platelets and stimulation were performed as described above. The final concentration of DMSO did not exceed 0.2% in incubation media.

Incubations were stopped by addition of 0.5 ml of ice-cold acetonitrile/methanol (1/1, vol/vol) containing 12.5 ng each of PGB_2 and 19-hydroxy-PGB_2 as internal standards, and the samples were stored at -20°C.

Activated platelet supernatants were prepared from a platelet suspension (4 x 10^8 cells/ml in HBSS-HEPES) containing 2 mM $CaCl_2$ and 0.5 mM $MgCl_2$. The cells were preincubated 5 min at 37°C and stimulated with 2.5 U/ml thrombin, 5 min at 37°C; the incubation media were then centrifuged at 2000 x g for 10 min at 4°C and the supernatants were collected. Supernatants (0.5ml) were used immediately for incubations with PMNL suspensions or denatured by addition of 2 vol of acetonitrile and stored at -20°C until analysis. In some experiments activated platelet supernatants were prepared in presence of 7 μM 5,8,11,14-eicosatetraynoic acid (ETYA) and 0.1 μM tiaprofenic acid, these two inhibitors being added during the preincubation period, i.e. 5 min prior to thrombin activation.

Analysis of 5-lipoxygenase products

The denatured samples were centrifuged at 2000 x g for 20 min to remove the precipitated material and the supernatants were analyzed without further treatment by reverse phase (RP) HPLC as described previously[26] with a minor modification of the mobile phase (methanol/acetonitrile/water, 23/23/54, vol/vol/vol was substituted for acetonitrile/water, 30/70, vol/vol) to improve the separation of the 20-hydroxy and 20-carboxy derivatives of the 5S,12S-DiHETE and LTB_4. The denatured samples initially containing HSA were evaporated down to a volume of ~ 1.4 ml at 40°C under a stream of nitrogen prior to RP-HPLC analysis. Briefly, samples were injected onto a Resolve C_{18} Radial Pak cartridge (5 x 100 mm, 5 µm particles) protected by Guard-Pak cartridges (silica and Resolve C_{18}, 5 µm particles) from Water's Millipore. Elution was performed at 1.5 ml/min. The lipoxygenase products were detected using fixed-wavelength UV photometers at 229 and 280 nm. Products were identified on the basis of their comigration with synthetic standards and specificity of absorption at either 229 nm or 280 nm. Products quantitation was done by comparing peak heights to calibrated standards of 20-hydroxy-LTB_4, LTB_4 and 15S-hydroxy-5,8,11,13(Z,Z,Z,E)-eicosatetraenoic acid (15-HETE), after correction for recovery using 19-hydroxy-PGB_2 and PGB_2.

Quantitation of arachidonic acid

Arachidonic acid was quantified in supernatants of thrombin-activated platelets using enzymatic and HPLC procedures. Briefly, the denatured platelet supernatants were centrifuged (2000 x g for 10 min) to remove the precipitated material. $[^{14}C(U)]$-arachidonic acid (1 nCi/sample) was added to the supernatants. Arachidonic acid was purified using the RP-HPLC system described previously[26], but using acetic acid instead of phosphoric acid in the mobile phases; the fractions containing arachidonic acid were collected and evaporated at 40°C under a stream of nitrogen. The residues were first dissolved in 50 µl of methanol; then 0.375 ml of 5 mM Na borate buffer pH 10 and 100 ng of 8,11,14(all cis)-eicosatrienoic acid as internal standard were added to each sample. One thousand units of soybean lipoxydase type I were then added and the reaction mixtures were incubated 30 min at 20°C to allow the conversion of arachidonic acid and the internal standard into the corresponding 15-hydroperoxyde derivatives which were subsequently reduced to 15-HETE and 15S-hydroxy-8,11,13(Z,Z,E)-eicosatrienoic acid (15-HETrE) by addition of 600 µg of $NaBH_4$ in 1ml of methanol. After 30 min at 20°C, the reaction was stopped by acidifying the reaction mixtures to pH 5 with acetic acid, and 15-HETE and 15-HETrE were analysed by RP-HPLC with UV detection at 229 nm. Arachidonic acid quantitation was performed by comparing peak heights of 15-HETE and 15-HETrE after correction for loss of $[^{14}C(U)]$-arachidonic acid in the course of the HPLC purification step. Recovery of arachidonic acid varied between 50 and 70%.

Statistical analysis

Statistical significance was evaluated by the Student t test for paired data.

RESULTS AND DISCUSSION

LTB$_4$ synthesis in coincubations of PMNL and intact platelets

We investigated LTB$_4$ synthesis in platelet/PMNL suspensions stimulated with 2.5 U/ml thrombin and/or 0.1 μM fMLP. A significant increase of LTB$_4$ synthesis was observed when the cell mixtures were stimulated with both stimuli in comparison to only one stimulus. When thrombin and fMLP were used to stimulate PMNL in absence of platelets LTB$_4$ synthesis was low (~4 pmol/8 x 10^6 cells), close to the limit of detection of the analytical system used (1-2 pmol) (table 1). These data clearly showed that platelets have the ability to promote LTB$_4$ synthesis in stimulated PMNL. The enhancement by platelets of LTB$_4$ synthesis required that both cell types be activated, since in the absence of a PMNL agonist LT synthesis is undetectable, and in the absence of thrombin, LT synthesis is comparable to that observed in PMNL alone. Similarly, thrombin-activated platelets enhanced LTB$_4$ synthesis in PMNL stimulated by 0.1 μM paf-acether (table 1). It is noteworthy that although platelets produce paf-acether, platelets/PMNL mixtures do not generate LTB$_4$ (in amounts detectable by HPLC) upon thrombin stimulation (table 1); this is likely explained by insufficient release of paf-acether by the activated platelets[27]. In these studies, the formation of LTB$_4$ was assessed from the cumulative amounts of LTB$_4$ (often undetectable) and of its metabolites, 20-hydroxy- and 20-carboxy-LTB$_4$, as LTB$_4$ is rapidly metabolized by ω-oxidation in PMNL suspensions[28,29].

Table 1. LTB$_4$ synthesis by activated PMNL in presence or in absence of platelets

Stimuli	LTB$_4$ synthesis (pmol/8x10^6 PMNL)	
	PMNL + Platelets	PMNL
Thrombin	< 2	n.d.
fMLP	5.3 ± 1.8	n.d.
Thrombin+fMLP	26.1 ± 1.3**	2.2 ± 0.9
paf	< 2	n.d.
Thrombin+paf	21 ± 8.5*	< 2

PMNL or platelet/PMNL mixtures were incubated with 0.1 μM fMLP and/or 2.5 U/ml thrombin, or 0.1 μM paf-acether and/or thrombin, for 10 min at 37°C in HBSS. In the cell mixture, the platelet:PMNL ratio was 25:1. LTB$_4$ was measured by RP-HPLC and the amounts indicated represent the sum of LTB$_4$ and of its ω-oxydation products (20-hydroxy-LTB$_4$ and 20-carboxy-LTB$_4$). Results are the mean ± SEM of 9 experiments (thrombin + fMLP), 3 experiments (fMLP) and 4 experiments (paf-acether ± thrombin). Incubations were carried out in triplicate in each experiments. ** $P < 0.001$ compared with PMNL stimulated with thrombin and fMLP; * $P < 0.05$ compared with PMNL stimulated with thrombin and paf. n.d., not determined.

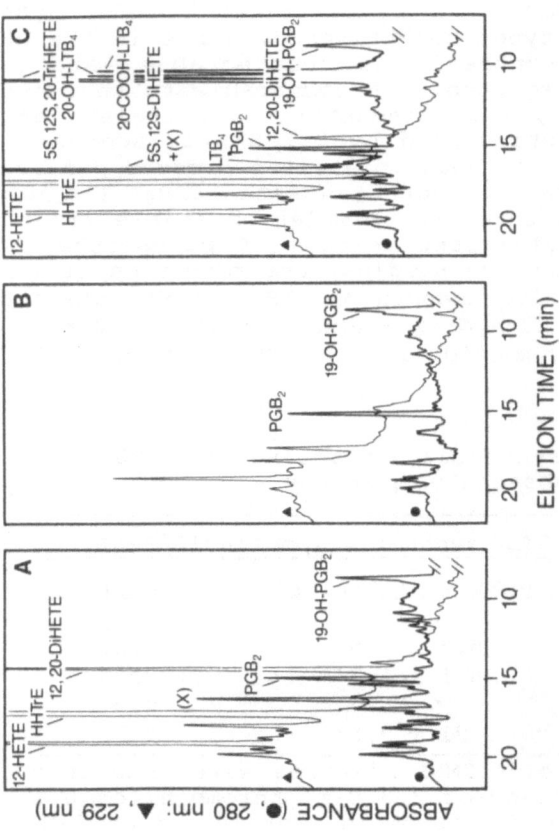

Fig. 1. RP-HPLC analysis of the arachidonic acid metabolites generated by rhGM-CSF-primed PMNL coincubated with platelets for 10 min at 37°C with 2.5 U/ml thrombin (A), 0.1 μM fMLP (B), 2.5 U/ml thrombin and 0.1 μM fMLP (C). PMNL were preincubated 60 min at 37°C in presence of 200 pM rhGM-CSF. The cells were incubated in 1 ml HBSS at densities of 8 × 10⁶ and 200 × 10⁶/ml for PMNL and platelets, respectively. The amount of internal standards added was 12.5 ng each of PGB₂ and 19-hydroxy-PGB₂. Attenuation settings of the UV photometers were 0.01 and 0.025 absorbance unit at full scale at 280 and 229 nm, respectively.

The effect of thrombin-activated platelets on LTB$_4$ synthesis was also investigated using PMNL primed by the hematopoietic growth factor, GM-CSF. GM-CSF is produced by monocytes, fibroblasts, lymphocytes and endothelial cells in a variety of conditions, for example in monocytes exposed to lipopoly-saccharides[30]. It is believed that GM-CSF plays an important role as modulator of inflammation and immune mechanisms. rhGM-CSF-treated or "primed" PMNL show increased fonctional responses to a second stimulation, such as enhanced superoxide anion formation and enzyme release in response to fMLP[31,32], increased phagocytosis[33] and also increased LT synthesis in response to soluble agonists (fMLP, C5a, paf-acether) and to the ionophore A23187[8,34,35,36].

The HPLC profiles of arachidonic acid metabolites generated by primed PMNL and platelet mixtures stimulated with 2.5 U/ml thrombin and/or 0.1 µM fMLP are shown in figure 1. When thrombin alone was added to the cells (Fig. 1A), platelets responded by producing the 12-lipoxygenase product, 12-HETE and the cyclooxygenase product HHTrE, while 5-lipoxygenase products (20-hydroxy- and 20-carboxy-LTB$_4$) were only detected in trace amounts. Some 12,20 DiHETE resulting from the ω-oxidation of 12-HETE by resting PMNL was produced. The absorption peak labeled "X" (280 nm) was tentatively identified as a 14,15-dihydroxy-5,8,10,12-eicosatetraenoic acid (14,15-DiHETE), a product of the 15-lipoxygenase[37,38] on the basis of its UV spectrum and comparison with the major 14,15-DiHETE isomer generated by platelets incubated with 15-HpETE (data not shown). Other small absorption peaks at 280nm, probably isomeric 14,15-DiHETEs and 8,15-DiHETEs were also present[37,38]. In this particular experiment, only traces of the metabolites of LTB$_4$ (20-hydroxy- and 20-carboxy-LTB$_4$) were detected when fMLP alone was the stimulus (Fig 1B). However in the several experiments performed (under the same experimental conditions but with different donors), LTB$_4$ synthesis varied from ~4 pmol to 45 pmol/8 x 10^6 cells (see next paragraph). When cells were stimulated with thrombin and fMLP (Fig. 1C), the formation of 5-lipoxygenase products by the PMNL was strikingly enhanced leading to significant accumulation of 20-hydroxy- and 20-carboxy-LTB$_4$. Although, as mentioned above, there was an important variability between different cell preparations in the responses to fMLP, enhancement of LTB$_4$ synthesis by the cell mixtures following platelet stimulation was always observed. The chromatogram (Fig. 1C) also showed the presence of the platelet products 12-HETE and HHTrE, as well as the products of the platelet/PMNL metabolic cooperation, 5S,12S-DiHETE which comigrated with 14,15-DiHETE (peak "X"), and 5S,12S,20-TriHETE the product of its ω-oxidation. A smaller amount of 12,20-DiHETE than in figure 1A was observed, probably originating from its competition with a better substrates as LTB$_4$ for ω-oxidation. The mechanisms of formation of these three products of platelet-PMNL interaction have been discussed previously[15,29,39]. In this study, the HPLC system used for the separation of lipoxygenase products enables the resolution of LTB$_4$ and 5S,12S-DiHETE (as well as their ω-oxidation products). Furthermore, the use of an high resolution (1.2 nm) photodiode array UV detector (Waters Millipore model 994) supported the identification of the compounds by the assessment of small differences in their UV

Table 2. LTB$_4$ synthesis by activated rhGM-CSF-treated PMNL in presence or in absence of platelets

Stimuli	LTB$_4$ synthesis (pmol/8x10^6 PMNL)	
	PMNL + platelets	PMNL
Thrombin	5.8 ± 1.8	n.d.
fMLP	37.6 ± 17.4	n.d.
Thrombin+fMLP	159 ± 34.5*	25.6 ± 10.1

rhGM-CSF-treated PMNL or platelet/PMNL (rhGM-CSF-treated) mixtures were incubated with 0.1 μM fMLP and/or 2.5 U/ml thrombin for 10 min at 37°C in HBSS buffer. PMNL were preincubated 60 min at 37°C in presence of 200 pM rhGM-CSF. LTB$_4$ was measured by RP-HPLC and the amounts indicated represent the sum of LTB$_4$ and of its ω-oxidation products. Results are the mean ± SEM of 6 experiments; in each experiment, incubations were performed in triplicate. *P< 0.01 compared with rhGM-CSF-treated PMNL stimulated with thrombin and fMLP. n.d., not determined.

spectra (data not shown), reflecting differences in the geometry of the conjugated triene units. The formation of lipoxins, reported recently in similar, but not identical experimental conditions[18], was not detectable in our studies, even when rhGM-CSF-treated PMNL were used and HPLC effluents were monitored at 302 nm for optimal detection. It seems likely that differences in the amount of cells used (lower in the present study, 4.5 and 2 times less PMML and platelets, respectively), account for this discrepancy. Similarly, LTC$_4$ derived from the transcellular metabolism of LTA$_4$ was not detected in this platelet-PMNL incubation system, probably owing to the absence of albumin in the incubation media and/or to insufficient PMNL activation induced by fMLP[17,40].

Table 2 sums up the results of the experiments performed with rhGM-CSF-treated PMNL. It clearly shows that stimulation of the cell mixture with both stimuli induced an average 8-fold increase in LTB$_4$ synthesis in comparison to the effect of fMLP alone. Using thrombin alone as a stimulus, LTB$_4$ synthesis was very low (~5 pmol/8 x 10^6 PMNL). In the present study, rhGM-CSF priming of PMNL led to a 6- fold, increase in fMLP-induced LTB$_4$ synthesis over level achieved in PMNL not previously exposed to rhGM-CSF (table 1 and 2). The results of our experiments with rhGM-CSF-treated PMNL indicate that stimulation of LTB$_4$ synthesis by activated platelets also occurs in primed PMNL, and that under these conditions, LTB$_4$ synthesis reaches levels strongly exceeding those measured in all other experimental conditions tested.

Mechanism underlying the stimulatory effect of activated platelets on LTB$_4$ synthesis

The first series of experiments was aimed at determining if factors responsible for the amplification of LTB$_4$ synthesis were released by platelets. We thus investigated LTB$_4$ synthesis by PMNL incubated in presence of activated platelet supernatants.

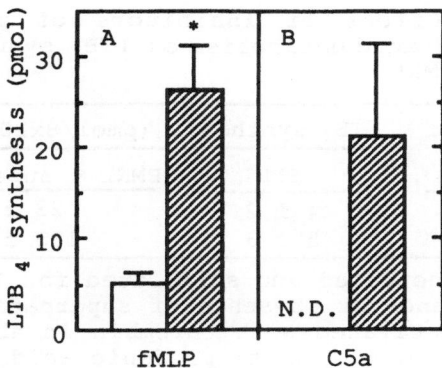

Fig 2. Synthesis of LTB$_4$ in PMNL stimulated by fMLP (A) or C5a (B), incubated in the absence (open bars) or presence (hatched bars) of thrombin-activated platelet supernatants for 10 min at 37°C Results are the mean ± SEM of 9 and 4 experiments for PMNL stimulated by fMLP and C5a, respectively. *P< 0.001. N.D., not detectable.

Supernatants of thrombin-activated platelets suspensions were obtained as described under "Methods". Figure 2A shows that the supernatants retained the ability to amplify LTB$_4$ synthesis. The addition of activated platelet supernatants to PMNL stimulated with fMLP caused a significant increase (5-fold or more) in LTB$_4$ synthesis by the PMNL, indicating that substances released by platelets upon activation by thrombin accounted for the potentiation of LTB$_4$ synthesis. Similar results were obtained when the complement fragment C5a was used as stimulus for PMNL instead of fMLP (Fig. 2B). The magnitude of the amplification of LTB$_4$ synthesis by the platelet supernatants was similar to that observed in the coincubation system (table 1 and Fig. 2).

In a second series of experiments, we used the 12-lipoxygenase and cyclooxygenase inhibitors ETYA and tiaprofenic acid to investigate whether products of platelet arachidonic acid metabolism might be involved in the stimulatory effect of activated platelet supernatants. Tiaprofenic acid is a potent inhibitor of the cyclooxygenase[41]. ETYA is an inhibitor of the 12- and 15-lipoxygenases and of the cyclooxygenase[42]. At the concentration of 7 μM and at the cell concentration used, ETYA showed minimal (< 10%) inhibitory activity on the PMNL 5-lipoxygenase (data not shown). These experiments were performed because 12-HpETE was previously identified as a mediator of the stimulatory effect of platelets on LTB$_4$ synthesis in a different in vitro model[19]; in addition thrombin-activated platelets release arachidonic acid metabolites of the cyclooxygenase and 12-lipoxygenase[43]. Supernatants were generated in presence or absence of ETYA and tiaprofenic acid. Table 3 shows that the presence of the inhibitors during activation of platelets with thrombin did not prevent the stimulatory effect of the platelet supernatants on LTB$_4$ synthesis by fMLP-stimulated PMNL, quite on the contrary, it strikingly enhanced the effect of platelets. These results suggested that platelet arachidonic acid metabolites were not

Table 3. Effect of inhibitors of platelets arachidonic acid metabolism on LTB4 synthesis by activated PMNL

	LTB4 synthesis (pmol/8x10^{6}PMNL)	
Inhibitors	PMNL	PMNL + supernatants
−	4 ± 2	23 ± 6
+	5 ± 3	153 ± 27

PMNL were incubated and stimulated for 10 min at 37°C in absence or presence of supernatants from platelets activated with thrombin in absence or in presence of 0.1 μM tiaprofenic acid and 7 μM ETYA. Values are means ± SEM of 3 experiments.

involved, and raised the possibility that arachidonic acid released by platelets was utilized for LT synthesis and accounted for the stimulatory effect of activated platelets. Measurements of arachidonic acid, 12-HETE and HHTrE levels in activated platelet supernatants generated in presence and absence of the inhibitors supported this hypothesis (table 4). Indeed, arachidonic acid and the two metabolites were present in activated platelet supernatants, and the inhibitors completely blocked the formation of 12-HETE and HHTrE, causing a dramatic increase in the level of free substrate.

Addition of arachidonic acid (0.1 μM to 0.5 μM) to fMLP-activated PMNL led to a dose-dependent increase in LTB$_4$ formation (Fig 3). The effect of arachidonic acid was significant at 0.1 μM, and at the concentration of 0.2 μM, corresponding to the average concentration measured in activated platelet supernatants (generated in absence of ETYA and tiaprofenic acid), LTB$_4$ synthesis was enhanced by ∼ 7-fold. Thus, exogenous arachidonic acid, at the concentration measured in activated platelet supernatants, enhanced LTB$_4$ synthesis in fMLP-stimulated cells to levels equivalent to those obtained with the activated platelet supernatants (table 1 and Fig 3); furthermore the profiles of products formed (mainly 20-hydroxy- and 20-carboxy-LTB$_4$) were identical in the two experimental settings. Therefore it can be concluded that arachidonic acid released by thrombin-activated platelets largely accounts for their stimulatory effect on LTB4 synthesis in the PMNL.

Table 4. Arachidonic acid, 12-HETE and HHTrE in thrombin-activated platelet supernatants

	picomoles/200x10^6 platelets	
products	Thrombin	Thrombin + inhibitors
AA	200 ± 25	3500 ± 790
12-HETE	1580 ± 400	< 40
HHTrE	1780 ± 160	< 20

Arachidonate (AA), 12 HETE and HHTrE released by thrombin-activated platelets in the presence or absence of inhibitors (tiaprofenic acid and eicosatetraynoic acid). Results are the mean values of 6 experiments ± SEM.

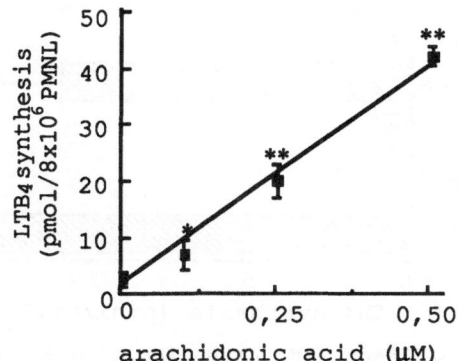

FIG. 3. Effect of exogenous arachidonic acid on the synthesis of LTB_4 (and metabolites) in PMNL incubated 10 min at 37°C in HBSS buffer in the presence of 0.1 μM fMLP. LTB_4, 20-hydroxy- and 20-carboxy-LTB_4 were measured by RP-HPLC. Results are the mean ± SEM of 3 experiments. Incubations were performed in triplicate in each experiments. * $P< 0,05$; ** $P< 0,005$

Besides arachidonic acid and its metabolites, thrombin-activated platelets release substances which could also contribute to increase LTB_4 synthesis in PMNL. In particular the adenine nucleotide ADP which is known to modulate superoxide anion production and intracellular Ca^{++} [12], and lyso-paf-acether which is acetylated to paf-acether by PMNL[27,44] and could enhance PMNL activation. Results shown in figure 4 however, demonstrate that ADP and lyso-paf-acether do not alter LTB_4 production in PMNL, even in presence of exogenous arachidonic acid. The platelet lipoxygenase product 12-HETE which might compete with arachidonic acid in reacylation processes (leading to increased availability of arachidonic acid) was also tested; the addition of up to 2 μM exogenous 12-HETE to PMNL activated with fMLP in the presence of 0.2 μM and 1 μM arachidonic acid did not alter LTB_4 synthesis (data not shown). Therefore, although it remains difficult to rule out that the above-mentioned coumpounds or other substances (such PF4 and PDGF) might contribute to the increased LTB_4 synthesis observed in platelet/PMNL coincubations, we propose that these products could only play a minor role in the model tested, as provision of substrate appears to account for most of the effect of activated platelets on LTB_4 synthesis.

Our finding that thrombin-activated platelets enhance LTB_4 synthesis in activated PMNL by providing substrate is in agreement with the current understanding of the regulation of LT synthesis in these cells. It is widely well accepted that the PMNL 5-lipoxygenase is present in an inactive form in these cells and that cell stimulation results in an enhanced activity of the enzyme[45]. It has also been observed that while PMNL stimulation with natural agonists (such as fMLP, C5a and paf-acether) induces 5-lipoxygenase activation and arachidonic acid release, the concomitant addition of exogenous substrate leads to a striking increase in the formation of LTB_4 and other 5-lipoxygenase products[3,5,7,47]. Thus, these observations clearly indicate that in

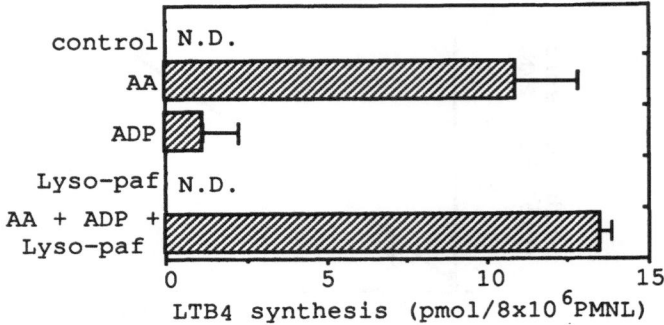

Fig. 4. Effect of 50 μM ADP, 0.5 μM lyso-paf, 0.2 μM exogenous substrate or combinations of the preceding compounds on LTB$_4$ synthesis in PMNL stimulated by 0.1 μM fMLP and incubated 10 min at 37°C in HBSS buffer. The results are from one experiment representative of 3 (incubations were done in triplicate and the values are the mean ± SD). N.D., not detectable.

PMNL activated by natural agonists, substrate availability is a limiting factor for LTB$_4$ synthesis.

In a previous study, we reported that the addition of platelets to PMNL suspensions incubated in presence of 60 μM arachidonic acid strongly enhanced the synthesis of LTB$_4$ by PMNL; this effect of platelets required the presence of arachidonic acid, was blocked by inhibition of the 12-lipoxygenase, and was clearly dependent on the production of the fatty acid hydroperoxide 12-HpETE by platelets[19]. However, LTB$_4$ synthesis was limited by 5-lipoxygenase activity, since substrate was present in excess; 12-HpETE likely acted directly on the 5-lipoxygenase, which is known to be activated by hydroperoxides[47]. In the current study, the 5-lipoxygenase was activated by fMLP, C5$_a$ or paf-acether, and substrate availability appeared to be the limiting factor of LTB$_4$ synthesis, while studies with inhibitors did not support a role for 12-HpETE.

Recent studies demonstrated the enhancement of lipoxins and paf-acether synthesis in PMNL coincubated with activated platelets lending further support to the implication of platelets in inflammatory processes[18,44]. The studies of Fiore and Serhan[21] and the present data also suggest that in physiopathological situations where PMNL and platelets are activated, in particular when PMNL are exposed to GM-CSF, platelet-PMNL interactions may promote the synthesis of another important lipid mediator of inflammation, LTB$_4$. It is interesting to point out the similarity in the mechanisms leading to enhanced LTB$_4$ synthesis (described herein) and enhanced paf-acether synthesis[44] which both involve transfer of their respective substrates, (i.e., arachidonic acid and lyso-paf-acether) from activated platelets to PMNL.

The biological significance of the present studies deserves a final comment. Indeed, the binding of arachidonic acid by serum albumin[48] makes it unlikely that transcellular metabolism could take place in a physiological environment, unless the cells

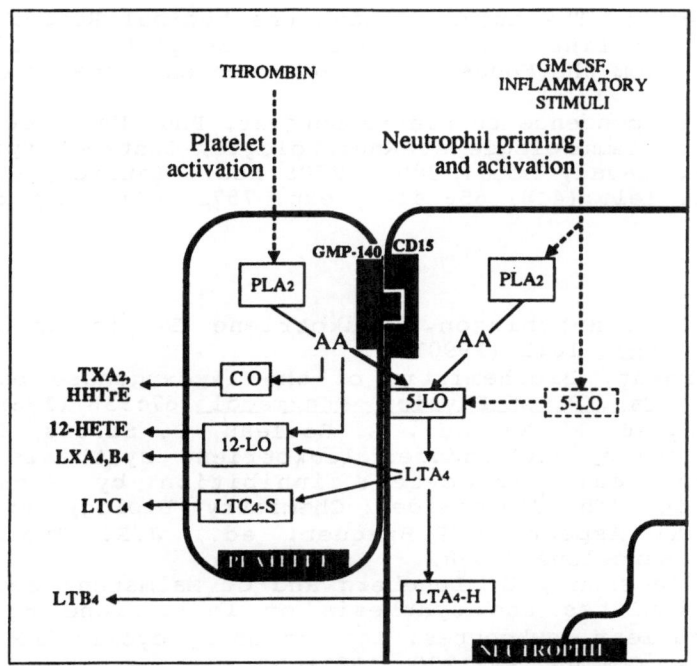

FIG. 5 Partial scheme of the transcellular metabolism of arachidonic acid by activated platelets and PMNL. Other interactions described previously (13-16) and leading to the formation of 5S,12S-DiHETE, 12,20-DiHETE and their w-oxidation products have not been depicted. These metabolic interactions are likely facilitated following adhesion of platelets to PMNL (see discussion); given the trapping of hydrophobic molecules by serum albumin[48], it is reasonable to speculate that efficient exchange of eicosanoids *in vivo* might occur in conditions of close cellular contact only. AA, arachidonic acid; CO, cylcooxygenase; LTA_4-H, LTA_4 hydrolase; LTC_4-S, LTC_4 synthase PLA_2, phospholipase A_2; TXA_2, thromboxane A_2, 5-LO, 5-lipoxygenase; 12-LO, 12-lipoxygenase.

involved are in direct contact. Figure 5 shows a scheme of the interactions in the metabolism between platelets and PMNL. In this regard, it is interesting that specific adherence of activated platelets to neutrophils and monocytes has been described[49]; furthermore, a platelet alpha granule membrane protein (GMP-140 or PADGEM), which mediates these cell interactions, has recently been characterized[50,51]. The CD15 antigene on human neutrophil could be a component of the GMP-140 ligand[52].. Obviously, attachment of platelets to PMNL could favor the recruitement of platelets at inflammatory sites and facilitate the occurence of metabolic or functional interactions between the two cell types. Further studies will address the question of the occurence of transcellular metabolism of eicosanoids between platelets and PMNL under physiological conditions and of the role of direct cells contact, as that mediated by GMP-140 in such metabolic cooperation.

Acknowledgement: The authors thank the Medical Research Council of Canada for financial support. P. Borgeat is holder of a scholarship from the Fonds de la recherche en santé du Québec.

Address correspondence to Pierre Borgeat, PhD, Unité de Recherche Inflammation, Immunologie et Rhumatologie, Centre Hospitalier de l'Université Laval, Local 9800, 2705 boul. Laurier, Québec, G1V 4G2, Canada. Tel.:(418) 656-4141, ext. 7572. FAX: (418)654-2765

REFERENCES

1.   A.W. Ford-Hutchinson, Leukotriene B4 in inflammation, Crit.Rev.Immunol. 10:1 (1990).
2.   P. Borgeat, Biochemistry of the lipoxygenase pathways in neutrophils, Can. J. of Physiol. Pharmacol. 67:936 (1989)
3.   P. Borgeat, M. Nadeau , G. Rouleau, P. Sirois, P. Braquet and P. Poubelle, PAF-induced leukotriene synthesis in human polymorphonuclear leukocytes: inhibition by Ginkgolide B (BN52021), in: "The Ginkgolides: Chemistry, Biology, Pharmacology and Clinical Aspects," P.Braquet, ed., J.R. Prous Science Publishers, Barcelone (1988).
4.   H.E. Claesson , U. Lundberg and C. Malmsten, Serum-coated zymosan stimulates the synthesis of leukotriene B4 in human polymorphonuclear leukocytes. Inhibition by cyclic AMP. Biochem. Biophys. Res. Commun. 99:1230 (1981).
5.   R.M. Clancy, C.A. Dahinden and T.E. Hugli, Arachidonate metabolism by human polymorphonuclear leukocytes stimulated by N-formyl-Met-Leu-Phe or complement component C5a is independent of phospholipase activation. Proc. Natl. Acad. Sci. USA 80:7200 (1983).
6.   A.H. Lin, D.R. Morton and R.R. Gorman, Acetyl glyceryl ether phosphorylcholine stimulates leukotriene B4 synthesis in human polymorphonuclear leukocytes J.Clin.Invest. 70:1058 (1982).
7.   H. Salari, P. Braquet, P. Naccache and P. Borgeat, Characterization of effect of N-formyl-methionyl-leucyl-phenylalanine on leukotriene synthesis in human polymorphonuclear leukocytes, Inflammation 9:127 (1985).
8.   S.R. McColl, E. Krump, P.H. Naccache, P. Poubelle, P.Braquet, M. Braquet and P. Borgeat, Granulocyte-macrophage colony-stimulating factor increases the synthesis of leukotriene B4 by human neutrophils in response to platelet-activating factor: enhancement of both arachidonic acid avaibility and 5-lipoxygenase activation, J. Immunol. in press.
9.   B. B. Weksler, Platelets, in: "Inflammation: Basic Principles and Clinical Correlates," J.I. Gallin , I.M. Goldstein & R. Snyderman, ed., Raven Press, New York (1988).
10.   T.F. Deuel, R.M. Senior and D. Chang, Platelet factor 4 is chemotactic for neutrophils and monocytes, Proc. Nat. Acad. Sci. USA. 78:4584 (1981).
11.   T.F. Deuel, R.M. Senior and J.S. Huang, Chemotaxis of monocytes and neutrophils to platelet derived growth factor, J. Clin. Invest. 69:1046 (1982).
12.   P.A. Ward, T.W. Cunningham, K.K. McCulloch, S.H. Phan, J. Powel and K.J. Johnson, Platelets enhancement of $O_2^{\dot{}}$ responses in stimulated human neutrophils, Lab.Inv. 58:37 (1988).
13.   M.A. Boogaert, O. Yamada, H.S. Jacob and C.F. Moldow, Enhancement of granulocytes-endothelial cell adherence and

granulocyte-induced cytotoxicity by platelet release products, Proc. Natl. Acad. Sci. USA. 79:7019 (1982).

14. A.J. Marcus, M.J. Broekman, L.B. Safier, H.L. Ullman, N. Islam, C.N. Serhan, L.E. Rutherford, H.M. Korchak and G. Weissman, Formation of leukotrienes and other hydroxy acids during platelet-neutrophil interactions in vitro, Biochem. Biophys. Res. Commun. 109:130 (1982).

15. P. Borgeat, B. Fruteau-de-Laclos, S. Picard, J. Drapeau, P. Vallerand and E.J. Corey, Studies on the mechanism of formation of the 5S,12S-dihydroxy-6,8,10,14(E,Z,E,Z) acid in leukocytes, Prostaglandins 23:713 (1982).

16. A.J. Marcus, L.B. Safier, H.L. Ullman, M.J. Broekman, N. Islam, T.D. Oglesby and R.R. Gorman, 12S,20-Dihydroxy-icosatetraenoic acid: A new icosanoid synthesized by neutrophils from 12S- hydroxyicosatetraenoic acid produced by thrombin- or collagen-stimulated platelets. Proc. Nat. Acad. Sci. USA 81:903 (1984).

17. J. Maclouf, F.A. Fitzpatrick and R.C. Murphy, Transcellular biosynthesis of eicosanoids, Pharmacol.Res. 21:1 (1989).

18. C.N. Serhan, K.A. Sheppard, Lipoxin formation during human neutrophil-platelet interactions - evidence for the transforma-tion of leukotriene-A4 by platelet 12-lipoxygenase in vitro, J. Clin. Invest. 85:772 (1990).

19. J. Maclouf, B. Fruteau de Laclos and P. Borgeat, Stimulation of leukotriene biosynthesis in human blood leukocytes by platelet-derived 12-hydroperoxy-icosatetraenoic acid. Proc. Natl. Acad. Sci. USA. 79:6042 (1982).

20. A. Del Maschio, J. Maclouf, E. Corvazier, M.J. Grange, P. Borgeat, Activated platelets stimulate human neutrophils functions, Nouv. Rev. Fr. Hematol. 27:275 (1985).

21. S. Fiore and C.N. Serhan, Formation of lipoxins and leukotrienes during receptor-mediated interactions of human platelets and recombinant human granulocyte/macrophage colony-stimulating factor-primed neutrophils, J. Exp. Med. 172:1451 (1990).

22. R. Palmantier and P. Borgeat, Thrombin-activated platelets amplify leukotriene B4 synthesis by neutrophils, Clinical and investigative medicine, 12:B7, Abst.#R-17 (1989).

23. R. Palmantier and P. Borgeat, Amplification of leukotriene B4 synthesis in neutrophils by thrombin activated platelets, FASEB j. 4:A2231, Abst.#3102 (1990).

24. M. Lagarde, P.A. Bryon, M. Guichardant, M. Dechavanne, A simple and efficient method for platelet isolation from their plasma, Thromb.Res. 17:581 (1980).

25. A. Boyum, Isolation of mononuclear cells and granulocytes from human blood. Isolation of mononuclear cells by one centrifugation, and of granulocytes by combining centrifugation and sedimentation at 1g. Scand. J. Clin. Lab. Invest. 21:77 (1968).

26. P. Borgeat and S. Picard, 19-Hydroxyprostaglandin B2 as an internal standard for on-line extraction-high-performance liquid chromatography analysis of lipoxygenase products. Anal. Biochem. 171:283 (1988).

27. M. Chignard, C. Lalau Keraly, D. Nunez, E. Coëffier and J. Benveniste, PAF-acether and platelets, in: "Platelets in biology and pathology III" MacIntyre & Gordon, ed., Elsevier Science Publisher, B.V., (1987).

28. W.S. Powell, Properties of leukotriene B4 20-hydroxylase from polymorphonuclear leukocytes. J. Biol. Chem. 259:3082 (1984).

29. J.A. Lindgren, G. Hansson, B. Samuelsson, Formation of novel hydroxylated eicosatetraenoic acids in preparations of human polymorphonuclear leukocytes. FEBS. Lett. 128:329 (1981).

30. C. Ruef and D.L. Coleman, Granulocyte-Macrophage Colony-Stimulating Factor: Pleitropic Cytokine With Potential Usefulness, Reviews of infectious diseases. 12:41 (1990).

31. R.H. Weisbart, L. Kwan, D.W. Golde, J.C. Gasson, Human GM-CSF primes neutrophils for enhanced oxidative metabolism in response to the major physiological chemoattractants. Blood. 69:18 (1987).

32. S.E. Kaufman, J.F. Dipersio, J.C. Gasson, Effects of Human GM-CSF on Neutrophil Degranulation In Vitro. Exp. Hematol. 17:800 (1989).

33. A.F. Lopez, D.J. Williamson, J.R. Gamble, C.G. Begley, J.M. Harlan, S.J. Klebanoff, A. Waltersdorph, G. Wong, S.C. Clark and M.A. Vadas, Recombinant human granulocyte-macrophage colony-stimulating factor stimulates in vitro mature human neutrophil and eosinophil function, surface receptor expression and survival, J.Clin.Invest. 78:1222 (1986).

34. D.S. Silberstein, W.F. Owen, J.C. Gasson, J.F. DiPersio and D.W. Golde, Enhancement of human eosinophil cytotoxicity and leukotriene synthesis by biosynthetic (recombinant) granulocyte-macrophage colony-stimulating factor. J. Immunol. 137:3290 (1986).

35. C.A. Dahinden, J. Zingg, F.E. Maly and A.L. De Weck, Leukotriene production in human neutrophils primed by recombinant human granulocyte/macrophage colony-stimulating factor and stimulated with the complement component C5a and FMLP as second signals. J. Exp. Med. 167:1281 (1988).

36. J.F. DiPersio, P.H. Naccache, P. Borgeat, J.C. Gasson, M-H. Nguyen and S.R. McColl, Characterization of the priming effects of human granulocyte-macrophage colony-stimulating factor on human neutrophil leukotriene synthesis, Prostaglandins. 36: 673 (1988).

37. R.L. Maas and A.R. Brash, Evidence for a lipoxygenase mechanism in the biosynthesis of epoxide and dihydroxy leukotrienes from 15(S)-hydroperoxyicosatetraenoic acid by human platelets and porcine leukocytes. Proc. Nat. Acad. Sci. USA. 80:2884 (1983).

38. P.Y-K. Wong, P. Westlund, M. Hamberg, E. Granstrom, P.H-W. Chao and B. Samuelsson, 15-Lipoxygenase in human platelets. J. Biol. Chem. 260:9162 (1985).

39. A.J. Marcus, L.B. Safier, H.L. Ullman, N. Islam, M.J. Broekman and C. von-Schacky, Studies on the mechanism of omega-hydroxylation of platelet 12-hydroxyeicosatetraenoic acid (12-HETE) by unstimulated neutrophils. J. Clin. Invest. 79:179 (1987).

40. J. Maclouf, R.C. Murphy and P.M. Henson, Transcellular Biosynthesis of sulfidopeptide leukotrienes during receptor-mediated stimulation of human neutrophil/platelet mixtures, Blood. 76:1838 (1990).

41. K. Schror, V. Neuhaus, B. Ahland, S. Sauerland, A. Kuhn, H. Darius and K. Bussmann, Actions of tiaprofenic acid on vascular prostacyclin biosynthesis and thromboxane and 12-HPETE formation of human platelets in vitro and ex vivo. Rheumatology. 7:88 (1982).

42. H. Salari, P. Braquet, P. Borgeat, Comparative effects of indomethacin, acetylenic acids, 15-HETE, nordihydroguaiaretic acid and BW755C on the metabolism of arachidonic acid in human leukocytes and platelets. Prostaglandins Leukotrienes Med. 13:53 (1984).

43. J.B. Smith, C. Dangelmaier and G. Mauco, Measurement of arachidonic acid liberation in thrombin-stimulated human platelets. Use of agents that inhibit both the cyclooxygenase and lipoxygenase enzymes. Biochim. Biophys. Acta. 835:344 (1985).

44. E. Coeffier, D. Delautier, J.P. Lecouedic, M. Chignard, Y. Denizot and J. Benveniste, Cooperation between platelets and neutrophils for PAF-acether (Platelet-Activating Factor) formation. J. Leukocyte Biol. 47:234 (1990).

45. P. Borgeat and B. Samuelsson, Arachidonic acid metabolism in polymorphonuclear leukocytes: effects of ionophore A23187. Proc. Natl. Acad. Sci. USA. 76:2148 (1979).

46. K.A. Haines, K.N. Giedd, A.M. Rich, H.M. Korchak and G. Weissmann, The leukotriene B4 paradox: neutrophils can, but will not, respond to ligand-receptor interactions by forming leukotriene B4 or its omega-metabolites. Biochem.J. 241:55 (1987).

47. A. Hatzelmann and V. Ullrich, Regulation of 5-lipoxygenase activity by the glutathione status in human polymorphonuclear leukocytes. Eur.J.Biochem. 169:175 (1987).

48. A.D. Purdon and A.K. Rao, Interaction of albumin, Arachidonic acid and prostanoids in platelets. Prostagland. Leuk. Essent. Fatty. 35:213 (1989).

49. T.W. Jungi, M.O. Spycher, U.E. Nydegger and S. Barandun, Platelet-leukocyte interaction: selective binding of thrombin-stimulated platelets to human monocytes, polymorphonuclear leukocytes, and related cell lines, Blood. 67:629 (1986).

50. E. Larsen, A. Celi, G.E. Gilbert, B.C. Furie, J.K. Erban, R. Bonfanti, D.D. Wagner, B. Furie, PADGEM protein - A receptor that mediates the interaction of activated platelets with neutrophils and monocytes, Cell. 59:305 (1989).

51. S.A. Hamburger, R.P. Mcever, GMP-140 Mediates Adhesion of Stimulated Platelets to Neutrophils, Blood. 75:550 (1990).

52. E. Larsen, T. Palabrica, S. Sajer, G.E Gilbert, D.D. Wagner, B.C. Furie and B. Furie, PADGEM-dependant adhesion of platelets to monocytes and neutrophils is mediated by a lineage-specific carbohydrate, LNF III (CD15), Cell. 63:467 (1990).

INTERACTION OF PLATELETS AND NEUTROPHILS IN THE GENERATION OF SULFIDOPEPTIDE LEUKOTRIENES

Robert C. Murphy[1], Jacques Maclouf[2], and Peter M. Henson[1]

[1]National Jewish Center for Immunology and Respiratory Medicine, 1400 Jackson Street, Denver, CO 80206; [2]INSERM Unit 150, Hopital Lariboisiere, 75475 Paris Cedex 10, FRANCE

INTRODUCTION

For approximately 50 years the mediator termed slow reacting substance of anaphylaxis (SRS-A) was suspected to play an important role in human allergic reactions, prolonged bronchoconstriction and asthma yet its chemical structure remained elusive (1,2). Details concerning the biosynthetic origin of this molecule as well as the regulatory mechanisms involved in controlling production and degradation of SRS-A were unknown. In 1979, the structure of SRS-A was elucidated (3) as a family of three novel compounds, having both a lipid portion derived from arachidonic acid and a peptide portion derived from glutathione (4). These molecules are now termed sulfidopeptide leukotrienes (leukotriene C_4, D_4, E_4), which differ in the number of amino acid residues resident in the peptide portion as either gamma-glutamylcysteinylglycine, cysteinylglycine, or cysteine respectively. During the past decade, a great deal of information has been obtained describing the biosynthesis of these molecules, the activation of phospholipase A_2 in liberating free arachidonic acid esterified to storage phospholipids (5), the importance of 5-lipoxygenase in generating the reactive intermediate leukotriene A_4 (6) and LTC_4 synthase which catalyzes the condensation of glutathione with leukotriene A_4 yielding LTC_4 (7). Furthermore, it is now recognized that sulfidopeptide leukotrienes can be synthesized in a variety of cells including mast cells (8), eosinophils (9), macrophages (10), and basophils (11). Interest in these molecules continues because of the potent biological activities which they possess including bronchoconstriction (12), vasoconstriction (13), and increased vascular permeability (14). Metabolism of LTC_4 is known to take place rapidly and includes sequential peptide cleavage reactions (leading to the sulfidopeptide leukotriene described above) as well as ω- and β-oxidation with ultimate elimination of metabolites into the urine (15).

In most studies concerning the biosynthetic origin of sulfidopeptide leukotrienes, emphasis has been placed upon the role of a single cell having phospholipase A_2, 5-lipoxygenase, and LTC_4 synthase capacity. However, the generation of these important class of lipid mediators can also take place through specific cell-cell interactions, most interestingly in the interaction between the human platelet and the human polymorphonuclear leukocyte. It is important to emphasize that independently these cells cannot produce sulfidopeptide leukotrienes, but in a coordinated fashion significant quantities of this specific lipid mediator can be produced following neutrophil activation.

Platelet-Neutrophil Interactions

An interaction between the platelet and neutrophil has been recognized for decades by investigators studying inflammatory and thrombotic processes (16-20). Intimate contact sites exist between these cells and this can be readily observed at the electron microscope level (eg ref 20 and Figure 1). However the studies of neutrophil-platelet interaction have tended to focus on the effects of neutrophils on the activation of platelets (17,19-22) or

Cell-Cell Interactions in the Release of Inflammatory Mediatiors
Edited by P. Y-K Wong and C.N. Serhan, Plenum Press, New York, 1991

91

more recently, on the alternative direction (17,23,24). Details concerning events regulating the actual adherence mechanisms did not receive as much attention. However the current emphasis on cellular adhesion proteins has generated a newfound interest in these processes. Both cell types express adhesion proteins on their surface that are enhanced in function or amount upon activation (25-27). At present there seem two likely candidates. Integrin receptors are present on both cell types (the CD11/18 family on the neutrophil and the GPIIb-IIIa protein on the platelet). Alternatively platelets express a key member of the selectin (LEC-CAM) family of lectins (PADGEM, GMP-140 or CD62) that appears to recognize carbohydrate structures on the neutrophil (28-30) . While a role for integrin receptors in neutrophil-platelet adhesion remains unclear, the platelet selectin does seem to be particularly involved in physical interactions with neutrophils (28-30).

Adhesion of these cells also presents an opportunity for optimal exchange of biochemical intermediates and transcellular metabolic events. While no direct experiments are available as yet that explore this possibility, it seems reasonable to suggest that physical adhesion may be a critical component in the transcellular biosynthesis of LTC_4 during platelet-neutrophil interactions.

While particular emphasis in this chapter will be placed upon the biosynthesis of sulfidopeptide leukotrienes, it should be mentioned that other lipid products are indeed synthesized during platelet-neutrophil interactions including dual lipoxygenase metabolites from the 5-lipoxygenase pathway (neutrophil-derived) and the 12-lipoxygenase pathway (platelet-derived). Thus, 5S, 12S,-diHETE (31) as well as 12,20-diHETE (32, 33) and further oxidation products (34) have been observed. Unfortunately, little is known concerning the biological role that these metabolites of transcellular events play in physiology or pathophysiology.

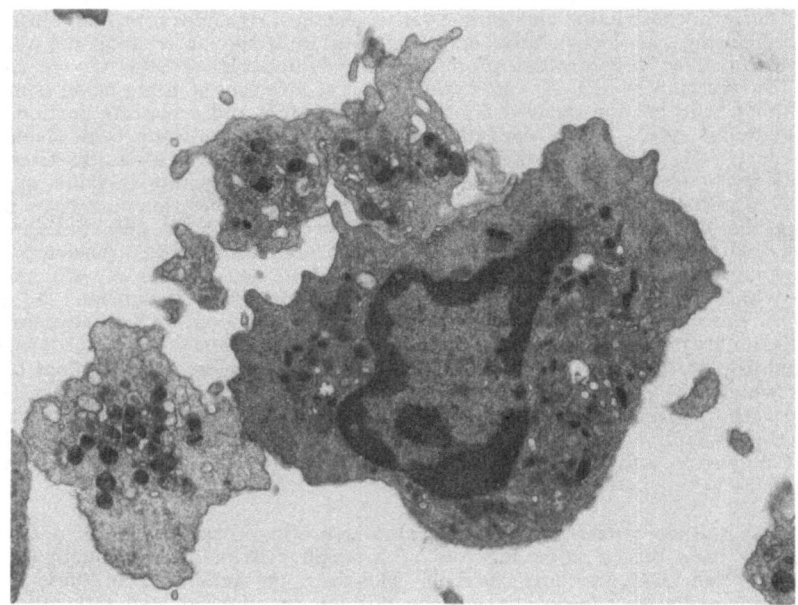

Figure 1 Electron micrograph showing platelet-neutrophil adherence following neutrophil stimulation with $10^{-7}M$ FMLP and $10^{-8}M$ PAF.

Synthesis and Release of LTA_4 from the Neutrophil

The first description of the enzyme 5-lipoxygenase (arachidonate: oxygenase 5-oxido reductase [EC 1.13.11.34]) was made in studies of arachidonate metabolism in the human polymorphonuclear leukocyte (35). A few years later it was found that the human neutrophil synthesized leukotriene B_4 from arachidonic acid (36) with intermediate formation of a chemically reactive epoxide leukotriene A_4. The neutrophil enzyme LTA_4 hydrolase could convert LTA_4 into LTB_4. The enzymatic steps involved in LTA_4 biosynthesis has been studied in great detail in the human neutrophil and some rather curious features have been revealed. When 5-lipoxygenase was first purified, it was also found to contain the activity to convert 5-hydroperoxyeicosatetraenoic acid into LTA_4, an enzymatic property previously termed as LTA_4 synthase (37). Thus, 5-lipoxygenase appears to have two catalytic activities, namely a monooxygenase as well as dehydratase. The 5-lipoxygenase enzyme was also found to be translocated to membranes and bound to an activating protein (38) following activation by ATP and calcium (39). Following this translocation process rapid inactivation takes place. Since the neutrophil also contains the enzyme LTA_4 hydrolase, it was widely felt that the product of LTA_4 biosynthesis within the human neutrophil was generation of the chemotactic leukotriene, LTB_4. However, in most studies of neutrophil-mediated biosynthesis of LTB_4 the production of several other LTA_4 products are typically observed (36). This is shown in Figure 2 with calcium ionophore (A23187) stimulated production of LTB_4 in isolated, purified human neutrophils. While there is an abundant production of leukotriene B_4, there is also an obvious production of other products derived from leukotriene A_4 which include the nonenzymatic products 5,12-diHETE and 5,6-di-HETE. This has always been a curious feature of neutrophil production of LTB_4.

In 1985 Dahinden, Hugli, and co-workers (40) described some rather interesting experiments suggesting that leukotriene A_4 can be released by human neutrophils following calcium ionophore stimulation. As seen in Figure 3, when neutrophils were stimulated with calcium ionophore in the presence of an albumin containing buffer, the supernatant could be treated with either methanol or ethanol to yield two novel metabolites either 12-O-methyl or 12-O-ethyl derivatives of LTB_4. These experiments showed that even after a 5 min incubation, LTA_4 was released into the supernatant. It was suggested that as much as 20-30% of the LTA_4 synthesized within the neutrophil by A23187 stimulation could be released from intact cells. The conclusion reached was that LTA_4 may be an important extracellular product of neutrophil stimulation. These authors went on to show that mast cells could take up exogenous LTA_4 and convert it into LTC_4 since such cells were known to contain LTC_4 synthase required for sulfidopeptide leukotriene biosynthesis. Even though cell mixing experiments were not performed, Dahinden *et al.* (40) speculated on the existence of cell-cell interaction in LTC_4 biosynthesis.

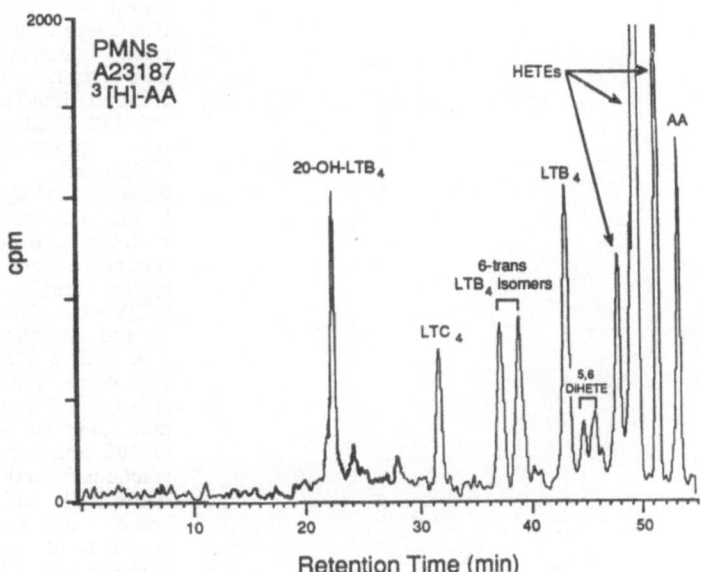

Figure 2 Reverse phase HPLC separation of the radioactive metabolites obtained following incubation of human neutrophils (5 x 10^6 cells/ml) with 5 μ M calcium ionophore A23187 following the addition of [^{3}H]arachidonic acid (10 μ Ci). 6-Trans-LTB_4 and 12-epi-6-trans-LTB_4 are nonenzymatic products derived from LTA_4 while LTB_4 formation requires LTA_4 hydrolase. Formation of nonenzymatic LTA_4 products indicate activation of 5-lipoxygenase, but escape of LTA_4 from the neutrophil. The production of LTC_4 suggests the presence of platelets (see text) and transcellular metabolism of neutrophil-derived LTA_4.

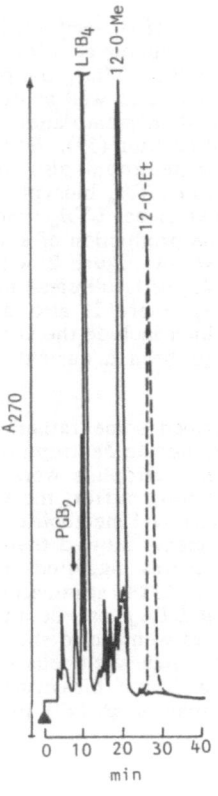

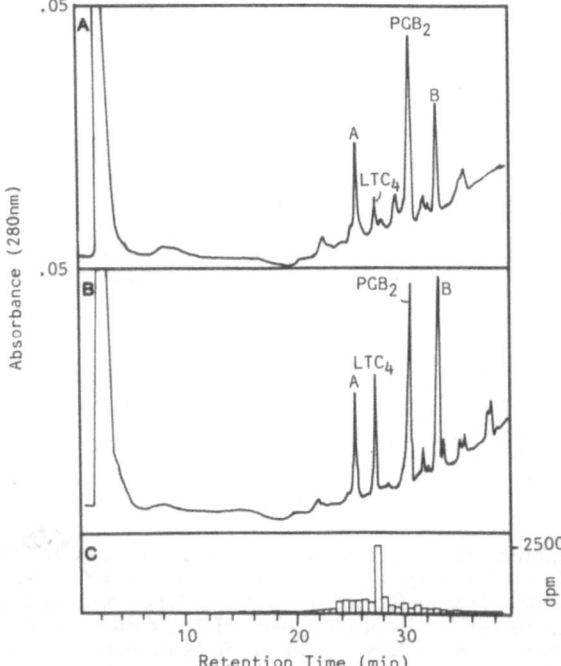

Figure 3 Reverse phase HPLC separation (taken from ref. 40) of the supernatant from 1 x 10^6 neutrophils following challenge for 5 min with 10 μ M calcium ionophore A23187, then treated with either methanol or ethanol (dotted trace). One nanomolar of prostaglandin B$_2$ (PGB$_2$) was added as internal standard. Products were detected following absorbance at 270 nm. The retention times (in minutes) of products include PGB$_2$ (7.8), 6-trans-LTB$_4$ (9.5), 12-epi-6-trans-LTB$_4$ (10.0), LTB$_4$ (10.8), 12-O-methyl-6-trans-LTB$_4$ isomers (19.2 and 19.7), and 12-O-ethyl-6-trans-LTB$_4$ isomers (27.2). (Reproduced with permission).

Figure 4 Reverse phase HPLC separation of extracts from neutrophil and neutrophil-platelet coincubations (taken from ref. 44). Washed platelets were prelabeled with [^{35}S]cysteine to form intracellular [^{35}S]glutathione. In the presence of 50 μ M cysteine, neutrophils were incubated alone (panel A) or in combination with [^{35}S]labeled platelets (panel B) and stimulated for 30 min with 5 μ M calcium ionophore A23187. The HPLC effluent was monitored for absorption at 280 nm. One minute fractions from the platelet-neutrophil coincubation were analyzed for radioactivity (panel C). Compounds identified include: A, 20-hydroxy-LTB$_4$; peak B, LTB$_4$ and a mixture of LTB$_4$ and 5(S),12(S)-dihydroxyeicosatetraenoic acid in panel B. PGB$_2$ (internal standard) and LTC$_4$. (Reproduced with permission).

When LTA_4 was first described as a product of 5-lipoxygenase and an intermediate in the formation of LTC_4 and LTB_4, emphasis was placed on the chemical reactivity of this allylic epoxide. A dilemma existed as to whether LTA_4 could exist outside of a cell for a sufficient length of time to be taken up by another cell. However, subsequent studies clearly established that LTA_4 could be stabilized by albumin when present even at modest concentrations (41). The uptake of intact LTA_4 was the second part of the dilemma. Studies of the export of both LTC_4 by human eosinophils (42) and export of LTB_4 from polymorphonuclear leukocytes (43) exploited the fact that an avid uptake of LTA_4 exists within certain cells. For example, even at $0°$ with incubations as short as 20 sec, LTA_4 could be efficiently taken up into the human neutrophil or eosinophil. Such an uptake processes has also been found to exist in the human platelet (Frank A. Fitzpatrick, unpublished observations). The uptake of LTA_4 by specific cells is still poorly understood, but is an essential feature of cell-cell interaction in the biosynthesis of leukotrienes.

LTC₄ Biosynthesis following Platelet-Neutrophil Interaction

Biochemical cooperation between human neutrophils and platelets was established by various studies where it was found that platelets could efficiently metabolize exogenous applied LTA_4 into a novel product which had an HPLC retention time, UV spectrum, as well as biological activity (guinea pig ileum), consistent with leukotriene C_4 (44). In order to establish the essential role of the neutrophil in providing LTA_4 for such a process and the platelet in providing glutathione, human neutrophils were incubated with platelets which had been preincubated with [^{35}S]cysteine to synthesize intracellular [^{35}S]glutathione. The results of this experiment are summarized as follows (Figure 4). Neutrophil preparations when stimulated in the absence of platelets yield a small amount of LTC_4 (Figure 4A) which has been recognized for over 10 years. We currently believe that the production of LTC_4 by neutrophil preparations may be due to the presence of other cells such as the eosinophil (present to a small extent in all neutrophil preparations) or alternatively contaminating platelets which are difficult to remove during neutrophil isolation. Following the addition of [^{35}S]-labeled platelets (3×10^8/ml) to neutrophils (1×10^7/ml) substantially enhanced the production of this sulfidopeptide leukotriene product as seen in the middle panel (Figure 4B). Furthermore, there was a co-elution of [^{35}S] labeled product with the LTC_4 (Figure 4C) confirming the production of LTC_4 from platelet-derived glutathione. When platelets were stimulated with calcium ionophore alone, there was no production of LTC_4. The production of LTC_4 by platelets was found in separate experiments to be dependent upon the platelet/neutrophil ratio as well being dependent upon calcium ionophore concentration. The synthesis of LTC_4 by platelet-neutrophil transcellular biosynthesis has subsequently been confirmed in other laboratories (45,46). Furthermore, treatment of platelets with aspirin, in concentrations which completely inhibit the production of thromboxane A_2, were found to have no effect on the capacity of the platelets to transform neutrophil-derived LTA_4 into LTC_4 (47).

Most of the investigations of the transcellular biosynthesis of neutrophil-derived LTA_4 by the platelet have involved the use of the calcium ionophore as a stimulatory agent to generate maximum amounts of LTA_4. However, this stimulus is generally believed to be rather nonspecific in that it activates many biochemical events both within the neutrophil as well as the platelet. An important question has been to ascertain whether or not LTA_4 transfer from neutrophil to platelet could occur under more physiologically relevant situation, such situations which involve a substantially milder stimulus that may occur *in vivo* following receptor-mediated stimulation with a chemotactic factor such as formyl-methionyl-leucyl-phenylalanine (FMLP) or phagocytosis induced by a particle such as opsonized zymosan. Using an experimental approach similar to that described above with [^{35}S] labeled glutathione in platelets, neutrophils (9×10^6 cells) and platelets (8×10^8 cells) were stimulated for 30 min at $37°$ with opsonized zymosan (40 particles/cell) and FMLP (1 μM) in the presence of labeled platelets (48). Following isolation and HPLC purification as seen in Figure 5, a very small amount of LTC_4 as well as LTB_4 could be observed in the HPLC effluent using the UV monitor set at 280 nm. However, there was a clear co-elution of a radioactive component ([^{35}S]) and an immunoreactive component to a specific antiserum at the appropriate retention time for LTC_4. These results confirmed that specific stimulatory conditions for the human neutrophil (namely phagocytosis) transcellular biosynthesis of LTC_4 could be observed to take place (37).

Activation of platelets does not appear to be required for LTA_4 processing into LTC_4. As seen in Figure 6, when platelets are stimulated with thrombin to synthesize thromboxane A_2 (as measured by the stable metabolite thromboxane B_2), there was no

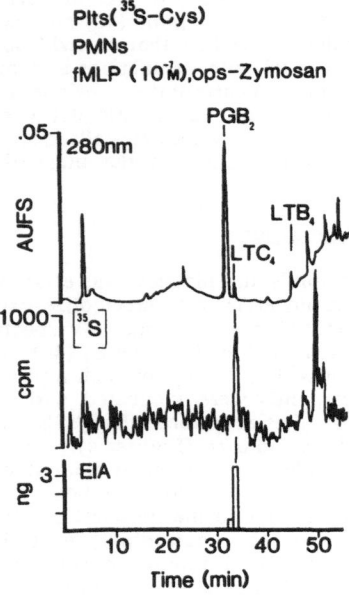

Plts(^{35}S–Cys)
PMNs
fMLP (10^{-7}M),ops–Zymosan

Figure 5 HPLC separation of the supernatant following receptor-mediated stimulation of a coincubation of human neutrophils and labeled platelets with FMLP (10^{-7} M) and opsonized zymosan for 5 min (taken from ref. 48). Platelets were prelabeled with [^{35}S]cysteine to form intracellular [^{35}S]glutathione. HPLC effluent was monitored at 280 nM (A) showing the elution of the internal standard PGB$_2$, a small peak for LTC$_4$, and LTB$_4$. Panel B shows the elution of radioactivity and the co-elution of a [^{35}S] labeled component with LTC$_4$. Panel C shows the enzyme immunoassay specific for LTC$_4$ which was carried out on 1 min fractions taken during the course of the HPLC run. The immunoreactive component eluted at exact retention time for the [^{35}S] labeled metabolite as well as the UV absorbing component at the retention time of LTC$_4$ (reproduced with permission).

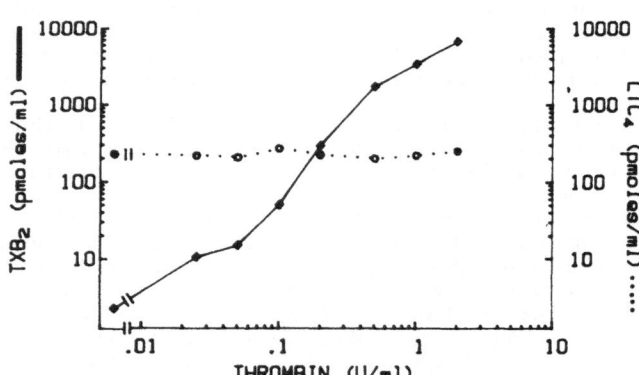

Figure 6 The effect of thrombin stimulation on the transformation of exogenously added LTA$_4$ by platelets into LTC$_4$. Platelets (6×10^8/ml) were stimulated with increasing concentrations of thrombin. After 10 min, LTA$_4$ (0.5 μ M) was added for 20 min. The supernatants were analyzed by enzyme immunoassay for thromboxane B$_2$ as well as LTC$_4$. These results show a constant transformation of LTA$_4$ into LTC$_4$, but a thrombin dose dependent increase in the production of thromboxane B$_2$ by stimulated platelets. (Taken from ref. 48 and reproduced with permission).

alteration in the capacity of these cells to transform LTA_4 into LTC_4 (48). This is consistent with the known activities of glutathione-S-transferases which are not believed to require an activation step. As mentioned above, the capacity of platelets to transform LTA_4 was not affected by inhibiting cyclooxygenase by aspirin. Thus, in those platelets which lost their capacity to make the vasoconstrictor TxA_2, there is still a possibility of producing a vasoconstrictor substance through the transcellular biosynthesis of LTC_4 from neutrophil derived LTA_4. These experiments also suggest that when platelets are aggregated (for example in a thrombus) they would still would retain their capacity to metabolize LTA_4. In the experiments illustrated in Figure 6, even when thrombin was present above 0.1 units/ml and platelets were fully aggregated, a constant amount of LTC_4 was still generated from exogenous LTA_4.

The above findings support the concept that transcellular biosynthesis of eicosanoids including leukotriene C_4 by platelet-neutrophil interactions is a viable alternative biosynthetic route for the generation of highly active lipid mediators. Yet, it is difficult to obtain experimental evidence for *in vivo* production of eicosanoids by this transcellular biosynthesis. In large part this is due to the complexities of cells as they are organized into tissues and the assignment of a synthetic cell for LTA_4 as well as the corresponding unique receptor cell. As a model to probe the potential for *in vivo* production of sulfidopeptide leukotrienes by transcellular biosynthesis, experiments have been carried out with whole blood (49). In these studies, neutrophils in blood were stimulated to undergo phagocytosis of zymosan particles and the production of eicosanoids quantitated. Individual cell populations in the blood were purified, reconstituted in buffer then stimulated with opsonized zymosan to probe the production capacity of various eicosanoids. As seen in Figure 7, production of LTE_4 in whole blood was substantial in response to zymosan, with production of 0.5 to 1 ng $LTE_4/10^6$ neutrophils. In these studies, LTE_4 production was measured since LTC_4 is rapidly metabolized to LTE_4 by peptidases in blood. This capacity to produce the sulfidopeptide leukotrienes could be duplicated when platelets and neutrophils from homologous blood were incubated together with opsonized zymosan. In contrast, when the neutrophil fraction was incubated alone, there was very minimal production of sulfidopeptide leukotrienes. It is also interesting to point out that the production of LTB_4 was substantially enhanced only in those preparations which contained red blood cells suggesting that the red blood cell plays an important role in the biosynthesis of leukotriene B_4 in this tissue model system. Even though the number of red blood cells greatly exceeded that of all other cells present, it was nonetheless possible for LTA_4 synthesized within the neutrophil to be converted into significant levels of sulfidopeptide leukotrienes by platelets following this physiological stimulus. It is likely that for such an efficient uptake of LTA_4 by platelets, it may be necessary to have cell-cell contact; thus little of the LTA_4 released by the neutrophil actually escapes the neutrophil-platelet complex. Furthermore, it is important to point out that the quantity of sulfidopeptide leukotrienes generated by transcellular biosynthesis in this model is close to nanomolar in concentration, quite compatible with significant biological activity for these mediators (50,51).

Lipid Mediator Network Revisited

In 1985 we argued for the existence of a lipid mediator network (52) operating *in vivo* during hypersensitivity reactions, for example in the lung, as part of the complex inflammatory process. We suggested the presence of such a network on the basis of three types of observation: 1. Simultaneous biosynthesis of numerous mediators with overlapping activities. 2. The ability of each member of the eicosanoid family (as well as PAF) to influence the synthesis of other mediators in these families through complex interactions at the receptor level. 3. A large overlap between the biologic activities of these mediators following cell activation (even though many have additional unique properties).

To these observations we now must add the concept of transcellular biosynthesis of lipid mediators, exemplified in the present discussion by generation of sulfidopeptide leukotrienes following neutrophil-platelet interactions. It seems likely that mobile inflammatory cells, as well as the cells of the blood vessels and tissues with which they interact, can cooperate in the synthesis (as well as catabolism) of most of the lipid mediators, thus adding further to the intricacy of the network. This type of extreme complexity and integrated cell cooperation would fit with the concept of a balanced network of positive and negative cell communication signals. Thus even though particular attempts may be made to alter the generation of specific lipid mediators, for example through the inhibition of cyclooxygenase and thus the reduced formation of thromboxane A_2 in platelets, nonetheless, vasoconstrictive substances would still be synthesized following

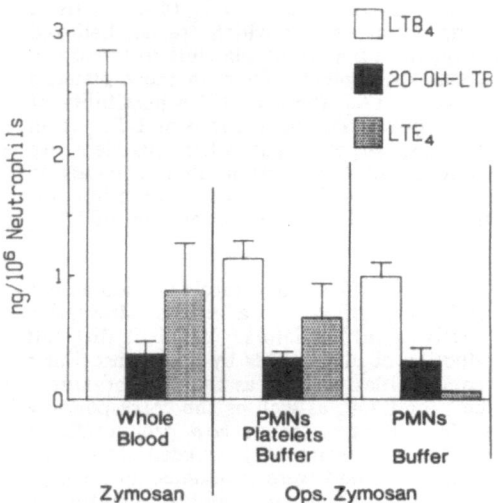

Figure 7 Formation of the lipid mediators LTB_4, 20-hydroxy-LTB_4 and LTE_4 in whole blood stimulated by zymosan (left panel), coincubation of platelets and neutrophils treated with opsonized zymosan (center panel) or neutrophils alone treated with opsonized zymosan (right panel). After a 5 min stimulation, the supernatant was partially purified by solid phase extraction and eicosanoid analyzed by combination HPLC and specific enzyme immunoassay. 20-Hydroxy-LTB_4 quantitation was carried out using gas chromatography/mass spectrometry and stable isotope dilution. (Taken from ref. 49 and reproduced with permission).

activation of neutrophils adhering to platelets to produce the vasoactive material LTC4 by cell-cell cooperation.

Thus, in order to understand the role that lipid mediators play in inflammatory processes it will be necessary to ascertain the overall biosynthesis and effects of numerous biologically active agents and the balance between them. This is also necessary when information is sought concerning the mechanisms by which inflammatory cells communicate with each other and with the surrounding tissues. While this might seem at first an impossible challenge, the availability of increasingly sophisticated analytic tools and of a new arsenal of specific antagonists is beginning to render the task open to rational investigation, both with isolated cell systems and even, in a preliminary fashion, in the whole body.

ACKNOWLEDGEMENTS

This work was supported, in part, by a grant from the National Institutes of Health (HL34303) and a grant from NATO (0264/89). The authors thank Dr. Elizabeth Hill for carrying out the experiment in Figure 1.

REFERENCES

1. C. H. Kelloway and E. R. Trethewie, The liberation of a slow-reacting smooth muscle-stimulating substance in anaphylaxis, J. Exptl. Physiol. 30:121 (1940).

2. R. P. Orange and K. F. Austen, Slow reacting substance of anaphylaxis, Adv. Immunol. 10:105 (1969).

3. R. C. Murphy, S. Hammarstrom, and B. Samuelsson, Leukotriene C: A slow reacting substance (SRS) from murine mastocytoma cells, Proc. Natl. Acad. Sci. USA 76:4275 (1979).

4. B. Samuelsson, P. Borgeat, S. Hammarstrom, and R. C. Murphy, Leukotrienes: A new group of biologically active compounds, in: "Advances in Prostaglandins and Thromboxane Research," B. Samuelsson, P. W. Ramwell, and R. Paoletti, eds., Raven Press, New York (1980).

5. C. C. Leslie, D. R. Voelker, J. Y. Channon, M. W. Wall, and P. T. Zelarney, Properties and purification of an arachidonoyl-hydrolyzing phospholipase A_2 from a macrophage cell line, RAW204.7, Biochim. Biophys. Acta 963:476 (1988).

6. P. Borgeat and B. Samuelsson, Arachidonic acid metabolism and polymorphonuclear leukocytes: Unstable intermediate in formation of dihydroxy acids, Proc. Natl. Acad. Sci. USA 76:3213 (1979).

7. T. Shimizu, Enzymes functional in the synthesis of leukotrienes and related compounds, Int. J. Biochem. 20:661 (1988).

8. B. Samuelsson, and C. D. Funk, Enzymes involved in the biosynthesis of leukotriene B_4, J. Biol. Chem. 264:19469 (1989).

9. A. Jorg, W. R. Henderson, R. C. Murphy, and S. J. Klebanoff, Leukotriene generation by eosinophils, J. Exp. Med. 155:390 (1982).

10. C. A. Rouzer, W. A. Scott, A. L. Hamill, and F. A. Cohn, Dynamics of leukotriene C production by macrophages, J. Exp. Med. 152:1236 (1980).

11. P. T. Peachell, L. M. Lichtenstein, and R. P. Schleimer, Inhibition by adenosine of histamine and leukotriene release from human basophils, Biochem. Pharm. 38:1717 (1989).

12. S. E. Dahlen, P. Hedqvist, S. Hammarstrom, and B. Samuelsson, Leukotrienes are potent constrictors of human bronchi, Nature 288:484 (1980).

13. N. A. Soter, R. A. Lewis, E. J. Corey, and K. F. Austen, Local effects of synthetic leukotrienes (LTC_4, LTD_4, LTE_4, and LTB_4) in human skin, J. Invest. Dermatol. 80:115 (1983).

14. Z. Marom, J. H. Shelhamer, M. K. Bach, D. R. Morton, and M. Kaliner, Slow reacting substances LTC_4 and D_4 increase the release of mucus from human airways *in vitro*, Am. Rev. Respir. Dis. 126:449 (1982).

15. A. Sala, N. Voelkel, J. Maclouf, and R. C. Murphy, LTE_4 elimination and metabolism in normal human subjects, J. Biol. Chem. 265:21771 (1990).

16. C. P. Page, Platelets as inflammatory cell, Immunopharmacol. 17:51 (1989).

17. P. M. Henson, Interactions between neutrophils and platelets, Editorial, Lab. Invest. 62:391 (1990).

18. J. Wester, J. J. Sixma, J. J. Geuze, and H. J. G. Heijin, Morphology of the hemostatic plug in human skin wounds. Transformation of the healing, Lab. Invest. 41:182 (1979).

19. P. M. Henson and C. G. Cochrane, Immunological induction of increased vascular permeability. I. A rabbit passive cutaneous anaphylactic reaction requiring complement, platelets and neutrophils, J. Exp. Med. 129:153 (1969).

20. P. M. Henson, Mechanisms of release of constituents from rabbit platelets by antigen-antibody complexes and complement II. Interaction of platelets with neutrophils, J. Immunol. 105:490 (1970).

21. A. Del Maschio, V. Evangelista, G. Rajtar, Z.-M. Chen, C. Cerletti, and G. De Gaetano, Platelet activation by polymorphonuclear leukocytes exposed to chemotactic agents, Am. J. Physiol. 258:H870 (1990).

22. M. A. Selak, M. Chignard, and J. B. Smith, Cathepsin G is a strong platelet agonist released by neutrophils, Biochem. J. 251:293 (1988).

23. D. Y. Tzeng, T. F. Deuel, J. S. Huang, R. M. Senior, L. A. Boxer, and R. L. Baehner, Platelet-derived growth factor promotes polymorphonuclear leukocyte activation, Blood 64:1123 (1984).

24. S. T. McGarrity, T. M. Hyers, and R. O. Webster, Inhibition of neutrophil functions by platelets and platelet-derived products: Description of multiple inhibitory properties, J. Leuk. Biol. 44:93 (1988).

25. S. M. Albelda and C. A. Buck, Integrins and other cell adhesion molecules, <u>FASEB J.</u> 4:2868-2880 (1990).

26. T. A. Springer, Adhesion receptors of the immune system, <u>Nature</u> 346:425 (1990).

27. T. A. Springer and L. A. Lasky, Sticky sugars for selectins, <u>Nature</u> 349:196 (1991).

28. E. Larsen, A. Celi, G. E. Gilbert, B. C. Furie, J. K. Erban, R. Bonfanti, D. D. Wagner, and B. Furie, PADGEM protein: a receptor that mediates the interaction of activated platelets with neutrophils and monocytes, <u>Cell</u> 59:305 (1989).

29. L. Corral, M. S. Singer, B. A. Macher, and S. D. Rosen, Requirement for sialic acid on neutrophils in a GMP-140 (PADGEM) mediated adhesive interaction with activated platelets, <u>Biochem. Biophys. Res. Commun.</u> 172:1349 (1990).

30. S. A. Hamburger and R. P. McEver, GMP-140 mediates adhesion of stimulated platelets to neutrophils, <u>Blood</u> 75:550 (1990).

31. J. Maclouf, B. Fruteau de Laclos, and P. Borgeat, Stimulation of leukotriene biosynthesis in human blood leukocytes by platelet-derived 12-hydroxyperoxy-eicosatetraenoic acids, <u>Proc. Natl. Acad. Sci. USA</u> 79:6042 (1982).

32. A. J. Marcus, L. B. Safier, H. L. Ullman, N. Islam, M. J. Broekman, N. Islam, T. D. Oglesby, and R. R. Gorman, 12S,20-Dihydroxyicosatetraenoic acid: A new eicosanoid synthesized by neutrophils from 12S-hydroxyicosatetraenoic acid produced by thrombin-or collagen-stimulated platelets, <u>Proc. Natl. Acad. Sci. USA</u> 81:903 (1984).

33. P. Wong, Y. K. Westlund, M. Hamberg, E. Granstrom, P. H. W. Chao, and B. Samuelsson, ω-Hydroxylation of 12-L-hydroxy-5,8,10,14-eicosatetraenoic acid in human polymorphonuclear leukocytes, <u>J. Biol. Chem.</u> 259:2683 (1984).

34. A. J. Marcus, L. B. Safier, H. L. Ullman, N. Islam, M. J. Broekman, J. R. Falck, S. Fischer, and C. von Schacky, Platelet-neutrophil interactions: (12S)-hydroxy-eicosatetraen-1,20-dioic acid: A new eicosanoid synthesized by unstimulated neutrophils from (12S)-20-dihydroxyeicosatetraenoic acid, <u>J. Biol. Chem.</u> 263:2223 (1988).

35. P. Borgeat, M. Hamberg, and B. Samuelsson, Transformation of arachidonic acid in dihomo-γ-linolenic acid by rabbit PMN leukocytes, <u>J. Biol. Chem.</u> 251:7816 (1976).

36. P. Borgeat and B. Samuelsson, Transformation of arachidonic acid by rabbit polymorphonuclear leukocytes, <u>J. Biol. Chem.</u> 254:2643 (1979).

37. T. Shimizu, O. Radmar, and B. Samuelsson, Enzyme with dual lipoxygenase activities catalyze leukotriene A_4 synthesis from arachidonic acid, <u>Proc. Natl. Acad. Sci. USA</u> 81:689 (1984).

38. R. A. F. Dixon, R. E. Diehl, E. Opas, E. Rands, P. J. Vickers, J. F. Evans, J. W. Gillard, and D. K. Miller, Requirement of a 5-lipoxygenase-activating protein for leukotriene synthesis, <u>Nature</u> 343:282 (1990).

39. C. A. Rouzer and S. Kargman, Translocation of 5-lipoxygenase to the membrane in human leukocytes challenged with ionophore A23187, <u>J. Biol. Chem.</u> 263:10980 (1988).

40. C. A. Dahinden, R. M. Clancy, M. Gross, J. M. Chiller, and T. E. Hugli, Leukotriene C_4 production by murine mast cells: Evidence of a role for extracellular leukotriene A_4, <u>Proc. Natl. Acad. Sci. USA</u> 82:6632 (1985).

41. F. A. Fitzpatrick, D. R. Morton, and M. A. Wynalda, Albumin stabilizes leukotriene A_4, <u>J. Biol. Chem.</u> 257:4680 (1982).

42. B. K. Lam, W. F. Owen, K. F. Austen, and R. J. Soberman, The identification of a distinct export step following the biosynthesis of leukotriene C_4 by human eosinophils, <u>J. Biol. Chem.</u> 264:12885 (1989).

43. B. K. Lam, W. F. Owen, K. F. Austen, and R. J. Soberman, The mechanism of leukotriene B_4 export from human polymorphonuclear leukocytes, J. Biol. Chem. 265:13438 (1990).

44. J. Maclouf and R. C. Murphy, Transcellular metabolism of neutrophil-derived leukotriene A_4 by human platelets, J. Biol. Chem. 263:174 (1988).

45. H. E. Claesson and J. Haeggstrom, Human endothelial cells stimulate leukotriene synthesis and convert granulocyte released leukotriene A_4 into leukotrienes B_4, C_4, D_4, and E_4, Eur. J. Biochem. 173:93 (1988).

46. C. Edenius, K. Heidvall, and J.A. Lindgren, Novel transformation of granulocyte-derived leukotriene A_4 into cysteinyl-containing leukotrienes by human platelets, Eur. J. Biochem. 178:81 (1988).

47. J. Maclouf, R. C. Murphy, and P. M. Henson, Transcellular sulfidopeptide leukotriene biosynthetic capacity of vascular cells, Blood 74:703 (1989).

48. J. Maclouf, R. C. Murphy, and P. M. Henson, Transcellular biosynthesis of sulfidopeptide leukotrienes during receptor-mediated stimulation of human neutrophil/platelet mixtures, Blood 76:1838 (1990).

49. A. Fradin, J. A. Zirrolli, J. Maclouf, L. Vausbinder, P. M. Henson, and R. C. Murphy, PAF and leukotriene biosynthesis in whole blood: A model for the study of transcellular arachidonate metabolism, J. Immunol. 143:3680 (1989).

50. R. A. Lewis, and K. F. Austen, The biologically active leukotrienes. Biosynthesis, metabolism, receptors, functions, and pharmacology, J. Clin. Invest. 73:889 (1984).

51. A. J. Marcus, Thrombosis and inflammation as multicellular processes: Pathophysiological significance of transcellular metabolism, Blood 76:1903 (1990).

52. R. C. Murphy and P. M. Henson, Mediator network, in: "Annales de l'Institut Pasteur/Immunologie," Institut Pasteur, ed., Institut Pasteur (1985).

RED CELL-NEUTROPHIL INTERACTIONS IN THE REGULATION

OF ACTIVE OXYGEN SPECIES AND LIPOXYGENASE PRODUCTS

Arnold Stern

New York University Medical Center
Department of Pharmacology
550 First Avenue
New York, NY 10016 USA

Recent evidence indicates that neutrophil eicosanoid metabolism is altered by the presence of another cell type. Possible mechanisms for effecting eicosanoid metabolism in neutrophils by other cell types are related to the capacity of these cell types to incorporate, deactivate or metabolize eicosanoids and their intermediates or provide or scavenge reactants that are utilized by neutrophils.

Red cells can interact with neutrophils to effect the accumulation of various substances generated by neutrophils. The regulation of the quantity of substances generated by the neutrophil could be influenced by the presence of hemoglobin and cytosolic and membrane enzymes in red cells. In early studies on neutrophil-red cell interactions, the red cell was used as a "target" for active oxygen generated by neutrophils(1).

Stimulated neutrophils produce H_2O_2, among other active oxygen species. When red cells are present, the extracellular concentration of H_2O_2 is diminished, most likely by scavenging of the H_2O_2 by red cell catalase and glutathione peroxidase(2). Additional evidence indicates that red cells may decrease utilization of H_2O_2 by myeloperoxidase, but this effect may depend on the availability of myeloperoxidase to compete with the red cell for H_2O_2(2).

Neutrophils produce HOCl through the interaction of myeloperoxidase and chloride ions and H_2O_2(3). The formation of HOCl by neutrophils was decreased by red cells and the accumulation of HOCl was affected by the activity of red cell catalase, indicating that the red cell could regulate the concentration of an oxidant such as HOCl in its environment by efficiently scavenging H_2O_2(4). Additionally, the red cell can reduce the formation of hydroxyl radical by this mechanism. The red cell does not efficiently scavenge superoxide.

Red cells can influence the formation of lipooxygenase products by neutrophils. In the coincubation of red cells and neutrophils, the production of the lipoxygenase product LTB_4

Cell-Cell Interactions in the Release of Inflammatory Mediators
Edited by P. Y-K Wong and C.N. Serhan, Plenum Press, New York, 1991

103

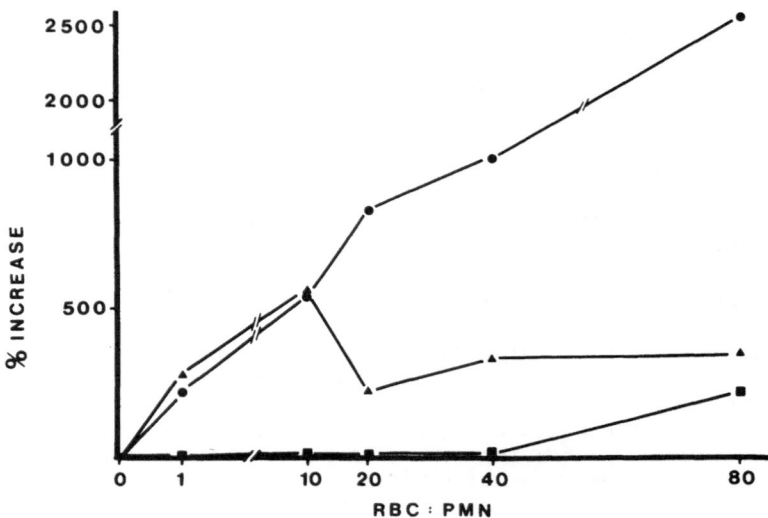

Figure 1. Percent increase in 5-lipoxygenase pathway products in
coincubation of red cells and neutrophils.
Neutrophils (20×10^6 cells/ml). Incubations took place
in PBS at 37°C for 20 minutes. Cells were stimulated
with A_{23187} (2.5μM). Adult fresh human blood was
prepared as previously reported(9). Neutrophils were
isolated by Ficoll-Hypague gradient centrifugation
(10). Lipoxygenase products were extracted and
quantitated as previously described(11). The
lipoxygenase products were separated by reverse phase
liquid chromatography utilizing an LKB dual-pump
system equipped with an Alter Ultrasphere-ODS (4.6 mm
x 25 cm) column, injector and solvent controller (LKB,
Brommer, Sweden) and a photodiode array rapid spectral
detector linked to an AT&T PC3600. Post run high
pressure liquid chromatography analyses were performed
with a 2140-202 Wavescan program and Nelson Analytical
3000 series chromatography data system. PGB_2 was used
as an internal standard. Recovery of 5-HETE in
neutrophils is 3.44 ng. Each value is the mean
percentage change of two samples. ■, LTB_4; ▲,
oxidation products of LTB_4; ●, 5-HETE.

was increased over that seen with neutrophils(5). When LTA_4 was incubated with red cells it was converted to LTB_4 implying that red cells contain epoxide hydrolase activity in their cytosol(7). This mechanism is similar to that observed in platelet-endothelial cell interactions. In that case, the activity of the enzyme, prostaglandin endoperoxide H_2, which is found in platelets, produced prostaglandin I_2 in endothelial cells through a transcellular mechanism. It has been suggested that stimulated neutrophils(6) release LTA_4 and/or 5-HPETE, both of which then enters red cells. In the case of 5-HPETE, it might undergo transformation by hemoglobin to LTA_4. By whatever mechanism LTA_4 appears in red cells, it is then converted to LTB_4 by epoxide hydrolase. Apparently, epoxide hydrolase in red cells behaves in a similar fashion as in neutrophils, since LTA_4 causes "suicide" inactivation of the enzyme in intact red cells(7,8).

Coincubation of neutrophils with increasing numbers of red cells indicated that LTB_4 production increases at high ratios of red cells to neutrophils (Figure). When other lipoxygenase products were analyzed it was observed that levels of 5-HETE rose significantly at low ratios of red cells to neutrophils and that this increase was 10 fold greater than LTB_4 at the highest red cell to neutrophil ratios. No increase occurred in 15-HETE or lipoxins in neutrophil-red cell interactions over that seen with neutrophils, implying that the behavior of red cells on neutrophils regarding lipoxygenase products of neutrophils is selective for the 5-lipoxygenase pathway. The precise mechanisms for this selectivity is not apparent, but may be related to the capacity of the red cell to regulate its environment(4,5,9,12).

Red cells play an important role in the regulation of neutrophil behavior by scavenging reduced products of oxygen metabolism, in particular hydrogen peroxide, thereby effecting both hydroxyl radical and HOCl formation and by regulating the formation of lipoxygenase pathway products. The potential for a transcellular interaction between red cells and neutrophils increases in a pathological process because this process causes stimulation of neutrophils to produce active oxygen and lipoxygenase products and because of the potential contact of neutrophils and red cells in areas of pathology.

The appearance of red cells and neutrophils in an inflammatory response could lead to an interaction between the cell types that could influence the 5-lipoxygenase pathway in the neutrophil. This would result in the accumulation of 5-lipoxygenase-derived products that provide further chemotactic activity. This activity could increase the inflammatory response and cause a progression of tissue damage. By diminishing both neutrophil infiltration and the occurrence of a neutrophil-red cell interaction, it may be possible to avoid certain pathological consequences. This was observed in studies on myocardial infarction, where in the reperfusion model, pharmacological intervention to decrease neutrophils in the affected area of the myocardium resulted in decreased infarct size (13). That modification of 5-lipoxygenase-derived product formation in the neutrophil by the red cell may represent an important addition to the inflammatory response awaits further studies.

Abbreviations

LTA$_4$ and LTB$_4$, leukotriene A$_4$ and B$_4$; HPETE, hydroperoxy-eicosatetraenoic acid; HETE, hydroxyeicosatetraenoic acid; PBS, phosphate buffered saline.

Acknowlegement

Supported by a grant from the National Institutes of Health, ES03425. Appreciation of Charles N. Serhan is acknowledged for providing facilities for conducting experiments on lipoxygenase pathway products and for helpful discussion.

Bibliography

1. Weiss, S.J. The role of superoxide in the destruction of erythrocyte targets by human neutrophils. J. Biol. Chem. 255, 9912-9917, 1980.

2. Test, S.T. and Weiss, S.J. Quantitative and temporal characterization of the extracellular H$_2$O$_2$ pool generated by human neutrophils. J. Biol. Chem. 259, 399-405, 1984.

3. Klebanoff, S.J. Myeloperoxidase-halide-hydrogen peroxide antibacterial system. J. Bacteriol. 95, 2131-2138, 1968.

4. Winterbourn, C.C. and Stern, A. Human red cells scavenge extracellular hydrogen peroxide and inhibit formation of hypochlorous acid and hydroxyl radical. J. Clin. Invest. 80, 1486-1491, 1987.

5. McGee, J.E. and Fitzpatrick, F.A. Erythrocyte-neutrophil interactions: Formation of leukotriene B$_4$ by transcellular biosynthesis. Proc. Natl. Acad. Sci. USA. 83, 1349-1353, 1986.

6. Marcus, A.J., Weksler, B.B., Jaffe, E.A. and Broekman, M.J. Synthesis of prostacyclin from platelet-derived endoperoxides by cultured human endothelial cells. J. Clin. Invest. 66, 979-986, 1980.

7. Fitzpatrick, F., Liggett, W., McGee, J., Bunting, S., Morton, D. and Samuelsson, B. Metabolism of leukotriene A$_4$ by human erythrocytes. J. Biol. Chem. 259, 11403-11407, 1984.

8. Orning, L., Jones, D.A. and Fitzpatrick, F.A. Mechanism-based inactivation of leukotriene A$_4$ hydrolase during leukotriene B$_4$ formation by human erythrocytes. J. Biol. Chem. 265, 14911-14916, 1990.

9. Stern, A. and Serhan, C.N. Human red cells enhance formation of 5-lipoxygenase-derived products by neutrophils. Free Rad. Res. Comm. 7, 335-339, 1989.

10. Boyum, A. Isolation of mononuclear cells and granulocytes from human blood. Scand. J. Clin. Lab. Invest. 21, Suppl. 97, 77-89, 1967.

11. Serhan, C.N. On the relationship between leukotriene and lipoxin production by human neutrophils; evidence for differential metabolism of 15-HETE and 5-HETE. Biochim. Biophys. Acta. <u>1004</u>, 158-168, 1989.

12. Sullivan, S.G. and Stern, A. Effects of physiologic concentrations of lactate, pyruvate and ascorbate on glucose metabolism in unstressed and oxidatively stressed human red blood cells. Biochem. Pharmacol. <u>32</u>, 2891-2902, 1983.

13. Romson, J.L., Hook, B.G., Kunkel, S.L. Abrams, G.D., Schork, M.A. and Lucchesi, B.R. Reduction of the extent of ischemic myocardial injury by neutrophil depletion in the dog. Circulation <u>67</u>, 1016-1023, 1983.

THE LIPOXIN BIOSYNTHETIC CIRCUIT AND THEIR ACTIONS

WITH HUMAN NEUTROPHILS

Stefano Fiore, Mark E. Brezinski, Kelly-Ann Sheppard, and
Charles N. Serhan*

Hematology Division
Brigham and Women's Hospital and Harvard Medical School
Boston, Mass

Introduction

Signal transduction in inflammatory cells is associated with the release and oxygenation of arachidonic acid by lipoxygenases (1,2). The lipoxins (LX) are a recent addition to the family of biologically active products generated from arachidonic acid collectively termed eicosanoids. Members of the LX series contain a conjugated tetraene structure (3) and display a unique spectrum of bioactivities which distinguish them from other eicosanoids (3,4). Along these lines, recent results from several laboratories (5-9) indicate that LXA_4** blocks some of the "pro-inflammatory" actions of leukotrienes. Taken together they suggest that LX may serve as *chalones* in inflammatory responses. Therefore, complete knowledge of their biosynthesis, temporal association of formation and relationship to other eicosanoids is essential to unraveling the functions of these tetraene-containing eicosanoids in inflammation as well as other physiologic events.

The biosynthesis of lipoxins from cellular sources of unesterified arachidonate can be triggered by initial oxygenation by either the 15-lipoxygenase (LO) or 5-LO followed by additional reactions (recently reviewed in 3). In addition to formation by a single cell type

---

* Address correspondence to Charles N. Serhan, Brigham and Women's Hospital, 75 Francis St., Boston, MA 02115

** Abbreviations: 8-trans-LXB_4, 5S,14R,15S-trihydroxy-6,8,10,12-trans-eicosatetraenoic acid; 11-trans-LXA_4, 5S,6R,15S-trihydroxy-7,9,11,13-trans-eicosatetraenoic acid; 15-HETE, 15S-hydroxy-5,8,11-cis-13-trans-eicosatetraenoic acid; 5-HETE, 5S-hydroxy-8,11,14-cis-6-trans-eicosatetraenoic acid; leukotriene B_4 (LTB_4), 5S,12R-dihydroxy-6,14-cis-8,10-trans-dihydroxyeicosatetraenoic acid; lipoxin A_4 (LXA_4), 5S,6R,15S-trihydroxy-7,9,13-trans-11-cis-eicosatetraenoic acid; lipoxin B_4 (LXB_4), 5S,14R,15S-trihydroxy-6,10,12-trans-8-cis-eicosatetraenoic acid; 12-HETE, 12S-hydroxy-5,8,14-cis-10-trans-eicosatetraenoic acid.

Cell-Cell Interactions in the Release of Inflammatory Mediators
Edited by P. Y-K Wong and C.N. Serhan, Plenum Press, New York, 1991

109

origin, the biosynthesis of eicosanoids can be the result of cell-cell interactions with the transcellular metabolism of key intermediates (10). An example of the latter is the production of eicosanoids via human platelet-neutrophil interactions (10-14), where platelets can utilize leukocyte-derived leukotriene (LT) A_4 to generate both peptido-LTs (13,14) and lipoxins (11,12). Human neutrophils (PMN) and platelets have served as a convenient model system for studying transcellular eicosanoid biosynthesis during cell-cell interactions (10-14) because these cell types are readily available by isolation from fresh venous blood and they display cell type specific distribution of the individual LOs involved (i.e. PMN 5-LO and 15-LO, platelet 12-LO). Each of the LO's plays an important role in the generation of eicosanoids.

To date, we have documented four independent routes with human peripheral blood cell types that can lead to the generation of lipoxins. They are:

(i)   15-lipoxygenation of arachidonic acid followed by transformation of 15-HETE by PMN to LX (15,16);

(ii)   acylation/deacylation of stored 15-HETE in inositol lipids of PMN (17);

(iii)   transformation of leukotriene A_4 by the platelet (11,12)/megakaryocyte 12-lipoxygenase to lipoxins (18); and

(iv)   formation and further metabolism of a 5(6)epoxytetraene intermediate (16,19).

The ionophore of divalent cations, $A_{23,187}$, is a useful tool in examining the biosynthesis of eicosanoids in intact cells, and in particular, transcellular pathways of formation because it (a) is not cell type specific in its actions, which enables activation of combined cell types and (b) it provokes the formation of individual eicosanoids in amounts that facilitate their isolation and structural elucidation by physical methods (2,10,14,20,21). By virtue of its ability to translocate divalent cations across cell membranes, it initiates a variety of responses that require an elevation in the intracellular levels of Ca^{2+}. In this regard, when added to human neutrophils and platelets, $A_{23,187}$ activates both the release of arachidonic acid from endogenous stores and lipoxygenation, which are two limiting events in the biosynthesis of lipoxygenase products by these cells. However, because $A_{23,187}$ activates cells by circumventing receptor-ligand interactions in both cell types, potential regulatory events between individual lipoxygenases which may be operative following receptor activation remain to be fully appreciated. Previous studies have shown that platelet-directed stimuli (i.e., thrombin) can initiate metabolic interactions between platelets and neutrophils (21). Thus, receptor-mediated activation of combined cell types may lead to both quantitative and qualitative differences in the profiles and individual eicosanoids generated by transcellular biosynthesis. To gain insight into the relationship between leukotrienes and lipoxins generated by receptor-mediated signals, we have determined the levels formed during co-incubations of platelets with PMN and have examined their routes of formation.

In this chapter we review recent evidence which indicates that (i) several distinct biosynthetic routes are operative in the generation of lipoxins by receptor-operated mechanisms which together contribute to the "LX-biosynthetic circuit" and (ii) LX stimulate selective events (i.e. phospholipid remodeling) in the activation sequence of human PMN (in vitro).

Lipoxin Formation during Cell-Cell Interactions

Simultaneous activation of PMN and platelets in suspension (1:10 cell ratio) with thrombin and fMLP leads to the formation of both LXA_4 and LXB_4 (11). In this system, one route of LX formation involves LTA_4, and evidence has been presented suggesting that platelet 12-LO and its ω-6-oxygenase activity can play a role in the formation of lipoxins from PMN-derived leukotriene A_4 (11,12). Because platelets can also utilize LTA_4 to generate LTC_4 (13,14), we determined the relationship between leukotrienes and lipoxins generated by PMN coincubated with platelets (1:100 cell ratio). When these cells were exposed to formylmethionyl-leucine-phenylalanine (fMLP) (10^{-7}M) and thrombin (0.1 U/ml), both tetraene-containing compounds and leukotrienes were generated in quantities which were consistent with earlier reports (22,23), namely that LTC_4 (which can be derived from platelets) (13,14) and ω-oxidation products of LTB_4 (formed by PMN) (22,23) registered as the dominant products in these incubation conditions. Since both electrochemical and UV detection with HPLC indicated that LX were formed in ng amounts, it was necessary to utilize electron-capture negative ion chemical ionization-mass spectrometry (NICI-MS) (24) (Fig. 1) to confirm the identity of the materials generated. To this end, materials were collected, converted to pentafluorobenzyl (PFB) ester derivatives, treated with bis(TMS)trifluoroacetamide (BSTFA), and analyzed by NICI-GC/MS as in ref. 24. The mass spectrum of the PFB ester, TMS (trimethylsilyl) ether of the material which coeluted with LXA_4 showed an intense anion at m/e 567 which corresponded to [M-PFB]$^-$, and ions of lower intensity (<5%) at m/e 477 and m/e 387 corresponding to the loss of TMSOH and 2(TMSOH). Together the physical characteristics of these products in electrochemical detection-ultraviolet-reverse phase high pressure liquid chromatography (ED-UV-RP-HPLC) and ions present in their mass spectra suggest that during coincubation platelets and PMN stimulated with fMLP and thrombin can generate LX.

The role of cell ratio and rank order of products generated via receptor-mediated formation of lipoxygenase derived eicosanoids was evaluated during coincubation of PMN and platelets (Fig. 2). Upon exposure to thrombin (0.1 U) and fMLP (10^{-7} M), the levels of LTBs (Σ LTB_4, ω-OH-LTB_4, and 20-COOH-LTB_4) and lipoxins (Σ LXA_4, LXB_4, and their all-trans isomers) were generated in excess of the amounts of peptido-LTs. The levels of peptido-LTs did not approach those observed for either the LXs or LTBs until the platelet count exceeded $\approx$10 per PMN (Fig. 2). At the cell ratio of 1:10, the increase observed for peptido-LTs appeared to follow a reciprocal decrement in the levels of LTBs. LTA_4 is the common precursor to both LTB_4 and LTC_4, and therefore it is likely that this shift to peptido-LTs represents the result of LTA_4 transcellular metabolism. The peptido-LTs measured in the coincubations were LTC_4, D_4, and E_4. Notably, both of these cell types have the ability to further metabolize LTC_4, once it is generated, into D_4 and E_4 as determined in a pulse chase experiment (Fig. 3). Here, addition of ^{3}H-LTC_4 to human platelets results in the appearance of labeled products that coelute with synthetic LTD_4 and E_4, respectively. A similar reciprocal relationship between LTB_4 and LTC_4 has been established with suspensions of platelets and granulocytes stimulated with ionophore $A_{23,187}$ (14). Thus, both cell types can contribute to the further metabolism of LTC_4 and the generation of bioactive LTD_4.

During coincubations stimulated with ionophore ($A_{23,187}$), platelet-derived arachidonate can serve as a precursor for PMN-derived LTB_4 and 5-HETE (10), and platelets can convert exogenous LTA_4 to lipoxins (11,12) as well as peptido-LTs (13,14). To determine if these two biosynthetic routes are operative during receptor-mediated eicosanoid formation by cell-cell interactions, labeled platelets were coincubated with PMN and the lipoxygenase products were analyzed for the content of [1-^{14}C]. A representative radiochromatogram of the mono-HETEs obtained after stimulation of co-incubations with fMLP (10^{-7} M) and thrombin (0.1 U) is given in Fig. 4. Both 5-HETE and 15-HETE carried [1-^{14}C]platelet-derived radiolabel. The results of these experiments illustrate that in incubation with thrombin platelets release [1-^{14}C]arachidonate that is utilized and transformed by both the 5- and 15-lipoxygenase of fMLP-activated PMN.

Further analysis revealed that peptido-LTs from these incubation conditions also carry [1-^{14}C] label (Fig. 5). These eicosanoids are generated from PMN-derived LTA_4 (13,14). Thus, it appears that during receptor-mediated cell-cell interaction platelet-derived [1-^{14}C]arachidonate is released and transformed by the PMN 5-LO to [1-^{14}C]-LTA_4, which is then transferred back to platelets for its conversion to labeled peptido-LTs (illustrated in Fig. 6).

Along these lines, evidence for at least two transcellular routes for lipoxin production has been obtained which involves the transfer of intermediates: one involves the transformation of exogenous 15-HETE by human PMN (25) and eosinophils (26), and the other involves the conversion of LTA_4 by human platelets (11,12). Lipoxins from the above incubation conditions also carried [1-^{14}C] label demonstrating that platelet-derived arachidonate contributes to their formation during cell-cell interactions that are initiated by receptor-mediated mechanisms. However, because both [1-^{14}C]15-HETE and [1-^{14}C]LTC_4 (which can serve as an indicator of labeled LTA_4 formation by PMN) were detected, it is not possible to conclude from these data which of the pathways predominates in the formation of lipoxins during PMN-platelet interactions. Nevertheless, the present findings provide evidence that both cell types (PMN and platelets) can generate lipoxins via transcellular routes (Fig. 6). During these interactions, agonist-induced generation of lipoxins by PMN appears to be enhanced by platelet-derived arachidonate, while the platelet generation of lipoxins is dependent upon leukocyte-derived LTA_4. Along these lines, we recently reported the levels of both LXs and LTs [LTC_4, LTD_4, LTB_4, and ω-oxidized LTB_4] generated from endogenous sources with PMN primed by recombinant granulocyte/macrophage colony-stimulating factor (GM-CSF) coincubated with platelets. In the absence of GM-CSF priming the ratio of LX to LT was ≈ 1 to 1.7, and with primed PMN the ratio shifted to $\approx 1:1$. Coincubation of [1-^{14}C]arachidonic acid (AA)-labeled platelets with primed PMN stimulated with thrombin and fMLP led to the production of 15-HETE, 5-HETE and lipoxins, which each carried ^{14}C-label. Thus, in addition to LTA_4 conversion to platelet-derived LXs, primed PMN can also transform platelet-derived [1-^{14}C]AA to generate both 5- and 15-LO products. Both of these routes may act in concert to generate LX. Thus, factors which lead to quantitative differences between LX and LT profiles appear

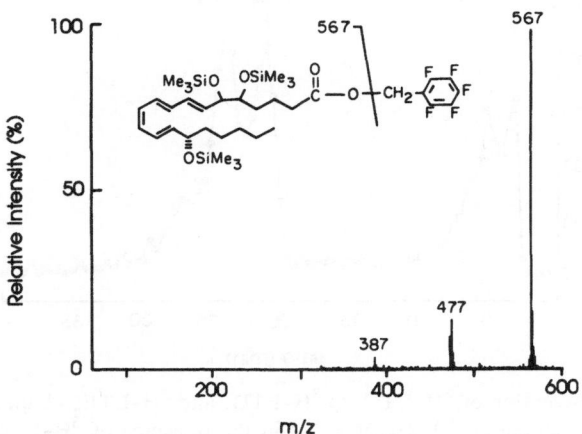

Fig. 1.   NICI mass spectrum of material formed during platelet-neutrophil coincubations. The material displaying chromatographic characteristics of LXs was collected after HPLC, converted to the pentafluorobenzyl (PFB) ester trimethylsilyl (TMS) ether and analyzed by NICI GC-MS (24). The formation of LXA_4 was confirmed by evidence of the intense anion at m/e 567 ([M-PFB]⁻) with ions of lower intensity at m/e 477 and 387 (due to the loss of TMSOH and 2(TMSOH) respectively) in its full scan spectrum as well as the physical characteristics of these products on ED-UV RP-HPLC.

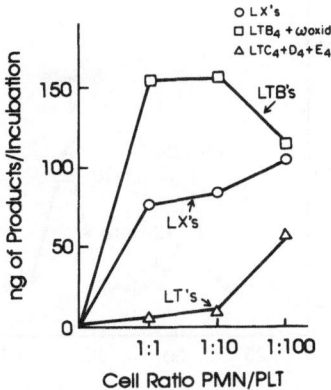

Fig. 2.   The role of cell ratio in the formation of lipoxins and leukotrienes by PMN-platelet coincubations stimulated with fMLP and thrombin. PMN (15×10^6 cells) were co-incubated with autologous platelets in 1 ml phosphate buffered saline (PBS) and stimulated with fMLP (10^{-7} M) and thrombin (0.1 U/ml) for 20 min at 37°C. Products were extracted and quantitated using combined ED-UV-HPLC in two separate mobil phase and RP-HPLC systems (27). LTBs denotes the sum of LTB_4, 20-COOH-LTB_4 and 20-OH-LTB_4, LXs are the sum of LXA_4, LXB_4, and their all-trans-isomers and LTs represents the sum of LTC_4, LTD_4, and LTE_4. The results are representative of three separate experiments.

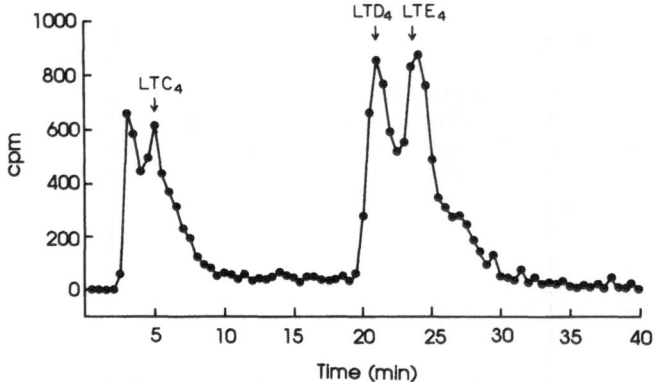

Fig. 3. Platelet conversion of ^{3}H-LTC$_4$ to ^{3}H-LTD$_4$ and ^{3}H-LTE$_4$. Human platelets (1.5 x 10^9) were incubated (20 min at 37°C) in the presence of ^{3}H-LTC$_4$ and 1 U/ml of thrombin. The incubation was terminated, extracted, and chromatographed using an RP-HPLC system (27) and the eluting material was collected (30 sec aliquots). Arrows denote the elution time of synthetic standards. Similar results were obtained with PMN incubations stimulated with fMLP. This profile is representative of 2 separate experiments.

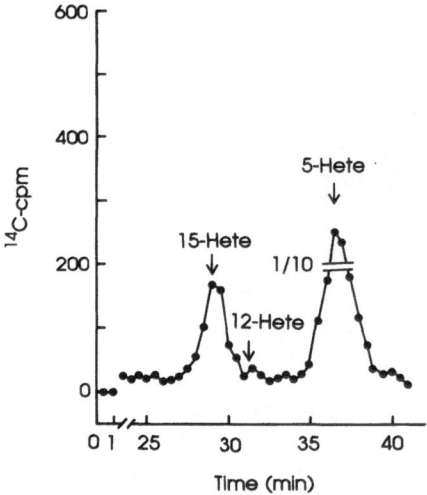

Fig. 4. RP-HPLC profile of [1-^{14}C]-labeled mono-HETEs from coincubation of unlabeled PMN with [1-^{14}C]arachidonate-labeled platelets exposed to fMLP and thrombin. Isolated platelets were labeled (10,27) with [1-^{14}C]-arachidonate, washed two times and incubated with autologous PMN (1:100, PMN/platelets) in PBS 1 ml with 15 x 10^6 PMN. Co-incubations were exposed to fMLP (10^{-6} M) and thrombin (0.1 U/ml) for 20 min at 37°C. Products were extracted and chromatographed with MeOH/H$_2$O/acetic acid (75:25:0.01, vol/vol/vol) as the mobile phase. Arrows denote co-chromatography of synthetic standards. Results are representative of three separate experiments.

to be an important determinant in the physiologic outcome of triggering these biosynthetic events.

Acylation-Deacylation of 15-HETE in Inositol Lipids

The levels of labeled 5-HETE generated in coincubations (Fig. 4) increased with GM-CSF primed PMN (27) consistent with the findings of Dahinden et al. (22). In contrast, conversion of platelet-derived arachidonic acid into labeled 15-HETE was similar with both primed PMN co-incubated with autologous platelets and naive peripheral blood PMN isolated and incubated with autologous platelets without intentional "priming." These findings raise the possibility that platelet derived C20:4 transformed to 15-HETE by PMNs may be subject to further transformations within PMN. This may be particularly relevant in view of the lack of evidence to support receptor operated formation of 15-HETE by isolated human neutrophils.

Although elevated levels of 15-HETE have been detected in various tissues, its role in inflammation and other physiologic responses remains of interest (28). Recent results indicate that 15-HETE injected into plaque lesions of psoriasis vulgaris in humans results in complete regression of the lesions (29). In addition, intraarticular administration of 15-HETE in canines with carragheenan-induced arthritis significantly reduces both the clinical severity and volume of effusates (30). 15-HETE can block both the formation of LTB_4 (31) and PGI_2 (32), and is a substrate for lipoxin biosynthesis by human neutrophils in vitro (33,34); however, the biochemical bases for the observed "anti-inflammatory" actions of 15-HETE in both in vivo and in vitro models remain to be elucidated. Stenson and Parker (35,39) first demonstrated that 5-HETE is incorporated into phospholipids and triglycerides of PMN, a finding which has been extended to a wide range of cell types (reviewed in 28). In view of the more recent observations with 15-HETE and its transformation products (29-34), experiments were undertaken to examine the following: Is 15-HETE incorporated into human PMN lipids in a profile similar to that of either AA or 5-HETE? Second, can either esterified 15-HETE or 5-HETE be mobilized upon activation of PMN and, if so, are they subject to lipoxygenation? Third, are functional responses of PMN altered by these events?

Exposure to $[^3H]$15-HETE results in the rapid incorporation of label into phospholipids of PMN (Fig. 7). Incorporation was time-dependent and associated with phosphatidylinositol (PI) ($\approx$20% of recovered label), while uptake into other phospholipid classes, CE, or triglyceride (TG) represented <4%. To determine whether labeled 15-HETE was esterified in the sn-2 position, labeled PI was isolated from PMNs by thin layer chromatography (TLC) and incubated with snake venom phospholipase A_2 (24°C, 60 min, pH 6.5). After extraction and reversed-phase HPLC, 98.9% of the radiolabel was released from PMN-derived PI, which coeluted with authentic 15(S)-HETE (n = 2). Together these findings indicate that PMNs rapidly esterify $[^3H]$15-HETE into the sn-2 of PI.

Mobilization of stored HETEs. Since PMNs stored 15-HETE in PI, it was of interest to determine whether PMNs can mobilize this source of 15-HETE upon activation. After labeling

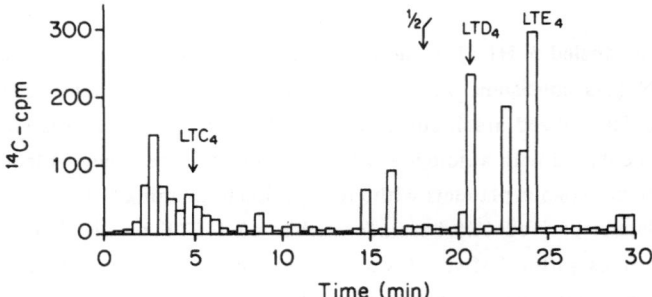

Fig. 5.   RP-HPLC profiles of [1-^{14}C]-labeled peptido-LTs from coincubation of unlabeled PMN with [1-^{14}C]arachidonate-labeled platelets exposed to fMLP and thrombin. Co-incubations were exposed to fMLP (10^{-7} M) and thrombin (0.1 U/ml) for 20 min at 37°C. Here, MeOH fractions from C18 Sep-pack extractions were chromatographed (27).   Arrows denote retention times of synthetic standards.   Results are representative of three separate experiments.

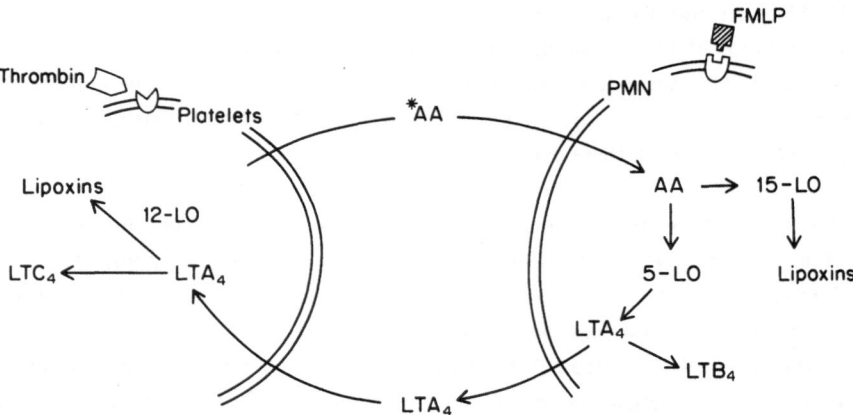

Fig. 6.   Proposed scheme of bidirectional routes operative in receptor-mediated biosynthesis of LT and LX during coincubation of PMN and [1-^{14}C]arachidonate labeled platelets.

with $[^3H]$15-HETE, PMNs were exposed to either selected agonists or vehicle alone (Fig. 8). fMLP activates PMNs via interaction with specific receptors, while $A_{23,187}$, phorbol 12-myristate 13-acetate (PMA) and AA each activate PMN responses by circumventing receptor-ligand interactions (36). PMNs exposed to either fMLP (100 nM), PMA (100 nM), or $A_{23,187}$ (5 μM) each showed a significant reduction in label content of PI.

We next determined whether agonist-induced release of esterified HETEs also promotes their oxygenation. In response to either $A_{23,187}$, PMA, or fMLP, PMNs deacylated 15-HETE and converted it to (5S,15S)-dihydroxy-6,13-trans-8,11-cis-eicosatetraenoic acid (5,15-DHETE), lipoxin A_4 (LXA$_4$), and lipoxin B_4 (LXB$_4$). The specific activities of 5,15-DHETE and deacylated 15-HETE were essentially equal while those of the lipoxins were lower, suggesting that additional sources of unlabeled substrate can contribute to their formation. In the absence of an agonist, neither $[^3H]$5,15-DHETE nor $[^3H]$lipoxins were detected (Table 1).

PMN also released and transformed esterified 5-HETE; however, a different profile of products was obtained (17). With all three stimuli, 5-HETE was converted to both its ω-oxidation products, 5,20-DHETE and 5,15-DHETE. In contrast to 15-HETE, none of the stimuli induced the formation of $[^3H]$-lipoxins from esterified $[^3H]$5-HETE. Thus, esterified sources of HETEs can be mobilized and transformed by PMN.

The impact of esterified 15-HETE on agonist-induced aggregation of PMN was examined (17), and in these experiments, threshold concentrations required to induce aggregation were determined in each case rather than monitoring relative changes in light transmittance since the latter may not represent an accurate value when comparing two distinct groups of cells (i.e. PMN vs. 15-HETE esterified PMN) with potentially different light scattering properties. For PMN exposed to fMLP, threshold concentration required to provoke aggregation was approximately two orders of magnitude higher for cells with esterified 15-HETE ($p < 0.05$). In contrast, neither $A_{23,187}$ nor PMA gave statistically significant differences in the threshold concentration required to induce aggregation. Also, when PMN were incubated with either AA (30 μM) or 5-HETE (30 μM) followed by washing, the concentrations of these stimuli required to induce aggregation were not significantly altered (17).

Formation of lipoxins by fMLP-stimulated labeled PMNs demonstrates a receptor-mediated generation of these eicosanoids from stores within a single cell type which depend upon a "first hit" event, namely the generation of 15-HETE by either PMN or neighboring cell types. Since LXA$_4$ is reported to (i) inhibit PMN responses to LTB$_4$ and fMLP (7), (ii) inhibit LTB$_4$-induced inflammation (6), and (iii) antagonize the actions of LTD$_4$ in glomerular hemodynamics (5), the acylation of one of its precursors (15-HETE) and agonist-induced formation described here may represent a scenario that can contribute to the production of lipoxins during tissue level events. Along these lines, LXA$_4$ has been identified in lavage fluids from patients with selected pulmonary disorders (8); however, the cellular origin of LXA$_4$ in the human bronchus is not known. 15-HETE is indeed a major product of airway epithelial cells (38); therefore the present results suggest that PMNs in such a local environment may rapidly remodel their inositol lipids

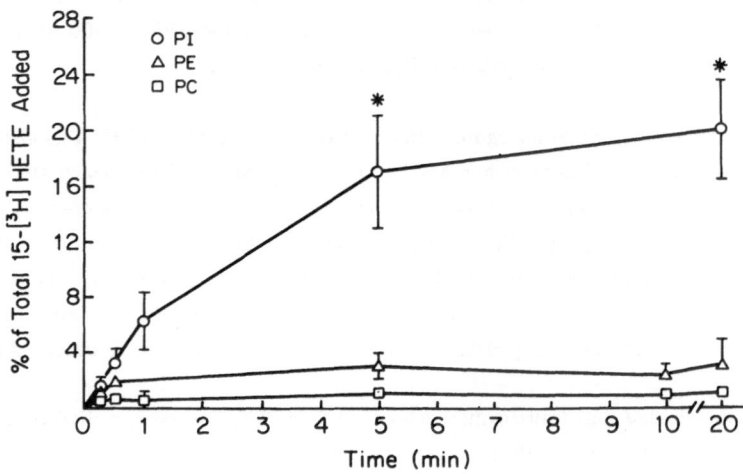

Fig. 7.   Time course of [³H]15–HETE incorporation into phospholipids. PMNs (30 x 10⁶ cells
per ml) were suspended in PBS (1 ml; 37°C, pH 7.45). Samples were extracted and
phospholipids were separated by two-dimensional TLC (17). Values represent the
means ± SE of four or five experiments. *, [³H]15–HETE incorporated with P < 0.05
when compared to the amounts associated with other phospholipid classes.

**Table 1. Stored 15-HETE in human PMNs: Agonist-induced release and
transformation**

Agonist	[³H]15-HETE-labeled PMNs, (cpm)			
	LXA₄	LXB₄	5,15-DHETE	15-HETE
A23187 (5 μM)	62	32	752	72
PMA (100 nM)	10	20	175	645
fMLP (100 nM)	15	15	45	1300
Vehicle	0	0	0	30

PMNs (3 x 10⁷ cells per ml) were labeled with [³H]15-HETE and 15-HETE (30
μM) for 20 min at 37°C. After washing in PBS, cells were exposed to
agonists or vehicle alone [0.4% EtOH (vol/vol)] and incubated for 20 min
at 37°C. Products were extracted and analyzed by reversed phase-HPLC
equipped with a rapid spectral detector. Values represent the average cpm
associated with PMN-derived products obtained for two representative
experiments. Products were identified by both coelution with synthetic
standards and characteristic UV spectra.

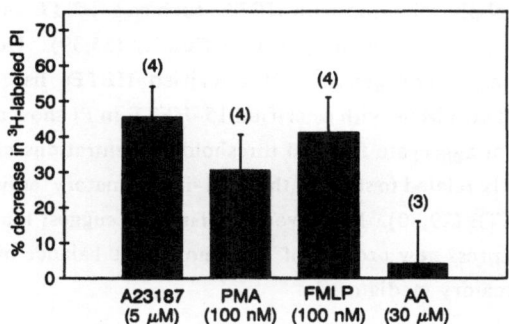

Fig. 8.   Agonist-induced release of [³H]15-HETE from PI. PMNs were labeled with [³H]15-HETE for 20 min. Stimuli or vehicle were added to 30 x 10⁶ PMNs per ml and the incubations continued for 20 min at 37°C. Incubations were stopped, products were extracted, and PI was isolated from each after two-dimensional TLC. Values represent the mean ± SE of three or four (numbers in parentheses) separate experiments.

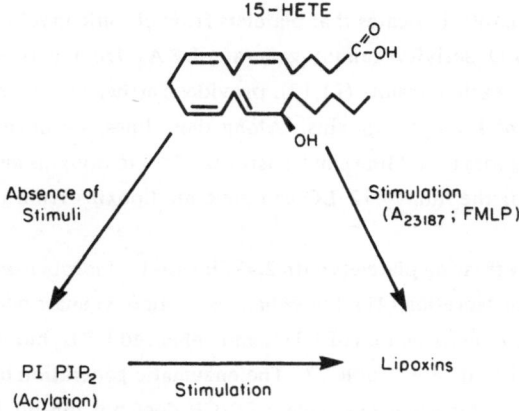

Fig. 9.   Acylation/deacylation cycle in inositol lipids of human PMN and transformation of stored 15-HETE.

119

with 15-HETE, which after stimulation can be mobilized and oxygenated to generate lipoxins. This form of "priming" and/or remodeling of membranes to give altered profiles of eicosanoids after a second challenge may be relevant in inflammation and other physiologic events (a hypothetical scheme is given in Fig. 9).

Remodeling of PMN phospholipid with HETEs such as 5-HETE and 12-HETE has been suggested to alter membrane characteristics such as fluidity (35,39). However, evidence for agonist-induced release and lipoxygenation of esterified HETEs has not been previously established. The finding that PMNs with esterified 15-HETE in PI show an impaired ability to generate LTB_4 and did not aggregate to usual threshold concentrations of fMLP suggests that these events may be causally related to some of the "anti-inflammatory" actions recently observed with injections of 15-HETE (29,30). Moreover, our findings suggest that cells can be primed by lipid remodeling to express new profiles of eicosanoids, the balance of which may regulate the actions of proinflammatory mediators.

Conversion of LTA_4 by Thrombin-activated Platelets

Platelet donation of agonist-induced released C20:4 could not account for the generation of LTC_4 in these incubations since human PMN generate LTB_4 rather than LTC_4 (1). Therefore, we monitored the relationship between lipoxin and peptido-LT production and assessed factors that alter their formation by platelets (Table 2). Thrombin-activated platelets generated LTC_4 + LTD_4 in amounts greater than those of LXA_4 + LXB_4 from exogenous LTA_4. Conversion of LTA_4 to lipoxins can involve the platelet 12-LO (11,12), while its conjugation to LTC_4 may involve a specific glutathione-S-transferase or platelet LTC_4 synthetase (13,14). In fact, recent evidence from Stenke et al. (40) indicates that platelets from chronic myeloid leukemia patients that do not display 12-LO activity cannot generate LXA_4 from exogenous LTA_4. This observation, together with earlier results (11,12), provides further evidence for the role of the 12-LO in the conversion of LTA_4 to lipoxins. Along these lines, we have found that the 12-LO from a human megakaryocyte cell line can transform LTA_4 to lipoxins and that a cos cell line transfected with cDNA for the human 12-LO can generate lipoxins from LTA_4 (18).

Depletion of GSH by treating platelets with 2,4-dinitro-1-chlorobenzene (DNCB) does not alter platelet aggregation or secretion (41), but enhances 12-lipoxygenase activity (42). Platelets treated with DNCB blocked the formation of LTC_4 and enhanced LXB_4 but not LXA_4 formation when compared with untreated cells (Table 2). The enzymatic generation of LTC_4 by platelets appears to be unlike that of macrophages, where DNCB does not inhibit LTC_4 formation by LTC_4 synthetase (43). Nitroprusside-treated platelets gave elevated levels of both LXA_4 and LXB_4 but generated less LTC_4. Since nitroprusside (10 μM) stimulates cGMP, inhibits Ca^{2+} mobilization, and blocks platelet aggregation (44), its ability to divert the fate of LTA_4 in favor of lipoxin production by platelets (Table 2) may be related to the vasodilatory actions of nitroprusside, since peptido-LTs are potent vasoconstrictors. Thus, the levels of both lipoxins and peptido-LTs generated by platelets may be altered by drugs that influence the glutathione status of platelets.

Neutrophil Responses to Lipoxins

Lipoxins have been shown to possess a number of intriguing biological actions which have proven to be specific for the stereochemistry of these compounds (recently reviewed in refs. 3 and 34). With respect to neutrophils, the first system examined (15), previous results have shown that LXA_4 displays selective actions with these cells by eliciting migration without triggering aggregation (45,46). In view of the inhibitory actions of lipoxin A_4 recently uncovered, we examined the impact of lipoxins in the stimulus-response coupling sequence of PMN which is well documented for these cells (see ref. 36) in order to determine which steps are specific LX targets.

Upon addition to labeled neutrophils, both LXA_4 and LXB_4 stimulated the release of esterified $[1-^{14}C]$-arachidonic acid in a dose- and time-dependent fashion. In each experiment, PMN obtained from individual donors were exposed in parallel to the chemotactic peptide fMLP as well as LXA_4 and LXB_4 for purposes of comparison. Previous results have shown that f-met-leu-phe is a potent stimulant for lipid remodeling (47-49). These findings with f-met-leu-phe are in agreement with those values previously reported both with respect to time-course and concentration. When compared at equal molar concentrations (10^{-7}M), time course experiments indicated that LXA_4, LXB_4, and f-met-leu-phe each stimulated a rapid release of $[1-^{14}C]$arachidonic acid that was evident within 5-30 sec following addition of stimuli. Maximal levels of unesterified $[1-^{14}C]$-arachidonic acid were observed at 15 sec following exposure of the cells to either LXA_4, LXB_4 or f-met-leu-phe (10^{-7}M). Here, unesterified $[1-^{14}C]$-arachidonic acid was identified, following extraction, by TLC and comigration with synthetic radiolabel-containing standard. A second phase of $[1-^{14}C]$arachidonic acid release was noted with each compound. Dose-response studies showed that both LXA_4 and LXB_4 at 10^{-7} M were essentially equipotent, while at concentrations below 10^{-7} M, LXA_4 was ineffective in stimulating the release of $[1-^{14}C]$-arachidonic acid. In addition, LXB_4 stimulated release of arachidonic acid at 10^{-8}M, which was comparable in magnitude to that induced by fMLP (10^{-8}M). The shapes of the dose-response curves obtained with lipoxins are consistent with those reported for other chemoattractants such as fMLP (49,50).

The extent of lipid remodeling was examined and the site of $[1-^{14}C]$-arachidonic acid release was determined utilizing dual radiolabeled neutrophils. The sn-2 positions of neutrophil phospholipids were labeled with $[1-^{14}C]$-arachidonic acid, and their sn-1 positions were labeled with $[^3H]$-palmitic acid as described by Walsh et al. (52) Following addition of optimal concentrations of either LXA_4, LXB_4, or fMLP (10^{-7}M), the changes in radiolabel content of phospholipid classes were determined after extraction and separation by two-dimensional TLC. LXA_4, LXB_4, and fMLP each induced extensive changes in the phospholipid classes of human PMN (Figure 10). Comparisons between the $[1-^{14}C]$ and $[^3H]$ label content of PL classes following addition of stimuli revealed that the major site of $[1-^{14}C]$arachidonic acid release induced by either LXA_4, LXB_4, or fMLP was from phosphatidylcholine (Figure 10). This is supported by the observed decrement in $[1-^{14}C]$ content in the phosphatidylcholine (PC) pool, which was not accompanied by changes in the $[^3H]$ content in PC (Fig. 10B). This finding suggests that lipoxins induce arachidonate release via activation of PLA_2, which has been

121

Table 2. Conversion of LTA$_4$ by thrombin-activated platelets: Relationship between lipoxins and peptidoleukotrienes

Incubations	LXB$_4$	LXA$_4$	LTC$_4$	LTD$_4$
Platelets +				
LTA$_4$ + thrombin 0.1 U/ml	15.9 ± 3.7	21.6 ± 7.9	111.8 ± 13.7	24.1 ± 9.1
DNCB (100 μM)	55.9 ± 6.4	23.7 ± 9.9	22.1 ± 6.2	21.0 ± 6.6
DNCB (500 μM)	149.3 ± 13.0	19.3 ± 7.6	21.0 ± 11.0	24.8 ± 8.4
Nitroprusside (10 μM)	31.7 ± 6.0	30.0 ± 1.8	14.5 ± 8.1	16.8 ± 5.3

Human platelets (1.5 x 10^9 cells/ml) were incubated (20 min, 37°C) with LTA$_4$ (20 μM), human albumin (0.1%), and thrombin (0.1 U/ml). Platelets were treated 30 min with DNCB or 3 min with nitroprusside at 37°C before addition of stimuli. Products were extracted and quantitated as described. Results are expressed in nanograms of product per incubation; mean ± SE of three separate experiments.

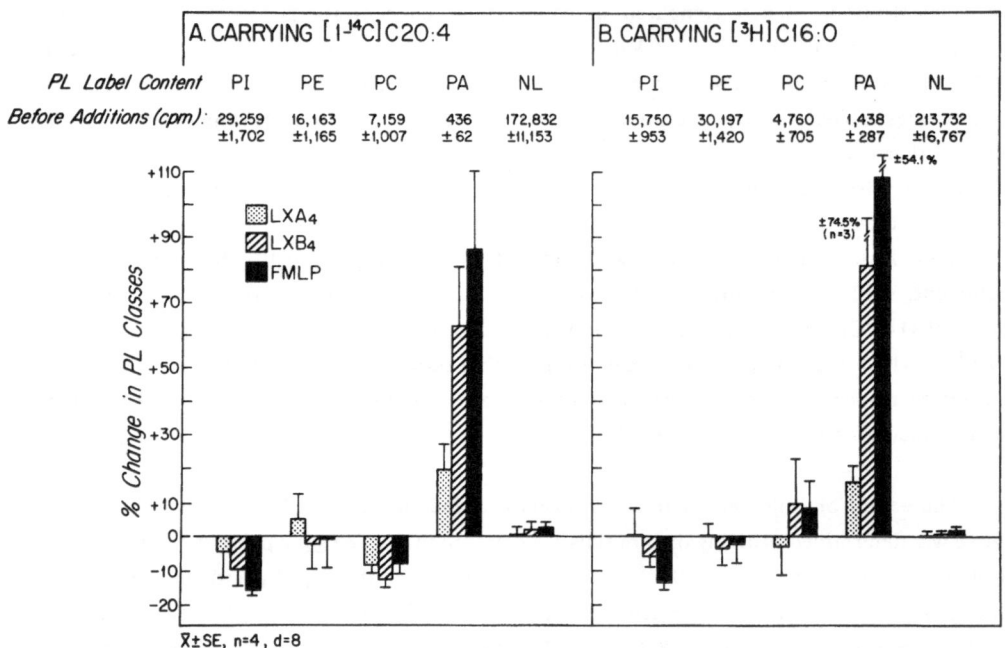

Fig. 10. Comparison between lipoxin and fMLP-induced changes in [1-^{14}C]C20:4 and [^{3}H]C16:0 content of PMN phospholipids. Following dual labeling of PMN with [1-^{14}C]C20:4 and [^{3}H]C16:0 as described (52), PMN (30 x 10^6 cells/ml) were warmed (5 min at 37°C) and incubated with 20 μl of either vehicle, LXA$_4$ (hatched bars), LXB$_4$ (empty bars) or fMLP (full bars) (10^{-7}M). Incubations were terminated at 10 sec by addition of 3.5 ml of chloroform/methanol, extracted and analyzed (4). Results are the mean ± S.E. of four separate experiments with duplicate determinations. Changes in [1-^{14}C]C20:4 (Panel A) and [^{3}H]C16:0 (Panel B) content in each phospholipid class are expressed as the percent difference obtained with PMN incubated with vehicle alone (EtOH final concentration less than 0.01% by volume). Radiolabel content within each PMN's phospholipid class determined before addition of agonist are expressed in cpm (mean ± S.E.).

documented with other stimuli, including fMLP, $A_{23,187}$ and zymogen (48-53). Following addition of stimuli, parallel changes in both labels ($[1-^{14}C]$ and $[^3H]$) were found in phosphatidic acid (PA), indicating that LXA_4 and LXB_4 each stimulate phosphatidic acid formation. In this respect, LXB_4 was more effective than LXA_4, and the extent of PA formation evoked by LXB_4 was similar to that obtained with fMLP. LXB_4 also gave parallel decrements in both labels in PI as did fMLP. In contrast, LXA_4 did not induce statistically significant change in labeled PI. Together these results suggest that LXB_4 can also activate PI metabolism, possibly by a phospholipase C mechanism, and that both LXA_4 and LXB_4 promote rapid phospholipid remodeling in human PMN.

A number of eicosanoids mediate their actions by interacting with receptors which couple to G proteins in the plasma membrane (reviewed in 73), and several chemoattractants mediate their actions on neutrophils via a pertussis toxin-sensitive step (23,47,54-56). Results in Table 3 indicate that prior treatment of $[1-^{14}C]$-arachidonate-labeled PMN with pertussis toxin completely inhibited the actions of LXA_4, LXB_4 and fMLP. The results obtained with fMLP are consistent with those previously reported with guinea pig neutrophils (47). In a second group of experiments the actions of the ß-oligomers of pertussis toxin, extracellular Ca^{2+}, and the isomers of LXA_4 and LXB_4 were assessed. Unlike pertussis toxin, treatment of labeled PMN with the ß-oligomers of the toxin did not significantly inhibit either fMLP or lipoxin-induced release of $[1-^{14}C]$-arachidonate (Table 4). The trans-isomers, e.g. 11-trans-LXA_4 and 8-trans-LXB_4, were ineffective in stimulating the release of $[1-^{14}C]$-arachidonate at equimolar concentrations to either of their native forms (10^{-7}M). In addition, prior exposure of the labeled cells to EGTA (5 mM) resulted in approx. 40.5% inhibition of fMLP-induced response while LXA_4-induced release of $[1-^{14}C]$-arachidonate was unaffected by chelation of extracellular Ca^{2+}.

PMN: Ca^{2+} mobilization, aggregation, and adhesion with vascular endothelial cells. Ca^{2+} mobilization is held to play a critical role in neutrophil activation and in the activation of the 5-lipoxygenase (36,57,69). Therefore, it was obligatory to determine if lipoxins stimulate increments in $[Ca^{2+}]_i$ with Fura-2-loaded PMN. LXA_4 and LXB_4 each stimulated a rapid yet relatively small increase in $[Ca^{2+}]_i$ which reached maximal levels within 10 sec following addition. Ca^{2+} mobilization was not evident at <100 nM of either LXA_4 or LXB_4. Both LXA_4 and LXB_4 were essentially equipotent in stimulating $[Ca^{2+}]_i$, while neither of their all-trans-isomers stimulated detectable increments in $[Ca^{2+}]_i$. However, neither compound ($10^{-8}-10^{-6}$M) stimulated PMN aggregation in the same concentration or temporal range found to activate arachidonate release. Although PMN did not aggregate in response to either LXA_4 or LXB_4, the same cells did release $[1-^{14}C]$20:4 in response to either compound (not shown). Since aggregation reflects homotypic adhesion between neutrophils, we also evaluated whether lipoxins stimulate heterotypic adhesion by monitoring their effects on adhesion of neutrophils to cultured human endothelial cell monolayers. LTB_4, ionophore, and PMA each stimulated the adhesion of labeled neutrophils to monolayers of endothelial cells. In contrast, neither LXA_4 nor LXB_4 ($10^{-6}-10^{-8}$M) stimulated neutrophil adhesion to endothelial cells. Together, these findings suggest that lipoxins do not stimulate adhesion of neutrophils.

Do lipoxins interact with LTB_4 receptors on human PMN? PMN possess high affinity receptors for LTB_4 that mediate its chemotactic activities (58-59). Lipoxins stimulated lipid

Table 3. Pertussis toxin sensitive release of $[1-^{14}C]C20:4$ from labeled human neutrophils following addition of either LXA_4, LXB_4 or fMLP

Incubations	$[1-^{14}C]C20:4$		Pertussis toxin treated cells	
	Δcpm	% increase	Δcpm	% inhibition
PMNs + LXA_4 10^{-7}M	200 ± 62	24	0.0 ± 0.0	100
PMNs + LXB_4 10^{-7}M	416 ± 52	49	0.0 ± 0.0	100
PMNs + fMLP 10^{-7}M	513 ± 90	60	0.0 ± 0.0	100

After labeling human neutrophils with $[1-^{14}C]C20:4$, cells (20 x 10^6 PMN/ml) were transferred in Hank's balanced salt solution without Ca^{++} and Mg^{++} containing 0.25% of fatty acid-free bovine serum albumin (BSA) incubated at 37°C for 90 minutes in the presence of pertussis toxin (2 μg/ml) and DNase Type I (50 U/ml). Next, cells were washed to remove the pertussis toxin and adjusted to 5 x 10^6 PMN/ml in PBS, pH 7.45 without BSA. Aliquots of labeled PMNs from the same cell preparations were taken and incubated in parallel in the absence of PTX. After treatment, cells were incubated for 5 min at 37°C followed by addition of either vehicle, LXA_4, LXB_4 or fMLP at 10^{-7}M (EtOH final concentration less than 0.01% v/v). All incubations were terminated at 10 sec and unesterified $[1-^{14}C]C20:4$ was extracted and quantified as described in Figure 1 (n=3). Results are the mean ± S.E. of a representative experiment with triplicate determinations for each experimental point.

Table 4. Release of $[1-^{14}C]C20:4$ from labeled human neutrophils: effects of pertussis toxin β oligomers, extracellular Ca^{2+} and stereoisomers[a]

Incubations 10^{-7}M	Δcpm, $[1-^{14}C]C20:4$		
	Δcpm	EGTA[b]	β-oligomers[c]
PMNs + fMLP	343.6 ± 29.1	204.5 ± 37.5[d]	299.0 ± 43.9
PMNs + LXA_4	169.0 ± 7.0	171.5 ± 33.0	166.2 ± 17.2
PMNs + 11-trans-LXA_4	7.3 ± 20.0		
PMNs + 8-trans-LXB_4	-72.8 ± 69.0		

[a]PMN were labeled as in Table 1.

[b]$EGTA_4$ (5 mM) + Mg^{2+} (1 mM) was added 20 sec before the addition of either fMLP or LXA_4, 10^{-7}M.

[c]Labeled PMNs (20 x 10^6 cells/ml) were treated with β oligomers of the pertussis toxin (5 μg/ml equimolar to the 2 μg/ml used for the holotoxin) under the same experimental conditions described in Table 1. Results are expressed as the mean ± SE of three separate experiments with 8 determinations.

[d]$P < 0.05$.

remodeling without provoking other neutrophil responses (Fig. 10 and Tables 1 and 2). Thus, it was possible that lipoxins act as partial agonists on the PMN LTB_4 receptor. To address this question, the impact of LXA_4, LXB_4, 15-HETE, and 5-HETE on the binding of $[^3H]$-LTB_4 was tested. Increasing concentrations of unlabeled LTB_4 blocked the specific binding of $[^3H]$-LTB_4 (1 nM) as previously reported (58). In contrast, neither LXA_4 nor LXB_4 effectively altered (10^{-10}-10^{-6} M) the binding of $[^3H]$-LTB_4; 15-HETE, which shares C_{15}-C_{20} identity with lipoxins, did partially block $[^3H]$-LTB_4 binding, albeit at high concentrations (10^{-6} M). These observations therefore suggest that LXA_4 and LXB_4 do not act on the LTB_4 receptor in human neutrophils. This situation with the PMN LTB_4 receptor is unlike that of the LTD_4 receptor of mesangial cells where LXA_4 acts as a partial agonist and can block both the in vivo and in vitro action of LTD_4 (see ref. 5 and vide infra).

LXA_4 is a highly stereospecific inducer of neutrophil chemotaxis that is active in the nanomolar range, but does not elicit aggregation within the same concentration range (4,15). LXA_4 has also been shown to be a chemokinetic agent (8,61) and can stimulate oxygen radical generation by PMN when they are exposed to concentrations in excess of 1 μM (4,62). Although lipoxins possess biological activities in several systems that are distinct from those of other eicosanoids (2,63), little information is available regarding the cellular events upon which they act. Thus, our finding that both LXA_4 and LXB_4 stimulate rapid lipid remodeling and release arachidonic acid without triggering the oxygenation of arachidonic acid or neutrophil adherence within the same temporal and concentration range may be relevant to the actions of these eicosanoids in inflammatory reactions.

Lipoxins A_4 and B_4 each stimulated the release of $[1-^{14}C]$arachidonic acid from labeled PMN in a dose- and time-dependent fashion. The magnitude and time course of arachidonic acid release induced by lipoxins were similar to those evoked by the chemotactic peptide f-met-leu-phe when the compounds were compared with PMN's from the same donors. The time course of release of $[1-^{14}C]$arachidonic acid proved to be a biphasic response with all three stimuli examined. The levels of unesterified $[1-^{14}C]$arachidonic acid were maximal within 15 sec and declined by 30 sec after addition of either LXA_4, LXB_4, or fMLP. The results obtained with fMLP are consistent with the rapid and extensive lipid remodeling previously reported with neutrophils exposed to this chemoattractant (48,49,64). The actions of LXA_4 and LXB_4 proved to be stereoselective since their all-trans-isomers were ineffective (Table 2). These results with LXA_4 and LXB_4 provide the first documentation indicating that lipoxins are potent stimulants for arachidonic acid release and lipid remodeling in human neutrophils (Fig. 10, Tables 1 and 2). Upon release, however, arachidonic acid was not transformed by either the 5- or 15-LO of human neutrophils and the lipoxins did not stimulate the formation of either LTB_4 or its ω-oxidation products. Thus, although lipoxins stimulated the release of arachidonic acid, it was not further transformed to other bioactive eicosanoids. These results with PMN are in sharp contrast to the actions of lipoxins with other cell types and tissues in which lipoxins can stimulate the release as well as transformation of arachidonic acid. For example, LXA_4 and LXB_4 stimulate endothelial cells to generate their primary arachidonic derived product, namely PGI_2 (65), while guinea pig lung strips exposed to LXA_4 generate thromboxane A_2 (66,67). These results are consistent with the notion that lipoxins stimulate the release of arachidonic acid but do not trigger the generation of appropriate signals at the levels required to activate the lipoxygenase pathways of human neutrophils. Activation of the 5-lipoxygenase in human leukocytes is regulated by a complex multicomponent system involving Ca^{2+}, ATP, and several nondialyzable factors (68,69). Therefore, our findings that LXA_4 and LXB_4 stimulate rapid but small increments in the levels of cytosolic Ca^{2+} as monitored with Fura-2 loaded cells (4) are consistent with the low levels of Ca^{2+} required to activate phospholipases, which may not be above the threshold required to trigger the oxygenation of arachidonic acid in these cells.

Unlike lipoxins, both leukotriene B_4 and 5-HETE are potent stimulants for elevating cytosolic Ca^{2+} (70,71), and leukotriene B_4 promotes neutrophil adherence (reviewed in 72). Neither LXA_4 nor LXB_4 stimulated aggregation, which is an index of homotypic adhesion (4). In addition, the lipoxins did not stimulate adhesion of neutrophils to cultured monolayers of human endothelial cells, indicating that these compounds do not stimulate heterotypic adhesion. Recently, Lee et al. (7) reported that neither LXA_4 nor LXB_4 stimulated neutrophil adhesion to endothelial cells. Our observations reported in ref. 4 are consistent with those of Lee et al. (7) with respect to adhesion. In addition, they provide further evidence that lipoxins display a profile of activities distinct from those of either LTB_4 or 5-HETE with human neutrophils.

Results from several laboratories have provided evidence for the role of G proteins in receptor-mediated signal transduction in neutrophils (36,47,54-56). Pertussis toxin catalyzes the ADP-ribosylation of a membrane protein ($\approx 41,000$ daltons) in guinea pig neutrophils that inhibits fMLP-induced release of $[1-^{14}C]$arachidonic acid (47). In addition, studies with rabbit neutrophils suggest that the receptors for fMLP, C5a, and leukotriene B_4 are each coupled to N_i-like proteins, which are important in mediating the functional responses of these chemoattractants (55). In a wide range of tissues, the eicosanoids (including prostaglandins, leukotrienes, and thromboxane) generally appear to exert their actions by interacting with specific receptors that are thought to be linked to G proteins (reviewed in 73). Prior treatment of $[1-^{14}C]$arachidonate-labeled PMN with pertussis toxin (PTX) completely inhibited fMLP as well as lipoxin-induced release of arachidonate (Table 1). In contrast, the ß-oligomers of pertussis toxin did not inhibit LX-induced release (Table 2). These findings with fMLP- and PTX-treated human neutrophils are consistent with those reported for both guinea pig and rabbit neutrophils (47,55) and they suggest a role for G proteins in lipoxin-induced release of arachidonic acid.

As mentioned above, PMN possess specific receptors for leukotriene B_4 (55,58-60); therefore, it was possible that the lipoxins might act through these receptors. Along these lines, LXA_4 has been shown to exert its actions in certain tissues by interacting with the same or a similar receptor site occupied by the cysteinyl-containing leukotrienes, namely LTC_4 and LTD_4 (5,74). Unlike LTC_4, LXA_4 does not contract the guinea pig ileum. Instead, it causes a dose-dependent inhibition of LTC_4-induced contractions in this tissue (74). Recently, LXA_4 and LTD_4 were shown to interact with a common site on mesangial cells, and LXA_4 can competitively antagonize both the cellular and hemodynamic actions of LTD_4 (5). However, neither LXA_4 nor LXB_4 alters the specific binding of LTB_4 to human PMN. Thus, although LXA_4 can interact with cysteinyl-containing leukotriene receptors in certain tissues (5,74), lipoxins do not appear to exert their actions on human PMN via binding with LTB_4 receptors (4). Whether lipoxins exert their actions on PMN by specific receptors remains to be demonstrated.

A number of soluble and insoluble PMN stimuli have been shown to initiate lipid remodeling and the release of arachidonic acid (48,49,52). Using the double-label technique with [^{3}H]arachidonic acid esterified into the sn-2 position and [^{14}C]palmitate esterified into the sn-1 position of distinct phospholipid classes, Walsh et al. (52) have found that oposonized zymosan and the ionophore $A_{23,187}$ stimulate arachidonic acid release predominantly via a phospholipase A_2 mechanism. Utilizing a similar strategy with [^{3}H]palmitate and [1-^{14}C]arachidonic acid labeling of neutrophils (52) followed by separation of phospholipids by two-dimensional TLC, we find that lipoxins stimulate the release of [1-^{14}C]arachidonic acid from predominantly phosphatidylcholine without provoking significant losses in the [^{3}H] content of phosphatidylcholine (Fig. 10). Similar results were obtained with fMLP, which are consistent with those of earlier reports (49,50). The results obtained with lipoxins are consistent with the role of a phospholipase A_2 in lipoxin-induced release of arachidonic acid. In addition, the finding that PA carried both labels following addition of either LXA_4, LXB_4, or fMLP suggests that, like fMLP (48,64), lipoxins stimulate the formation of PA (Fig. 10). Several recent reports have provided evidence suggesting that, in addition to the well-established phospholipase C mechanism and PI cycle, neutrophils may also generate PA by means of a phospholipase D (64,75). The complete mechanism of lipoxin-induced PA formation was not determined in the present study and thus remains the subject of further experiments. Nevertheless, since the levels of PA were elevated following exposure to lipoxins and neither compound stimulates adhesion (4), it appears that these two events, namely lipid remodeling and adhesion, can be dissociated in cells exposed to lipoxins. Together, the results to date demonstrate that lipoxins exert a selective profile of actions with human neutrophils (in vitro) and that lipid remodeling events induced by lipoxins can be dissociated from both heterotypic and homotypic adhesion of these cells. Moreover, in view of these results (Fig. 10, ref. 4), it is possible that lipoxins may be helpful agents in identifying critical steps in signal transduction pathways of human neutrophils.

Conclusions. Stimulatory as well as counterregulatory actions have been documented with lipoxins (reviewed in 76). The chalone activities of these eicosanoids are of particular interest because the natural mediators that can down-regulate inflammation have remained to be elucidated. Natural "endogenous-inhibitors" are likely to exist in the local milieu that can control key proinflammatory events because inflammation is itself a self-limited response that when controlled locally is of paramount value to the host. Several laboratories including this one have provided evidence to support this role for LX inflammation. In the nanomolar range, LXA_4 can antagonize both the cellular and in vivo actions of leukotriene D_4 in renal microcirculation. Prior exposure of glomerular mesangial cells to LXA_4 blocks both the binding of leukotriene D_4 to its receptor and the generation of inositol-triphosphate in mesangial cells which can produce LXA_4 (5). Along these lines, the recent results by Lee et al. (7) have shown that preincubation of human neutrophils with LXA_4 inhibits (IC_{50} 10^{-8} M) their subsequent chemotactic responses to either LTB_4 (10^{-7} M) or fMLP (10^{-7} M). Preincubation with LXA_4 also inhibited agonist-induced Ca^{2+} mobilization and hydrolysis of phosphoinositides (7). Lipoxin A_4 has been found to inhibit LTB_4-induced plasma leakage and leukocyte migration in vivo in the hamster cheek pouch model of inflammation (6). In view of these findings, our results that lipoxins stimulate rapid remodeling of neutrophil lipids and small increases in $[Ca^{2+}]_i$ which return to baseline within seconds suggest that these events may be related to the inhibitory responses of neutrophils observed following exposure of the cells to a second challenge.

The profile of bioactions documented thus far for LXA_4 (4-9; reviewed in 3) and its identification in human bronchoalveolar fluids (8), as well as in the intracoronary blood of patients ≈ 10 sec following angioplasty (77), suggest that the generation of LX may play a functional role (in vivo in humans). Thus, the production of physiologically relevant levels (ng) of LX by receptor-mediated activation of PMN interacting with platelets by at least two separate routes, namely one which involves the donation of platelet-arachidonate to the 15-LO of PMN and the other PMN-derived LTA_4 transformation by platelets (bidirectional formation), may be important in both inflammation and interaction of these cells with the vessel wall. Our finding that esterified stores of HETE can be deacylated upon cell activation and are transformed suggests that cells can be primed by lipid remodeling to express new profiles of eicosanoids, the balance of which may regulate the actions of pro-inflammatory mediators. Thus, together the triggering of individual LX biosynthetic circuits which may be amplified or enhanced during cell-cell interactions as well as interaction between LOs of individual cell types may represent a chalone type regulatory component in the inflammatory response.

Acknowledgments

We thank Mary Halm Small for skillful preparation of this chapter and Dr. Santosh Nigam for helpful discussions during his sabbatical in this laboratory. These studies were supported in part by National Institutes of Health grants AI-26714 and GM-38765. C.N.S. is a recipient of an Established Investigator Award from the American Heart Association and is a Pew Scholar in the Biomedical Sciences. Dr. Fiore is the recipient of American Heart Association, Massachusetts Affiliate postdoctoral fellowship no. 13-407-890.

REFERENCES

1. B. Samuelsson, Leukotrienes: Mediators of immediate hypersensitivity reactions and inflammation, Science 220:568 (1983).
2. B. Samuelsson, S.-E. Dahlén, J.Å. Lindgren, C.A. Rouzer, and C.N. Serhan, Leukotrienes and lipoxins: Structures, biosynthesis, and biological effects, Science 237:1171 (1987).
3. C.N. Serhan, Lipoxins: eicosanoids carrying intra- and intercellular messages, J. Bioenerget. Biomembr. 23:105 (1991).
4. S. Nigam, S. Fiore, F.W. Luscinskas, and C.N. Serhan, Lipoxin A_4 and lipoxin B_4 stimulate the release but not the oxygenation of arachidonic acid in human neutrophils: Dissociation between lipid remodeling and adhesion, J. Cell. Physiol. 143:512 (1990).
5. K.F. Badr, D.K. DeBoer, M. Schwartzberg, and C.N. Serhan, Lipoxin A_4 antagonizes cellular and in vivo action of leukotriene D_4 in rat glomerular mesangial cells: Evidence for competition at a common receptor, Proc. Natl. Acad. Sci. USA 86:3438 (1989).
6. P. Hedqvist, J. Raud, U. Palmertz, J. Haeggström, K.C. Nicolaou, and S.-E. Dahlén, Lipoxin A_4 inhibits leukotriene B_4-induced inflammation in the hamster cheek pouch, Acta Physiol. Scand. 137:571 (1989).

7. T.H. Lee, C.E. Horton, U. Kyan-Aung, D. Haskard, A.E.G. Crea, and B.W. Spur, Lipoxin A_4 and lipoxin B_4 inhibit chemotactic responses of human neutrophils stimulated by leukotriene B_4 and N-formyl-L-methionyl-L-leucyl-L-phenylalanine, Clin. Sci. 77:195 (1989).

8. T.H. Lee, A.E.G. Crea, V. Gant, B.W. Spur, B.E. Marron, K.C. Nicolaou, E. Reardon, M. Brezinski, and C.N. Serhan, Identification of lipoxin A_4 in the bronchoalveolar lavage fluid obtained from patients with pulmonary disease, Am. Rev. Respir. Dis. 141:1453 (1990).

9. B.M. Grandordy, H. Lacroix, E. Mavoungou, S. Krilis, A.E.G. Crea, B.W. Spur, and T.H. Lee, Lipoxin A_4 inhibits phosphoinositide hydrolysis in human neutrophils, Biochem. Biophys. Res. Commun. 167:1022 (1990).

10. A.J. Marcus, M.J. Broekman, L.B. Safier, H.L. Ullman, N. Islam, C.N. Serhan, L.E. Rutherford, H.M. Korchak, and G. Weissmann, Formation of leukotrienes and other hydroxy acids during platelet-neutrophil interactions in vitro, Biochem. Biophys. Res. Commun. 109:130 (1982).

11. C.N. Serhan and K.-A. Sheppard, Lipoxin formation during human neutrophil-platelet interactions: evidence for the transformation of leukotriene A_4 by platelet 12-lipoxygenase in vitro, J. Clin. Invest. 85:772 (1990).

12. C. Edenius, J. Haeggström, and J.A. Lindgren, Transcellular conversion of endogenous arachidonic acid to lipoxins in mixed human platelet-granulocyte suspensions, Biochem. Biophys. Res. Commun. 157:801 (1988).

13. J.A. Maclouf and R.C. Murphy, Transcellular metabolism of neutrophil-derived leukotriene A_4 by human platelets: a potential cellular source of leukotriene C_4, J. Biol. Chem. 263:174 (1988).

14. C. Edenius, K. Heidvall, and J.A. Lindgren, Novel transcellular interaction: conversion of granulocyte-derived leukotriene A_4 to cysteinyl-containing leukotrienes by human platelets, Eur. J. Biochem. 178:81 (1988).

15. C.N. Serhan, M. Hamberg, and B. Samuelsson, Lipoxins: novel series of biologically active compounds formed from arachidonic acid in human leukocytes, Proc. Natl. Acad. Sci. USA 81:5335 (1984).

16. C.N. Serhan, K.C. Nicolaou, S.E. Webber, C.A. Veale, S.-E. Dahlén, T.J. Puustinen, and B. Samuelsson, Lipoxin A: stereochemistry and biosynthesis, J. Biol. Chem. 261:16340 (1986).

17. M. Brezinski and C.N. Serhan, Selective incorporation of 15-HETE in phosphatidylinositol of human neutrophils: agonist-induced deacylation and transformation of stored hydroxyeicosanoids, Proc. Natl. Acad. Sci. USA 87:6248 (1990).

18. K.-A. Sheppard, S. Greenberg, C. Funk, M. Romano, and C.N. Serhan, Lipoxin generation by human megakaryocyte-induced 12-lipoxygenase, Biochim. Biophys. Acta, in press.

19. T. Puustinen, S.E. Webber, K.C. Nicolaou, J. Haeggström, C.N. Serhan, and B. Samuelsson, Evidence for a 5(6)-epoxytetraene intermediate in the biosynthesis of lipoxins in human leukocytes: Conversion into lipoxin A by cytosolic epoxide hydrolase, FEBS Lett. 207:127 (1986).

20. J.A. Maclouf, B.F. de Laclos, and P. Borgeat, Stimulation of leukotrienes biosynthesis in human blood leukocyte by platelet-derived 12-hydroperoxy-icosatetraenoic acid, Proc. Natl. Acad. Sci. USA 79:6042 (1982).

21. A.J. Marcus, L.B. Safier, H.L. Ullman, M.J. Broekman, N. Islam, T.D. Oglesby, and R.R. Gorman, 12S,20-Dihydroxyicosatetraenoic acid: A new icosanoid synthesized by neutrophils from 12S-hydroxyicosatetraenoic acid produced by thrombin- or collagen-stimulated platelets, Proc. Natl. Acad. Sci. USA 81:903 (1984).

22. C.A. Dahinden, J. Zingg, F.E. Maly, and A.L. de Weck, Leukotriene production in human neutrophils primed by recombinant human granulocyte/macrophage colony-stimulating factor and stimulated with the complement component C5a and fMLP as second signals, J. Exp. Med. 167:1281 (1988).

23. S.R. McColl, E. Krump, P.H. Naccache, and P. Borgeat, Enhancement of human neutrophil leukotriene synthesis by human granulocyte-macrophage colony-stimulating factor, Agents Actions 27:465 (1989).

24. D.A. Brezinski, and C.N. Serhan, Characterization of lipoxins by combined gas chromatography and electron-capture negative ion chemical ionization mass spectrometry: formation of lipoxin A_4 by stimulated human whole blood, Biol. Mass Spectrom. 20:45 (1991).

25. C.N. Serhan, On the relationship between leukotriene and lipoxin production by human neutrophils: evidence for differential metabolism of 15-HETE and 5-HETE, Biochim. Biophys. Acta 1004:158 (1989).

26. C.N. Serhan, U. Hirsch, J. Palmblad, and B. Samuelsson, Formation of lipoxin A by granulocytes from eosinophilic donors. FEBS Lett. 217:242 (1987).

27. S. Fiore and C.N. Serhan, Formation of lipoxins and leukotrienes during receptor-mediated interactions of human platelets and recombinant human granulocyte/macrophage colony-stimulating factor-primed neutrophils, J. Exp. Med. 172:1451 (1990).

28. A.A. Spector, J.A. Gordon, and S.A. Moore, Hydroxyeicosatetraenoic acids (HETES), Prog. Lipid Res. 27:271 (1988).

29. K. Fogh, H. Søgaard, T. Herlin, and K. Kragballe, Improvement of psoriasis vulgaris after intralesional injections of 15-hydroxyeicosatetraenoic acid (15-HETE), J. Am. Acad. Dermatol. 18:279 (1988).

30. K. Fogh, E.S. Hansen, T. Herlin, V. Knudsen, T.B. Henriksen, H. Ewald, C. Bünger, and K. Kragballe, 15-Hydroxy-eicosatetraenoic acid (15-HETE) inhibits carrgeenan-induced experimental arthritis and reduces synovial fluid leukotriene B_4 (LTB_4), Prostaglandins 37:213 (1989).

31. J.Y. Vanderhoek, R.W. Bryant, and J.M. Bailey, Inhibition of leukotriene biosynthesis by the leukocyte product 15-hydroxy-5,8,11,13-eicosatetraenoic acid, J. Biol. Chem. 255:10064 (1980).

32. B.N.Y. Setty and M.J. Stuart, 15-Hydroxy-5,8,11,13-eicosatetraenoic acid inhibits human vascular cyclooxygenase: potential role in diabetic vascular disease, J. Clin. Invest. 77:202 (1986).

33. C.N. Serhan, On the relationship between leukotriene and lipoxin production by human neutrophils: Evidence for differential metabolism of 15-HETE and 5-HETE, Biochim. Biophys. Acta 1004:158 (1989).

34. S.-E. Dahlén and C.N. Serhan, 1991, Lipoxins: Bioactive lipoxygenase interaction products, in: "Lipoxygenases and Their Products," A. Wong and S. Crooke, eds., Academic Press, San Diego.

35. W.F. Stenson and C.W. Parker, 12-L-hydroxy-5,8,10,14-eicosatetraenoic acid, a chemotactic fatty acid, is incorporated into neutrophil phospholipids and triglyceride, Prostaglandins 18:285 (1979).

36. E.L. Becker, J.C. Kermode, P.H. Naccache, R. Yassin, J.J. Munoz, M.L. Marsh, C.-K. Huang, and R.I. Sha'afi, Pertussis toxin as a probe of neutrophil activation, Fed. Proc. Fed. Am. Soc. Exp. Biol. 45, 2151 (1986).

37. J.J. Murray, A.B. Tonnel, A.R. Brash, L.J. Roberts II, P. Gosset, R. Workman, A. Capron, and J.A. Oates, Release of prostaglandin D_2 into human airways during acute antigen challenge, N. Engl. J. Med. 315:800 (1986).

38. M.J. Holtzman, A. Pentland, N.L. Baenziger, and J.R. Hansbrough, Heterogeneity of cellular expression of arachidonate 15-lipoxygenase: implications of biological activity, Biochim. Biophys. Acta 1003:204 (1989).

39. W.F. Stenson and C.W. Parker, Metabolism of arachidonic acid in ionophore-stimulated neutrophils: esterification of a hydroxylated metabolite into phospholipids, J. Clin. Invest. 64:1457 (1979).

40. L. Stenke, B. Näsman-Glaser, C. Edenius, J. Samuelsson, J. Palmblad and J.Å. Lindgren, 1990, Lipoxygenase products in myeloproliferative disorders: increased leukotriene C_4 and decreased lipoxin formation in chronic myeloid leukemia, in: "Advances in Prostaglandin, Thromboxane, and Leukotriene Research," Vol. 21, B. Samuelsson et al., eds, Raven Press, New York.

41. T.D. Hill, J.G. White, and G.H.R. Rao, The influence of glutathione depleting agents on human platelet function, Thromb. Res. 53:457 (1989).

42. T.D. Hill, J.G. White, and G.H.R. Rao, Role of glutathione and glutathione peroxidase in human platelet arachidonic acid metabolism, Prostaglandins 38:21 (1989).

43. M. Abe and T.E. Hugli, Characterization of leukotriene C_4 synthetase in mouse peritoneal exudate cells, Biochim. Biophys. Acta 959:386 (1988).

44. R.O. Morgan and A.C. Newby, Nitroprusside differentially inhibits ADP-stimulated calcium influx and mobilization in human platelets, Biochem. J. 258:447 (1989).

45. J. Palmblad, H. Gyllenhammar, B. Ringertz, C.N. Serhan, B. Samuelsson, and K.C. Nicolaou, The effects of lipoxin A and lipoxin B on functional responses of human granulocytes, Biochem. Biophys. Res. Commun. 145:168 (1987).

46. S. Fiore, M. Romano, and C.N. Serhan, 1990, Lipoxin and leukotriene production during receptor-activated interactions between human platelets and cytokine-primed neutrophils, in: "Advances in Prostaglandin, Thromboxane, and Leukotriene Research," Vol. 21, B. Samuelsson et al., eds, Raven Press, New York.

47. G.M. Bokoch and A.G. Gilman, Inhibition of receptor-mediated release of arachidonic acid by pertussis toxin, Cell 39:301 (1984).

48. C.N. Serhan, M.J. Broekman, H.M. Korchak, A.J. Marcus, and G. Weissmann, Endogenous phospholipid metabolism in stimulated neutrophils. Differential activation by fMLP and PMA, Biochem. Biophys. Res. Commun. 107:951 (1982).

49. E.M. Wynkoop, M.J. Broekman, H.M. Korchak, A.J. Marcus, and G. Weissmann, Phospholipid metabolism in human neutrophils activated by N-formyl-methionyl-leucyl-phenylalanine, Biochem. J. 236:829 (1986).

50. A. Sellmayer, Th. Strasser, and P.C. Weber, Differences in arachidonic acid release, metabolism and leukotriene B_4 synthesis in human polymorphonuclear leukocytes activated by different stimuli, Biochim. Biophys. Acta 927:417 (1987).

51. S. Nigam, S. Nodes, G. Cichon, E.J. Corey, and C.R. Pace-Asciak, Receptor-mediated action of hepoxilin A_3 releases diacylglycerol and arachidonic acid from human neutrophils, Biochem. Biophys. Res. Commun. 171:944 (1990).

52. C.E. Walsh, B.M. Waite, M.J. Thomas, and L.R. DeChatelet, Release and metabolism of arachidonic acid in human neutrophils, J. Biol. Chem. 256:7228 (1981).

53. F.H. Chilton and R.C. Murphy, Remodeling of arachidonate-containing phosphoglycerides within the human neutrophil, J. Biol. Chem. 261:7771 (1986).

54. N. Okamura, M. Uchida, T. Ohtsuka, M. Kawanishi, and S. Ishibashi, Diverse involvements of Ni protein in superoxide anion production in polymorphonuclear leukocytes depending on the type of membrane stimulants, Biochem. Biophys. Res. Commun. 130:939 (1985).

55. D.E. Feltner, R.H. Smith, and W.A. Marasco, Characterization of the plasma membrane bound GTPase from rabbit neutrophils. I. Evidence for an N_i-like protein coupled to the formyl peptide, C5a, and leukotriene B_4 chemotaxis receptors, J. Immunol. 137:1961 (1986).

56. S. Mong, G. Chi-Rosso, J. Miller, K. Hoffman, K.A. Razgaitis, P. Bender, and S.T. Crooke, Leukotriene B_4 induces formation of inositol phosphates in rat peritoneal polymorphonuclear leukocytes, Molec. Pharmacol. 30:235 (1986).

57. P. Borgeat and B. Samuelsson, Arachidonic acid metabolism in polymorphonuclear leukocytes: Effects of ionophore A23,187, Proc. Natl. Acad. Sci. USA 76:2148 (1979).

58. A.H. Lin, P.L. Ruppel, and R.R. Gorman, Leukotriene B_4 binding to human neutrophils, Prostaglandins 28:837 (1984).

59. S. Mong, G. Chi-Rosso, J. Miller, K. Hoffman, K.A. Razgaitis, P. Bender, and S.T. Crooke, Leukotriene B_4 induces formation of inositol phosphates in rat peritoneal polymorphonuclear leukocytes, Mol. Pharmacol. 30:235 (1986).

60. D.W. Goldman and E.J. Goetzl, Selective transduction of human polymorphonuclear leukocyte functions by subsets of receptors for leukotriene B_4, J. Allergy Clin. Immunol. 74:373 (1984).

61. B. Spur, C. Jacques, A.E. Crea, and T.H. Lee, 1988, Lipoxins of the 5-series derived from

eicosapentaenoic acid, in: "Lipoxins: Biosynthesis, Chemistry, and Biological Activities, P.Y.-K. Wong and C.N. Serhan, eds, Plenum Press, New York, Vol. 229 in Advances in Experimental Medicine and Biology.

62. J. Palmblad, H. Gyllenhammar, and B. Ringertz, 1988, Effects of lipoxins A and B on functional responses of human granulocytes, in: "Lipoxins: Biosynthesis, Chemistry, and Biological Activities," P.Y.-K. Wong and C.N. Serhan, eds, Plenum Press, New York, Vol. 229 in Advances in Experimental Medicine and Biology.

63. C.N. Serhan and B. Samuelsson, 1988, Lipoxins: A new series of eicosanoids (biosynthesis, stereochemistry, and biological activities), in "Lipoxins: Biosynthesis, Chemistry, and Biological Activities,", P.Y.-K. Wong and C.N. Serhan, eds, Plenum Press, New York, Vol. 229 in Advances in Experimental Medicine and Biology.

64. D.E. Agwu, L.C. McPhail, R.L. Wykle, and C.E. McCall, Mass determination of receptor-mediated accumulation of phosphatidate and diglycerides in human neutrophils measured by Coomassie blue staining and densitometry, Biochem. Biophys. Res. Commun. 159:79 (1989).

65. M.E. Brezinski, M.A. Gimbrone, Jr., K.C. Nicolaou, and C.N. Serhan, Lipoxins stimulate prostacyclin generation by human endothelial cells, FEBS Lett. 245:167 (1989).

66. E. Wikström, P. Westlund, K.C. Nicolaou, and S.-E. Dahlén, Lipoxin A_4 causes generation of thromboxane A_2 in the guinea-pig lung, Agents Actions 26:90 (1989).

67. S.-E. Dahlén, Biological activities of lipoxins, in: "New Trends Lipid Mediators Research," Vol. 3, U. Zor, Z. Naor, and A. Danon, eds., Karger, Basel, in press.

68. C.A. Rouzer and B. Samuelsson, Reversible, calcium-dependent membrane association of human leukocyte 5-lipoxygenase, Proc. Natl. Acad. Sci. USA 84:7393 (1987).

69. T. Puustinen, M.M. Scheffer, and B. Samuelsson, Regulation of the human leukocyte 5-lipoxygenase: Stimulation by micromolar Ca^{2+} levels and phosphatidylcholine vesicles, Biochim. Biophys. Acta 960:261 (1988).

70. C.N. Serhan, A. Radin, J.E. Smolen, H. Korchak, B. Samuelsson, and G. Weissmann, Leukotriene B_4 is a complete secretagogue in human neutrophils: A kinetic analysis, Biochem. Biophys. Res. Commun. 107:1006 (1982).

71. J.T. O'Flaherty and J. Nishihira, 5-hydroxyicosatetraenoate promotes Ca^{2+} and protein kinase C mobilization in neutrophils, Biochem. Biophys. Res. Commun. 148:575 (1987).

72. R. Snyderman and E.J. Goetzl, Molecular and cellular mechanisms of leukocyte chemotaxis, Science 213:830 (1981).

73. W.L. Smith, The eicosanoids and their biochemical mechanisms of action, Biochem. J. 259:315 (1989).

74. S.-E. Dahlén, L. Franzén, J. Raud, C.N. Serhan, P. Westlund, E. Wikström, T. Björck, H. Matsuda, S.E. Webber, C.A. Veale, T. Puustinen, J. Haeggström, K.C. Nicolaou, and B. Samuelsson, 1988, Actions of lipoxin A_4 and related compounds in smooth muscle preparations and on the microcirculation in vivo, in: "Lipoxins: Biosynthesis, Chemistry and Biological Activities," P.Y.-K. Wong and C.N. Serhan, eds., Plenum Press, New York, Vol. 229 in Advances in Experimental Medicine and Biology.

75. J. Balsinde, E. Diez, and F. Mollinedo, Phosphatidylinositol-specific phospholipase D: A pathway for generation of a second messenger, Biochem. Biophys. Res. Commun. 154:502 (1988).

76. C.N. Serhan, Components of the arachidonic acid signalling cascade: a brief update and hypothesis, in: "Advances in Rheumatology and Inflammation," Eular Verlag, Basel, in press.

77. D.A. Brezinski, R.A. Nesto, and C.N. Serhan, Angioplasty triggers intracoronary release of leukotrienes and lipoxin A_4: impact of aspirin therapy, submitted.

HEPOXILINS MODULATE SECOND MESSENGER SYSTEMS IN THE HUMAN NEUTROPHIL

[1]Cecil R. Pace-Asciak and [2]Santosh Nigam

[1]Research Institute, Hospital for Sick Children,
555 University Avenue, Toronto, Canada M5G 1X8
[2]Eicosanoid Research, Department of Gynaecological
Endocrinology, Free University of Berlin, Germany

SUMMARY

In this chapter, we will review recent findings which implicate the hepoxilins as modulators of second messenger systems in the human neutrophil. We have shown that the hepoxilins affect calcium homeostasis in the cell and that they stimulate the release of arachidonic acid and diradylglycerol but not inositol phosphate indicating a mode of action for these 12-lipoxygenase metabolites that is independent of phospholipase C activation. In fact lipid analyses indicate that the phospholipid affected by the hepoxilins is phosphatidyl choline, and that this phospholipid is hydrolyzed by a phospholipase D. These findings indicate that the hepoxilins, which are formed by the platelet as well as the neutrophil, may affect neutrophil activation through a potential cell-cell interaction in the circulation or at pathologic sites to initiate or potentiate the inflammatory process.

FORMATION OF THE HEPOXILINS

Hepoxilins are hydroxy-epoxide derivatives of arachidonic acid formed from 12-HPETE through an intramolecular rearrangement catalyzed by heme proteins[1-3]. The hepoxilin terminology was introduced in 1984 in an attempt to combine important structural features (__H__ydroxy and __epox__ide) with their first reported biological activity (Insu__lin__ secretion)[4]. Two hepoxilins are formed, i.e. 8-hydroxy-11,12-epoxyeicosa-5Z, 10E, 14Z-trienoic acid (hepoxilin A$_3$) and 10-hydroxy-11,12-epoxyeicosa-5Z, 8Z, 14Z-trienoic acid (hepoxilin B$_3$). The latter was originally reported formed from activated platelets by Jones et al[5] while evidence implicating the formation of the former was obtained by Walker et al [6] who isolated the 10, 11, 12-trihydroxy product. Both hepoxilins were isolated from rat lung in the hydroxy-epoxide form in 1983[1]. Since then the hepoxilins have been isolated from a variety of tissues and cells including rat pancreatic islets[7], rat liver[8], rat brain[9], Aplysia neurons[10], human neutrophils[11], rat aorta[12], rat skin[13], and cultured mouse neuroblastoma cells[13]. A recent report has shown that hepoxilin B$_3$ (presumably also hepoxilin A$_3$ is also formed but not isolated) has been extracted from the marine red algae[14]. While hepoxilin A$_3$ is acid labile and hydrolyzes to the trihydroxy product, hepoxilin B$_3$ is much less sensitive to hydrolysis. Figure 1 describes the structures of the primary hepoxilins and their metabolites, their pathways of formation and the enzymes involved.

Cell-Cell Interactions in the Release of Inflammatory Mediators
Edited by P. Y-K Wong and C.N. Serhan, Plenum Press, New York, 1991

133

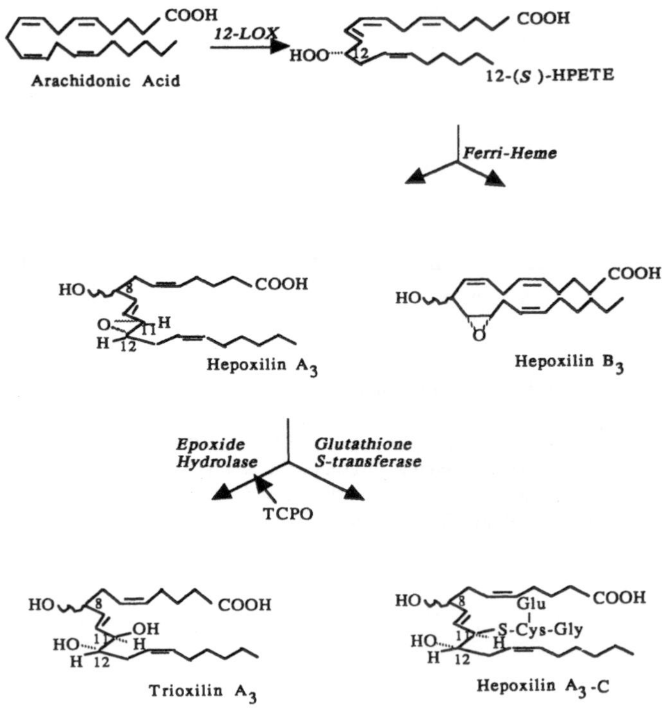

Fig. 1. Scheme depicting structures of the hepoxilins including the biosynthetic and metabolic pathways described to date.

METABOLISM OF THE HEPOXILINS

The hepoxilins are metabolized through two distinct pathways (see Fig. 1). First, an epoxide hydrolase has been identified and purified[15] which converts the epoxide moiety to a dihydroxy moiety to produce a trihydroxy metabolite which is termed 'trioxilin' i.e. trioxilin A_3 and trioxilin B_3. Second, a glutathione S-transferase-catalyzed conjugation of the epoxide moiety with glutathione has been identified which forms a metabolite termed hepoxilin A_3-C[8]. Hepoxilin B_3 is less reactive to both enzyme pathways than hepoxilin A_3. Hepoxilin A_3-C is formed by a specific isozyme of glutathione S-transferase isolated from rat liver, the Y_{b2} homodimer[16]. We have shown that rat brain has the capacity to form hepoxilin A_3-C, and this can be readily observed when the epoxide hydrolase pathway is blocked through use of trichloropropene oxide, TCPO[17]. Neither of the two metabolizing pathways for hepoxilins appear to be present in human neutrophils in detectable amounts[13].

BIOLOGICAL ACTIONS OF THE HEPOXILINS

To date biological actions reported for the hepoxilins appear to have as their basis the stimulation of intracellular calcium. Indeed the hepoxlins were earlier shown to stimulate the transport of calcium across membranes[18], and this was recently confirmed in the neutrophil where hepoxilins were shown to cause an increase in intracellular free calcium both through transport from the extracellular medium as well as through the mobilization from intracellular stores of calcium[11]. Other calcium-mediated actions of hepoxilins are in the stimulation of the release of insulin from pancreatic islets, in the presynaptic inhibition of neurotransmitter release as evidenced from electrophysiological studies in the Aplysia[10] and the mammalian brain[19] as well as from recent biochemical studies in the mammalian brain[20]. Initially the hepoxilins were shown to enhance the secretion of insulin from pancreatic islets that were stimulated with glucose[4]. But this effect has since been shown to be more potent at low glucose concentrations suggesting that the hepoxilins may indeed play a more important role in the basal secretion of insulin[13]. Studies in the Aplysia indicate that the hepoxilins mimic the FMRFamide-induced presynaptic release of 5-HT, while in the mammalian studies hepoxilins were shown to possess synaptic actions[17,19] and to inhibit the release of tritiated norepinephrine from brain slices that were prelabelled with the neurotransmitter[20]. Additionally hepoxilins appear to modify K^+-channel activity, as the recent report by Belardetti has shown in the Aplysia[21]. In this study, inside-out patches of Aplysia neurons containing a specific type of channel activity, the serotonin-type of K^+-channel, responded to the hepoxilins with an increase in the probability of opening, while the hepoxilin precursor, 12-HPETE, had no effect. The reason for the lack of effect of 12-HPETE may lie in its lack of conversion into the hepoxilins by this cytosol-free patch and also indicates that a hepoxilin generating system exists in the cytosol. Indeed heme-proteins present in the cytosol are capable of converting 12-HPETE into the hepoxilins[1-3].

ACTIONS OF HEPOXILINS ON THE NEUTROPHIL

We have recently shown that hepoxilin A_3 is capable of exerting calcium-related actions on the neutrophil[11]. Through the fluorescent dye technique, Dho et al showed that hepoxilin A_3 is capable of raising intracellular calcium through a presumed receptor-mediated fashion as the hepoxilin effect was blocked by pertussis toxin. It was also shown that the response to hepoxilin A_3 was biphasic, only the first phase being independent of extracellular calcium. From these studies it can be inferred that the first phase results from the mobilization of calcium from intracellular stores and is therefore unaffected by extracellular calcium, while the second phase is due to a leakage or transport of calcium from the extracellular medium. One possibility for the rise in intracellular calcium may be that the hepoxilins activate the formation of IP_3 leading to a mobilization of calcium from intracellular stores. However, measurement of total IP in hepoxilin-activated neutrophils indicated that this was not increased (see below). Further studies showed that hepoxilin A_3 caused the release of arachidonic acid and diradylglycerol from neutrophils, and that this release is more evident in FMLP activated neutrophils[22] (Table 1). These results suggest that hepoxilins may activate phospholipases in the cell and that the rise in calcium is brought about through a phospholipase C-independent (IP_3-independent) mechanism. The hepoxilins additionally bring about changes in intracellular pH with an initial acidification of the neutrophil followed by a prolonged alkalinization[11]. The hepoxilins bring about shape changes in neutrophils qualitatively similar to those induced by chemotactic peptides[11].

Mechanism of action of hepoxilins in the neutrophil

Source of diradylglycerol. Diradylglycerol can be formed through two independent pathways, one directly from phosphatidyl inositol via a phospholipase C dependent pathway together with the formation of inositol phosphate (IP_3)[23], the other through the hydrolysis of phosphatidyl choline via a phospholipase D dependent pathway through the formation of phosphatidic acid and its subsequent hydrolysis via phospatidic acid phosphohydrolase[24, 25]. Measurements of total IP in the neutrophil indicated that this second messenger was not increased in the presence of hepoxilins suggesting that a phospholipase C-type of activity was not involved. Analysis of the phospholipid pools in arachidonic acid-prelabelled neutrophils indicated that hepoxilins cause a decrease in the retention of label mostly in the phosphatidyl choline pool and to a much lesser extent in phosphatidyl inositol or other phospholipid pools with an accumulation of label in phosphatidic acid (Table 2). During hepoxilin activation, the level of choline was elevated confirming that a phospholipase D-type of phosphatidyl choline hydrolysis may have taken place. Indeed, in support of this finding was the accumulation of label in phosphatidic acid (Table 2). Further confirmation of the phosphatidic acid pathway for the formation of diradylglycerol via the phosphatidic acid phosphohydrolase enzyme was obtained through use of propranolol, previously shown to inhibit this enzyme[25]. Use of propranolol resulted in the appearance of elevated levels of phosphatidic acid in hepoxilin-activated

Table 1. Time-dependent release of diradylglycerol and arachidonic acid from ^{14}C-labelled human neutrophils by hepoxilin A$_3$ (HxA$_3$)[a].

Time (sec)	Diradylglycerol (% increase over FMLP) +HxA$_3$ (n=5)	Arachidonic acid (% increase over FMLP) +HxA$_3$ (n=5)
10	0	8±1
30	8±2	40±3
60	16±4	68±4
120	34±5	52±11
300	40±5	49±5

[a]Neutrophils were prelabeled with arachidonic acid and challenged in the presence or absence of FMLP (100nM) for the determined time in the presence of hepoxilin A$_3$ (3μM). Data represents the mean±SD.

Table 2. Release of ^{14}C-arachidonic acid from phosphatidyl choline during exposure to hepoxilin A$_3$.

Condition	Phospholipid type (% change in CPM)[a]			
	PA	PE	PC	PI
Vehicle	0	0	0	0
+ FMLP	+37±5	+2±5	-19±1	-12±4
+ FMLP+ hepoxilin A$_3$	+55±5	+2±4	-30±1	-16±3

[a] Human neutrophils were labelled with ^{14}C-arachidonic acid for 15 min at 37° in PBS buffer and exposed to FMLP (100nM) and HxA$_3$ (3μM) during 2 min. Positive sign relates to an accumulation of label, negative sign relates to the disappearance of label. Data represents the mean±SD, n=3 separate experiments.

Table 3. Propranolol inhibits phosphatidic acid phosphohydrolase resulting
in an elevation of phosphatidic acid at the expense of diradylglycerol
during hepoxilin A_3 activation of human neutrophils[a]

Condition	Phosphatidic acid (% of control)	Diradylglycerol (% of control)
Hepoxilin A_3 (3µM)	100±0	100±0
" + propranolol (100µM)	121±16	87±7
Hepoxilin A_3 (3µM)		
" + FMLP (100nM)	147±13	152±9
" + propranolol (100µM)	192±14	108±11

[a]Incubations were carried out in the presence of propranolol (100µM) for
2 min at 37°C using neutrophils labelled with ^{14}C-arachidonic acid. Data
represents the mean±SE, n=3 separate experiments.

neutrophils with a concomittant decrease in the level of diradylglycerol confirming that the
source of diradylglycerol was indeed phosphatidic acid which was formed through
hydrolysis of phosphatidyl choline with phospholipase D (Table 3).

Source of arachidonic acid. The release of arachidonic acid can be derived from the
hydrolysis of phospholipids by phospholipase A_2 , or from neutral lipids by the various
glyceride lipases[26,27]. The above experiments with propranolol resulted in the lowering of
the levels of diradylglycerol and a corresponding lowering of the levels of arachidonic acid
(not shown). These experiments clearly indicated that arachidonic acid is derived from
diradylglycerol through a diradylglycerol lipase.

Further metabolism of released arachidonic acid. That the actions of hepoxilins were
not mediated via products that may be formed from the released arachidonic acid was
investigated through use of 'selective' blockers of prostaglandin and lipoxygenase product
formation including the leukotrienes. Indeed NDGA, ETYA, and SKF-525A were
ineffective in blocking the actions of the hepoxilins in increasing the release of arachidonic
acid and diradylglycerol[13].

CONCLUSIONS

The present results point to a novel action of the hepoxilins on the neutrophil. Hepoxilins
stimulate the release of arachidonic acid as well as diradylglycerol through a phospholipase
C-independent mechanism (Fig. 2). Evidence to date points to an activation of a
phospholipase D-type of enzyme in the neutrophil which utilizes phosphatidyl choline as
substrate to generate phosphatidic acid and choline. Phosphatidic acid in turn is
dephosphorylated to diradylglycerol through activation of phosphatidic acid
phosphohydrolase, and arachidonic acid is subsequently released from diradylglycerol
through the activation of diradylglycerol lipase. Hepoxilins also raise the intracellular
concentration of calcium through an IP_3-independent mechanism. The blockade of the
above mentioned hepoxilin actions by pertussis toxin suggests that hepoxilins may act
directly to release calcium from intracellular stores through activation of their own receptor
although we do not have any additional evidence yet of the existence of such a receptor.
Indeed such actions of the hepoxilins are supported by their formation in these cells as well
as other cells in the circulation (e.g. platelets), so that neutrophil activation by hepoxilins
may occur at pathologic sites to initiate or potentiate the inflammatory process.

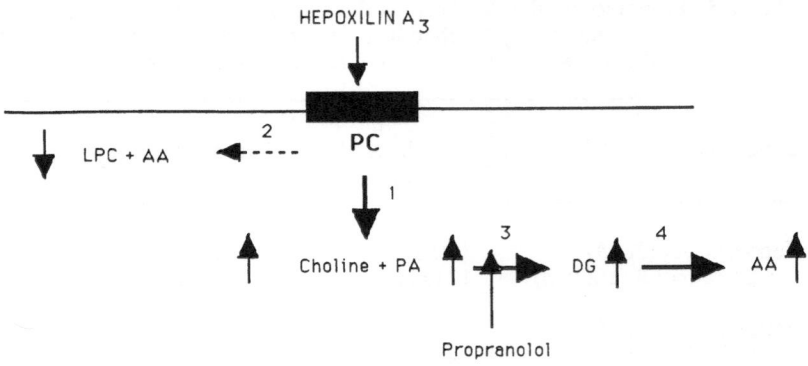

Fig. 2. Scheme depicting the action of hepoxilin A3 on the neutrophil resulting in the activation of a phospholipase D-mediated reslease of diradylglycerol (DG) and arachidonic acid (AA). PA = phosphatidic acid, PC = phosphatidyl choline, LPC = lysophosphatidyl choline.

REFERENCES

1. C. R. Pace-Asciak, E. Granstrom, and B. Samuelsson, Arachidonicacid epoxides: Isolation and structure of two hydroxy epoxide intermediates in the formation of 8 11 12 trihydroxy eicosatrienoic acid and 10 11 12 Trihydroxy eicosatrienoic acid, J. Biol. Chem., **258**: 6835 (1983).
2. C. R. Pace-Asciak, Hemoglobin- and hemin-catalyzed transformation of 12L-hydroperoxy-5,8,10,14-eicosatetraenoic acid, Biochim. Biophys. Acta., **793**: 485 (1984).
3. C. R. Pace-Asciak, Arachidonic acid epoxides: demonstration through oxygen-18 labeled oxygen gas studies of an intramolecular transfer of the terminal hydroxyl group of 12s hydroperoxy eicosa-5 8 10 14-tetraenoic-acid to form hydroxy epoxides, J. Biol. Chem., **259**: 8332 (1984).
4. C. R. Pace-Asciak, and J. M. Martin, Hepoxilin, A new family of Insulin secretagogues formed by intact rat pancreatic islets, Prostagl. Leukotriene and Med., **16**: 173 (1984).
5. R. L. Jones, P. J. Kerry, N. L. Poyser, I. C. Walker, and N.H. Wilson, The identification of trihydroxy eicosatrienoic acids as products from the incubation of arachidonic acid with washed blood platelets, Prostaglandins, **16**: 583 (1978).
6. I. C. Walker, R. L. Jones, P. J. Kerry, and N. H. Wilson, An epoxy-hydroxy product from arachidonate, Adv. Prostagl. Thromb. Res., **6**: 107 (1980).
7. C. R. Pace-Asciak, J. M. Martin, E. J. Corey, and W.-G. Su, Endogenous release of hepoxilin A3 from isolated perifused pancreatic islets of Langerhans, Biochem. Biophys. Res. Commun., **128**: 942 (1985).
8. C. R. Pace-Asciak, O. Laneuville, M. Chang, C. C. Reddy, W.-G. Su, and E. J. Corey, New products in the hepoxilin pathway: isolation of 11-glutathionyl hepoxilin A3 through reaction of hepoxilin A3 with glutathione S-transferase, Biochem. Biophys. Res. Commun., **163**: 1230 (1989).

9. C. R. Pace-Asciak, Formation and metabolism of hepoxilin A_3 by the rat brain, Biochem. Biophys. Res. Commun., 151: 493-498 (1988).

10. D. Piomelli, E. Shapiro, R. Zipkin, J. H. Schwartz, and S. J. Feinmark, Formation and action of 8-hydroxy-11,12-epoxy-5,9,14-icosatetraenoic acid in Aplysia: A possible second messenger in neurons, Proc. Natl. Acad. Sci. (USA, 86: 1721 (1989).

11. S. Dho, S. Grinstein, E. J. Corey, W. G. Su, and C. R. Pace-Asciak, Hepoxilin A_3 induces changes in cytosolic calcium, intracellular pH and membrane potential in human neutrophils, Biochem. J., 266: 63 (1990).

12. O. Laneuville, E. J. Corey, R. Couture, and C. R. Pace-Asciak, Hepoxilin A_3 (HxA3) is formed by the rat aorta and is metabolized into HxA3-C, a glutathione conjugate, Biochim. Biophys. Acta, 1991, in press.

13. C. R. Pace-Asciak, unpublished.

14. M. F. Moghaddam, W. H. Gerwick, and D. L. Ballantine, Discovery of the mammalian insulin release modulator, hepoxilin B3, from the tropical red algae Platysiphonia miniata and Cottoniella filamentosa, J. Biol. Chem., 265: 6126 (1990).

15. C. R. Pace-Asciak, and W.-S. Lee, Purification of hepoxilin epoxide hydrolase from rat liver, J. Biol. Chem., 264: 9310 (1989).

16. O. Laneuville, M. Chang, C. C. Reddy, E. J. Corey, and C. R. Pace-Asciak, Isozyme specificity in the conversion of hepoxilin A_3 into a glutathionyl hepoxilin (HxA3-C) by the Y_{b2} subunit of rat liver glutathione S-transferase. J. Biol. Chem., 1991, in press.

17. C. R. Pace-Asciak, O. Laneuville, W.-G. Su, E. J. Corey, N. Gurevich, P. Wu, and P. L. Carlen, A glutathione conjugate of hepoxilin A_3: Formation and action in the rat central nervous system, Proc. Natl. Acad. Sci. USA., 87: 3037 (1990).

18. L. O. Derewlany, C. R. Pace-Asciak and I. C. Radde, Hepoxilin A, hydroxyepoxide metabolite of arachidonic acid stimulates transport of calcium-45 across the guinea-pig visceral yolk sac, Can. J. Physiol. Pharmacol., 62: 1466 (1984).

19. P. L. Carlen, N. Gurevich, P. H. Wu, Su-W-G., E. J. Corey, and C. R. Pace-Asciak, Actions of arachidonic acid and hepoxilin A_3 on mammalian CA1 neurons, Brain Res., 497: 171 (1989).

20. C. R. Pace-Asciak, L. Wong, and E. J. Corey, Hepoxilin A_3 blocks the release of norepinephrine from rat hippocampal slices, Biochem. Biophys. Res. Commun., 1990, in press.

21. F. Belardetti, W. B. Campbell, J. R. Falck, G. Demontis, and M. Rosolowsky, Products of heme-catalyzed transformation of the arachidonate derivative 12-HPETE open S-type K^+ channels in Aplysia, Neuron, 3: 497 (1989).

22. S. Nigam, S. Nodes, G. Cichon, E. J. Corey, and C. R. Pace-Asciak, Receptor-mediated action of hepoxilin A_3 releases diacylglycerol and arachidonic acid from human neutrophils, Biochem. Biophys. Res. Commun., 171: 944 (1990).

23. M. J. Berridge, and R. F. Irvine, Inositol triphosphate, a novel second messenger in cellular signal transduction, Nature, 312: 315 (1984).

24. M. M. Billah, S. Eckel, T. J. Mullman, R. W. Egan, and M. I. Siegel, Phosphatidyl choline hydrolysis by phospholipase D determines phosphatidate and diglyceride levels in chemotactic peptide-stimulated human neutrophils, J. Biol. Chem., 264: 17069 (1989).

25. J. Exton, Signaling through phosphatidyl choline breakdown, J. Biol. Chem., 265: 1, (1990).

26. S. Nakashima, K. Nagata, K. Ueda, Y. Nozawa, Stimulation of arachidonic acid release by guanine nucleotide in saponin-permeabilized neutrophils: evidence for involvement of CTP-binding protein in phospholipase A_2 activation, Arch. Biochem. Biophys., 261: 375 (1988).

27. W. Tao, F.P. Molski, and R. I. Sh'afi, Arachidonic acid rlease in rabbit neutrophils, Biochem. J., 257: 633 (1989).

REGULATED SECRETION IN VASCULAR ENDOTHELIUM

Bruce M. Ewenstein[*][§], Brian C. Jacobson[*], and Kimberly A. Birch[‡]

[*]Division of Hematology and [‡]Department of Pathology, Brigham & Women's Hospital, Boston, MA 02115; and Departments of [*]Medicine and [‡]Pathology, Harvard Medical School, Boston, MA 02115

Introduction

Eukaryotic cells secrete proteins through both regulated and constitutive pathways (1). In the regulated pathway, exocytosis of stored material is coupled to an external stimulus which produces transient elevations in $[Ca^{2+}]_i$, cAMP or other secondary messengers (2,3). Secretion is ultimately accomplished by the translocation of the exocytotic granule to the subcortical regions of the cell and the fusion of the cellular plasma membrane with the membranes of the secretory organelle. The process of regulated secretion has at least two distinct consequences. First, the rapid release of the contents of the storage granule generates locally high concentrations of peptides, proteins and other bioactive compounds in the vicinity of the secretory cell. These products may promote cell-cell interaction through the activation of target cells or the crosslinking of surface receptors. Second, the translocation of secretory granule membrane proteins to the plasma membrane of the secretory cell may provide new or additional binding sites for cellular adhesion.

Far from its earlier conception as a simple anatomical barrier, vascular endothelium has come to be viewed as a "dynamic partner" in the multiple, complex interactions involving cellular constituents of blood (4). The past several years have witnessed a growing appreciation of the significance of regulated endothelial cell (EC) secretion in these interactions. This article will review what is known about the process of regulated secretion in EC and its likely contributions to cell-cell interactions in the vasculature.

von Willebrand factor (vWF)

Although EC secrete a variety of proteins, only vWF [and possibly platelet-derived growth factor (5) and protein S (6)] is rapidly released in response to agonist stimulation.   vWF is a

---

[§]to whom correspondence should be addressed.

Cell-Cell Interactions in the Release of Inflammatory Mediators
Edited by P. Y-K Wong and C.N. Serhan, Plenum Press, New York, 1991

complex glycoprotein which is synthesized exclusively by EC (7) and by megakaryocytes (8) and circulates in plasma as a series of polymers (multimers) of M_r varying from 0.5 to 20 x 10^6. In vivo, vWF serves both as a stabilizing "carrier protein" of coagulation factor VIII, with which it is noncovalently complexed, and as a platelet-subendothelial "molecular bridge" at sites of vessel wall damage (reviewed in 9). Quantitative or qualitative defects in circulating vWF result in von Willebrand's disease, the most common inherited bleeding disorder (reviewed in 10).

vWF Biosynthesis

Circulating vWF is chiefly composed of identical subunits (M_r 220,000) of 2050 amino acids (11). Each subunit is derived from a pre-pro-polypeptide of 2813 amino acids consisting of a 22 amino acid signal peptide and a 741 amino acid propeptide (12-15). The biosynthesis of vWF has recently been reviewed (16) and will only be briefly outlined here. Following the co-translational cleavage of the signal peptide and the addition of 13 high mannose N-linked oligosaccharide chains, the 2791 amino acid (M_r 275,000) pro-vWF forms dimers which are covalently linked through an unknown number of C-terminal disulfide bonds (17). The dimerization of the pro-vWF polypeptide chain is believed to occur in the rough endoplasmic reticulum (18-20). In the Golgi apparatus the high mannose glycans are processed to the mature form and up to ten O-linked oligosaccharide chains are added to the protein backbone (11). Propeptide cleavage and multimerization of dimeric vWF subunits are the final steps in vWF biosynthesis and are thought to occur chiefly in the trans Golgi compartment and in nascent secretory granules (16,21). Although these steps occur at approximately the same time, they are likely to be independent events as propeptide cleavage can be inhibited by site-specific mutagenesis without affecting multimerization (22,23), and multimerization can be inhibited by intracellular alkalinization without affecting pro-vWF cleavage (24).

It has been suggested that the propeptide facilitates the proper alignment of adjacent pro-vWF dimers (17) and several lines of evidence suggest that this portion of the molecule is critical for vWF multimer formation. Verweij et al. (22) demonstrated that truncated recombinant vWF lacking the propeptide failed to multimerize when expressed in heterologous cells under conditions in which full-length vWF formed multimers. Wise et al. (23) further demonstrated that the ability of similarly truncated vWF proteins to multimerize could be restored by the co-expression of the pro-vWF polypeptide even when it was not covalently linked to the mature vWF subunit. Finally, in a series of cell-free experiments, Mayadas and Wagner (25) observed that dimers of pro-vWF, but not of fully processed vWF, were able to form high molecular weight polymers under acidic conditions and that monoclonal antibodies directed against the propeptide blocked this process. The cleaved vWF propeptide is co-localized with the mature vWF protein in the secretory granule (Weibel-Palade body) and circulates in the plasma as a noncovalently-linked dimer which has been termed von Willebrand antigen II (vW Ag II) 26-29).

Constitutive Secretion of vWF

In culture, EC secrete vWF along both constitutive and regulated pathways (Fig. 1). Constitutive secretion appears to be directly linked to vWF biosynthesis as it is completely blocked by the protein synthesis inhibitor, cycloheximide (30). The vWF secreted along this pathway consists of predominantly smaller multimeric forms which contain a variable quantity of uncleaved (pro-vWF) subunits. Thus, although both multimerization and propeptide cleavage can occur, neither process is favored in the constitutive pathway.

In vitro, nearly 95% of newly synthesized vWF is secreted along the constitutive pathway, the remainder being targeted to the Weibel-Palade body and the regulated pathway (31) (see below). In vitro studies of other secretory cells have demonstrated that constitutive secretion is a default pathway for proteins normally destined for packaging in exocytotic granules (1). The fraction of vWF targeted to the constitutive pathway may thus be a consequence of an imbalance between the biosynthesis of vWF and that of other components of the Weibel-Palade body. This concept is supported by observations that variations in EC culture conditions may lead to increased numbers of Weibel-Palade bodies (32) or enhanced rates of basal secretion of the larger vWF multimeric forms (33). The relative contribution of the constitutive pathway to vWF plasma levels in vivo remains to be determined.

Regulated Secretion of vWF

The presence of a clinically important storage pool of vWF was first suggested by the observations that a variety of stimuli including exercise, adrenalin (34), desmopressin, and the desmopressin analog, 1-desamino-8-D-arginine-vasopressin (DDAVP) (35) rapidly increase plasma vWF levels and ameliorate clinical bleeding (36). Human umbilical vein EC in vitro rapidly secrete vWF in response to a number of physiologic agonists including thrombin (30,37,38)(Fig. 1), histamine (39), vascular permeability factor (40), fibrin (41,42) and complement membrane attack complexes (43) and thus provide a convenient experimental system

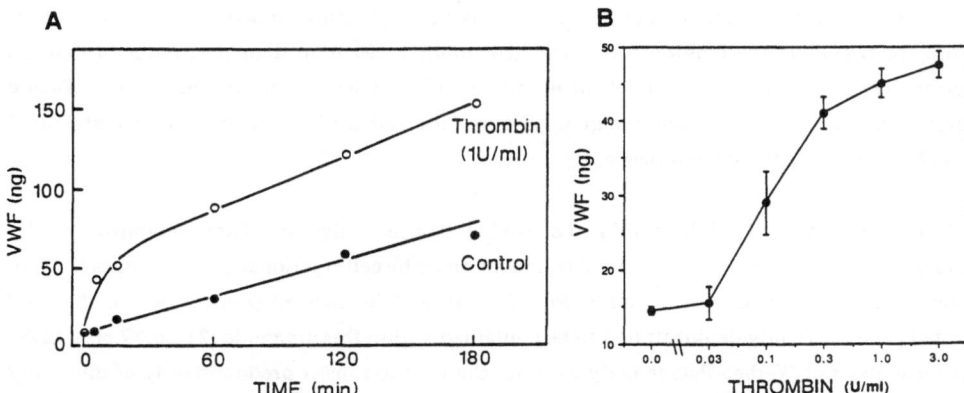

Fig. 1.   Constitutive and thrombin-stimulated vWF secretion. A, Cultured EC constitutively secrete vWF (•). The addition of human α-thrombin results in the rapid release of preformed vWF (○) followed by a return to near basal levels of secretion. B, Thrombin-mediated secretion is dose-dependent ($EC_{50} \approx 0.2$ U/ml). Data depicts total vWF released in 20 min. from approximately 2×10^5 passaged human umbilical vein EC. Results are expressed as the mean (± S.E.M.) of triplicate determinations of vWF measured by an inhibition ELISA (21).

in which to study the process of regulated exocytosis. Proteolytically active thrombin is required to stimulate vWF secretion and prior treatment of the agonist with diisopropyl fluorophosphate (DIP), D-Phe-Pro-Arg-chloromethyl ketone (PPACK), or hirudin abolishes its secretogogue activity (37,39,41). Thrombin has been shown to bind to several classes of EC surface proteins (44,45) but the receptor responsible for vWF secretion has not been defined. Both histamine-induced elevations of $[Ca^{2+}]_i$ and vWF secretion are mediated through H_1 receptors as these responses can be inhibited by the H_1 blocker, pyrilamine, but not by the H_2 antagonist, cimetidine (39,46). Interestingly, vasopressin analogs such as DDAVP which cause rapid and marked elevations of vWF in vivo, fail to induce vWF secretion in cultured EC (47,48), suggesting that their clinical effects may not be mediated by direct EC stimulation.

Concentrations of agonists which produce maximal vWF release result in the depletion of the intracellular pool of Weibel-Palade bodies. Full regeneration of these granules following thrombin or phorbol ester stimulation requires up to 48 hours (49) and is not accompanied by either a compensatory increase in de novo vWF synthesis or in an alteration in the fraction of vWF which is targeted to the regulated pathway of secretion (50). These in vitro findings are also reflected in the clinical observation that patients become refractory to DDAVP treatment following as few as one or two exposures to the drug and require several days before a hemostatic response is restored (36).

The Weibel-Palade Body

vWF is stored in and secreted from two principal sources: vascular EC and the α-granule of the platelet. The vascular vWF storage organelle is now recognized to be the Weibel-Palade body, a rod-shaped organelle unique to endothelium (51)(Fig. 2). First described in large blood vessels, Weibel-Palade bodies have since been found in virtually all vascular and lymphatic endothelial cells (9) and may be increased in atherosclerotic blood vessels (52) and neovascular tissue surrounding certain tumors (53). In the electron microscope, Weibel-Palade bodies are seen to be 0.1 μm in diameter and up to 4 μm in length, and are surrounded by a unit membrane. Within the organelle are longitudnally arranged tubules, 150 Å in diameter, embedded in an electron-dense matrix (51). From electron microscopic studies, it appears that Weibel-Palade bodies originate from the trans Golgi apparatus (54) and are eventually found distributed throughout the endothelial cytoplasm.

The presence of vWF within the Weibel-Palade body was first demonstrated by immunoelectron microscopy (55-57) and later confirmed by cell fractionation techniques (21,59). To date, fully processed vWF (subunit M_r 220,000) and its cleaved propeptide are the only secreted proteins firmly demonstrated to be contained within this organelle (21,26,59,60). SDS-agarose analysis of Weibel-Palade body vWF has shown it to consist predominantly of unusually large multimeric forms (21). Thus it is most likely that the unique morphology of the Weibel-Palade body derives from the highly polymerized state of its principle secretory product. The precise conditions within the organelle which facilitate multimerization are not known but both high protein concentration (61) and low pH (25) have been shown to promote vWF

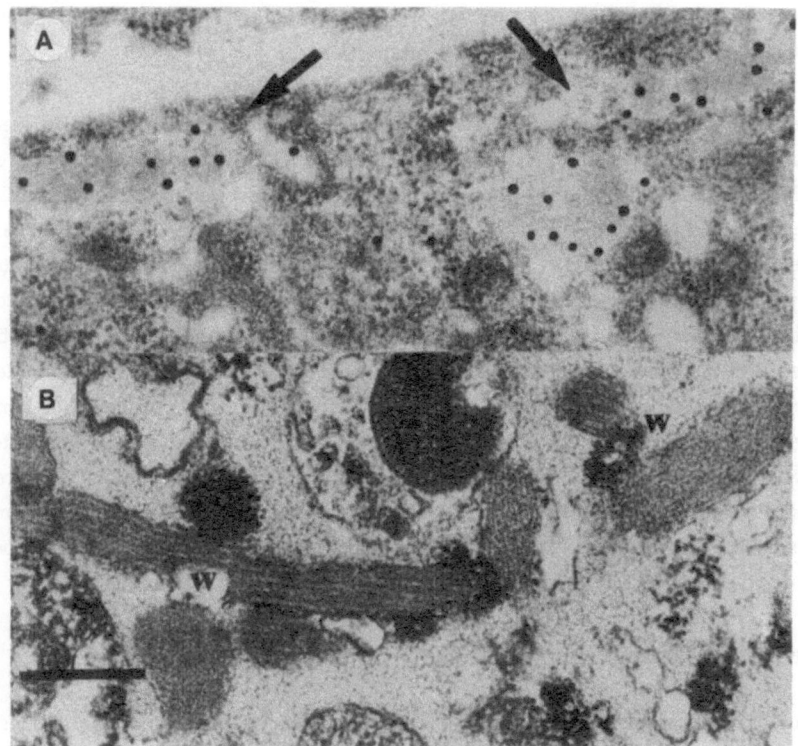

Fig. 2.   Electron micrographs of Weibel-Palade bodies. A, Weibel-Palade bodies (→) in cultured human umbilical vein EC stained with colloidal gold particles coupled to anti-vWF antibody. B, Weibel-Palade bodies (w) purified by Percoll gradient fractionation (21). The internal lamellar structure believed to result from the highly polymerized vWF contained within the organelle is apparent. Immuno-electron microscopy was performed by Dr. Michael J. Warhol at Brigham & Women's Hospital (Boston, MA).

polymerization in vitro. The vWF stored within the Weibel-Palade body and that released in response to agonist stimulation are similar both in multimeric size distribution and in the absence of pro-VWF forms (21,31). These findings lend strong support to the concept that the Weibel-Palade body is the source of agonist-stimulated vWF secretion.

More recently, CD62, an integral membrane protein of the platelet α-granule, alternatively named GMP140 and PADGEM (62,63), was also shown, using both immunoelectron microscopic (64) and immunofluorescence (65) techniques, to be associated with the Weibel-Palade body. Structurally, GMP140/PADGEM is a 789 amino acid glycoprotein consisting of an NH_2-terminal domain homologous to Ca^{2+}-dependent lectins, followed by an EGF-like domain, nine tandem consensus repeats similar to those in complement regulatory proteins, a transmembrane domain, and a short cytoplasmic tail (66). Of particular significance is the fact that GMP140/PADGEM shares overall structural homology and extensive sequence identity with two other cell adhesion molecules, ELAM-1 (67) and the leukocyte peripheral lymph node homing receptor (68), and thus belongs to a newly described family of lectin-containing receptors ("selectins" or "LECCAMs") that promote selective cell-cell interactions.

Following agonist stimulation of EC, Weibel-Palade bodies are rapidly translocated to the cell surface (69). As measured by the appearance of GMP140 (70) and direct electron microscopic observation (21), the fusion of organelle and plasma membranes occurs rapidly and reaches near maximal levels by 3-5 minutes following agonist stimulation. The discharge of vWF into the surrounding medium takes longer and "release patches" of cell surface associated vWF can be seen by immunofluorescence to persist for up to several hours (31,41). The association of vWF with the external surface of the endothelial cell has yet to be fully explained but may be mediated either by the endothelial vitronectin receptor (71), or by an as yet unidentified constituent of the Weibel-Palade body itself. In contrast, the vWF propeptide, which co-localizes with fully processed vWF in the Weibel-Palade body, does not remain associated with EC after stimulation (29). A significant fraction of stimulated vWF secretion is also directed to the basolateral surface (72). Moreover, vWF isolated from the subendothelial matrix of cultured EC consists primarily of high molecular weight forms, suggesting that Weibel-Palade bodies are the principle source of this protein (73).

Stimulus-secretion Coupling

Regulated exocytosis is presumed to be under the influence of second messengers generated upon agonist stimulation. One pathway of second messenger generation is the activation of phosphatidylinositol-specific phospholipase C which results in the generation of both inositol 1,4,5-trisphosphate (IP_3) and 1,2-diacylglyercol (DAG) from phosphatidylinositol 4,5-bisphosphate (PIP_2) (74). Thrombin, the ultimate serine protease activated by the coagulation cascade, has been shown to activate phospholipase C in cultured human umbilical vein EC (75,76). It is generated in active form on the surface of stimulated platelets and on endothelial membranes at sites of inflammation (77,78). Thrombin thus serves as an important and relevant paradigm for the study of stimulus-secretion coupling in vWF secretion.

In EC, thrombin stimulation leads to a rapid increase in cytosolic free calcium concentration ($[Ca^{2+}]_i$) followed (in the presence of thrombin concentrations greater than 0.1 U/ml) by a prolonged plateau phase. The initial rise in $[Ca^{2+}]_i$ is preceded by, and is thought to be a consequence of, IP_3-mediated release of internal Ca^{2+} stores while the plateau phase is associated with the influx of extracellular Ca^{2+} (75,76,79). The importance of Ca^{2+} as a second messenger in endothelial vWF secretion is suggested by a number of experimental observations. First, the Ca^{2+} ionophore, A23187, which elevates $[Ca^{2+}]_i$, also induces vWF release (30). Second, other agonists which induce vWF release, such as histamine (39), vascular permeability factor (40) and complement membrane attack complexes (43), also raise $[Ca^{2+}]_i$.

The importance of Ca^{2+} in thrombin stimulus-secretion coupling in EC has been directly demonstrated by the inhibition of vWF secretion in the presence of extracellular Ca^{2+} chelation (38) and the close correlation between thrombin-stimulated elevations of $[Ca^{2+}]_i$ and vWF release (80). In many cellular responses, elevations of $[Ca^{2+}]_i$ are mediated through calmodulin (CaM).

146

That this may also be true for thrombin-stimulated vWF secretion is suggested by the finding that the CaM inhibitor, W-7, produces a dose-dependent inhibition of vWF release in response to the agonist (Fig. 3). In contrast, no inhibition is observed with the less potent W-7 analog, W-5. These data suggest that CaM plays an obligatory role in transducing the Ca^{2+} signal in thrombin-mediated vWF release.

Ca^{2+}/CaM may regulate secretion at multiple sites along the secretory pathway. One attractive target currently under investigation in our laboratory is the CaM-dependent enzyme, myosin light chain kinase (MLCK). In vitro, MLCK has been shown to cause the selective phosphorylation of myosin light chains and is believed to be a key regulator of myosin contractile activity in vivo (reviewed in 81). The phosphorylation of myosin light chains by MLCK correlates with the exocytotic process in a number of other cell types (82,83) and more recently in EC retraction (84), a process with which secretion may share certain common features. Support for a role of MLCK in thrombin-mediated vWF secretion comes from the finding that ML-7, a selective inhibitor of MLCK, also causes a dose-dependent inhibition of thrombin-stimulated vWF secretion (Fig. 3).

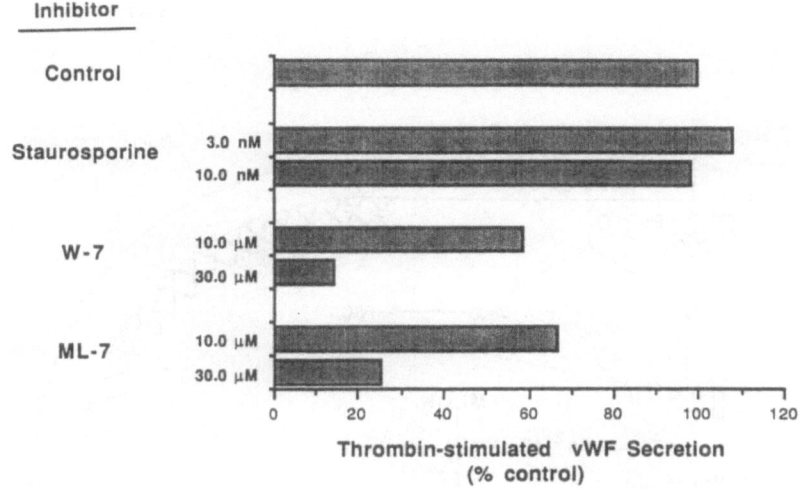

Fig. 3.   Effect of protein kinase inhibitors on thrombin-stimulated vWF secretion. EC were preincubated in the presence or absence of the indicated inhibitor for 15 min. prior to the addition of thrombin (1 U/ml). The quantity of vWF released in 15 min. following thrombin addition was measured by an inhibition ELISA (21). Results are means of triplicate determinations.

The DAG generated by thrombin-stimulated PIP_2 hydrolysis is thought to be an endogenous activator of protein kinase C (PKC). A possible role for PKC as a mediator of vWF secretion is suggested by the observation that the PKC activator, PMA, which does not cause elevations of $[Ca^{2+}]_i$ in human umbilical vein EC, is a potent secretogogue of vWF (21,38).

147

Thrombin has been shown to induce the translocation of PKC from cytosolic to membrane fractions in bovine pulmonary artery EC (85) and to stimulate the phosphorylation of a myristoylated, alanine-rich C kinase substrate (MARCKS) (86) in human umbilical vein EC (87). However, a role for PKC in thrombin-mediated secretion in EC has not been established. Indeed, staurosporine, a potent inhibitor of PKC, exerted no net effect on thrombin-mediated vWF release (Fig. 3) suggesting that PKC activation does not play a primary role in the signal transduction system leading to vWF secretion.

Taken together, these data provide the basis for a working model of thrombin stimulation in which elevations of $[Ca^{2+}]_i$ provide the dominant signal for vWF secretion. This signal is transduced through CaM-dependent processes which likely include the stimulation of an MLCK-dependent translocation of the Weibel-Palade body to the inner surface of the cell (Fig. 4). This translocation may also depend upon the presence of an intact microtubule system (88). Additional levels of control are likely to be exerted by the subcortical actin network which has been proposed to serve as a barrier to regulated exocytosis in other cell types (89). PKC likely modulates thrombin-mediated vWF release in a number of ways including the regulation of thrombin-stimulated $[Ca^{2+}]_i$ responses (76), but appears not to be of primary importance in the secretory process.

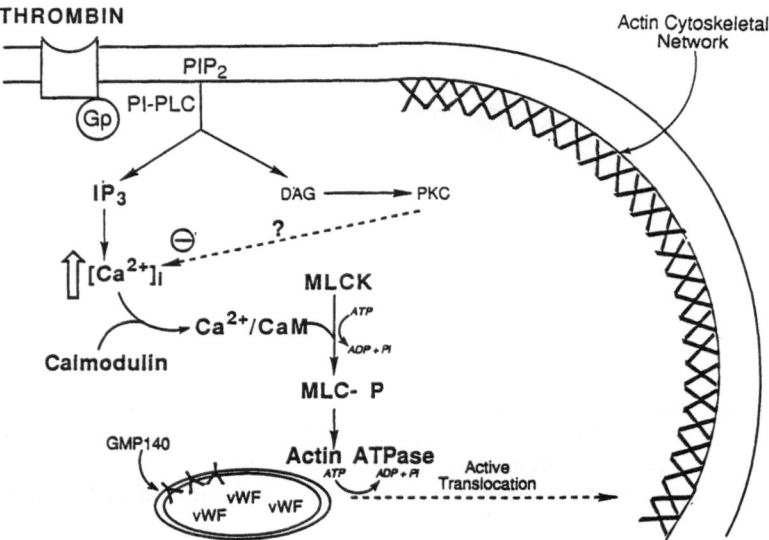

Fig. 4.   Hypothetical model of thrombin stimulus-secretion coupling in vascular EC. Thrombin is proposed to generate increases in $[Ca^{2+}]_i$ principally through the generation of IP_3 through the hydrolysis of PIP_2 (74). The Ca^{2+} signal is transduced through a calmodulin (CaM)-dependent pathway possibly involving myosin light chain kinase (MLCK) leading to the phosphorylation of myosin light chains (MLC-P) and the active translocation of the Weibel-Palade body to the subcortical regions of the cell.

Implications for Cell-Cell Interactions

Our appreciation of the range of physiologic processes modulated by regulated EC exocytosis continues to expand. Weibel-Palade bodies are the storage pool for the unusually large multimeric forms of vWF which bind with the greatest avidity to platelet receptors and vascular subendothelium (90-92). The vWF released from these granules is thus ideally suited to promote hemostasis by enhancing platelet-subendothelial adhesion in the vicinity of stimulated endothelium. Through the rapid surface expression of GMP140/PADGEM, the process of regulated exocytosis may also play a key role in the early phases of leukocyte adhesion to vascular endothelium. Previous studies have shown that neutrophils (and monocytes) bind to purified GMP140/PADGEM, a process which does not require that the leukocytes be metabolically active (93-95). More significantly, anti-GMP140 antibodies block leukocyte adhesion to stimulated EC in vitro (93). The observation that GMP140/PADGEM is heavily concentrated in small veins and venules (64) further supports its proposed role, although other protein (reviewed in 96) and phospholipid mediators (97) of leukocyte-EC adhesion are clearly also important in the full expression of the inflammatory response.

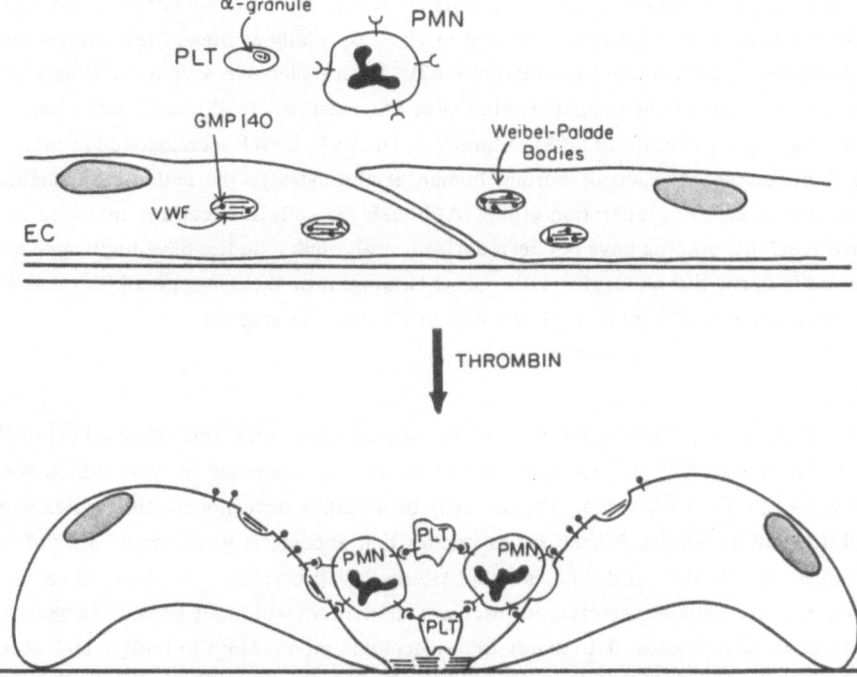

Fig. 5.   A schematic representation of the action of thrombin on vascular EC resulting in cell retraction, the deposition of large multimeric forms of vWF; and the translocation of GMP140 (PADGEM) to the cell surface. Thrombin also stimulates the surface expression of GMP140/PADGEM on activated platelets promoting a complex interaction of polymorphonuclear cells (PMN), platelets, EC, and the subendothelial matrix.

In vivo, it is likely that platelet and neutrophil adhesion in response to EC stimulation are mutually interactive events. For example, it has been appreciated for some time that products of activated platelets can enhance neutrophil activation (98) and that platelets can mediate the transcellular metabolism of neutrophil-derived eicosanoid lipid mediators (99,100). More recently, it has been demonstrated that GMP140/PADGEM promotes the binding of neutrophils and monocytes to activated platelets (101,102), experiments which provide a unifying view of the early hemostatic and inflammatory responses to tissue injury. In one possible scenario (Fig. 5), thrombin, generated on the surface of activated platelets, is proposed to stimulate a number of different cell types including neutrophils, platelets and nearby EC. The release of high molecular weight vWF from stimulated EC would facilitate the binding of additional platelets to exposed subendothelium while the expression of GMP140/PADGEM on the surfaces of both the platelet and EC would promote polymorphonuclear cell adhesion. Positive feedback loops may then be generated leading to further cellular activation. For example, peroxides, such as those produced by activated neutrophils, have recently been shown to stimulate additional GMP140/PADGEM expression and, in turn, promote leukocyte-EC adhesion (103). Additional platelet-derived products may also stimulate further cellular activation and adhesion.

Recently, interest has focused on the role of red blood cell (RBC)-EC interactions in the pathophysiology of sickle cell vasoocclusive crises. Sickle cell RBC are abnormally adherent to vascular endothelium in vitro (104-106) and in vivo, especially at sites of low shear stress such as the immediate post-capillary circulation (107). High molecular weight multimers of vWF, such as those secreted by the regulated pathway of EC, markedly facilitate the adhesion of sickle cell RBC under flow conditions (108). Moreover, DDAVP, a vWF secretagogue in intact blood vessels, induces the adhesion of normal human erythrocytes to the endothelial surface of a perfused microvascular preparation (109). Although the cellular receptors involved in vWF-mediated RBC-EC binding have not been defined, preliminary studies have implicated both Gp Ib-like and integrin-like molecules (110). These findings raise the strong possibility that Weibel-Palade body-derived vWF plays a pivitol role in RBC-EC interactions.

Conclusions

Through the regulated secretion of highly multimerized vWF and the rapid expression of GMP140/PADGEM, EC interact with circulating cellular elements in ways which were not anticipated a few short years ago. The diversity of naturally occuring agonists which stimulate the mobilization of Weibel-Palade bodies implies that exocytosis in EC may be involved in a wide range of physiological and pathophysiological processes. If true, then a fuller understanding of regulated secretion in vascular endothelium will likely provide further insights into each of these processes. Ultimately, pharmacologic intervention to inhibit vWF secretion, GMP140 expression, or the subsequent cellular adhesive events mediated by these molecules, may prove useful in the management of a variety of clinically important processes such as thrombosis, inflammation and vasoocclusive disease.

Acknowledgments

We thank Mary Halm Small for assistance in the preparation of this manuscript and Drs. Robert J. Wise and Jordan S. Pober for helpful discussions. These studies were supported in part by National Institutes of Health grants PO1-HL36028 and RO1-HL36003, and American Heart Association, Massachusetts Affiliate, grant 13-535-889.

REFERENCES

1.	T.L. Burgess and R.B. Kelly, Constitutive and regulated secretion of proteins, <u>Annu. Rev. Cell Biol</u>. 3:243 (1987).

2.	J.F. Harper, Stimulus-secretion coupling:	Second messenger-regulated exocytosis, <u>Advances in Second Messenger and Phosphoprotein Research</u>, 22:193 (1988).

3.	T.J. Rink and D.E. Knight, Stimulus-secretion coupling: A perspective highlighting the contributions of Peter Baker, <u>J. Exp. Biol</u>. 139:1 (1988).

4.	M.A. Gimbrone Jr. and M.P. Bevilacqua, 1988, Vascular endothelium:	Functional modulation at the blood interface, <u>in</u>:	"Endothelial Cell Biology," N. Simionescu and M. Simionescu, eds., Plenum Press, New York.

5.	J.M. Harlan, P.J. Thompson, R.R. Ross, and D.F. Bowen-Pope, Alpha-thrombin induces release of platelet-derived growth factor-like molecule(s) by cultured human endothelial cells, <u>J. Cell. Biol</u>. 103:1129 (1986).

6.	D.M. Stern, J. Brett, K. Harris, and P. Nawroth, Participation of endothelial cells in the protein C-protein S anticoagulant pathway:	the synthesis and release of protein S, <u>J. Cell Biol</u>. 102:1971 (1986).

7.	E.A. Jaffe, L.W. Hoyer, R.L. Nachman, Synthesis of antihemophilic factor antigen by cultured human endothelial cells, <u>J. Clin. Invest</u>. 52:2757 (1973).

8.	R.L. Nachman, R. Levine, and E.A. Jaffe, Synthesis of factor VIII antigen by cultured guinea pig megakaryocytes, <u>J. Clin. Invest</u>. 60:914, (1977).

9.	R.I. Handin and D.D. Wagner, 1989, Molecular and cellular biology of von Willebrand factor, <u>in</u>: "Progress in Hemostasis and Thrombosis," Vol. 9, B.S. Coller, ed., Saunders, Philadelphia.

10.	T.S. Zimmerman and Z.M. Ruggeri, von Willebrand disease, <u>Hum. Pathol</u>. 18:140 (1987).

11.	K. Titani, S. Kumar, K.Takio, L.H. Ericsson, R.D. Wade, K. Ashida, K.A. Walsh, Michael W. Chopek, J.E.Sadler, and K. Fujikawa, Amino acid sequence of human von Willebrand factor, <u>Biochemistry</u> 25:3171 (1986).

12.	D. Ginsburg, R.I. Handin, D.T. Bonthron, T.A. Donlon, G.A.P. Bruns, S.A. Latt, and S.H. Orkin, Human von Willebrand factor (vWf): Isolation of complementary DNA (cDNA) clones and chromosomal localization, <u>Science</u> 228:1401 (1985).

13.	D.C. Lynch, T.S. Zimmerman, C.J. Collins, M. Brown, M.J. Morin, E.H. Ling, and D.M. Livingston, Molecular cloning of cDNA for human von Willebrand factor: Authentication by a new method, <u>Cell</u> 41:49 (1985).

14.	J.E. Sadler, B.B. Shelton-Inloes, J.M. Sorace, J.M. Harlan, K. Titani, and E.W. Davie, Cloning and characterization of two cDNAs coding for human von Willebrand factor, <u>Proc. Natl. Acad. Sci. USA</u> 82:6394 (1985).

15.	C.L. Verweij, C.J.M. deVries, B. Distel, A.-J. van Zonneveld, A.G. van Kessel, J.A. van Mourik, and H. Pannekoek, Construction of cDNA coding for human von Willebrand factor using antibody probes for colony screening and mapping of the chromosomal gene, <u>Nucleic Acids Res</u>. 13:4699 (1985).

16.	D.D. Wagner, Cell biology of von Willebrand factor, <u>Annu. Rev. Cell Biol</u>. 6:217 (1990).

17.	D.D. Wagner, S.O. Lawrence, B.M. Ohlsson-Wilhelm, P.J. Fay, and V.J. Marder, Topology and order of formation of interchain disulfide bonds in von Willebrand factor, <u>Blood</u> 69:27 (1987).

18. D.D. Wagner and V.J. Marder, Biosynthesis of von Willebrand protein by human endothelial cells: Processing steps and their intracellular localization, <u>J. Cell Biol</u>. 99:2123 (1984).

19. D.D. Wagner, T. Mayadas, M. Urban-Pickering, B.H. Lewis, and V.J. Marder, Inhibition of disulfide bonding of von Willebrand protein by monensin results in small, functionally defective multimers, <u>J. Cell Biol</u>. 101:112 (1985).

20. M.C. Roarke, D.D. Wagner, V.J. Marder, and L.A. Sporn, Temperature-sensitive steps in the secretory pathway for von Willebrand factor in endothelial cells, <u>Eur. J. Cell Biol</u>. 48:337 (1989).

21. B.M. Ewenstein, M.J. Warhol, R.I. Handin, and J.S. Pober, Composition of the von Willebrand factor storage organelle (Weibel-Palade body) isolated from cultured human umbilical vein endothelial cells, <u>J. Cell Biol</u>. 104:1423 (1987).

22. C.L. Verweij, M. Hart and H. Pannekoek, Expression of variant von Willebrand factor (vWf) cDNA in heterologous cells: Requirement of the propolypeptide in vWF multimer formation, <u>EMBO J</u>. 6:2885 (1987).

23. R.J. Wise, D.D. Pittman, R.I. Handin, R.J. Kaufman, and S.H. Orkin, The propeptide of von Willebrand factor independently mediates the assembly of von Willebrand multimers, <u>Cell</u> 52:229 (1988).

24. D.D. Wagner, T. Mayadas, and V.J. Marder, Initial glycosylation and acidic pH in the Golgi apparatus are required for multimerization of von Willebrand factor, <u>J. Cell Biol</u>. 102:1320 (1986).

25. T. Mayadas and D.D. Wagner, In vitro multimerization of von Willebrand factor is triggered by low pH: Importance of the propolypeptide and free sulfhydryls, <u>J. Biol. Chem</u>. 264:13497 (1989).

26. D.R. McCarroll, E.G. Levin, and R.R. Montgomery, Endothelial cell synthesis of von Willebrand antigen II, von Willebrand factor and von Willebrand factor/von Willebrand antigen II complex, <u>J. Clin. Invest</u>. 75:1089 (1985).

27. R.R. Montgomery and T.S. Zimmerman, von Willebrand's disease antigen II. A new plasma and platelet antigen deficient in severe von Willebrand's disease, <u>J. Clin. Invest</u>. 61:1498 (1978).

28. P.J. Fay, Y. Kawai, D.D. Wagner, D. Ginsburg, D. Bontron, B.M. Ohlsson-Wilhelm, S.I. Charin, G.N. Abraham, R.I. Handin, S.H. Orkin, R.R. Montgomery, and V.J. Marder, Propolypeptide of von Willebrand factor circulates in blood and is identical to von Willebrand factor antigen II, <u>Science</u> 232:995 (1986).

29. D.D. Wagner, P.J. Fay, L.A. Sporn, S. Sinha, S.O. Lawrence, and V.J. Marder, Divergent fates of von Willebrand factor and its propolypeptide (von Willebrand antigen II) after secretion from endothelial cells, <u>Proc. Natl. Acad. Sci. USA</u> 84:1955 (1987).

30. C. Loesberg, M.D. Gonsalves, J. Zandbergen, C. Willems, W.G. van Aken, H.V. Stel, J.A. van Mourik, and P.G. deGroot, The effect of calcium on the secretion of factor VIII-related antigen by cultured human endothelial cells, <u>Biochim. Biophys. Acta</u>. 763:160 (1983).

31. L.A. Sporn, V.J. Marder, and D.D. Wagner, Inducible secretion of large biologically potent von Willebrand factor multimers, <u>Cell</u> 46:185 (1986).

32.   P.K. Nguyen and P.L. Bockensted, Induction of von Willebrand factor synthesis in human umbilical vein endothelial cells by sodium butyrate, Blood 76;431a (1990).

33.   H.M. Tsai, R.L. Nagel, V.B. Hatcher, and I.I. Sussman, Multimeric composition of endothelial cell-derived von Willebrand factor, Blood 73:2074 (1989).

34.   C.R.M. Prentice, C.D. Forbes, and S.M. Smith, Rise of factor VIII after exercise and adrenalin infusion measured by immunological and biological techniques, Thromb. Res. 1:493 (1972).

35.   Z.M. Ruggeri, P.M. Mannucci, R. Lombardi, A.B. Federici, and T.S. Zimmerman, Multimeric composition of factor VIII/von Willebrand factor following administration of DDAVP: Implications for pathophysiology and therapy of von Willebrand's disease subtypes, Blood 59:1272 (1982).

36.   P.M. Mannucci, 1986, Desmopressin (DDAVP) for treatment of disorders of hemostasis, in: "Progress in Hemostasis and Thrombosis", Vol. 8, B.S. Coller, ed., Grune & Stratton, New York.

37.   J.D. Levine, J.M. Harlan, L.A. Harker, M.L. Joseph, and R.B. Counts, Thrombin-mediated release of factor VIII antigen from human umbilical vein endothelial cells in culture, Blood 60:531 (1982).

38.   P.G. deGroot, M.D. Gonsalves, C. Loesberg, M.R. van Buul-Wortelboer, W.G. van Aken, and J.A. van Mourik, Thrombin-induced release of von Willebrand factor from endothelial cells is mediated by phospholipid methylation, J. Biol. Chem. 259:13329 (1984).

39.   K.K. Hamilton and P.J. Sims, Changes in cytosolic Ca^{2+} associated with von Willebrand factor release in human endothelial cells exposed to histamine. J. Clin. Invest. 79:600 (1987).

40.   T.A. Brock, H.F. Dvorak, and D.R. Senger, Tumor-secreted vascular permeability factor increases cytosolic Ca^{2+} and von Willebrand factor release in human endothelial cells, Am. J. Pathol. 138:213 (1991).

41.   J.A. Ribes, C.W. Francis, and D.D. Wagner, Fibrin induces release of von Willebrand factor from endothelial cells, J. Clin. Invest. 79: 117 (1987).

42.   J.A. Ribes, F. Ni, D.D. Wagner, and C.W. Francis, Mediation of fibrin-induced release of von Willebrand factor from cultured endothelial cells by the fibrin beta chain. J. Clin. Invest. 84:435 (1989).

43.   R. Hattori, K.K. Hamilton, R.P. McEver, and P.J. Sims, Complement proteins C5b-9 induce secretion of high molecular weight multimers of endothelial von Willebrand factor and translocation of granule membrane protein GMP-140 to the cell surface, J. Biol. Chem. 264:9053 (1989).

44.   B.J. Awbrey, J.C. Hoak, and W.G. Owen, Binding of human thrombin to cultured human endothelial cells, J. Biol. Chem. 254:4092 (1979).

45.   P. Lollar and W.B. Owen, Evidence that the effects of thrombin on arachidonate metabolism in cultured human endothelial cells are not mediated by a high affinity receptor, J. Biol. Chem. 255:8031 (1980).

46.   D. Rotrosen and J.I. Gallin, Histamine type I receptor occupancy increases endothelial cytosolic calcium, reduces F-actin, and promotes albumin diffusion across cultured endothelial monolayers, J. Cell Biol. 103:2379 (1986).

47.   F.M Booyse, A.J. Quarfoot, J. Chediak, M.B. Stemerman, and T. Maciag, Characterization

and properties of cultured human von Willebrand umbilical vein endothelial cells, <u>Blood</u> 58:788 (1981).

48.   E.G.D. Tuddenham, J. Lazarchich, and L.W. Hoyer, Synthesis and release of factor VIII by cultured human endothelial cells, <u>Brit. J. Haematol</u>. 47:617 (1981).

49.   J.H. Reinders, R. C. Vervoorn, C.L. Verweij, J.A. van Mourik, and P.G. deGroot, Perturbation of cultured human vascular endothelial cells by phorbol ester or thrombin alters the cellular von Willebrand factor distribution, <u>J. Cell Physiol</u>. 133:79 (1987).

50.   T. Mayadas, D.D. Wagner, and P.J. Simpson, von Willebrand factor biosynthesis and partititoning between constitutive and regulated pathways of secretion after thrombin stimulation, <u>Blood</u> 73:706 (1989).

51.   E.R. Weibel and G.E. Palade, New cytoplasmic components in arterial endothelia, <u>J. Cell Biol</u>. 23:101 (1964).

52.   A.A. Trillo and R.W. Prichard, Early endothelial changes in experimental primate atherosclerosis, <u>Lab. Invest</u>. 41:294 (1979).

53.   P. Kumar, A. Erroi, A. Sattar, and S. Kumar, Weibel-Palade bodies as a marker for neovascularization induced by tumor and rheumatoid angiogenesis factors, <u>Cancer Res</u>. 45:4339 (1985).

54.   A. Sengel and P. Stoebner, Golgi origin of tubular inclusions in endothelial cells, <u>J. Cell Biol</u>. 44:223 (1970).

55.   D.D. Wagner, J.B. Olmsted, and V.J. Marder, Immunolocalization of von Willebrand protein in Weibel-Palade bodies of human endothelial cells, <u>J. Cell Biol</u>. 95:355 (1982).

56.   M. Hormia, V.P. Lehto, and I. Virtanen, Intracellular localization of factor VIII-related antigen and fibronectin in cultured human endothelial cells: Evidence for divergent routes of intracellular translocation, <u>Eur. J. Cell Biol</u>. 33:217 (1984).

57.   M.J. Warhol and J.M. Sweet, The ultrastructural localization of von Willebrand factor in endothelial cells, <u>Am. J. Pathol</u>. 117:310 (1984).

58.   H. Kagawa, S. Fujimoto, H. Ueda, and K. Hamasaki, Immunocytochemical localization of factor VIII-related antigen in human umbilical vein, <u>J. VOEH</u> 7:365 (1985).

59.   J.H. Reinders, P.G. de Groot, M.D. Gonsalves, J. Zandbergen, C. Loesberg, and J.A. van Mourik, Isolation of a storage and secretory organelle containing von Willebrand protein from cultured human endothelial cells, <u>Biochim. Biophys. Acta</u> 804:361 (1984).

60.   J.H. Reinders, P.G. de Groot, J. Dawes, N.R. Hunter, H.A.A. van Heugten, J. Zandbergen, M.D. Gonsalves, and J.A. van Mourik, Composition of secretion and subcellular localization of von Willebrand protein with that of thrombospondin and fibronectin in cultured human vascular endothelial cells, <u>Biochim. Biophys. Acta</u> 884:306 (1985).

61.   J. Loscalzo, M. Fisch, and R.I. Handin, Solution studies of the quaternary structure and assembly of human von Willebrand factor, <u>Biochemistry</u> 24:4468 (1985).

62.   R.P. McEver and M.N. Martin, A monoclonal antibody to a membrane glycoprotein binds only to activated platelets, <u>J. Biol. Chem</u>. 259:9799 (1984).

63.   S. Hsu-Lin, C.L. Berman, B.C. Furie, D. August, B. Furie, A platelet membrane protein expressed during platelet activation and secretion. Studies using a monoclonal antibody specific for thrombin-activated platelets, <u>J. Biol. Chem</u>. 259:9121 (1984).

64.   R.P. McEver, J.H. Beckstead, K.L. Moore, L. Marshall-Carlson, and D.F. Bainton, GMP-140, a platelet alpha-granule membrane protein, is also synthesized by vascular endotheial cells

and is localized in Weibel-Palade bodies, J. Clin. Invest. 84:92 (1989).

65.   R. Bonfanti, B.C. Furie, B. Furie, and D.D. Wagner, PADGEM (GMP-140) is a component of Weibel-Palade bodies of human endothelial cells, Blood 73:1109 (1989).

66.   G.I. Johnston, R.G. Cook, and R.P. McEver, Cloning of GMP-140, a granule membrane protein of platelets and endothelium: sequence similarity to proteins involved in cell adhesion and inflammation, Cell 56:1033 (1989).

67.   M.P. Bevilacqua, S. Stengelin, M.A. Gimbrone, Jr, and B. Seed, Endothelial leukocyte adhesion molecule 1: An inducible receptor for neutrophils related to complement regulatory proteins and lectins, Science 243:1160 (1989).

68.   L.A. Lasky, M.S. Singer, T.A. Yednock, D. Dowbenko, C. Fennie, H. Rodriguez, T. Nguyen, S. Stachel, and S.D. Rosen, Cloning of a lymphocyte homing receptor reveals a lectin domain, Cell 56:1045 (1989).

69.   J.M. McNiff and J. Gil, Secretion of Weibel-Palade bodies observed in extra-alveolar vessels of rabbit lung, J. Appl. Physiol. 54:1284 (1983).

70.   R. Hattori, K.K. Hamilton, R.D. Fugate, R.P. McEver, and P.J. Sims, Stimulated secretion of endothelial von Willebrand factor is accompanied by rapid redistribution to the cell surface of the intracellular granule membrane protein GMP-140, J. Biol. Chem. 264:7768 (1989).

71.   L.A. Fitzgerald, I.F. Charo, and D.R. Phillips, Human and bovine endothelial cells synthesize membrane proteins similar to human platelet glycoproteins IIb and IIIa, J. Biol. Chem. 260:10893 (1985).

72.   L.A. Sporn, V.J. Marder, and D.D. Wagner, Differing polarity of the constitutive and regulated secretory pathways for von Willebrand factor in endothelial cells, J. Cell. Biol. 108:1283 (1989).

73.   S.H. Tannenbaum, M.E. Rick, B. Shafer, and H.R. Gralnick, Subendothelial matrix of cultured endothelial cells contains fully processed high molecular weight von Willebrand factor, J. Lab. Clin. Med. 113:372 (1989).

74.   M.J. Berridge, Inositol trisphosphate and diacylglycerol as second messengers, Biochem. J. 220:345 (1984).

75.   E.A. Jaffe, J. Grulich, B.B. Weksler, G. Hampel, and K. Watanabe, Correlation between thrombin-induced prostacylin production and inositol trisphosphate and cytosolic free calcium levels in cultured human endothelial cells, J. Biol. Chem. 262:8557 (1987).

76.   T.A. Brock and E.A. Capasso, Thrombin and histamine activate phospholipase C in human endothelial cells via a phorbol ester-sensitive pathway, J. Cell Physiol. 136:54 (1988).

77.   D.M. Stern, P.P. Nawroth, W. Kisiel, D. Handley, M. Drillings, and J. Bartos, A cogulation pathway on bovine aortic segments leading to the generation of Factor Xa and thrombin, J. Clin. Invest. 74:1910 (1984).

78.   D.M. Stern, P.P. Nawroth, D. Handley, and W. Kisiel, An endothelial cell-dependent pathway of coagulation, Proc. Natl. Acad. Sci. USA 82:2523 (1985).

79.   T.J. Hallam, R. Jacob, and J.E. Merritt, Evidence that agonists stimulate bivalent-cation influx into human endothelial cells, Biochem. J. 255:179 (1988).

80.   K.A. Birch, G.B. Zavoico, J.S. Pober, and B.M. Ewenstein, A direct role of calcium and a regulatory role of protein kinase C (PKC) in thrombin-stimulated von Willebrand factor (vWF) secretion by human umbilical vein endothelial cells (EC), FASEB J. 4:481A (1990).

81.   K.E. Kamm and J.T. Stull, The function of myosin and myosin light chain kinase

phosphorylation in smooth muscle, Ann. Rev. Pharmacol. Toxicol. 25:593 (1985).

82. M. Inagaki, S. Kawamoto, and H. Hidaka, Serotonin secretion from human platelets may be modified by Ca^{2+}-activated phospholipid-dependent myosin phosphorylation, J. Biol. Chem. 259:14321 (1984).

83. M. Saitoh, M. Naka, and H. Hidaka, The modulatory role of myosin light chain phosphorylation in human platelet activation, Biochem. Biophys. Res. Commun. 140:280 (1986).

84. R.B. Wysolmerski and D. Lagunoff, Involvement of myosin light-chain kinase in endothelial cell retraction, Proc. Natl. Acad. Sci. USA 87:16 (1990).

85. J.J. Lynch, T.J. Ferro, F.A. Blumenstock, A.M. Brockenauer, and A.B. Malik, Increased endothelial albumin permeability mediated by protein kinase C activation, J. Clin. Invest. 85:1991 (1990).

86. J.E. Niedel and P.J. Blackshear, 1986, Protein kinase C, in: "Phosphoinositides and Receptor Mechanisms," J.W. Putney, ed., Alan R. Liss.

87. B.C. Jacobson, J.S. Pober, J.W. Fenton, and B.M. Ewenstein, Thrombin and Histamine rapidly stimulate the phosphorylation of the Myristoylated alanine-rich C kinase substrate (MARCKS) in human umbilical vein endothelial cells: Evidence for distinct patterns of protein kinase activation, submitted.

88. S. Sinha and D.D. Wagner, Intact microtubules are necessary for complete processing, storage and regulated secretion of von Willebrand factor by endothelial cells, Eur. J. Cell Biol. 43:377 (1987).

89. D. Aunis and M.-F. Bader, The cytoskeleton as a barrier to exocytosis in secretory cells, J. Exp. Biol. 139:253 (1988).

90. A.B. Federici, R. Bader, S. Pagani, M.L. Colibretti, L. De Marco, and P.M. Mannucci, Binding of von Willebrand factor to glycoproteins Ib and IIb/IIIa complex: affinity is related to multimeric size, Brit. J. Haematol. 73:93 (1989).

91. L.A. Sporn, V.J. Marder, and D.D. Wagner, von Willebrand factor released from Weibel-Palade bodies binds more avidly to extracellular matrix than that secreted constitutively, Blood 69:1531 (1987).

92. J.L. Moake, N.A. Turner, N.A. Stathopoulos, L.H. Nolasco, and J.D. Hellums, Involvement of large plasma von Willebrand factor multimers and unusually large von Willebrand factor multimers and unusually large von Willebrand factor forms derived from endothelial cells in shear-stress induced platelet aggregation, J. Clin. Invest. 78:1456 (1986).

93. J.-G. Geng, M.P. Bevilacqua, K.L. Moore, T.M. McIntyre, S.M. Prescott, J.M. Kim, G.A. Bliss, G.A. Zimmerman, and R.P. McEver, Rapid neutrophil adhesion to activated endothelium mediated by GMP-140, Nature 343:757 (1990).

94. J.R. Gamble, M.P. Skinner, M.C. Berndt, and M.A. Vadas, Prevention of activated neutrophil adhesion to endothelium by soluble adhesion proteins GMP140, Science 249:414 (1990).

95. K.L. Moore, A. Varki, and R.P. McEver, GMP-140 binds to a glycoprotein receptor on human neutrophils: Evidence for a lectin-like interaction, J. Cell. Biol. 112:491 (1991).

96. L. Osborn, Leukocyte adhesion to endothelium in inflammation, Cell 62:3 (1990).

97. G.A. Zimmerman, T.M. McIntyre, M. Mehra, and S.M. Prescott, Endothelial cell-associated platelet-activating factor: a novel mechanism for signaling intercellular adhesion, J. Cell Biol. 110:529 (1990).

98.   M.A. Boogaerts, G. Vercelotti, C. Roelant, S. Malbrain, R.L. Verwilghen, and H.S. Jacob, Platelets augment granulocyte aggregation and cytotoxicity: Uncovering of their effects by improved cell separation techniques using Percoll gradients, Scand. J. Haematol. 37:229 (1986).

99.   A.J. Marcus, M.J. Broekman, L.B. Safier, H.L. Ullman, N. Islam, C.N. Serhan, L.E. Rutherford, H.M. Korchak, and G. Weissmann, Formation of leukotrienes and other hydroxy acids during platelet-neutrophil interactions in vitro, Biochem. Biophys. Res. Commun. 109:130 (1982).

100.   J.A. Maclouf and R.C. Murphy, Transcellular metabolism of neutrophil-derived leukotriene A_4 by human platelets. A potential cellular source of leukotriene C_4, J. Biol. Chem. 263:174 (1988).

101.   E. Larsen, A. Celi, G.E. Gilbert, B.C. Furie, J.K. Erban, R. Bonfanti, D.D. Wagner, and B. Furie, PADGEM protein: A receptor that mediates the interaction of activated platelets with neutrophils and monocytes, Cell 59:305 (1989).

102.   S.A. Hamburger and R.P. McEver, GMP-140 mediates adhesion of stimulated platelets to neutrophils, Blood 75:550 (1990).

103.   K.D. Patel, G.A. Zimmerman, S.M. Prescott, R.P. McEver, and T.M. McIntyre, Oxygen radicals induce human endothelial cells to express GMP-140 and bind neutrophils, J. Cell Biol. 112:749 (1991).

104.   R.P. Hebbel, M.A.B. Boogaerts, J.W. Eaton, and M.H. Steinberg, Erythrocyte adherence to endothelium in sickle-cell anemia: A possible determinant of disease severity, N. Engl. J. Med. 302:992 (1980).

105.   N. Mohandas and E. Evans, Sickle erythrocyte adherence to vascular endothelium: Morphologic correlates and the requirement for divalent cations and collagen-binding proteins, J. Clin. Invest. 76:1605 (1985).

106.   G.A. Barabino, L.V. McIntire, S.G. Eskin, D.A. Sears, and M. Udden, Endothelial cell interactions with sickle cell, sickle trait, mechanically injured, and normal erythrocytes under controlled flow, Blood 70:152 (1987).

107.   D.K. Kaul, M.E. Fabry, and R.L. Nagel, Microvascular sites and characteristics of sickle cell adhesion to vascular endothelium in shear flow conditions: Pathophysiological implications. Proc. Natl. Acad. Sci. USA 86:3356 (1989).

108.   T.M. Wick, J.L. Moake, M.M. Udden, S.G. Eskin, D.A. Sears, and L.V. McIntire, Unusually large von Willebrand factor multimers increase adhesion of sickle erythrocytes to human endothelial cells under controlled flow, J. Clin. Invest. 80:905 (1987).

109.   H.M. Tsai, I.I. Sussman, R.L. Nagel, and D.K. Kaul, Desmopressin induces adhesion of normal human erythrocytes to the endothelial surfaces of a perfused microvascular preparation, Blood 75:251 (1990).

110.   T.M. Wick, J.L. Moake, M.M. Udden, and L.V. McIntire, Unusually large (UL) vWF multimers bind to GPIb-like and integrin receptors on sickle and young non-sickle RBC and on endothelial cells (EC): A mechanism for sickle and other young RBC adhesion to EC, Blood 72:76a (1988).

ROLE OF EICOSANOIDS AND THE CYTOKINE NETWORK IN TRANSMEMBRANE

SIGNALING IN VASCULAR CELLS

Kenneth B. Pomerantz* and David P. Hajjar

Departments of Medicine*, Pathology, and Biochemistry and the
National Institutes of Health Specialized Center of Research in
Thrombosis, Cornell University Medical College, New York, NY

INTRODUCTION

The concept that atherosclerosis is an inflammatory response to injury is based on observations that cells of monocytic origin, including macrophages and T-cells populate the developing atherosclerotic lesion. This is in addition to the well-characterized smooth muscle cell infiltrate [1]. Since endothelial cell activation or injury is a prerequisite for monocyte adhesion and diapedesis, the histologic evidence described above support the hypothesis that endothelial cell injury is an important initial event in the development of the inflammatory lesion. Activation or injury may thus initiate a series of process that lead to intimal hyperplasia and cholesteryl ester deposition within the macrophage and smooth muscle cell.

Under normal conditions, the vascular endothelium promotes thromboresistance, maintains vascular tone through the secretion of: a) vasoconstrictors (including endothelin, endothelial cell contracting factors 1 and 2, and angiotensin II), and b) vasodilators (prostacyclin [PGI_2], endothelial cell-derived relaxing factor [EDRF], and endothelial cell hyperpolarizing factor [EDHF]). These factors maintain arterial smooth muscle cells in the contractile phenotype, and influence smooth muscle cell proliferation by the constituitive secretion of heparin. In addition, endothelial cells normally restrict permeability of large macromolecules (including plasma low density lipoproteins [LDL]), preventing local cholesterol accumulation.

In response to injury or activation, endothelial cells become increasingly permeable to LDL, have higher replicative rates, develop pro-coagulant properties by increasing the synthesis and/or expression of thrombogenic vs thrombolytic proteins, and express surface glycoproteins which promote the adhesion of neutrophils, monocytes, and platelets. Activated endothelium may also promote the transition of smooth muscle cells from the contractile to synthetic phenotype (promoting smooth muscle cell proliferation) and alter lipoprotein and cholesterol metabolism (resulting in net cholesterol accretion). These findings are supported by the observation that cholesteryl esters accumulated to a greater extent under re-endothelialized than de-endothelialized areas of denuded rabbit aorta in vivo [2,3].

Cell-Cell Interactions in the Release of Inflammatory Mediators
Edited by P. Y-K Wong and C.N. Serhan, Plenum Press, New York, 1991

The above observations suggest that humoral mediators derived from injured endothelial cells or inflammatory cells interacting with other arterial cells may promote macrophage insudation, smooth muscle cell proliferation, and subsequent connective tissue and lipid deposition [4]. Principal mediators implicated in the mediation of these events collectively include growth factors and cytokines. Importantly, these agents also stimulate the release of numerous secondary messengers, including eicosanoids. The eicosanoid profile of cells comprising the vascular wall constitute a wide array of products, which have functionally diverse, and, commonly, antagonistic biological properties toward their agonists. Thus, in addition to transducing membrane associated events, eicosanoids also have important paracrine activity that is modulated by other biological response modifiers. The purpose of this review is to summarize, and focus on the role of eicosanoids in mediating or modulating cellular proliferation and cholesterol trafficking in an inflammatory milieu, such as the athersclerotic vessel wall. While other membrane signaling processes, such as G-protein stimulus-coupling is of significant importance in signal tranduction, this is an extensive area that is adaquately covered elsewhere.

INTRACELLULAR REGULATION OF ARTERIAL CHOLESTEROL METABOLISM

Cellular cholesterol content is dependent upon the activity of the processes of <u>cholesterol delivery</u> (by cholesterol donors such as LDL, β-VLDL, and modified LDL), <u>metabolism</u> (lysosomal and cytoplasmic cholesteryl ester hydrolysis, cholesterol esterification, and trafficking to the cell membrane), and <u>efflux</u> (by plasma cholesterol acceptors, including high density lipoprotein (HDL), by its receptor). Delivery of cholesterol to vascular cells is mediated by the LDL (apo B) receptor, or by a receptor recognizing modified forms of LDL, called the scavenger receptor. Intracellular cholesteryl ester processing is mediated by enzymes which promote cholesteryl ester hydrolysis in the lysosome and in the cytoplasm by cholesteryl esterases, esterification by ACAT, and synthesis via HMG-CoA reductase. Clearly, the regulation of these processes will determine the cholesterol and cholesteryl ester content of smooth muscle cells and macrophages, while dysregulation of these processes may result in lipid accumulation. LDL receptor activity is regulated by cellular cholesterol levels, acting through sterol-mediated down-regulation of genes whose products control the processing of LDL-cholesteryl esters (the LDL receptor) and <u>de novo</u> cholesterol synthesis (HMG-CoA reductase) [5]. Since LDL receptor activity varies inversely with cellular cholesterol levels [6], LDL receptor-independent pathways of cholesterol uptake assume relative importance in cholesterol accumulation and eventual foam cell development. Macrophages accumulate cholesteryl esters by a scavenger receptor, whose gene has recently been cloned [7], and whose expression is not down-regulated by cholesterol [6]. Putative physiological ligands for the scavenger receptor include oxidized LDL, LDL-heparin-fibronectin aggregates, and LDL complexes.

Conversely, cholesterol efflux from smooth muscle cells and macrophages is affected by plasma cholesterol acceptor proteins, the most efficacious of which is HDL [8]. HDL promotes net cholesterol efflux from vascular smooth muscle cells, macrophages, and endothelium by interaction with specific receptors. Once cholesterol is incorporated into HDL, it is esterified to form core cholesteryl esters by association with the plasma enzyme lecithin cholesteryl acyl transferase (LCAT). Furthermore HDL cholesteryl esters exchange with LDL cholesteryl esters by facilitated transfer via the plasma cholesteryl ester transfer protein.

160

Thus, the ultimate control of vascular cholesterol metabolism relies on the modulation of cholesterol influx and egress, which may ultimately be regulated in turn by mediators derived from endothelial and/or inflammatory cells and plasma that directly interact with endothelium to alter lipoprotein permeability and vascular cholesterol metabolism.

EVIDENCE IMPLICATING GROWTH FACTORS AND CYTOKINES IN ATHEROGENESIS

There is accumulating evidence that growth factors and cytokines derived from endothelial cells modulate smooth muscle and macrophage function.

Growth factor and cytokine synthesis by vascular cells. Growth factors, including platelet-derived growth factor (PDGF), acidic and basic fibroblast growth factor (FGF), and transforming growth factor-beta (TGF-β) are synthesized by vascular cells. Interestingly, the expression of biological activity of these growth factors are quite diverse. PDGF is secreted in biologically active form by endothelial cells, and is inducible by endotoxin, thrombin, and TGF-β. However, FGF is not secreted, but is localized either intracellularly, or bound to extracellular matrix. Therefore, expression of FGF activity requires either cell injury or exposure of the extracellular matrix to heparin or heparinases. Finally, TGF-β is secreted into the extracellular milieu in a latent form, and must be activated (by plasmin or cathepsin D). In this regard, endothelial and smooth muscle cells in co-culture produce activated TGF-β while endothelial and smooth muscle cells cultured alone produce only the latent form.

Cytokines, including interleukin-I (IL-1), tumor necrosis factor (TNF), and granulocyte/macrophage colony stimulating factor (GM-CSF), are synthesized by inflammatory cells and activated arterial endothelial and smooth muscle cells. Cytokine secretion is autoinducible, although cytokine secretion may be induced by other mechanisms, such as by exposure to endotoxin and other activators.

Biological Effects of Growth Factors and Cytokines on Vascular Cells

Growth factors. All known growth factors stimulate smooth muscle cell proliferation (Table 1). In this regard, the most-studied growth factor is PDGF, which stimulates smooth muscle cell proliferation [9], migration [10], chemotaxis, [11] and chemokinesis [12]. The importance of this growth factor as a mediator in smooth muscle cell hyperplasia is evidenced by the fact that there is increased production of PDGF by proliferative smooth muscle cells after vascular injury [13]. In addition, PDGF messenger RNA (mRNA) transcripts are increased in atheromatous [14] but not normal regions of aorta [15]. Smooth muscle cells derived from atheromatous lesions are also more proliferative [16]. Interestingly, macrophages newly recruited into atherosclerotic lesions possess significant amounts of PDGF-B chain protein and mRNA, which is reduced after the macrophages have subsequently accumulated lipid [17].

In addition to PDGF, other growth factors synthesized by vascular cells mediate smooth muscle cell proliferation, but usually by indirect means. FGF promotes proliferation in endothelial and smooth muscle cells [18] through the induction of PDGF-A chain, and early response genes, such as c-myc, c-jun, and c-fos [19]. TGF-β, on the other hand, has been shown to inhibit, and stimulate smooth muscle cell proliferation. The mechanisms of these divergent effects of TGF-β on smooth muscle cell proliferation are multifactorial, and depend on the following: a) cell-density. TGF-β inhibits sparse cultures of smooth muscle cells, but stimulates proliferation in confluent cultures of smooth muscle cells; and b) dose - TGF-β augments proliferation at low concentrations by inducing PDGF release, but causes inhibition at higher concentrations by down-regulation of the PDGF receptor. Finally, it has been recently suggested that the inhibitory effects of TGF-β is augmented by heparin.

Growth factors are important in the progression of intimal hyperplasia following arterial cell injury. Thus, novel methods of prevention of intimal hyperplasia should be considered in the future as important therapeutic modalities in both the slowing of the progression of intimal hyperplasia, in atherosclerosis, and for the prevention of intimal hyperplasia following saphenous vein coronary artery transplantation.

These growth factors have also been implicated in the regulation of cholesterol metabolism. PDGF [20] and FGF [21] promote the binding and uptake of LDL in smooth muscle cells. In fibroblasts, PDGF increases LDL receptor gene transcription [22], increases fluid phase endocytosis in the absence of a storage pool of cellular cholesterol [23], and renders the cell dependent on cholesterol synthesis to affect the proliferative response to PDGF [24]. PDGF can stimulate cholesteryl ester hydrolase activity [25]; and TGF-β activates LDL receptor activity and the cholesteryl ester cycle in arterial smooth muscle cells [26]. The stimulation of LDL receptor activity by growth factors may be due to the requisite need for cholesterol to provide membrane biogenesis accompanying the proliferative response to these agents. Presently, there is no information on the effects of growth factors on the activity or expression of the scavenger receptor whose role in proliferation is unclear, but whose role in the accumulation of intracellular cholesteryl ester by the macrophage is considered to be of major importance.

Table 1. Effect of growth factors on smooth muscle cell proliferation and the cholesteryl ester cycle.

Agents	Proliferation	LDL$_R$[a]	aLDL$_R$[b]	HMG CoA-Reductase	CEH[c]	ACAT
PDGF	I	I	N/A	I	I	N/A
FGF	I	I	N/A	N/A	N/A	N/A
TGF-β	V	I	N/A	N/A	D	N/A

[a] - LDL receptor, [b] - acetylated LDL (scavenger) receptor, [c] - cholesteryl ester hydrolase, I - increased effect, D - decreased effect, V - biphasic effect, with increased, and decreased effects being observed, N/A - not available.

Cytokines. The fact that cytokines may modulate the atherogenic response to injury is based on a number of in vivo studies. Firstly, TNF, GM-CSF, and IL-2 lower serum cholesterol in humans [27]. Secondly, gamma-interferon inhibits atherogenesis in rabbits [28]. The mechanism providing cytokines with anti-atherogenic effects is unclear, but it has been postulated that these cytokines may promote cholesterol deposition and clearance by the spleen and other reticulo-endothelial cell beds, thus depleting the plasma of LDL-derived cholesteryl ester, thereby reducing net cholesterol load to the arterial vasculature.

Accordingly, there have been numerous _in vitro_ studies which evaluate the biological activities of the known cytokines, with special reference to endothelial, smooth muscle cell, and macrophage function. These agents exhibit a variety of effects, depending on the cell type, and their degree of activation. These data are summarized below:

On endothelial cells, cytokines generally promote a procoagulant surface, promote leukocyte-endothelial cell interactions, and have variable effects of endothelial cell proliferation. IL-1 promotes monocyte and neutrophil chemotaxis and diapedesis into the vessel wall [29], and TNF causes increased endothelial cell permeability to LDL [30] Gamma-interferon and GM-CSF act in a synergistic manner to induce procoagulant activity of endothelial cells [31], and promote lymphocyte adhesion to endothelial cells by induction of HLA-DR antigens [32]. However, the effects of cytokines on endothelial cell proliferation appear to depend on their activation state. On quiescent endothelial cells, IL-1 causes proliferation, and interferon and TNF may be cytotoxic, while on activated endothelium, interferon, TGF-β, and IL-1 uniformly inhibit endothelial cell proliferation [33].

On smooth muscle cells, IL-1 induces proliferation which is due to induction of PDGF-AA homodimer [34]. Interestingly, the effect of IL-1 on smooth muscle cell proliferation is antagonized by gamma-interferon, which inhibits smooth muscle cell proliferation [35]. The ability of cytokines to stimulate growth may be indirect; TNF and IL-1 have been shown to stimulate smooth muscle cell proliferation by increasing the number of FGF receptors [36]. Thus, growth factors may amplify their own response by stimulating their own production, and the production of secondary growth factors.

Unfortunately, there is surprisingly little information on the effects of cytokines on vascular cholesterol metabolism; studies of cytokines on cholesterol metabolism have been restricted to macrophages. Cytokines appear to down-regulate receptor-mediated uptake of LDL. In one study, products of activated lymphocytes decreased the activity of the LDL and scavenger receptors on monocytes [37]. In other studies, bacterial lipopolysacchride, whose biological effects appear to be mediated principally through the generation of TNF, have been shown to down-regulate the macrophage scavenger receptor [38]. Yet, significance of these findings is unclear, since decreased cellular clearance of native and modified LDL may provoke increased interstitial LDL levels, with commensurate increases in plasma LDL and net decreased intracellular net cholesteryl ester deposition.

In summary, the apparent anti-atherogenic effects of cytokines when evaluated in the context of whole animal studies appear to be paradoxical to the apparent inflammatory effects of these agents when evaluated in the context of _in vitro_ studies. Interpretation of these results suggests that the actions of cytokines on other areas, such as the reticulo-endothelial system, (liver, etc.) may be of greater biological importance in the clearance of LDL than the apparent inflammatory effects of these agents on arterial endothelium and subintimal cells.

EVIDENCE THAT EICOSANOIDS ARE INVOLVED IN ATHEROGENESIS

The prospect that eicosanoids may be involved in atherogenesis is based on numerous pathophysiologic data demonstrating that atherosclerosis is accompanied by altered eicosanoid metabolism. These data have been reviewed recently [39]. Briefly, most studies have demonstrated that diet-induced atherosclerosis is accompanied by a reduction in

PGI_2 synthesis [40,41]. The significance of these findings is suggested by the observation that carbacyclin (a stable PGI_2 analog) decreased arterial cholesterol content [42], implicating PGI_2 as an important regulator of vascular cholesterol content.

The most obvious aspect of the significance of reduced eicosanoid-synthetic capacity in atherosclerosis is the fact that PGI_2 is a major vasodilator and inhibitor of platelet aggregation, and loss of PGI_2-synthetic activity may predispose to enhanced vasoconstriction and platelet aggregation. However, the role of PGI_2 and other eicosanoids in cell-cell communication and modulation of the effects of growth factors and cytokines may now be of increased importance in the understanding of the pathogenesis of the atherosclerotic lesion. It now becomes relevant to discuss in detail the eicosanoid profile of the major cell type associated with the atherosclerotic lesion, and to discuss the effects of these eicosanoids on arterial cell function.

EICOSANOIDS AND ARTERIAL CELL FUNCTION

Eicosanoids are synthesized by all vascular tissue, and have potent autocrine and paracrine activity. Therefore, the net effect of eicosanoids on cell function will be determined by the type and distribution of cell types found in both the normal areas, and in areas associated with plaque. This section summarizes the major autocrine and paracrine effects of eicosanoids on the principal cell types found in normal and atherosclerotic vessels.

Endothelium. The effects of eicosanoids on endothelial cell function are summarized herein and in Table 2:

Eicosanoid biosynthesis. Endothelial cells normally secrete PGI_2 as its principal eicosanoid metabolite, with lesser amounts of $PGF_{2\alpha}$, PGE_2, and, to a lesser extent, hydroxy- [43] and epoxy-eicosatetraenoic acids [44]. Endothelial cells can convert neutrophil-derived leukotriene (LT) A_4 to LTC_4 [45], which can also stimulate endothelial cell PGI_2 synthesis [46]. Endothelial cells can also incorporate 12-HETE into cellular phospholipids [47] which can subsequently inhibit PGI_2 synthesis [48].

Proliferation. The direct effects of PGI_2 on endothelial cell proliferation are unknown. LTC_4, a product of macrophages, and endothelial cells as a result of cell-cell communication with neutrophils, stimulates angiogenesis by bovine endothelial cells [46], with PGE_2 and LTB_4 having lesser, but still significant effects [49]. Endothelial cells secrete less 15- and 12-HETE, but these HETEs, which may be synthesized by adjacent macrophages, may be greater importance in generating their biological effects. 15- and 12-HETE stimulate endothelial cell proliferation, which appears to be mediated by diacyl glycerol kinase inhibition.

Adhesion. Eicosanoids are apparently extremely important in mediating adhesion of inflammatory cells to endothelium. In general, cyclooxygenase products (the major eicosanoid metabolites generated by endothelial cells) inhibit Ia antigen expression on endothelial cells, while lipoxygenase inhibitors decrease leukocyte adherence to gamma-interferon treated endothelial cells [50]. It appears that lipoxygenase and epoxygenase products of arachidonic acid promote increase adhesiveness of endothelial cells for a variety of cell types. Thus, LTB_4, a product or neutrophils and macrophages, causes hyperadhesiveness of neutrophils for endothelial cells [51], and increases neutrophil migration through intact endothelial cell monolayers [52]. Lipoxygenase products mediate increase tumor cell adhesion to endothelial cells via the well characterized glycoprotein IIb/IIIa receptor system [53]. Lipoxygenase products also mediate arachidonic acid-induced neutrophil adhesion to endothelial cell monolayers [54]. Finally 14,15 epoxyeicosatetraenoic acid stimulates the adhesion of monocyte/macrophages to endothelial cells [55]. Thus, these data demonstrate that under conditions of cell-cell interaction, lipoxygenase-generated eicosanoids mediate neutrophil/macrophage adherence to endothelium, and may be of major importance in the sequence of macrophage diapedesis into the subendothelial space and subsequent alterations in macrophage function.

Table 2. Effect of Eicosanoids on Endothelial Cell Function.

Eicosanoid	Proliferation	Adhesion	Permeability
PGI_2	N/A[a]	D	0
PGE_2	I	D	0
$PGF_{2\alpha}$	I	D	0
LTC_4	I	N/A	D
LTB_4	I	I	D
HETES[b]	I	I	D
14,15-EET	N/A	I	N/A

[a]: Symbols are defined as follows: N/A = Not Available; I = increase; D = decrease, 0 = no effect.
[b]: Includes 5-, 12-, and 15-HETE.

Permeability. Lipoxygenase products also enhance endothelial cell permeability after thrombin-induced pulmonary embolism [56]. In an _in vitro_ study, increased endothelial cell permeability observed after calcium ionophore (A-23187) or phorbol acetate was not abolished by cyclooxygenase inhibitors, suggesting that prostaglandins do not mediate permeability changes due to these agonists. However, physiological agonists including histamine and thrombin, which are known to increase endothelial cell permeability, did not alter basal permeability in these cells [57]. These data imply that there is endothelial cell specificity in agonist induced alterations in permeability.

Smooth Muscle Cells. The effects of the various eicosanoids on smooth muscle cell phenotype, chemotaxis, proliferative capacity, and cholesterol metabolism have been reviewed recently [39], and are summarized below and in Table 3.

Eicosanoid biosynthesis. Arterial smooth muscle cells synthesize principally PGI_2, and PGE_2, with lesser amounts of $PGF_{2\alpha}$ [58]. These cells also synthesize 15- and 11-HETE, but to a much lesser extent [59]. Interestingly, smooth muscle cells can also convert neutrophil-derived LTA_4 to LTC_4 [60], suggesting that under conditions of acute endothelial desquamation, neutrophil/smooth muscle cell interactions may result in the synthesis of a potent vasoconstrictor. Smooth muscle cells synthesize quantitatively very little lipoxygenase products relative to cyclooxygenase products, the net effect of cyclooxygenase-generated eicosanoids predominate. However, under conditions or cell-cell interaction, as when occurs when neutrophils, macrophages, and platelets interact with smooth muscle cells, the biological effects of lipoxygenase-generated eicosanoids may predominate.

Phenotype. Unfortunately, there is little information on the role
of eicosanoids in the regulation of smooth muscle cell phenotype, but
has been a topic in a number of recent reviews [39,61]. In an early
study, it was shown that exogenously added PGE_1, which behaves similarly to PGI_2, stimulates the transition from contractile to synthetic
phenotype [62], although modulated arterial smooth muscle cells synthesized more PGI_2 than their contractile counterparts [63]. These data
suggest that PGI_2 may promote an inflammatory phenotype. More recently, it was reported that the leukotrienes, including LTC_4, LTD_4, and
LTE_4 increase the rate of transition from the contractile to synthetic
phenotype [64]. These data suggest although PGI_2 and related eicosanoids
may promote the transition from the contractile to synthetic phenotype, the increased synthesis of PGI_2 may serve to modulate the proliferative response to mitogenic factors.

Table 3. Effects of Eicosanoids on Smooth Muscle Cell Function.

Eicosanoid	Phenotype	Migration	Proliferation	CE Metabolism CEH[a]	ACAT[b]
PGI_2	I	N/A	V[c]	I	0
PGE_2	N/A	N/A	V	0	D
$PGF_{2\alpha}$	N/A	N/A	0	0	0
HETE	N/A	I	V	I	N/A
LTB_4	I	I	I	N/A	N/A
LTC_4	I	I	I	N/A	N/A

[a]: Acid and Neutral Cholesteryl esterases; [b] - Acyl CoA cholesterol
acyl transferase; [c] - Variable effects, with increased, decreased, or
no effect being observed, depending on conditions (see text).

Migration/chemotaxis. 12-HETE and 15-HETE, synthesized to a greater
extent by neutrophils and platelets, are potent chemotactic eicosanoids for smooth muscle cells; 5-HETE is not active in this regard [65]
LTB_4 (a neutrophil lipoxygenase product) is also chemotactic but not
chemokinetic for arterial smooth muscle cells [66]. These data support
the premise that under conditions of acute arterial injury when neutrophils and platelets can interact with smooth muscle cells, communication between these cells types via lipoxygenase-derived eicosanoids
serve to promote smooth muscle cell migration and chemotaxis.

Proliferation. PGI_2 inhibits proliferation of arterial smooth
muscle cells in vivo, presumably by inhibiting platelet activation
(with resultant release of PDGF and other mitogens) [67], and in vitro
by inhibiting DNA synthesis [68]. However, opposite results were observed by Thyberg et al, who reported that PGI_2 and PGE_2 stimulated
DNA synthesis in serum-free conditions, but inhibited DNA synthesis in
the presence of mitogens, such as PDGF [61], suggesting a biphasic role
for PGI_2 in the regulation of smooth muscle proliferation. PGE_1, more
so than PGE_2 inhibited proliferation induced by PDGF [69]. In in vitro
studies, LTB_4 stimulated arterial cell proliferation, suggesting that
LTB_4 is important in mediating the vascular inflammatory response to
injury [66]. The ability of LTB_4 to stimulate DNA synthesis in arterial
smooth muscle cells may be mediated through cyclooxygenase products,
since indomethacin and aspirin blocked LTB_4-induction of DNA synthesis
[70], or through induction of phospholipase activity as observed with

LTD_4 [71]. The effects of HETEs on smooth muscle cell proliferation are variable with either no effect [66] or inhibition [72] of smooth muscle cell proliferation being reported. These data support the concept that specific endothelial and smooth muscle cell-derived eicosanoids can antagonize the response to mitogens. Conversely, eicosanoids derived from inflammatory cells, especially leukotrienes, may promote smooth muscle cell proliferation. We interpret these findings to imply that a reduction of arterial eicosanoid synthesis relative to the synthesis of eicosanoids generated by inflammatory cells may predispose to increased smooth muscle cell proliferation.

Cholesterol Metabolism. PGI_2, some of its metabolites, and 12-HETE stimulated those enzymes responsible for the hydrolysis of lysosomal (LDL-derived) and cytoplasmic cholesteryl esters in intact arterial smooth muscle cells [73,74]. The importance in endothelial cell modulation of smooth muscle cell cholesteryl ester hydrolytic activities was established in our laboratory. In smooth muscle cells co-cultured with endothelial cells or with endothelial cell conditioned medium, the stimulation of lysosomal cholesteryl ester hydrolase activity was eicosanoid-dependent [25]. Furthermore, 12,20-diHETE (a product of transcellular metabolism of platelet-derived 12-HETE by neutrophils) also stimulated lysosomal cholesteryl esterase activity by similar mechanisms [75]. PGE_2, but not PGI_2, 6-keto-$PGF_{1\alpha}$, or 6-keto-PGE_1 inhibited ACAT activity [76] in smooth muscle cells. These data demonstrate that endothelial cell-derived eicosanoids may have significant paracrine activity to regulate cholesterol content in smooth muscle cells. Unfortunately, there is no information on the effects of eicosanoids on the regulation of the apoB (LDL) or scavenger receptors on smooth muscle cells.

Macrophages. Although macrophages are not normally present to a significant degree in the subintimal space, these cells do accumulate during and after endothelial cell injury, and assume major importance in the early generation of the fatty streak. Unfortunately, it has not been feasible to obtain subintimal macrophages in large numbers for studies of eicosanoid metabolism. Therefore, most studies have employed either peripheral blood monocytes, monocytes derived from either alveolar or peritoneal exudate, or immortalized human or murine cell lines of monocyte/macrophage origin. Thus, application of data derived from these sources to the subintimal macrophage can only be inferred.

Eicosanoid biosynthesis. The spectrum of arachidonic acid metabolites synthesized by macrophages is dependent upon numerous factors, including the species, tissue of origin, nature of the stimulus, the degree of activation, and the influence of calcium. **Peripheral blood monocytes** synthesize principally cyclooxygenase products in response to the calcium ionophore, A-23187 [77]. **Alveolar type II macrophages** synthesize PGE_2, with lesser amounts of thromboxane A_2, $PGF_{2\alpha}$, PGD_2, and HHT in response to lipopolysacchride; in response to A-23187, lipoxygenase products, including 5-HETE [77] and LTC_4 [78] were the preferred products, with little or no cyclooxygenase products being formed [79]. In a more definitive report, Rouzer et al demonstrated that murine alveolar macrophages synthesized principally leukotrienes C_4, D_4, and E_4, and other cyclooxygenase products as described above, while interstitial pulmonary macrophages synthesize principally LTC_4 [80]. **Human peritoneal macrophages** synthesize LTB_4 and 5-HETE, with lesser amounts of thromboxane B_2, 12-HETE and PGI_2, while murine peritoneal macrophages synthesize mainly TxB_2, 12-HETE [81], and other HETE's including 5-,8-, and 15-HETE in response to exogenous arachidonate [81]. In response to zymosan, murine peritoneal macrophages synthesize principally LTC_4 and lesser amounts of PGE_2 and PGI_2 [81]. The mechanisms resulting in altered eicosanoid biosynthesis as a function of differentiation may be due to changes in protein kinase C-sensitive pathways [82] or calcium flux. When calcium availability is relatively low, cyclooxygenase products are elaborated in preference to lipoxygenase products, whereas when calcium levels are relatively high, as during conditions of maximal stimulation, LTC_4 is the predominant eicosanoid produced, with lesser or even diminished prostaglandin production being observed [83]. Eicosanoid production by macrophages is also regulated by eicosanoids. Prostaglandin production is irreversibly inhibited by 12- and 15-hydroperoxyeicosatetrenoic acids, but reversibly inhibited by the corresponding 12- and 15-HETE's [84], and PGE_2 behaves as a negative regulator of macrophage eicosanoid synthesis at the level of phospholipase A_2 [85].

In summary, these data suggest that the uncommitted monocyte synthesizes principally cyclooxygenase products, but, as a result of tissue-specific differentiation, synthesize mostly lipoxygenase products. Since lipoxygenase products are clearly pro-inflammatory, the differentiated macrophage is not capable of modulating the inflammatory response to injury, and is now primed to mediate stimulating processes leading to smooth muscle cell hyperplasia and net cholesteryl ester deposition. It remains to be determined what factor(s) causes the cyclooxygenase-to-lipoxygenase transition.

Phenotype. Endogenously synthesized eicosanoids modulate Ia expression on macrophages [86]. 5,8,11,14 - Eicosatetraynoic acid (ETYA), a competitive inhibitor of cyclooxygenase and lipoxygenase, induces the differentiation of U937 monocyte macrophages, suggesting that endogenously synthesized eicosanoids inhibit the differentiation of this monocytic cell line [87]. The mechanisms by which cyclooxygenase and lipoxygenase products modulate macrophage phenotype are currently unknown.

Cholesterol metabolism. PGE_2, a major cyclooxygenase product generated by macrophages, inhibits cholesteryl ester accumulation, presumably by inhibiting ACAT activity [88]. Importantly, PGI_2 and PGE_1 reduced LDL receptor activity and cholesterol synthesis in leukocytes [89], and implies that these eicosanoids restrict LDL entry into these cells. Although PGI_2 is synthesized by macrophages, it is not a major macrophage cyclooxygenase product. Thus, in communication with endothelial cells, as in an early response to injury, endothelial cell-derived PGI_2 may be of significant importance in the modulation of cholesteryl ester accumulation. However, this concept has yet to be developed experimentally. In contrast to cyclooxygenase products, lipoxygenase inhibitors will prevent cholesterol accumulation in macrophages exposed to acetylated LDL, suggesting that macrophage HETE or leukotriene production may promote cholesterol deposition [90].

In summary, uncommitted monocytes synthesize principally cyclooxygenase products, whose biological activities antagonize the effects of vasoconstrictors, promote smooth muscle cell contractile phenotype, inhibit the cell proliferation, and foster net cholesteryl ester hydrolysis and the potential for net cholesterol efflux. However, committed or differentiated macrophages synthesize principally pro-inflammatory lipoxygenase products, including leukotrienes and 5-HETE, whose properties in aggregate increase permeability, promote inflammatory cell chemotaxis, favor cellular proliferation and the potential for net cholesteryl ester retention. These results remain to be corroborated with macrophages derived from atherosclerotic lesions.

Growth Factors. The fact that soluble mediators could modify eicosanoid metabolism was first implied by the observation that exposure of
fibroblasts to macrophage products produced a persistent increase in
PGE_2 production [91]. Later studies with purified materials demonstrated
that growth factors had a multiplicity of effects on eicosanoid generation. It was first demonstrated that PDGF stimulated eicosanoid
production by activating phospholipase A_2 [92]. Later studies also
proved that PDGF promoted eicosanoid generation by stimulation of
transcription and translation of cyclooxygenase and PGI_2 synthase
genes [93]. PDGF is synergistic with serotonin in stimulating smooth
muscle cell PGI_2 synthesis [94]. EGF also stimulates smooth muscle cell
PGI_2 generation [95]. However, PDGF does not stimulate endothelial cell
eicosanoid production [96], presumably because large vessel endothelial
cells do not possess receptors for PDGF. However, the effect of FGF on
endothelial cell eicosanoid biosynthesis is apparently different from
the effects of other growth factors, in that FGF was shown to inhibit
[97] endothelial cell eicosanoid production by reducing the cellular
content of cyclooxygenase and PGI_2 synthase.

Numerous studies have now demonstrated that not only do growth
factors augment eicosanoid biosynthesis, but that eicosanoids generated in response to growth factors may mediate or modulate the cellular
responses to growth factor stimulation. Firstly, eicosanoids appear to
antagonize the mitogenic effects of growth factors, including PDGF [61].
Prostaglandins were demonstrated to be negative regulators in TNF and
EGF-induced fibroblast proliferation [98]. Secondly, eicosanoids also
appear to modulate the chemotactic effect of PDGF on smooth muscle
cells, since it was demonstrated that eicosatetraynoic acid (ETYA, a
dual cyclooxygenase/lipoxygenase inhibitor) but not indomethacin (a
cyclooxygenase inhibitor), inhibited PDGF-induced cell migration [12].
Our laboratory has also demonstrated that eicosanoids modulate the
effects of growth factors on cellular cholesterol metabolism. Endothelial cell-derived PDGF stimulated ACEH activity in arterial smooth
muscle cells. Eicosanoid-dependency was established when aspirin- or
ETYA treatment of endothelial cells abolished endothelial cell stimulation of ACEH activity in smooth muscle cells [25]. In summary, eicosanoids derived from endothelial and smooth muscle cells appear to
antagonize the proliferative response to growth factors. This has
important implications for the progression of the hyperplastic response following arterial injury, in that a loss of eicosanoid-
synthetic capacity may provoke unregulated smooth muscle cell proliferation and exaggerated cholesteryl ester deposition.

Cytokines. IL-1, an important cytokine derived from activated endothelial cells and macrophages augments endothelial cell [99], macrophage
[100], and smooth muscle cell [101] eicosanoid synthesis by direct phospholipase activation [99] and by increased transcription and translation of cyclooxygenase [102]. Other phospholipase activators, such as
bradykinin and thrombin are synergistic with cytokines in this effect
[103,104]. Other cytokines, including TGF-β [105], and IL-2 [106] also
stimulate endothelial cell eicosanoid biosynthesis which, in similarity with the mechanism of eicosanoid generation in response to other
agonists, is dependent upon induction of cyclooxygenase. In fact, TNF
and IL-1 are additive [107] or synergistic [108] in stimulating eicosanoid
synthesis. Thus, two important features of cytokine-induction of
eicosanoid biosynthesis should be considered. Firstly, the mechanisms
of the stimulatory effect of cytokines on arterial eicosanoid production are similar to the mechanisms by which growth factors augment
eicosanoid biosynthesis (nascent protein synthesis). These data sug-

gest that induction of eicosanoid-synthetic enzymes by cytokines and growth factors are a general feature of these groups of biologically active molecules. However, an alternative explanation is equally probable. Since cytokines also stimulate the production of growth factors (as described above), the stimulation of eicosanoid biosynthesis by cytokines may be indirect (i.e., mediated by growth factors). This may serve as a mechanism of biological amplification of the negative modulatory role of eicosanoids in the regulation of growth factor- and cytokine-induced cellular activation.

There is now numerous data to demonstrate that eicosanoids also modulate the mitogenic effects of cytokines on vascular and non-vascular cells. For instance, in smooth muscle cells, IL-1-induced proliferation was augmented in the presence of cyclooxygenase inhibitors, and inhibited by exogenous PGE_1 and PGE_2. These findings demonstrate that endogenously synthesized eicosanoids are anti-proliferative or serve to control the proliferative effect of IL-1 [109]. Endogenously synthesized eicosanoids also mediate the expression of the IL-1 receptor following exposure of fibroblasts to IL-1 [110]. In similar studies, TNF, EGF, and TGF-β stimulated fibroblast proliferation, which was augmented in the presence of cyclooxygenase inhibitors, and reduced in the presence of exogenously added PGD_2, PGE_2, and $PGF_{2\alpha}$ [98]. Thus, in mesenchymal cells such as smooth muscle cells and fibroblasts, eicosanoids serve to antagonize the proliferative effects of cytokines.

Eicosanoids also regulate numerous aspects of macrophage cell function, including cytokine gene expression and ultimate release, and differentiation. Firstly, endotoxin (whose biological activity is mediated through TNF) stimulated IL-1 production, whose synthesis was augmented by indomethacin, but inhibited by exogenous PGE_2. These data demonstrated that PGE_2 down-regulate endotoxin-induced IL-1 secretion [111]. However, there appears to be stimulus-specificity, since zymosan-induced IL-1 secretion was augmented by indomethacin and blocked by nordihydroguarietic acid (a lipoxygenase inhibitor), suggesting that lipoxygenase, and not cyclooxygenase products mediate zymosan-induced IL-1 production [111]. Secondly, phorbol ester upregulates TNF gene expression through the synthesis of LTB_4; the upregulation of TNF gene expression by phorbol ester is blunted by PGE_2 [112]. This mechanism may represent part of a positive feed-back loop since LTB_4 can stimulate IL-1 and TNF release [113]. In addition to stimulating PGE_2, TNF also stimulates colony-stimulating factor - 1 (CSF-1). Since exogenous PGE_2 inhibits CSF-1 production, this eicosanoid plays a negative regulatory role in the modulation of CSF-1 release induced by TNF [114].

In summary, cytokines induce eicosanoid production in vascular cells in a manner similar to that observed with growth factors. Importantly, cyclooxygenase-generated eicosanoids appear to antagonize the effects of cytokine activation, while lipoxygenase-derived eicosanoids serve to mediate or amplify the effects of cytokine activation. Clearly, then, the net influence of cytokine effects in vascular cells will ultimately depend on eicosanoid-modulation of their biological activities. This concept is outlined in Figure 1.

SIGNAL TRANSDUCTION MECHANISMS IN GROWTH FACTOR/CYTOKINE-INDUCED EICOSANOID SYNTHESIS: INTRACELLULAR REGULATION OF ARTERIAL CELL FUNCTION

It is now well established that eicosanoids are one of numerous second messenger systems synthesized in response to, and modulate the effects of humoral stimulation. The impact of eicosanoids on the net response to growth factor/cytokine regulation is therefore dependent

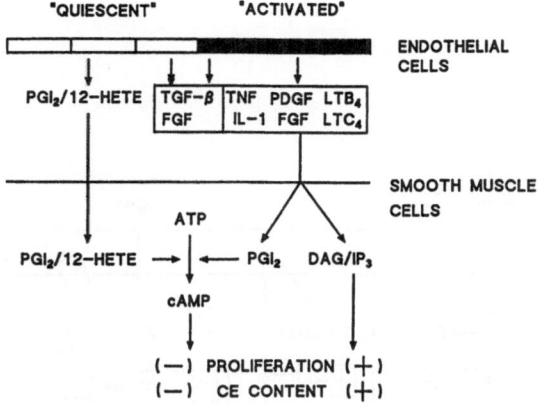

Figure 1. Alterations in smooth muscle cell function as a result of
endothelial cell injury or activation. Under normal conditions, endothelial cell and smooth muscle cell-derived eicosanoids, including PGI_2, PGE_2, and 12-HETE, and growth factors,
including TGF-β, and FGF. The net result of eicosanoid and
growth factor release by endothelial cells in the quiescent
state is to maintain smooth muscle cells in a non-proliferative state, and maintain low cholesteryl ester content. Under
conditions of injury, endothelial cell and monocyte release of
cytokines, growth factors, and leukotrienes, which stimulate
monocyte adhesion and diapedesis. In addition, the net effect
of cytokine and growth factor production is the stimulation of
smooth muscle cell proliferation. Furthermore, in the absence
or reduction in vascular eicosanoids, the mitogenic effects of
growth factors and eicosanoids are unmodulated, leading to
unrestricted smooth muscle cell growth and cholesteryl ester
deposition.

upon the regulation of eicosanoid biosynthesis (as described above),
spectrum and quantity of eicosanoids elaborated in response to various
stimuli, and the influence of eicosanoids on other principal second
messenger systems.

For example, interaction of a peptide hormone with its receptor
results in activation of numerous intracellular proteins. The mechanism by which this occurs is principally by phosphorylation by the
calcium and phospholipid-dependent protein kinase, also known as
protein kinase C, which itself is activated by a tryosine kinase
activity intrinsic to the peptide hormone receptor. This mechanism has
been elucidated in detail for the signal transduction of PDGF upon
interaction with its receptor [115]. Protein kinase C is activated in
endothelial cells by endotoxin, TNF, and IL-1 [116]. Activated protein
kinase C then may phosphorylate numerous proteins, including phospholipase C. Phosphatidyl inositol is the preferred substrate for this
enzyme, which hydrolyzed to 1,2-diacyl glycerol, and 1,4,5 inositol
trisphosphate. Diacylglycerol itself has numerous biological activities, including stimulation of protein kinase C as part of a positive
feedback to stimulate eicosanoid biosynthesis [117]. Diacylglycerol is
also a cofactor for adenylate cyclase, which when phosphorylated,
stimulates the ATP to 3'-5'-cyclic adenosine monophosphate (cyclic
AMP) conversion. In addition, inositol phosphates activate calcium
channels; calcium permits the association of soluble phospholipase A_2

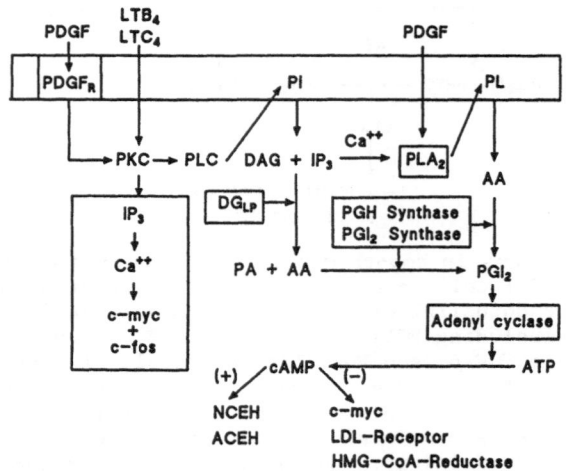

Figure 2. Signal transduction mechanisms in growth factor/cytokine-
induced eicosanoid biosynthesis: regulation of arterial cell
function. For the purposes of this discussion, the effects of
PDGF is outlined here. PDGF binds to its receptor (PDGF$_R$),
which results in the activation of its intrinsic tyrosine
kinase activity, which in turn activates protein kinase C
(PKC) via phosphorylation. Activated PKC in turn phosphorylates phospholipase C (PLC), which, in its active form, hydrolyzes membrane associated phosphatidyl-inositol (PI) to dia-
cyl-glycerol (DAG) and inositol-tris-phosphate (IP$_3$). DAG is
further hydrolyzed by diglyceride lipase (DG$_{LP}$) to produce
phosphatidic acid (PA) and arachidonate (AA). AA is also
derived from IP$_3$-mediated release of calcium from intracellular stores and subsequent activation of phospholipase A$_2$
(PLA$_2$). In some cells, PDGF can also stimulate PLA$_2$ directly,
leading to AA release. AA is metabolized by PGH synthase and
PGI$_2$ synthase to PGI$_2$, which directly activates adenyl cyclase, and subsequent production of cyclic AMP (cAMP). Increased cAMP is associated with decreased cellular proliferation, and upregulation of cholesteryl ester hydrolytic (NCEH
and ACEH) activities. These effects serve to antagonize the
primary effects of PDGF, which include activation of early
response genes (c-myc and c-fos), as outlined in the outlined
box at left. Importantly, eicosanoids derived from inflammatory cells, including leukotrienes can directly stimulate smooth
muscle cell proliferation and eicosanoid biosynthesis.

172

will cell membranes, suggesting that membrane association is a requisite for phospholipase activation and arachidonate release [118], and subsequent eicosanoid biosynthesis. Indeed, protein kinase C activation occurs in parallel with phospholipase A_2 activation [119].

Eicosanoid biosynthesis is under stringent metabolic control, and only a small percentage of released arachidonic acid is actually converted to oxygenated metabolites. Eicosanoid generation is limited by the activity of acyl-CoA lysophosphatide acyltransferase [117] and cyclic AMP; cyclic AMP itself inhibits further autocoid release in a negative feedback fashion by inhibition of calcium movements, and by activation of the cyclic AMP-dependent protein kinase [120]. In fact, the biological activity of the principal eicosanoids, particularly PGI_2, appears to be in the regulation of other second messenger systems, most importantly, cyclic AMP. PGI_2 and PGE_1 potently stimulate cyclic AMP synthesis in vascular tissue [121]. All of the biological activity of PGI_2 can be ascribed to its ability to stimulate cyclic AMP production. The significance of this observation lies in the observation that the most important features of atherogenesis are manifestations of a loss of cyclic AMP function. This important concept is summarized in Figure 2.

The significance of these observations lies in the fact that cyclic AMP inhibits smooth muscle cell proliferation [122] and agents which elevate cyclic AMP, such as interferon [123], PGE_1 [124] inhibit smooth muscle cell proliferation by this mechanism. In fibroblasts, TNF and IL-1 increase cyclic AMP [116], which is followed by a reduction in the expression of c-myc mRNA and inhibition of thymidine incorporation [125]. Furthermore, while thrombin stimulates endothelial cell proliferation [126] and PGI_2 synthesis, agents which stimulate cyclic AMP inhibit the induction of c-sis induction by thrombin and TGF-β [126], and inhibit thrombin-induced PDGF-B chain synthesis [127]. Thus, agents which stimulate cyclic AMP may be doing so via PGI_2 generation, and loss of PGI_2-synthetic activity would conceivably result in unregulated proliferation in response to mitogens such as TNF and IL-1.

The regulation of cellular cholesterol metabolism in peripheral tissue is also under cyclic nucleotide regulation. We have demonstrated that lysosomal (acid) cholesteryl ester hydrolase activity in arterial smooth muscle cells is activated by cyclic AMP [73,76]. In addition, we and others have identified a neutral cholesteryl ester hydrolase activity in smooth muscle cells [74] and in macrophages [128], whose activity is stimulated by eicosanoids through activation of the cyclic AMP-dependent protein kinase. Importantly, the upregulation of the LDL receptor gene, LDL receptor expression [129] and HMG-CoA reductase [130] is also dependent upon activation of protein kinase C [129]. However, PGI_2 reduces the number of LDL receptors [89] and sterol synthesis [131] in macrophages, suggesting that there may be independent regulation of LDL receptor and HMG CoA reductase by protein kinase C and agents, such as PGI_2, that stimulate cyclic AMP production.

CONCLUSIONS

The data summarized in this review demonstrate that under normal conditions, eicosanoids derived from the blood vessel wall, including PGI_2 and related eicosanoids, contribute to the barrier function of intact endothelium, inhibit the permeation of large macromolecules such as LDL, inhibit monocyte adhesion and diapedesis, and modulate numerous aspects of smooth muscle cell, including maintenance of the contractile phenotype, inhibition of smooth muscle cell proliferation, and maintenance of low cellular cholesteryl ester content. Following acute inflammatory injury, interaction of inflammatory cells with the vascular endothelium promote monocyte adhesion with subsequent alteration of endothelium to a more pro-coagulant phenotype, increased permeability to macromolecules such as LDL. Subacute or chronic injury may then exacerbate these events, leading to smooth muscle cell proliferation, and potentially, increased cholesteryl ester deposition. These processes clearly occur via the release of, and response to potent inflammatory mediators, including cytokines, growth factors, and eicosanoids, which are elaborated by activated endothelium, macrophages, and smooth muscle cells. The roles of each cytokines and growth factors in the atherogenic response to injury appear to be fairly complex; some of these agents appear to be pro-atherosclerotic (PDGF, FGF, EGF, IL-1), while others appear to be anti-atherosclerotic (TGF-b, TNF, interferon, GM-CSF), even though the synthesis of each is induced by similar stimuli, such as endotoxin. This basic concept is summarized in Figure 1.

Furthermore, the data also suggest that eicosanoids synthesized by the normal blood vessel wall, such as PGI_2 and PGE_2, act to antagonize the influence of pro-atherogenic factors, and mediate to some degree the influence of anti-atherogenic factors, by maintaining the endothelium in a quiescent state, characterized by a non-thrombotic, non-adhesive surface maintaining a low degree of permeability to circulating macromolecules, as well as maintain the arterial smooth muscle cells in the contractile phenotype, inhibit proliferation, and promote low cell cholesterol content. This is in contrast to HETEs and leukotrienes, including LTB_4 and LTC_4, which appear to have pro-atherogenic properties.

It is thus clear that intact PGI_2-synthetic capacity is important in the modulation of pro-atherogenic stimuli. Thus, factors which inhibit the vascular production of PGI_2, such as high dose-aspirin, or dietary maneuvers which inhibit PGI_2 generation, may result in depressed vascular eicosanoid production. Secondly, since the biological activities of PGI_2 and related eicosanoids are mediated through the generation of cyclic AMP, the importance of the eicosanoid/cyclic AMP axis becomes manifest when eicosanoid/cAMP production is reduced (Figure 1).

Finally, it should be noted that other humoral factors implicated in the regulation of endothelial cell, smooth muscle cell, and macrophage proliferative capacity and cholesterol metabolism, alter the cellular responses to cytokine and growth factor activation. Furthermore, these humoral factors may modify cytokine, growth factor or eicosanoid production and effects. Plasma factors implicated in this regard include native LDL and modified forms of LDL (oxidized LDL, for example), HDL, heparin, EDRF, and marine lipids (eicosapentaenoic acid). Thus, regulation of the cytokine/growth factor network by eicosanoids may represent an important aspect of arterial response to injury, and in the progression of intimal hyperplasia and cholesteryl ester deposition in a setting of inflammatory cell activation.

ACKNOWLEDGEMENTS

This work was funded by the National Institutes of Health (NIH) SCOR in Thrombosis (HL-18828), and NIH Grant HL-39701. The authors would like to thank Drs. Rosemary Kraemer and Andrew Nicholson for their critical review of this manuscript.

REFERENCES

1. P. Davies, Biology of Disease: Vascular cell interactions with special reference to the pathgenesis of atherosclerosis, Lab. Invest. 55:5 (1986).
2. D. Falcone, D. Hajjar, and C. Minick, Enhancement of cholesterol and cholesteryl ester accumulation in re-endothelialized aorta, Amer. J. Pathol. 99:81 (1980).
3. D. Hajjar, D. Falcone, S. Fowler, and C. Minick, Endothelium modifies the altered metabolism of the injured aortic wall, Amer. J. Pathol. 102:28 (1981).
4. R. Cotran, and J. Pober, Effects of cytokines on vascular endothelium: their role in vascular and immune injury, Kidney Int. 35:969 (1989).
5. P. Dawson, S. Hofmann, D. van de Westhuyzen, T. Sudhof, M. Brown, and J. Goldstein, Sterol-dependent repression of low density lipoprotein receptor promotor mediated by 16-base pair sequence adjacent to binding site for transcription factor Sp1, J. Biol. Chem. 263:3372 (1988).
6. M. Brown, and J. Goldstein, A receptor-mediated pathway for cholesterol homeostasis, Science 232:34 (1986).
7. T. Kodama, M. Freeman, L. Rohrer, J. Zabrecky, P. Matsudaira, and M. Krieger, Type I macrophage scavenger receptor contains alpha-helical and collagen-like coiled coils, Nature 343:531 (1990).
8. O. Stein, J. Vanderhoek, and Y. Stein, Cholesterol content and sterol synthesis in human skin fibroblasts and rat aortic smooth muscle cells exposed to lipoprotein-depleted serum and high density apoprotein/phospholipid mixtures, Biochim.Biophys.Acta 431:347 (1976).
9. R. Ross, J. Glomset, B. Kariya, and L. Harker, A platelet-dependent serum factor that stimulates the proliferation of arterial smooth muscle cells in vitro, Proc. Natl. Acad. Sci. USA 71:1207 (1974).
10. J. Fingerle, R. Johnson, A. Clowes, M. Majesky, and M. Reidy, Role of platelets in smooth muscle proliferation and migration after vascular injury in rat carotid artery, Proc. Natl. Acad. Sci. USA 86:8412 (1989).
11. G. Grotendorst, H. Seppa, H. Kleinman, and G. Martin, Attachment of smooth muscle cells to collagen and their migration toward platelet-derived growth factor, Proc. Natl. Acad. Sci. USA 78:3669 (1981).
12. J. Nakao, H. Ito, W. Chang, Y. Koshihara, and S. Murota, Aortic smooth muscle cell migration caused by platelet-derived growth factor is mediated by lipoxygenase product(s) of arachidonic acid, Biochem. Biophys. Res. Comm. 112:866 (1983).
13. C. Walker, D. Bowen-Pope, R. Ross, and M. Reidy, Production of platelet-derived growth factor-like molecules by cultured arterial smooth muscle cells accompanies proliferation after arterial injury, Proc. Natl. Acad. Sci. USA 83:7311 (1986).
14. J. Wilcox, K. Smith, L. Williams, S. Schwartz, and D. Gordon, Platelet-derived growth factor mRNA detection in human atherosclerotic plaques by in situ hybridization, J. Clin. Invest. 82:1134 (1988).
15. T. Barrett, and E. Benditt, sis (platelet-derived growth factor B chain) gene transcript levels are elevated in human atherosclerotic lesions compared to normal artery, Proc. Natl. Acad. Sci. USA 84:1099 (1987).

16. K. Pietila, and T. Nikkari, Enhanced growth of smooth muscle cells from atherosclerotic rabbit aortas in culture, <u>Athero</u> 36:241 (1980).

17. R. Ross, J. Masuda, E. Raines, A. Gown, S. Kutsuda, M. Sasahara, L. Malden, H. Masuko, and H. Sato, Localization of PDGF-B chain in macrophages in all phases of atherogenesis, <u>Science</u> 248:1009 (1990).

18. M. Klagsbrun, and E. Edelman, Biological and biochemical properties of fibroblast growth factors. Implications for the pathogenesis of atherosclerosis, <u>Arterio.</u> 9:269 (1989).

19. C. Gay, and J. Winkles, Heparin-binding growth factor-I stimulation of human endothelial cells induces platelet-derived growth factor A-chain gene expression, <u>J. Biol. Chem.</u> 265:3284 (1990).

20. A. Chait, R. Ross, J. Albers, and E. Bierman, Platelet-derived growth factor stimulates activity of low density lipoprotein receptors, <u>Proc. Natl. Acad. Sci. USA</u> 77:4084 (1980).

21. P. Davies, and C. Kerr, Modification of low density lipoprotein metabolism by growth factors in cultures of vascular endothelial cells and human skin fibroblasts, <u>Biochim.Biophys.Acta</u> 712:26 (1982).

22. T. Mazzone, K. Basheerruddin, L. Ping, S. Frazer, and G. Getz, Mechanism of growth-related activation of the low density lipoprotein receptor pathway, <u>J. Biol. Chem.</u> 264:1787 (1989).

23. P. Davies, and R. Ross, Mediation of pinocytosis in cultured arterial smooth muscle and endothelial cells by platelet-derived growth factor, <u>J. Cell Biol.</u> 79:663 (1978).

24. K. Suzuki, M. Hara, A. Kitani, M. Haragai, K. Norioka, K. Kondo, F. Hirata, N. Sakata, M. Kawakami, M. Kawagoe, and H. Nakamura, Augmentation of LDL receptor activities on lymphocytes by interleukin-2 and anti-CD3 antibody: a flow cytometric analysis, <u>Biochim.Biophys.Acta</u> 1042:352 (1990).

25. D. Hajjar, A. Marcus, and K. Hajjar, Interactions of arterial cells: studies on the mechanisms of endothelial cell modulation of cholesterol metabolism in co-cultured smooth muscle cells, <u>J. Biol. Chem.</u> 262:6976 (1987).

26. A. Nicholson, and D. Hajjar, Transforming growth factor-beta: effects on smooth muscle cell growth and cholesterol metabolism, <u>FASEB J.</u> 3:A612 (1989). (Abstract)

27. B. Feinberg, R. Kurzrock, M. Talpaz, M. Blick, and S. Saks, A phase-I trial of intravenously-administered recombinant tumor necrosis factor in cancer patients, <u>J. Clin. Oncol.</u> 6:1328 (1988).

28. A. Wilson, R. Schaub, R. Goldstein, and P. Kuo, Suppression of aortic atherosclerosis in cholesterol-fed rabbits by purified rabbit interferon, <u>Arterio.</u> 10:208 (1990).

29. T. Issekutz, Effects of six different cytokines on lymphocyte adherence to microvascular endothelium and in vivo lymphocyte migration in the rat, <u>J.Immunol.</u> 144:2140 (1990).

30. K. Shinjo, S. Tsuda, T. Hayama, T. Asahi, and H. Kawaharada, Increase in permeability of human endothelial cell monolayer by recombinant human lymphotoxin, <u>Biochem. Biophys. Res. Comm.</u> 162:1431 (1989).

31. S. Zuckerman, and Y. Surprenant, Induction of endothelial cell/macrophage procoagulant activity: synergistic stimulation by gamma interferon and granulocyte-macrophage colony stimulating factor, <u>Thomb. Haemo.</u> 61:178 (1989).

32. J. Masuyama, N. Minato, and S. Kano, Mechanisms of lymphocyte adhesion to human vascular endothelial cells in culture. T-lymphocyte adhesion to endothelial cell HLD-DR antigens induced by gamma interferon, <u>J. Clin. Invest.</u> 77:1596 (1986).

33. C. Hicks, S. Breit, and R. Penny, Response of microvascular endothelial cells to biological response modifiers, Immunol. Cell Biol. 67:271 (1989).

34. E. Raines, S. Downer, and R. Ross, Interleukin-1 mitogenic activity for fibroblasts and smooth muscle cells is due to PDGF-AA, Science 243:393 (1989).

35. G. Hansson, M. Hellstrand, L. Rymo, L. Rubbia, and G. Gabbiani, Interferon-gamma inhibits both proliferation and expression of differentiation-specific alpha-smooth muscle actin in arterial smooth muscle cells, J. Exptl. Med. 170:1595 (1989).

36. H. Sawada, M. Kan, and W. McKeehan, Opposite effects of mono-kines (interleukin-1 and tumor necrosis factor) on proliferation and heparin-binding (fibroblast) growth factor binding to human aortic endothelial and smooth muscle cells, In Vitro 26:213 (1990).

37. A. Fogelman, J. Seager, M. Haberland, M. Hokom, R. Tanaka, and P. Edwards, Lymphocyte-conditioned medium protects human mono-cyte-macrophages from cholesteryl ester accumulation, Proc. Natl. Acad. Sci. USA 79:922 (1982).

38. B. Van Lentin, A. Fogelman, J. Seager, E. Ribi, M. Haberland, and P. Edwards, Bacterial endotoxin selectively prevents the expression of scavenger-receptor activity on human monocyte-macrophages, J.Immunol. 134:3718 (1985).

39. K. Pomerantz, and D. Hajjar, Eicosanoids in regulation of arte-rial smooth muscle cell phenotype, proliferative capacity, and cholesterol metabolism, Arterio. 9:413 (1989).

40. J. Wang, Y. Lu, Z. Guo, E. Zhen, and F. Shi, Lipid peroxides, glutathione peroxidase, prostacyclin and cell cycle stages in normal and atherosclerotic Japanese quail arteries, Athero 75:219 (1989).

41. R. Gryglewski, E. Kosta-Trabka, A. Deminska-Kiec, and R. Korbut, Prostacyclin and atherosclerosis - experimental and clinical approaches, Adv. Expt. Med. Biol. 243:21 (1988).

42. S. Akopov, A. Orekhov, V. Tertov, K. Khashimov, E. Gabrielyan, and V. Smirnov, Stable analogues of prostacyclin and throm-boxane A_2 display contradictory influences on atherosclerotic properties of cells cultured from human aorta, Athero 72:245 (1988).

43. E. Jaffe, Cell biology of endothelial cells, Hum. Pathol. 18:234 (1987).

44. K. Pritchard, P. Wong, and M. Stemerman, Atherogenic concentra-tions of low density lipoprotein enhance endothelial cell generation of epoxyeicosatrienoic acid products, Amer. J. Pathol. in press:(1990).

45. S. Feinmark, and P. Cannon, Endothelial cell leukotriene C_4 synthesis results from intercellular transfer of leukotriene A_4 synthesized by polymorphonuclear leukocytes, J. Biol. Chem. 261:16466 (1986).

46. M. Clark, D. Littlejohn, S. Mong, and S. Crooke, Effect of leukotrienes, bradykinin, and calcium ionophore (A-23187) on bovine endothelial cells: release of prostacyclin, Prostag-landins 31:157 (1986).

47. S. Moore, L. Prokuski, P. Figard, A. Spector, and M. Hart, Murine cerebral microvascular endothelium incorporate and metabolize 12-hydroxyeicosatetraenoic acid, J. Cell. Physiol. 137:75 (1988).

48. S. Hong, T. Carty, and D. Deykin, Tranylcypromine and 15-hydrop-eroxyarachidonate affect arachidonic acid release in addition to inhibition of prostacyclin synthesis in calf aortic en-dothelial cells, J. Biol. Chem. 255:9538 (1980).

49. T. Kanayasu, J. Nakao-Hayashi, N. Asuwa, I. Mirota, T. Ishii, H. Ito, and S. Murota, Leukotriene C_4 stimulates angiogenesis in bovine carotid artery endothelial cells in vitro, <u>Biochem. Biophys. Res. Comm.</u> 159:572 (1989).

50. D. Leszczynski, and P. Hayry, Eicosanoids are regulatory molecules in gamma-interferon-induced endothelial antigenicity and adherence for leukocytes, <u>FEBS Letters</u> 242:383 (1989).

51. J. Palmblad, P. Lindstrom, and R. Lerner, Leukotriene B_4-induced hyperadhesiveness of endothelial cells for neutrophils, <u>Biochem. Biophys. Res. Comm.</u> 166:848 (1990).

52. T. Casale, and M. Abbas, Comparison of leukotriene B_4-induced neutrophil migration through different cellular barriers, <u>American Journal of Physiology</u> 258:C639 (1990).

53. K. Honn, I. Grossi, L. Fitzgerald, L. Umbarger, C. Diglio, and J. Taylor, Lipoxygenase products regulate IRGpIIb/IIIa receptor adhesion of tumor cells to endothelial cells, subendothelial matrix and fibronectin, <u>Proc. Soc. Exptl. Biol. Med.</u> 189:130 (1988).

54. M. Buchanan, M. Vazquez, and M. Gimbrone, Arachidonic acid metabolism and the adhesion of human polymorphonuclear leukocytes to cultured vascular endothelial cells, <u>Blood</u> 62:889 (1983).

55. K. Pritchard, R. Tota, M. Stemerman, and P. Wong, 14,15-Epoxyeicosatetraenoic acid promotes endothelial cell dependent adhesion of human monocytic tumor U937 cells, <u>Biochem. Biophys. Res. Comm.</u> 167:137 (1990).

56. M. Perlman, A. Johnson, W. Jubiz, and A. Malik, Lipoxygenase products induce neutrophil activation and increase endothelial permeability after thrombin-induced pulmonary microembolism, <u>Circ. Res.</u> 64:62 (1989).

57. J. Gudgeon, and W. Martin, Modulation of arterial permeability: studies on an in vitro model, <u>Br. J. Pharmacol.</u> 98:1267 (1989).

58. K. Pomerantz, and D. Hajjar, Eicosanoid metabolism in cholesterol-enriched arterial smooth muscle cells: reduced arachidonate release with concomitant decrease in cyclooxygenase products, <u>J. Lipid. Res.</u> 30:1219 (1989).

59. J. Bailey, R. Bryant, J. Whiting, and K. Salata, Characterization of 11-HETE and 15-HETE, together with prostacyclin, as major products of the cyclooxygenase pathway in cultured rat aorta smooth muscle cells, <u>J. Lipid. Res.</u> 24:1419 (1983).

60. S. Feinmark, and P. Cannon, Vascular smooth muscle cell leukotriene C_4 synthesis: requirement for transcellular leukotriene A_4 metabolism, <u>Biochim.Biophys.Acta</u> 922:125 (1987).

61. J. Thyberg, U. Hedin, M. Sjolund, L. Palmberg, and B. Bottger, Regulation of differentiated properties and proliferation of arterial smooth muscle cells, <u>Arterio.</u> 10:966 (1990).

62. M. Sjolund, J. Nilsson, L. Palmberg, and J. Thyberg, Phenotype modulation of primary cultures of arterial smooth muscle cells. Dual effect of prostaglandin E_1, <u>Differentiation</u> 27:158 (1984).

63. J. Larrue, D. Daret, J. Demond-Henri, C. Allieres, and H. Bricaud, Prostacyclin synthesis in proliferative aortic smooth muscle cells. A kinetic in vivo and in vitro study, <u>Athero</u> 50:63 (1984).

64. L. Palmberg, H. Claesson, and J. Thyberg, Effect of leukotrienes on phenotypic properties and growth of arterial smooth muscle cells in primary culture, <u>J. Cell Sci.</u> 93:403 (1989).

65. J. Nakao, T. Ooyama, H. Ito, W. Chang, and S. Murota, Compara-

tive effect of lipoxygenase products of arachidonic acid on rat aortic smooth muscle cell migration, <u>Athero</u> 44:339 (1982).

66. J. Hirusumi, A. Nomoto, Y. Ohkubo, C. Sekiguchi, S. Mutoh, I. Yamaguchi, and H. Aoki, Inflammatory responses in cuff-induced atherosclerosis in rabbits, <u>Athero</u> 64:243 (1987).

67. H. Sinzinger, T. Zidek, P. Fitscha, J. OGrady, O. Wagner, and J. Kaliman, Prostaglandin I_2 reduces activation of human arterial smooth muscle cells in vivo, <u>Prostaglandins</u> 33:915 (1987).

68. Y. Uehara, T. Ishimitsu, K. Kimura, M. Ishii, T. Ikeda, and T. Sugimoto, Regulatory effects of eicosanoids on thymidine uptake by vascular smooth muscle cells of rats, <u>Prostaglandins</u> 36:847 (1988).

69. J. Nilsson, and A. Olsson, Prostaglandin E_1 inhibits DNA synthesis in arterial smooth muscle cells stimulated with platelet-derived growth factor, <u>Athero</u> 53:77 (1984).

70. L. Palmberg, H. Claesson, and J. Thyberg, Leukotrienes stimulate initiation of DNA synthesis in cultured arterial smooth muscle cells, <u>J. Cell Sci.</u> 88:151 (1987).

71. M. Clark, D. Littlejohn, T. Conway, S. Mong, S. Steiner, and S. Crooke, Leukotriene D_4 treatment of bovine aortic endothelial cells and murine smooth muscle cells in culture results in an increase in phospholipase A_2 activity, <u>J. Biol. Chem.</u> 261:10713 (1986).

72. D. Smith, A. Willis, and I. Mahmud, Eicosanoid effects on cell proliferation in vitro: relevance to atherosclerosis, <u>Prosta. Leuko. Med.</u> 16:1 (1984).

73. D. Hajjar, B. Weksler, D. Falcone, J. Hefton, K. Tack-Goldman, and C. Minick, Prostacyclin modulates cholesteryl ester hydrolytic activity by its effect on cyclic adenosine monophosphate in rabbit aortic smooth muscle cells, <u>J. Clin. Invest.</u> 70:479 (1982).

74. D. Hajjar, C. Minick, and S. Fowler, Arterial neutral cholesteryl esterase. A hormone-sensitive enzyme distinct from the lysosomal enzyme, <u>J. Biol. Chem.</u> 258:192 (1983).

75. D. Hajjar, A. Marcus, and O. Etingin, Platelet-Neutrophil-Smooth Muscle Cell Interactions: Lipoxygenase-derived mono- and dihydroxy acids activate cholesteryl ester hydrolysis by the cyclic AMP dependent protein kinase cascade, <u>Biochem.</u> 28:8885 (1989).

76. D. Hajjar, and B. Weksler, Metabolic activity of cholesteryl esters in aortic smooth muscle cells is altered by prostaglandins I_2 and E_2, <u>J. Lipid. Res.</u> 24:1176 (1983).

77. M. Balter, G. Toews, and M. Peters-Golden, Different patterns of arachidonate metabolism in autologous blood monocytes and alveolar macrophages, <u>J. Immunol.</u> 142:602 (1989).

78. G. Cott, J. Westcott, and N. Voelkel, Protaglandin and leukotriene production by alveolar type II cells in vitro, <u>Am. J. Physiol.</u> 258:L179 (1990).

79. G. Brown, M. Monick, and G. Hunninghake, Human alveolar macrophage arachidonic acid metabolism, <u>Am. J. Physiol.</u> 254:C-809 (1988).

80. C. Rouzer, W. Scott, A. Hammill, and Z. Cohn, Synthesis of leukotriene C and other arachidonic acid metabolites by mouse pulmonary macrophages, <u>J. Exptl. Med.</u> 155:720 (1982).

81. U. Schade, H. Holl, and E. Rietschel, Metabolism of exogenous arachidonic acid by mouse peritoneal macrophages, <u>Prostaglandins</u> 34:401 (1987).

82. M. Peters-Golden, R. McNish, J. Brieland, and J. Fantone, Diminished protein kinase C-activated arachidonate metabolism accompanies rat macropahge differentiation in the lung, J.Immunol. 144:4320 (1990).

83. V. Kaever, H. Pfannkuche, K. Wessel, and K. Resch, The ratio of macrophage prostaglandin and leukotriene synthesis is determined by the intracellular free calcium level, Biochem. Pharmacol. 39:1313 (1990).

84. J. Humes, E. Opas, M. Galavage, D. Soderman, and R. Bonney, Regulation of macrophage eicosanoid production by hydroperoxy- and hydroxy-eicosatetraenoic acids, Biochem. J. 233:199 (1986).

85. C. Kadiri, J. Masliah, M. Bachelet, B. Vargftig, and G. Bereziat, Phospholipase A_2-mediated release of arachidonic acid in stimulated guinea pig alveolar macrophages: interaction with lipid mediators and cyclic AMP, J. Cell. Biochem. 40:157 (1989).

86. C. Tripp, A. Wyche, E. Unanue, and P. Needleman, The functional significance of the regulation of macrophage Ia expression by endogenous arachidonate metabolites in vitro, J.Immunol. 137:3915 (1986).

87. F. Ondrey, K. Anderson, D. Hoeltgen, and J. Harris, Differentiation of U937 cells induced by 5,8,11,14 - eicosatetraynoic acid, a competitive inhibitor of arachidonic acid metabolism, Expt. Cell Res. 179:477 (1988).

88. N. Morisaki, T. Kanzaki, M. Kitahara, Y. Saito, and S. Yoshida, Inhibitory effect of prostaglandin E_2 on cholesterol ester accumulation in macrophages, Biochem. Biophys. Res. Comm. 137:461 (1986).

89. W. Krone, A. Klass, H. Nagele, B. Behnke, and H. Greten, Effect of prostaglandins on LDL receptor activity and cholesterol synthesis in freshly isolated human mononuclear leukocytes, J. Lipid. Res. 29:1663 (1988).

90. J. Schroeff, L. Havekes, A. Weerheim, J. Emeis, and B. Vermeer, Suppression of cholesteryl ester accumulation in cultured human monocyte-derived macrophages by lipoxygenase inhibitors, Biochem. Biophys. Res. Comm. 127:366 (1985).

91. J. Korn, Fibroblast prostaglandin E_2 synthesis. Persistance of an abnormal phenotype after short-term exposure to mononuclear cell products, J. Clin. Invest. 71:1240 (1983).

92. A. Habernicht, M. Goerig, J. Grulich, D. Rothe, R. Gronwald, U. Loth, G. Schettler, B. Kommerell, and R. Ross, Human platelet-derived growth factor stimulates prostaglandin synthesis by activation and by rapid de novo synthesis of cyclooxygenase, J. Clin. Invest. 75:1381 (1985).

93. M. Goerig, A. Habenicht, W. Zeh, P. Salbach, K. Burkhard, D. Rothe, W. Nastainszyk, and J. Glomset, Evidence for coordinate, selective regulation of eicosanoid synthesis in platelet-derived growth factor-stimulated 3T3 fibroblasts and in HL-60 cells induced to differentiate into macrophages or neutrophils, J. Biol. Chem. 263:19384 (1988).

94. S. Coughlin, M. Moskowitz, H. Antoniades, and L. Levine, Serotonin receptor-mediated stimulation of bovine smooth muscle cell prostacyclin synthesis and its modulation by platelet-derived growth factor, Proc. Natl. Acad. Sci. USA 78:7134 (1981).

95. J. Blay, and M. Hollenberg, Epidermal growth factor stimulation of prostacyton produciton by cultured aortic smooth muscle cells: requirement for increased cellular calcium levels, J. Cell. Physiol. 139:524 (1989).

96. A. Ristimaki, O. Ylikorkala, and L. Viinikka, Effect of growth factors on human vascular endothelial cell prostacyclin production, _Arterio._ 10:653 (1990).

97. B. Weksler, Heparin and acidic fibroblast growth factor interact to decrease prostacyclin synthesis in human endothelial cells by affecting both prostaglandin H synthase and prostacyclin synthase, _J. Cell. Physiol._ 142:514 (1990).

98. T. Hori, S. Kashiyama, M. Hayakawa, S. Shibamoto, M. Tsujimoto, N. Oku, and F. Ito, Possible role of prostaglandins as negative regulators in growth stimulation by tumor necrosis factor and epiderminal growth factor in human fibroblasts, _J. Cell. Physiol._ 141:275 (1989).

99. F. Breviairio, P. Proserpio, F. Bertocchi, M. Lampugnani, A. Mantovani, and E. Dejana, Interleukin-1 stimulates prostacyclin production by cultured human endothelial cells by increasing mobilization and conversion, _Arterio._ 10:129 (1990).

100. C. Albrightson, N. Baenziger, and P. Needleman, Exaggerated human vascular cell prostaglandin biosynthesis mediated by monocytes: role of monokines and interleukin I, _J.Immunol._ 135:1872 (1985).

101. H. Bull, M. Rustin, J. Spaull, J. Cohen, E. Wilson-Jones, and P. Dowd, Pro-inflammatory mediators induce sustained release of prostaglandin E_2 from human dermal microvascular endothelial cells, _Br. J. Dermatol._ 122:153 (1990).

102. M. Rustin, H. Bull, and P. Dowd, Effect of human recombinant interleukin - 1α on release of prostacyclin from human endothelial cells, _Br. J. Dermatol._ 120:153 (1989).

103. R. Burch, J. Connor, and J. Axelrod, Interleukin 1 amplifies receptor-mediated activation of phospholipase A_2 in 3T3 fibroblasts, _Proc. Natl. Acad. Sci. USA_ 85:6306 (1988).

104. L. O'Neill, and G. Lewis, Interleukin-1 potentiates bradykinin- and TNF-alpha-induced PGE_2 release, _Europ. J. Pharmacol._ 166:131 (1989).

105. A. Ristamaki, Transforming growth factor alpha stimulates prostacyclin production by cultured human vascular endothelial cells more potently than epidermal growth factor, _Biochem. Biophys. Res. Comm._ 160:1100 (1989).

106. K. Frasier-Scott, H. Hatzakis, D. Seong, C. Jones, and K. Wu, Influence of natural and recombinant interleukin 2 on endothelial cell arachidonate metabolism: induction of de novo synthesis of prostaglandin H synthase, _J. Clin. Invest._ 82:1877 (1988).

107. H. Endo, T. Akahoshi, and S. Kashiwazaki, Additive effects of IL-1 and TNF on induction of prostacyclin synthesis in human vascular endothelial cells, _Biochem. Biophys. Res. Comm._ 156:1007 (1988).

108. J. Pfeilschifter, W. Pignat, K. Vosbeck, and F. Marki, Interleukin-1 and tumor necrosis factor synergistically stimulate prostaglandin synthesis and phospholipase A_2 release from rat renal mesangial cells, _Biochem. Biophys. Res. Comm._ 159:385 (1989).

109. P. Libby, S. Warner, and G. Friedman, Interleukin I: a mitogen for human vascular smooth muscle cells that induces the release of growth-inhibitory prostanoids, _J. Clin. Invest._ 81:487 (1988).

110. T. Akahoshi, J. Oppenheim, and K. Matsushima, Interleukin 1 stimulates its own receptor expression on human fibroblasts through the endogenous production of prostaglandin(s), _J. Clin. Invest._ 82:1219 (1988).

111. S. Kunkel, and S. Chensue,   Arachidonic acid metabolites regulate interleukin-1 secretion, <u>Biochem. Biophys. Res. Comm.</u> 128:892 (1985).

112. J. Horiguchi, D. Spriggs, K. Imamura, R. Stone, R. Luebbers, and D. Kufe,   Role of arachidonic acid metabolism in transcriptional induction of tumor necrosis factor gene expression by phorbol ester, <u>Mol. Cell. Biol.</u> 9:252 (1989).

113. L. Gagnon, L. Filion, C. Dubois, and M. Rola-Pleszczynski, Leukotrienes and macrophage activation: augmented cytotoxic activity and enhanced interleukin 1, tumor necrosis factor, and hydrogen peroxide production, <u>Agents and Actions</u> 26:142 (1989).

114. M. Sherman, B. Weber, R. Datta, and D. Kufe,   Transcriptional and posttranscriptional regulation of macrophage-specific colony stimulating factor gene expression by tumor necrosis factor. Involvement of arachidonic acid metabolites, <u>J. Clin. Invest.</u> 85:442 (1990).

115. L. Williams,   Signal transduction by the platelet-derived growth factor, <u>Science</u> 243:1564 (1989).

116. Y. Zhang, J. Lin, Y. Yip, and J. Vilcek,   Enhancement of cAMP levels and of protein kinase activity by tumor necrosis factor and interleukin 1 in human fibroblasts: role in the induction of interleukin 6, <u>Proc. Natl. Acad. Sci. USA</u> 85:6802 (1988).

117. H. Pfannkuche, V. Kaever, and K. Resch,   A possible role of protein kinase C in regulating prostaglandin synthesis of mouse peritoneal macrophages, <u>Biochem. Biophys. Res. Comm.</u> 139:604 (1986).

118. J. Channon, and C. Leslie,   A calcium-dependent mechanism for associating a soluble arachidonoyl-hydrolyzing phospholipase A_2 with membrane in the macropahge cell line RAW 264.7, <u>J. Biol. Chem.</u> 265:5409 (1990).

119. J. Balsinde, B. Fernandez, and E. Diez,   Regulation of arachidonic acid release in mouse peritoneal macrophages. The role of extracellular calcium and protein kinase C, <u>J.Immunol.</u> 144:4298 (1990).

120. B. Undem, T. Torphy, D. Goldman, and F. Chilton,   Inhibition by adenosine 3':5'-monophosphate of eicosanoid and platelet-activating factor biosynthesis in the mouse PT-18 mast cell, <u>J. Biol. Chem.</u> 265:6750 (1990).

121. M. Huang, and G. Drummond,   Adenylate cyclase in cerebral microvessels: effect of guanine nucleotides, adenosine, and other agonists, <u>Molecul. Pharm.</u> 16:462 (1979).

122. K. Southgate, and A. Newby,   Serum-induced proliferation of rabbit aortic smooth muscle cells from the contractile state is inhibited by 8-Br-cAMP but not 8-Br-cGMP, <u>Athero</u> 82:113 (1990).

123. Y. Fukumoto, Y. Kawahara, K. Kariya, S. Araki, H. Fukuzaki, and Y. Takai,   Independent inhibition of DNA synthesis by protein kinase C, cyclic AMP and interferon alpha/beta in rabbit aortic smooth muscle cells, <u>Biochem. Biophys. Res. Comm.</u> 157:337 (1988).

124. R. Stout,   Cyclic AMP: a potent inhibitor of DNA synthesis in cultured arterial endothelial and smooth muscle cells, <u>Diabetologia</u> 22:51 (1982).

125. N. Heldin, Y. Paulsson, K. Forsberg, C. Heldin, and B. Westermark,   Induction of cyclic AMP synthesis is followed by a reduction in the expression of c-myc messenger RNA and inhibition of ^{3}H-thymidine incorporation in human fibroblasts, <u>J. Cell. Physiol.</u> 138:17 (1989).

126. T. Daniel, V. Gibbs, D. Milfray, and L. Williams, Agents that increase cAMP accumulation block endothelial c-sis induction by thrombin and transforming growth factor-b, <u>J. Biol. Chem.</u> 262:11893 (1987).

127. W. Kavanaugh, G. Harsh, N. Starksen, C. Rocco, and L. Williams, Transcriptional regulation of the A and B chain genes of platelet-derived growth factor in microvascular endothelial cells, <u>J. Biol. Chem.</u> 263:8470 (1988).

128. J. Khoo, E. Mahoney, and D. Steinberg, Neutral cholesterol esterase activity and its enhancement of cAMP-dependent protein kinase, <u>J. Biol. Chem.</u> 256:12659 (1981).

129. J. Auwerx, A. Chait, G. Wolfbauer, and S. Deeb, Involvement of second messengers in regulation of the low-density lipoprotein receptor gene, <u>Mol. Cell. Biol.</u> 9:2298 (1989).

130. J. Auwerx, A. Chait, and S. Deeb, Regulation of the low density lipoprotein receptor and hydroxymethylglutaryl coenzyme A reductase genese by protein kinase C and a putative negative regulatory protein, <u>Proc. Natl. Acad. Sci. USA</u> 86:1133 (1989).

131. W. Krone, P. Kaczmarczyk, D. Muller-Wieland, and H. Greten, The prostacyclin analog iloprost and prostaglandin E_1 suppress sterol synthesis in freshly isolated human mononuclear leukocytes, <u>Biochim.Biophys.Acta</u> 835:154 (1985).

LIPOXINS INHIBIT MICROVASCULAR INFLAMMATORY ACTIONS OF LEUKOTRIENE B_4

Johan Raud, Ulla Palmertz, Sven-Erik Dahlén, and Per Hedqvist

Department of Physiology I and Institute of Environmental Medicine, Karolinska Institutet Stockholm, Sweden

INTRODUCTION

The lipoxins constitute a new class of arachidonic acid derivatives which may be produced by for example leukocytes, or by leukocytes interacting with platelets (*cf*. Dahlén and Serhan 1990). The biosynthesis of lipoxins is the result of interactions between different lipoxygenases, and the lipoxins characteristically possess a conjugated tetraene structure (Serhan *et al*., 1984; *cf*. Dahlén and Serhan 1990). The two major compounds, lipoxin A_4 and B_4 (LXA$_4$, 5S,6R,15S-trihydroxy-7,9,13-*trans*-11-*cis*-eicosatetraenoic acid; LXB$_4$, 5S,14R,15S-trihydroxy-6,10,12-*trans*-8-*cis*-eicosatetraenoic acid), display a number of biological actions, indicating their involvement as modulators of events such as intracellular signal transduction, immunological defence and inflammation (*cf*. Dahlén and Serhan 1990).

In studies of the microcirculation in the hamster cheek pouch by intravital microscopy, we found that LXA$_4$ and one of its precursors, 15-HPETE (15S-hydroperoxy-5,8,11-*cis*-13-*trans*-eicosatetraenoic acid), are vasodilators without the ability to cause increased vascular permeability or leukocyte adherence/emigration (Dahlén *et al*., 1987). Further studies disclosed that LXA$_4$ can inhibit the plasma leakage and leukocyte emigration induced by local challenge in the cheek pouch with the chemoattractant LTB$_4$ (Hedqvist *et al*., 1989a; 1989b; 1990). In this chapter we summarize our findings with LXA$_4$ as a modulator of acute inflammation, and compare some of the anti-inflammatory activities of LXA$_4$ with actions of LXB$_4$ and 15-HETE (15S-hydroxy-5,8,11-*cis*-13-*trans*-eicosatetraenoic acid).

METHODS

As previously described in detail (*cf*. Björk *et al*., 1984; Raud 1989), the left cheek pouch of sodium pentobarbital anaesthetized male Syrian hamsters was everted and

Cell-Cell Interactions in the Release of Inflammatory Mediators
Edited by P. Y-K Wong and C.N. Serhan, Plenum Press, New York, 1991

185

prepared for intravital microscopy under continuous super-
fusion with a bicarbonate buffer maintaining physiological
temperature, pH and gas tensions. Lipoxin A_4, LXB_4, and 15-
HETE were administered topically (included in superfusate
around the cheek pouch), either together with LTB_4 or
histamine for 5 min, or in combination with prostaglandin E_2
(PGE_2) for 15 min, starting 5 min before challenge with LTB_4
or histamine. Plasma extravasation was quantitated using
FITC-dextran (fluorescein isothiocyanate-conjugated dextran,
25 mg/100 mg b.w., i.v.) as a tracer, and assessed by
counting of fluorescent spots per cm^2 of cheek pouch area at
five min intervals. Leukocyte actions and arteriolar
diameters were visualized on a TV monitor via a videocamera
and stored on videotape for subsequent quantitation. The
number of endothelium adherent (endothelial) and emigrated
leukocytes was counted around venules with an inner diameter
of $\approx$ 15 μm (cf. Raud 1989). Leukotriene B_4 and 15-HETE were
from Biomol (Plymouth Meeting, PA, USA) and PGE_2 from Upjohn
(Kalamazoo, MI, USA). Lipoxins A_4 and B_4 were synthesized as
described (Nicolaou et al., 1985). Stock solutions of LTB_4,
LXA_4, and LXB_4 (free acids) in ethanol were stored at -80°C
and the concentration and purity of the compounds was checked
before use by RP-HPLC and UV-spectrometry. FITC-dextran
(mol.wt. 150,000) and histamine dihydrochloride were
purchased from Sigma. Data are expressed as mean
values ± SEM, using Student's t-test for statistical
evaluations.

RESULTS

    In line with previous observations (Björk et al., 1982;
Raud 1989), topical challenge of the cheek pouch with LTB_4
(300 nM, 5 min) provoked plasma leakage from both postcapillary and larger venules. This response was almost abolished
when LTB_4 was administered in the presence of LXA_4 (3 μM)
(Fig. 1, upper panel). Likewise, although to a lesser
degree, the effect of LTB_4 was reduced by LXB_4 (3 μM) and 15-
HETE (3 μM) (Fig. 1, upper panel). On the other hand, LXA_4
did not affect the leukocyte-independent postcapillary
leakage of plasma evoked by histamine (10 μM) (Fig. 1, lower
panel).

    Because the plasma leakage evoked by LTB_4, in contrast
to histamine, is dependent upon accumulation and diapedesis
of leukocytes (Björk et al., 1982), experiments were performed to evaluate the influence of LXA_4 on leukocyte adherence
to the vascular endothelium and the ensuing emigration of
leukocytes into the tissue. The vasodilator PGE_2 (30 nM),
which per se does not cause increased vessel permeability or
leukocyte diapedesis (cf. Raud 1989), was used to enhance the
inflammatory response to LTB_4 (30 nM). In this case, the
leakage response to LTB_4 was also substantially depressed by
LXA_4 (300 nM) (Fig. 2). Moreover, the number of emigrated
leukocytes was reduced in the presence of LXA_4 (Fig. 3, upper
panel), as determined 40 min after challenge with LTB_4.
Interestingly, the initial (after 5 min) LTB_4-induced
adherence of leukocytes to the venular endothelium, a process
required for emigration to occur (Lindbom et al., 1990), was
not significantly attenuated by LXA_4 (Fig. 3, lower panel),

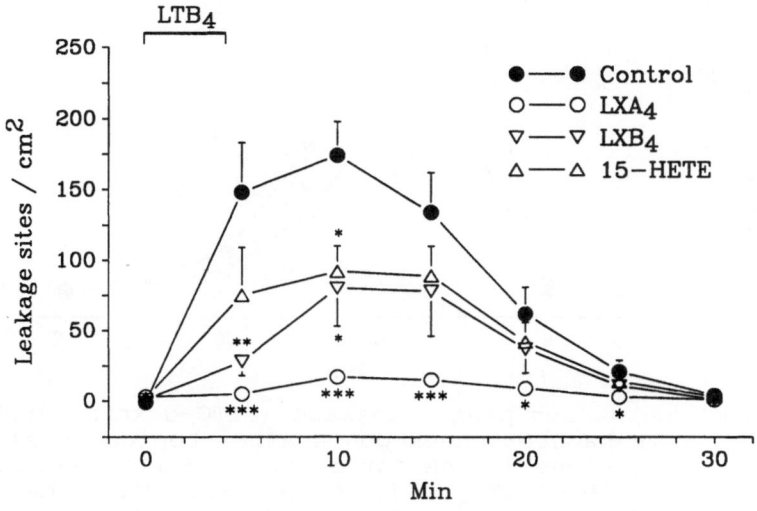

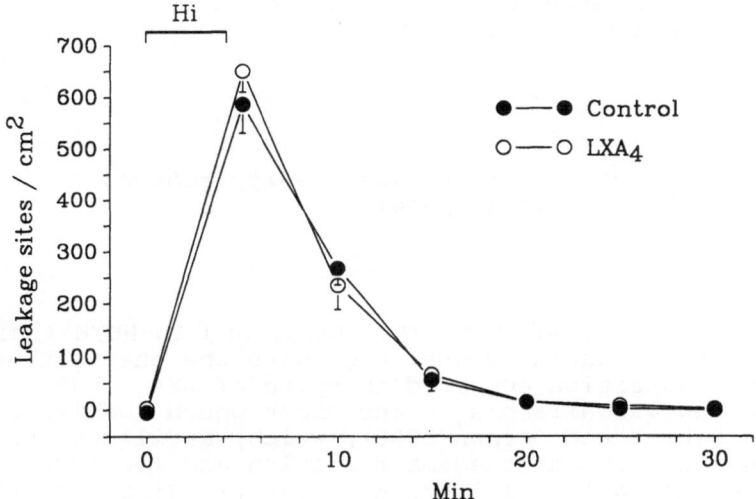

Fig.1. <u>Upper panel</u>: Time course for plasma leakage (FITC-dextran leakage sites) in hamster cheek pouch after topical challenge with LTB_4 alone (300 nM for 5 min, n = 12), and LTB_4 in the presence of LXA_4 (3 μM, n = 6), 15-HETE (3 μM, n = 5), or LXB_4 (3 μM, n = 5). <u>Lower panel</u>: Time course for plasma leakage in hamster cheek pouch after topical challenge with histamine alone (10 μM, for 5 min, n = 11), and histamine in the presence of LXA_4 (3 μM, n = 9). Mean values ± SEM. * = $P < 0.05$, ** = $P < 0.01$, *** = $P < 0.001$ *vs.* control.

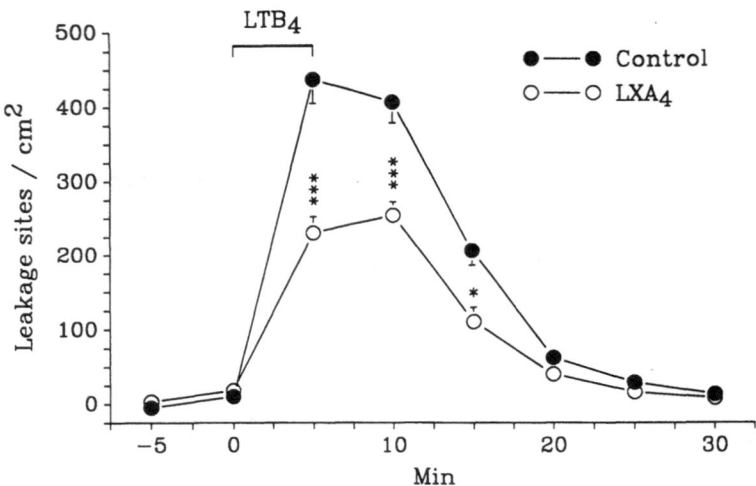

Fig.2. Time course for plasma leakage (FITC-dextran leakage sites) in hamster cheek pouch after topical challenge with LTB_4 alone (30 nM for 5 min, n = 11), and LTB_4 in the presence of LXA_4 (300 nM, n = 8). Experiments performed in the presence of PGE_2 (30 nM from -5 to +10 min). Mean values ± SEM. * = P < 0.05, *** = P < 0.001 *vs.* control.

while the leukocyte adherence was clearly reduced at later time points (Fig. 3, lower panel).

DISCUSSION

We have described that LXA_4, LXB_4, and 15-HETE inhibit plasma leakage induced by challenge with the chemoattractant LTB_4. The inhibition occurred in spite of LXA_4, LXB_4, and 15-HETE being vasodilators in the cheek pouch (Dahlén *et al.*, 1987; *cf.* Dahlén and Serhan 1990), which, as illustrated for PGE_2, generally enhances edema formation and leukocyte emigration (*cf.* Williams 1983; Raud 1989). Since the LXA_4-induced inhibition of plasma leakage evoked by LTB_4 appeared to correlate with suppression of the leukocyte emigration, we suggest that LXA_4 interferes with mechanisms for leukocyte-dependent alterations of microvascular permeability. This is further supported by the fact that the leukocyte-independent extravasation of plasma induced by histamine was unaffected by LXA_4. Future studies will show if our *in vivo* findings relate to the results of Lee *et al.*, (1989), documenting that LXA_4 and LXB_4 can inhibit LTB_4- and f-met-leu-phe-induced neutrophil chemotaxis *in vitro*. Our observation that LXA_4 did not inhibit the initial adherence of leukocytes to the endothelium may suggest an action by LXA_4 at target sites other than leukocyte adhesion-molecules or the LTB_4-receptor. In this context, LXA_4 does not inhibit LTB_4-induced contrac-

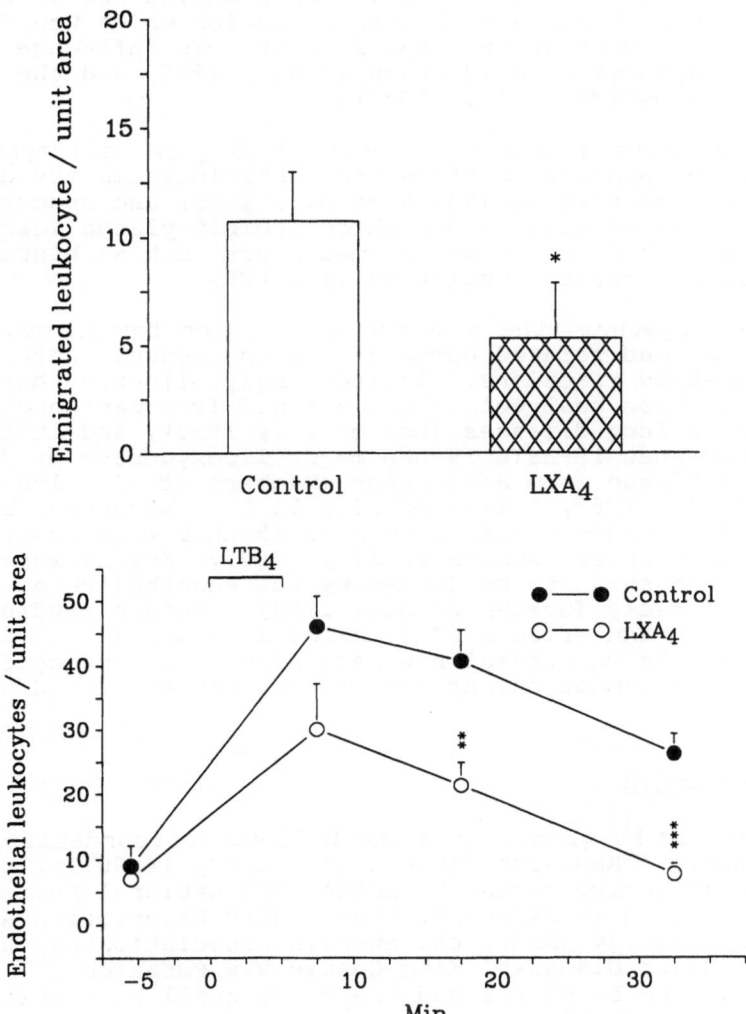

Fig.3. Upper panel: Increase in number of emigrated
leukocytes in hamster cheek pouch 40 min after topical
challenge with LTB_4 alone (30 nM for 5 min, n = 10),
and LTB_4 in the presence of LXA_4 (300 nM, n = 7).
Lower panel: Changes in the number of endothelial
leukocytes in hamster cheek pouch after topical
challenge with LTB_4 alone (30 nM for 5 min, n = 11),
and LTB_4 in the presence of LXA_4 (300 nM, n = 8).
Experiments performed in the presence of PGE_2 (30 nM
from -5 to +10 min). Mean values ± SEM. * = P < 0.05,
** = P < 0.01, *** = P < 0.001 vs. control.

tions of the guinea pig lung strip *in vitro* (Dahlén, S.-E., unpublished observation), further indicating that LXA_4 is not an LTB_4-receptor antagonist. As an alternative, the lipoxins may interfere with intracellular events during the process of leukocyte emigration, a mechanism of action which would be in line with the observations that lipoxins can influence protein phosphorylation (Hansson *et al.*, 1986) and the Golgi apparatus (Ramstedt *et al.*, 1987).

The apparent selective effects of LXA_4 on leukocyte-dependent inflammation differs from anti-inflammatory drugs such as corticosteroids (Björk *et al.*, 1985) and nedocromil sodium (Dahlén *et al.*, 1989), which inhibit plasma leakage induced by both directly acting mediators such as histamine as well as chemotactic mediators like LTB_4.

Taken together, the observations prompt for an evalua-tion of LXA_4 and related compounds as endogenous inhibitors of inflammatory reactions. Interestingly, lipoxins have been isolated in bronchoalveolar lavage fluid from patients with inflammatory lung diseases (Lee *et al.*, 1990), and it is well established that 15-HETE is the major lipoxygenase product in human lung tissue from asthmatics (Hamberg *et al.*, 1980; Dahlén *et al.*, 1983; Kumlin *et al.*, 1990). Moreover, bronchi of asthmatics had elevated levels of 15-HETE when compared with non-asthmatics (Kumlin *et al.*, 1990). Furthermore, 15-HETE has been shown to be formed by the endothelium of human umbilical vessels (Gorman *et al.*, 1985). Future studies on the formation and effects of lipoxins in experimental and clinical models may reveal new mechanisms for leukocyte actions in inflammation, as well as new targets for drug development.

ACKNOWLEDGEMENTS

Supported by grants from the Wallenberg Foundation, the Swedish Medical Research Council (14X-4342, 14X9071), the Institute of Environmental Medicine, the National Environment Protection Board (5324060-2), the Swedish Association Against Asthma and Allergy (RmA), the Swedish Association Against Chest and Heart diseases, King Gustav V:s Research Foundation, the L. Hierta and the M. Bergvall Foundations, and Karolinska Institutet.

REFERENCES

Björk, J., Goldschmidt, T., Smedegard, G., and Arfors, K.-E., 1985, Methylprednisolone acts at the endothelial cell level reducing inflammatory responses, Acta Physiol. Scand., 123:221.

Björk, J., Hedqvist, P., and Arfors, K.-E., 1982, Increase in vascular permeability induced by leukotriene B4 and the role of polymorphonuclear leukocytes, Inflammation, 6:189.

Björk, J., Smedegård, G., Svensjö, E., and Arfors, K.-E., 1984, The use of the hamster cheek pouch for intravital microscopy studies of microvascular events *Prog. appl. Microcirc.*, 6:41.

Dahlén, S.-E., Björck, T., Kumlin, M., Sydbom, A., Raud, J., Palmertz, U., Franzén, L., Grönneberg, R., and Hedqvist P., 1989, Dual Inflammatory action of Nedocromil sodium on antigen-induced inflammation, *Drugs*, 37(Suppl. I):63.

Dahlén, S.-E., Hansson, G., Hedqvist, P., Björck, T., Granström, E., and Dahlén, B., 1983, Allergen challenge of lung tissue from asthmatics elicits bronchial contraction that correlates with the release of leukotriene C_4, D_4, and E_4, *Proc. Natl. Acad. Sci. USA*, 80:708.

Dahlén, S.-E., Raud, J., Serhan, C.N., Björk, J., and Samuelsson, B., 1987, Biological activities of Lipoxin A include lung strip contraction and dilation of arterioles *in vivo*, *Acta Physiol. Scand.*, 130:643.

Dahlén, S.-E. and Serhan, C.N., 1991, Lipoxins: Bioactive lipoxygenase interaction products, in: "Lipoxygenases in Inflammation," A. Wong and S.T. Crooke, ed., Academic Press, San Diego, p. 235.

Gorman, R.R., Oglesby, T.D., Bundy, G.L. and Hopkins, N.L., 1985, Evidence for 15-HETE synthesis by human umbilical vein endothelial cells, *Circulation*, 72:708.

Hamberg, M., Hedqvist, P., and Rådegran, K., 1980, Identification of 15-hydroxy-5,8,11,13-eicosatetraenoic acid (15-HETE) as a major metabolite of arachidonic acid in human lung, *Acta Physiol. Scand.*, 110:219.

Hansson, A., Serhan, C.N., Haeggström, J., Ingelman-Sundberg, M., Samuelsson, B., and Morris, J., 1986, Activation of protein kinase C by lipoxin A and other eicosanoids. Intracellular action of oxygenation products of arachidonic acid, *Biochem. Biophys. Res. Commun.*, 134:1215.

Hedqvist, P., Dahlén, S.-E., Raud, J., Lindbom, L., Thuresson-Klein, Å., and Nicolaou, K.C., 1989a, Aspects of eicosanoids in inflammation, in: "Prostanoids and Drugs," B. Samuelsson, F. Berti, G.C. Folco, and G.P. Velo, ed., Plenum Press, New York, p. 169.

Hedqvist, P., Raud, J., Palmertz, U., Haeggström, J., Nicolaou, K.C., and Dahlén, S.-E., 1989b, Lipoxin A_4 inhibits leukotriene B_4-induced inflammation in the hamster cheek pouch, *Acta Physiol. Scand.*, 137:571.

Hedqvist, P., Raud, J., and Dahlén, S.-E., 1990, Microvascular actions of eicosanoids in the hamster cheek pouch, *Adv. Prostaglandin Thromboxane Leukotriene Res.*, 20:153.

Kumlin, M., Hamberg, M., Granström, E., Björck, T., Dahlén, B., Matsuda, H., Zetterström, O., and Dahlén, S.-E., 1990, 15(S)-hydroxyeicosatetraenoic acid is the major arachidonic acid metabolite in human bronchi: Association with airway epithelium, Arch. Biochem. Biophys., 282:254.

Lee, T.H., Crea, A.E.G., Gant, V., Spur, B.W., Marron, B.E., Nicolaou, K.C., Reardon, E., Brezinski, M., and Serhan, C.N., 1990, Identification of Lipoxin A_4 and its relationship to the sulfidopeptide leukotrienes C_4, D_4, and E_4 in the bronchoalveolar lavage fluid obtained from patients with selected pulmonary diseases, Am. Rev. Respir. Dis., 141:1453.

Lee, T.H., Horton, C.E., Kyan-Aung, U., Haskard, D., Crea, A.E.G., and Spur, B.W., 1989, Lipoxin A_4 and B_4 inhibit chemotactic responses of human neutrophils stimulated by leukotriene B_4 and N-formyl-L-methionyl-L-leucyl-phenylalanine, Clin. Sci., 77:195.

Lindbom, L., Lundberg, C., Prieto, J., Raud, J., Nortamo, P., Gahmberg, C.G., and Patarroyo, M., 1990, Rabbit leukocyte adhesion molecules CD11/CD18 and their participation in acute and delayed inflammatory responses and leukocyte distribution in vivo, Clin. Immunol. Immunopathol. 57:105.

Nicolaou, K.C., Veale, C.A., Webber, S.E., Katerinopoulus, K., 1985, Stereocontrolled total synthesis of Lipoxins A, J. Am. Chem. Soc., 107:7515.

Ramstedt, U., Serhan, C.N., Nicolaou, K.C., Webber, S.E., Wigzell, H., and Samuelsson, B., 1987, Lipoxin A-induced inhibition of natural killer cells: Studies on stereospecificity and mode of action, J. Immunol., 138:266.

Raud J., 1989, Intravital microscopic studies on acute mast cell-dependent inflammation, Acta Physiol. Scand., 135(Suppl. 578):1.

Serhan, C.N., Hamberg, M., and Samuelsson, B., 1984, Lipoxins: Novel series of biologically active compounds formed from arachidonic acid in human leukocytes, Proc. Natl. Acad. Sci. USA., 81:5335.

Williams, T.J., 1983, Interactions between prostaglandins, leukotrienes and other mediators of inflammation, Br. Med. Bull., 39:239.

PAF AND TNFα INTERACTIONS IN THE PATHOPHYSIOLOGY OF SEPTIC SHOCK

Reuven Rabinovici, Tian Li Yue, Jerome Vernick, and Giora Feuerstein

Department of Surgery, Jefferson Medical College, Philadelphia, PA 19107 (R.R. and J.V.) and Department of Cardiovascular Pharmacology, SmithKline Beecham Pharmaceuticals, King of Prussia, PA 19406 (T-L.Y. and G.F.)

INTRODUCTION

The septic shock syndrome is a medical emergency caused in part by the release into the circulation of bacterial lipopolysaccharide (LPS) endotoxin. The evolving concept in the pathophysiology of septic shock suggests that endotoxin induces its detrimental effects indirectly through the production of multiple mediators among which platelet activating factor (PAF) and tumor necrosis factor-α (TNFα) play a major role. The role of PAF and TNFα in endotoxemia was inferred from the following observations: 1) many of the manifestations of septic shock such as hypotension, hemoconcentration, thrombocytopenia and leukopenia can be produced by PAF and TNFα (Table 1); 2) elevated levels of PAF and TNFα were found in various septic conditions (Table 2); and 3) PAF antagonists and anti-TNFα neutralizing antibodies were shown to convey significant protection in animal models of septic shock (Table 3).

Since it seems that LPS exerts its biological effects in part through PAF and TNFα it is possible that interactions between these mediators further contribute to the mechanisms involved in the pathogenesis of septic shock. The purpose of this chapter is to give a brief overview on the interrelationships between PAF and TNFα in the septic state and to summarize the in vitro and in vivo data that link these two mediators in the pathophysiology of the septic shock syndrome.

IN VITRO STUDIES ON PAF/TNFα INTERACTIONS

Several in vitro studies in cell systems suggested a possible link between PAF and TNFα. Camussi et al. (1) observed that murine (1-10 ng/ml) and human (10-50 ng/ml) TNFα stimulated PAF production by cultured rat peritoneal macrophages, rat polymorphonuclear neutrophils, and by human vascular endothelial cells cultured

Table 1. Biological responses to administration of endotoxin, PAF and TNFα. ↑ increase; ↓ decrease; 0 no effect; ? unknown; * dependent on phase of septic shock

	Endotoxin	PAF	TNFα
blood pressure	↓	↓	↓
heart rate	↑	↑	↑
cardiac output	↑ or ↓ *	↓	0 or ↓
peripheral resistance	↓ or ↑ *	↓	0 or ↑
platelets	↓	↓	↓
leukocytes	↓	0	↓
hematocrit	↑	↑	↑
TXB_2	↑	↑	?
$6\text{-keto-PGF}_{1\alpha}$	↑	0	?
epinephrine	↑	↑	↑
norepinephrine	↑	↑	↑
cortisol	↑	↑	↑
lactate	↑	↑	↑
capillary permeability	↑	↑	↑
complement activation	↑	?	?
lethality	↑	↑	↑
DIC	↑	?	?

Table 2.   Serum TNFα (A) and PAF (B) levels in endotoxic shock. IV - intravenous, IA - intraarterial, IP - intraperitoneal, bac-live bacteriae, REF - reference, n.p. - not published

A.

ENDOTOXIN	DOSE/MODE	SPECIES	TNF	REF
n.p.	400 µg/mice IP	mice	6.2×10^3 U/ml at 60 min	(34)
E. coli	0.1 mg/kg IV	rat	1.6×10^5 U/ml at 90-120 min	(35)
E. coli	5 mg/kg IV	rat	3.5×10^5 U/ml at 60-90 min	(36)
E. coli	1 mg/kg IV	rat	9.5×10^3 U/ml at 90 min	(37)
S. minessota	10 µg/rabbit IV	rabbit	2.5×10^3 U/ml at 45-100 min	(38)
E. coli	30 µg/animal IV	rabbit	9.5×10^3 U/ml at 120 min	(39)
E. coli	1.2×10^{11} bac/kg IA	baboon	1.7×10^3 U/ml at 120 min	(40)
E. coli	10^{11}-10^{12} bac/kg IV	baboon	810 U/ml at 90 min	(41)
E. coli	4 ng/kg IV	human	10 U/ml at 120 min	(42)
E. coli	20 µg/kg IV	human	14 U/ml at 90 min	(41)

B.

ENDOTOXIN	DOSE/MODE	SPECIES	PAF	REF
E. coli	50 mg/kg IV	rat	4.7 nM at 10 min	(43)
S. enteritdis	20 mg/kg IP	rat	13.7 ng/ml at 20 min	(21)
S. typhosa	0.5 mg/kg IV	rat	8.3 ng/ml at 180 min	(16)
S. aureus	-	septic	4.1 nM	(44)
P. mirabilis	-	children	2.2 nM	
H. Influenzae	_		1.3 nM	
K. pneumonia	_		1.8 nM	

Table 3. Summary of PAF antagonists (A) and neutralizing anti-TNFα antibodies (B) which protect against lethality in animals exposed to endotoxic shock. C- control (endotoxin alone), T- treatment with PAF antagonist or anti-TNFα antibody, Δ- change in mortality between the control and the treatment groups. Ref - reference, IV - intravenous, IP - intraperitoneal, SC - subcutaneous, np - not published, Ab - antibody.

A.

Antagonist/Ab	Endotoxin	Species	Mortality C	T	Δ	Ref
CV 3988 10 mg/kg, IV, 5 min before and 50 min after LPS	S. enteritdis 20 mg/kg, IP	rat	6/6 at 24 hrs	4/11	-64%	(21)
CV 3988 10 mg/kg, IV, with LPS	E. coli 5 mg/kg, IV	rat	8/12 at 20 hrs	7/24	-38%	(45)
BN 52021 5 mg/kg, SC, 15 min before LPS	S. enteritidis 5 mg/kg, IV	rat	np at 24 hrs	np	-45%	(46)
WEB 2086 10 mg/kg, PO, 60 min before LPS, and 10 mg/kg, IV, with LPS	E. coli 7.5 mg/kg, IV	rat	18/18	0/12	-100%	(47)
BN 50739 10 mg/kg, IP, 30 min before LPS	E. coli 14.4 mg/kg, IV	rat	9/15	0/11	- 60%	(20)
BN 52021 5 mg/kg, IV, 15 min before LPS	E. coli 1.0 mg/kg, IV	dog	5/5 at 72 hrs	0/5	-100%	(48)

B.

Antagonist/Ab	Endotoxin	Species	Mortality C	T	Δ	Ref
polyclonal rabbit anti murine TNFα 200 μl, IP 1.5 hrs before LPS	E. coli 400 μg/kg, IP	mice	7/14 at 7 d	0/14	-100%	(49)
polyclonal goat anti human TNFα, 8 ml, IV 45 min before LPS	S. minnesota 10 μg/kg, IV	rabbit	5/5 at 24 hrs	0/5	-100%	(50)
monoclonal murine anti human TNFα, F(ab') fragments, 10 mg/kg, IV	E. coli 1.2×10^{11} bac/kg,	baboon	6/6 at 48 hrs	0/3	-100%	(51)

from the umbilical cord vein. A possible mechanism of this interaction could be the activation by TNFα of 1-0-alkyl-2(R)-lyso-glycero-3-phosphocholine:acetyl CoA acetyltransferase (1-4), a key enzyme in the biosynthesis of PAF. Interestingly, PAF (5-50 ng/ml for 1-4 hr) did not induce appreciable release of TNFα into the culture medium of rat peritoneal macrophages. In contrast, Hayashi et al. (5) reported that PAF (10^{-5}M) can release TNFα from guinea pig peritoneal macrophages with maximal activity at 8 hr. Also, Bonavida et al. (6,7), using a sensitive radio-immunoassay, demonstrated that PAF facilitates in a dose-dependent manner the production of TNFα by peripheral blood derived monocytes. TNFα activity was present initially, but declined after 24 hr. Additional evidence that PAF induces TNFα production was provided by Rola-Pleszcynski et al. (8) who used cell cultures of human and rodent lympocytes and macrophages. Other studies linked PAF and TNFα-induced endothelial cell injury. For example, PAF antagonists attenuated priming of PMNs by TNFα, including oxygen radicals and leukotrienes production (9).

To date, only one in vitro study is available where PAF/TNFα interactions were evaluated in relationship to endotoxin challenge (10). In this study, pre- or post-treatment with the new PAF antagonist RP 55778, was reported to inhibit TNFα production by isolated murine macrophages exposed to Salmonella enteritidis endotoxin. Also, high concentrations of RP 55778 added simultaneously to endotoxin, completely inhibited the TNFα mRNA signal.

IN VIVO STUDIES ON PAF/TNFα INTERACTIONS

Possible PAF/TNFα relationships were documented in several in vivo non-septic tissue injury models. For example, in guinea pigs superfused with TNFα, a subsequent injection of a low-dose PAF into the mesenteric vasculature resulted in increased thrombosis. This synergistic effect was inhibited by pretreatment with the PAF antagonist BN 52021 or anti-TNFα antibodies (Bourgain and Braquet, in preparation). Additional evidence to support interactions between PAF and TNFα was obtained in studies where TNFα (0.5-1.0 mg/kg, i.v.) increased PAF synthesis by the intestinal tissue in rats (11). Also, relationship between PAF and TNFα has been recorded in rats where pretreatment with TNFα amplified the extravasation induced by a low and non-active dose of PAF (12). Recently, combined administration of non-active doses of PAF and TNFα caused a significant loss of glycosaminoglycans in cartilage tissue (13) and the administration of intrapulmonary TNFα triggered local PAF production in rat immune complex alveolitis (14).

Several studies investigated the complicated interactions between PAF and TNFα in models of endotoxemia. Heuer and Weber (15) demonstrated that pretreatment of mice with either S. typhosa endotoxin or TNFα resulted in a synergistic increase in PAF-induced mortality whereas when administered alone at similar doses, both TNFα and endotoxin did not effect mortality. Other investigators (16) reported that endogenously-produced TNFα (primed by zymosan) increased the hypotensive response and bowel injury induced by LPS in rats and that pretreatment with the PAF antagonist SRI 63-441 markedly ameliorated the hypotension and tissue injury. These authors also demonstrated that following LPS injection, zymosan-primed rats had higher serum PAF and TNFα levels than nonprimed animals. In contrast, Myers et al. (17) observed no beneficial effect of

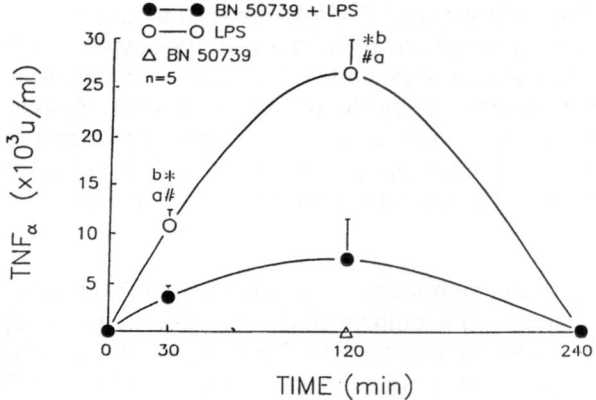

Fig. 1.   Effect of pre-treatment with BN 50739 (10 mg/kg) on TNFα (n=5) response to LPS in rats. *P<0.05, #P<0.01. a - vs. baseline, b - vs. other group.

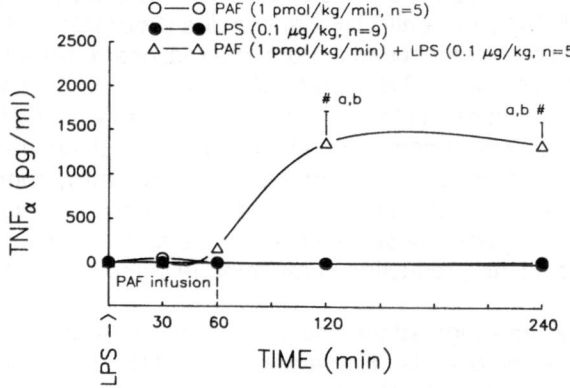

Fig. 2.   Priming effect of PAF on TNFα response to endotoxin in the rat. LPS-lipopolysaccharide; a - vs. basal value; b - vs. all other groups.

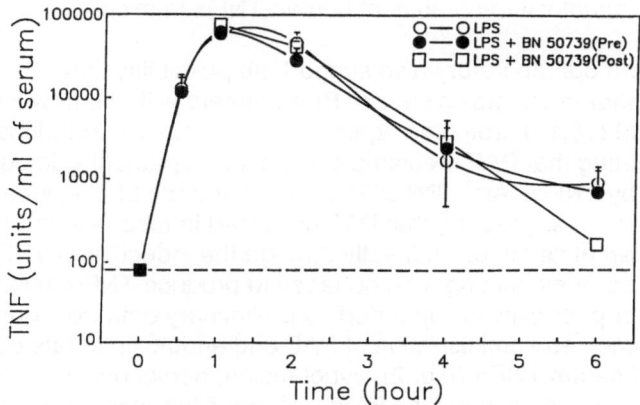

Fig. 3.   Effect of pretreatment (-30 min) or post-treatment (+15 min) with BN
50739 (10 mg/kg, i.p.) on TNFα response to endotoxin (50 µg/kg, i.v.)
in the rabbit (n=9). LPS - lipopolysaccharide.

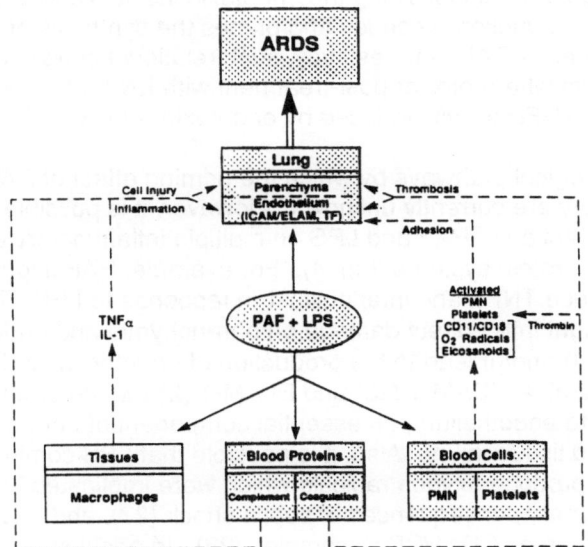

Fig. 4.   A scheme of the possible mechanisms by which the priming effect of
PAF facilitates LPS-induced lung injury. We hypothesize that lung
injury results from the action of PAF and LPS on multiple inflammatory
cells and plasma protein systems: 1) Primed macrophages challenged
with LPS can produce TNFα and interleukin-1 which in turn inflict injury
on pulmonary cells and lead to the production of endothelial surface
adhesion molecules; 2) PMN primed by PAF and exposed to LPS are
activated to produce oxygen radicals and express adhesion molecules;
3) Platelets and coagulation factors are primed to produce a
procoagulant state; 4) PAF primes for LPS-induced activation of the
complement system; and 5) LPS and PAF in combination might injure
pulmonary cells directly through yet unknown mechanisms. For more
details see text.

pre- or post-treatment with the PAF antagonist BN 52021 on the lethality of combined low dose endotoxin/recombinant human TNFα in mice.

Recent data from our laboratory also support the possibility that PAF and TNFα interact with each other in endotoxic shock. Pretreatment with the new PAF antagonist BN 50739 (18,19) protected against recombinant TNFα-induced mortality in the rat (20), indicating that PAF mediates at least some pathophysiological processes initiated by TNFα. Also, BN 50739 attenuated the LPS-induced elevation of plasma TNFα (Fig. 1), suggesting that PAF produced in response to LPS stimulation exerts part of its effects indirectly through the induction of TNFα synthesis and release or by priming macrophages to produce TNFα in response to endotoxin. The latter possibility is supported by preliminary data from our laboratory showing that the combined administration of PAF and endotoxin in rats caused marked elevation of serum TNFα (Fig. 2), hypotension, hemoconcentration, leukopenia, thrombocytopenia, and lung injury, whereas intravenous infusion of subthreshold doses of PAF or endotoxin alone failed to elicit these changes.

Taken together, these interactions suggest a positive feedback loop between PAF and TNFα. Moreover, since the plasma levels of PAF peak as early as 20 min after the administration of LPS (21), whereas endotoxin-induced elevation of plasma TNFα peaks later at 90-120 min following the endotoxic insult (Table 2), it is conceivable that PAF production precedes and primes the synthesis of TNFα which, in turn, acts later to release PAF. Interestingly, such relationship was not demonstrated in rabbits where pre- or post-treatment with BN 50739 did not effect the increase in serum TNFα activity induced by endotoxin (Fig. 3) (22).

The pathophysiological pathways by which the priming effect of PAF facilitate LPS-induced lung injury are currently unknown. However, it is possible that lung injury results from the action of PAF and LPS on multiple inflammatory cellular elements and plasma protein systems (Fig. 4). For example, PAF might prime macrophages to produce TNFα and interleukin-1 in response to LPS. These humoral mediators might inflict direct damage on parenchymal and endothelial pulmonary cells (23,24) and/or lead to the production of endothelial surface adhesion molecules such as ICAM-1 (25) and ELAM-1 (26) which promote leukocyte adherence to endothelium, an essential component of neutrophil-mediated vascular and tissue injury. Also, it is possible that LPS combined with PAF activate PMN to produce oxygen radicals which were implicated in the development of pulmonary consequences of septic shock (27), and express adhesion molecules such as CD11/CD18 complex (28). In addition, LPS administered with PAF can possibly activate platelets to promote disseminated intravascular coagulation (DIC) (29) and TXA_2 production (20,30), two known phenomena of sepsis. Activation of the complement system, a process involved in producing the syndrome of septic shock (31,32), might be another mechanism leading to lung injury. PAF might also prime the activation of the contact and coagulation cascade by endotoxin and therefore facilitate the procoagulant state (DIC) observed in septic shock (33). Finally, LPS and PAF might injure pulmonary cells directly through yet unknown mechanisms.

CONCLUSION

PAF and TNFα are both produced in response to endotoxin challenge and act directly on target cells to evoke some of the pathophysiological consequences of septic shock. In addition, an increasing body of evidence indicates that these two mediators interact to amplify each other production and to enhance their biological effects. Recent in vivo data suggests priming mode relationship between PAF and TNFα, i.e., PAF primes for endotoxin-induced TNFα production, lung injury and cardiovascular shock. The priming action of PAF possibly represents an evolving concept that septic shock results from interactions among minuscule doses of multiple mediators. However, the relative contribution of PAF/TNFα relationship to the overall pathophysiological mechanisms of the septic shock syndrome are still obscure and should await further investigations.

REFERENCES

1. G. Camussi, F. Bussolino, G. Salvido, and C. Baglioni, J Exp Med 166:1390 (1987).
2. F. Bussolino, F. Breviario, N. Aglietta, F. Sanavio, A. Bosia, and E. Dejana, Biochim Biophys Acta 927:43 (1987).
3. F. Bussolino, F. Breviario, C. Tetta, M. Aglietta, A. Mantovani, and E. Dejana, J Clin Invest 77:2027 (1986).
4. F. Bussolino, G. Camussi, and C. Baglioni, J Biol Chem 263:11856 (1988).
5. H. Hayashi, I. Kudo, T. Kato, and K. Inoue, in: "Platelet-activating factor and diseases," K. Saito, and D. J. Hanahan, eds., Int Med Publishers, Tokyo (1989).
6. B. Bonavida, J. M. Mencia-Huerta, and P. Braquet, J Lipid Mediators (in press).
7. B. Bonavida, J. M. Mencia-Huerta, and P. Braquet, J Allergy Appl Immunol (in press).
8. M. Rola-Pleszczyski, J. Bosse, E. Bissonnette, and C. Dubois, Prostaglandins 35:802 (1988).
9. M. Paubert-Braquet, P. Koltz, M. O. Longchamp, J. Guilbaud, D. Hosford, and P. Braquet, J Lipid Mediators (in press).
10. A. Floch, A. Bousseau, E. Hetier, F. Floch, P-E. Bost, and I. Cavero, J Lipid Mediators 1:349 (1989).
11. X-M. Sun, and W. Hsueh, J Clin Invest 81:1328 (1988).
12. M. G. Sirois, G. E. Plante, J. J. Oppenheim, P. Braquet, and P. Sirois, J Lipid Mediators (in press).
13. D. Howat, F. Desa, C. Chander, A. Moore, and D. A. Willoughby, J Lipid Mediators (in press).
14. J. S. Warren, P. A. Barton, D. M. Mandel, and K. Matrosic, Lab Invest 63:764 (1990).
15. H. Heuer, and K. H. Webber, Prostaglandins 35:814 (1988).
16. X-M. Sun, W. Hsueh, and G. Torre-Amione, Am J Pathol 136:949 (1990).
17. A. K. Myers, J. W. Robey, and R. M. Price, Br J Pharmacol 99:499 (1990).
18. T-L. Yue, R. Rabinovici, M. Farhat, and G. Feuerstein, J Lipid Mediators 3:13 (1991).
19. T-L. Yue, R. Rabinovici, M. Farhat, and G. Feuerstein, Prostaglandins 39:469 (1990).
20. R. Rabinovici, T-L. Yue, M. Farhat, E. F. Smith III, M. Slivjak, and G. Feuerstein, J Pharmacol Exp Ther 255:256 (1990).
21. S. W. Chang, C. O. Fedderson, P. M. Henson, and N. F. Voelkel, J Clin Invest 79:1498 (1987).

22. T-L. Yue, M. Farhat, R. Rabinovici, P. Y. Perera, S. N. Vogel, and G. Feuerstein, J Pharmacol Exp Ther 254:976 (1990).
23. K. E. Stephens, A. Ishizaka, J. W. Larrick, and T. A. Raffin, Am Rev Respir Dis 137:1364 (1988).
24. D. C. Hocking, P. G. Phillips, T. J. Ferro, and A. Johnson, Circ Res 67:68 (1990).
25. M. L. Dustin, R. Rothlein, A. K. Bhan, C. A. Dinarello, and T. A. Springer, J Immunol 137:245 (1986).
26. M. P. Bevilacqua, M. E. Wheeler, J. S. Pober, W. Fiers, D. L. Mendrick, R. S. Cotran, and M. A. Gimbrone, in: "Leukocyte Emigration and its Sequelae," H. Z. Movat, ed., Karger, Basel (1987).
27. H. Shennib, R. C. Chiu, D. S. Mulder, G. K. Richards, and J. Prentis, Am Rev Resp Dis 130:444 (1984).
28. N. B. Vedder, B. W. Fouty, R. K. Winn, J. M. Harlan, and C. L. Rice, Surgery 106:509 (1989).
29. D. C. Morrison and R. J. Ulevitch, Am J Pathol 93:527 (1978).
30. J. A. Cook, W. C. Wise, and P. V. Halushka, J Clin Invest 65:227 (1980).
31. W. W. Spink and J. Vick, J Exp Med 114:501 (1961).
32. W. W. Spink, R. B. Davis, R. Potter, and S. Chartrand, J Clin Invest 43:696 (1964).
33. C. G. Cochrane, in: "Handbook of Endotoxin," L. B. Hinshaw, ed., Elsevier, Amsterdam, New York, Tokyo (1985).
34. G. F. Evans and S. H. Zuckerman, Agents and Actions 26:329 (1989).
35. G. Feuerstein, J. M. Hallenbeck, B. Vanatta, R. Rabinovici, P. Y. Perara, and S. N. Vogel, Circ Shock 30:265 (1990).
36. A. Wagga, Clin Immunol Immunopathl 45:348 (1987).
37. G. J. Bagby, J. J. Thompson, L. A. Wilson, and R. M. Hazel, Circ Shock 28:385 (1989).
38. J. C. Mathison, E. Wolfson, and R. J. Ulevitch, J Clin Invest 81:1925 (1988).
39. B. Beutler, I. W. Milsark, and A. Cerami, J Immun 135:3972 (1985).
40. K. J. Tracey, Y. Fong, D. J. Hesse, K. R. Manouge, A. T. Lee, G. C. Kuo, S. F. Lowry, and A. Cerami, Nature 330:662 (1987).
41. D. G. Hesse, K. J. Tracey, Y. Fong, K. R. Manouge, M. A. Palladino, A. Cerami, T. Shires, and S. F. Lowry, Surg Gynecol Obstet 166:147 (1988).
42. H. R. Michie, K. R. Manouge, D. R. Spriggs, A. Revhaug, S. O'Dwyer, C. A. Dinarello, A. Cerami, S. M. Wolff, and D. W. Wilmore, N Engl J Med 318:1481 (1988).
43. T. Doebber, M. Wu, J. Robbins, B. Choy, M. Chang, and T. Shen, Biochem Biophys Res Commun 127:799 (1985).
44. F. Bussolino, M. G. Porcellini, L. Varese, and A. Bosia, Throm Res 48:619 (1987).
45. Z. Terashita, Y. Imura, K. Nishikawa, and S. Sumida, Eur J Pharmacol 109:257 (1985).
46. A. Etienne, F. Hecquet, C. Soulard, B. Spinnewyn, F. Clostre, and P. Braquet, Agents and Actions 17:368 (1985).
47. J. Casals-Stenzel, Eur J Pharmacol 135:117 (1987).
48. J. Fletcher, M. Earnest, J. Moore, A. Disimone, and N. Abumard, Circ Shock 27:359 (1989).
49. B. Beutler, I. W. Milsark, and A. Cerami, Science 229:869 (1985).

50. J. C. Mathison, E. Wolfson, and R. J. Ulevitch, J Clin Invest 81:1925 (1988).
51. K. J. Tracey, Y. Fong, D. J. Hesse, K. R. Manouge, A. T. Lee, G. C. Kuo, S. F. Lowry, and A. Cerami, Nature 330:662 (1987).

LTB4 AND PAF IN THE CYTOKINE NETWORK

Marek Rola-Pleszczynski

Immunology Division, Department of Pediatrics, Faculty of
Medicine, University of Sherbrooke, Sherbrooke QC
Canada, J1H 5N4

I.   INTRODUCTION

In most instances when the host is asked to mount a specific immune response, the initial event or the subsequent response, or both, are associated with non-specific inflammation. In recent years, with the identification of numerous molecular and cellular components of the inflammatory response, investigators have initiated studies of potential interactions between the latter and the protagonists of more specific immunological reactions. In particular, soluble mediators of inflammation, produced by phagocytes, endothelial cells or nerves have been studied in regard to their possible modulation of lymphocyte and monocyte-macrophage functions. In the present chapter, we will focus on a group of lipid molecules, the leukotrienes (LTs) and platelet-activating factor (PAF), and their potential role in modulation of the immune response at the cytokine level.

LTs are derived from the oxidative metabolism of arachidonic acid[1-3]. While sharing some of the myotropic properties of the other leukotrienes, LTB_4 has been found to exert very strong leukocytotropic activities. It is one of the most powerful chemokinetic[4] and chemotactic[5] agents and it can induce neutrophil aggregation[4], degranulation[6], hexose uptake[7] and enhanced binding to endothelial cells[8]. It can also induce cation fluxes[9], augment cytoplasmic calcium concentrations from intracellular pools[10,11] and activate phosphatidylinositol hydrolysis[12,13].

LTB_4 is rapidly synthesized by phagocytic cells, principally neutrophilic PMN[3] and alveolar macrophages[14], upon challenge with a variety of stimuli such as microbial pathogens[15], toxins, aggregated immunoglobulin[16], particulate material[17] or ionophores, especially when exogenous arachidonic acid is available. Its synthesis can also be induced by platelet-derived 12-HPETE[18] or PAF[19,20].

Effective cell separation techniques suggest that LTB_4 is not a product of lymphocytes[21-23], in spite of several earlier reports to the contrary[24,25]. Under a variety of situations, LTB_4 seems to be found in lymphocyte cultures

Cell-Cell Interactions in the Release of Inflammatory Mediators
Edited by P. Y-K Wong and C.N. Serhan, Plenum Press, New York, 1991

only in the presence of contaminating monocytes[22]. Recently, however, human T and B cells and lymphocytic cell lines have been shown to convert LTA_4 to LTB_4, while failing to produce LTB_4 following stimulation with ionophore alone[26]. That the gene(s) involved in LT synthesis can be derepressed is suggested however by the observed production of LTC_4 by a human hybridoma formed by fusing CD8+ T cells with CEM-6 lymphoma (CD4+) cells[27].

LTB_4 has been detected in significant concentrations in inflammatory synovial exudates[28], psoriatic skin lesions[29], peritonitis[30] and inflammatory bowel disease[31]. In an animal model of colitis, local production of LTB_4 and LTC_4 correlates with inflammatory cell infiltrates[32].

PAF is also a potent mediator of inflammation, being released during numerous and varied inflammatory, allergic or immune reactions and participating in their development. The discovery of selective PAF receptor antagonists has contributed immensely to the definition of a role for PAF in many disease processes.

Because both PAF and leukotrienes can derive from the same membrane phospholipid, alkyl-arachidonyl-glycero-phosphocholine[33], they may be produced simultaneously, or in sequence, and some of the effects of PAF appear to be mediated by subsequent LT production.

Cellular sources of PAF are numerous and include various leukocyte populations, endothelial cells and probably other cell types[34]. Resting lymphocytes are unable to produce PAF, possibly due to a lack of acetyl transferase[35]. However, several T and B cell lines[36] and large granular lymphocytes[37] can be activated to produce PAF under certain circumstances. Many of the cells or tissues which produce PAF are themselves targets of PAF-induced bioactions, with the potential for amplification loops.

Globally, LTB_4 and PAF thus appear to be potent pro-inflammatory agents. Several years ago, we and others became interested in their potential immunoregulatory functions (reviewed in 38 and 39). It became progressively evident that LTB_4 and PAF could effect many regulatory functions through interactions with a variety of lymphocyte and other cell populations, as presented in the next sections.

II. LTB_4 AND CYTOKINE PRODUCTION

Many of the effects of LT on immune and inflammatory cell functions could be amplified or modulated by interactions of these mediators with accessory cells, such as monocytes or macrophages and the cytokines they produce. Since these phagocytic cells can also produce LTB_4 and LTC_4 themselves, these mediators can play a role of feedback regulators or internal messengers for cytokine production.

A. IL 1

Indirect evidence that lipoxygenase metabolites of arachidonic acid would possibly be involved in IL 1 production was provided by use of metabolic inhibitors of lipoxygenation[40]. The direct effects of LT on production of IL 1 by human peripheral blood monocytes were also examined. Using the thymocyte proliferation assay to assess bioactive IL1, we previously reported that LTB_4 enhanced production of IL1-like material by lipopolysaccharides (LPS)-stimulated monocytes twofold to threefold[41]. The most efficient concentrations of LTB_4 were 10^{-8} to 10^{-7}M. LTD_4 also enhanced IL 1 production, but to a lesser extent than LTB_4. Adherence-purified, but otherwise unstimulated, human monocytes could also be induced to produce IL 1-like material in response to LTB_4. Similarly, IL 1 production by monocytes stimulated with the known IL 1 inducers muramyl dipeptide, silica, or zymosan was also enhanced by LTB_4[41]. Inhibition of cyclooxygenase with use of indomethacin during IL 1 production by LPS-treated monocytes enhanced thymocyte response to IL 1, but LTB_4 further enhanced IL 1 production when added to indomethacin-treated monocyte cultures.

LTB_4 can also affect IL 1 production indirectly, via T lymphocyte activation: human T cells pulsed with LTB_4 modulate IL 1 production by human monocytes by secreting IFN-γ[42]. In fact, we found that LTB_4-pulsed T cells were capable of inducing a suppression of lymphocyte proliferation if allowed to interact with monocytes, but that this suppression was reversed to an enhancing effect when monocytes were treated with the cyclooxygenase inhibitor indomethacin. Furthermore, LTB_4-pulsed T cells released a soluble factor that would mediate both effects. This factor was found to be IFN-γ, because affinity-purified IFN-γ could reproduce the effects, and a rabbit polyclonal anti-serum to human IFN-γ could block the activities of supernatants from LTB_4-pulsed T cells[42].

At the molecular level, LTB_4 induced, within 30 min, nuclear transcription of IL1β mRNA. Surprisingly, we failed to show any augmented production of immuno-reactive IL 1α or IL 1β following incubation of monocytes with LTB_4[43]. Since the thymocyte proliferation assay is not specific for IL 1, being sensitive to IL 6, IL 2, IL 4, and IL 7, and since monocyte supernatants may contain IL 6 in addition to IL 1, we investigated whether LTB_4 could preferentially stimulate IL 6 production and thus enhance thymocyte proliferation in the LAF assay in the absence of augmented IL 1 production.

B. IL 6

As indicated above, LTB_4 enhanced the production of IL 1-like material by human monocytes, as measured in the thymocyte costimulation assay[41]. Since that report, the advent of enzyme immunoassays and cDNA probes for cytokines allowed us to reexamine this phenomenon: LTB_4 preferentially induced IL 6 mRNA accumulation and IL 6 protein release as assessed by ELISA and the B9 cell bioassay[43]. In contrast, minimal IL 1 mRNA or protein was induced by LTB_4 above control values either in the absence or presence of muramyl dipeptide. Supernatants of LTB_4-treated monocytes consistently showed enhanced thymocyte costimulatory activity and this was abrogated by 75-80% by anti-IL 1 antibody. Baseline production of IL 1 appeared however to be sufficient for a synergistic stimulation of thymocytes in the presence of IL 6. Our results now help clarify that LTB_4 stimulates preferentially IL 6 production and that the observed LTB_4-

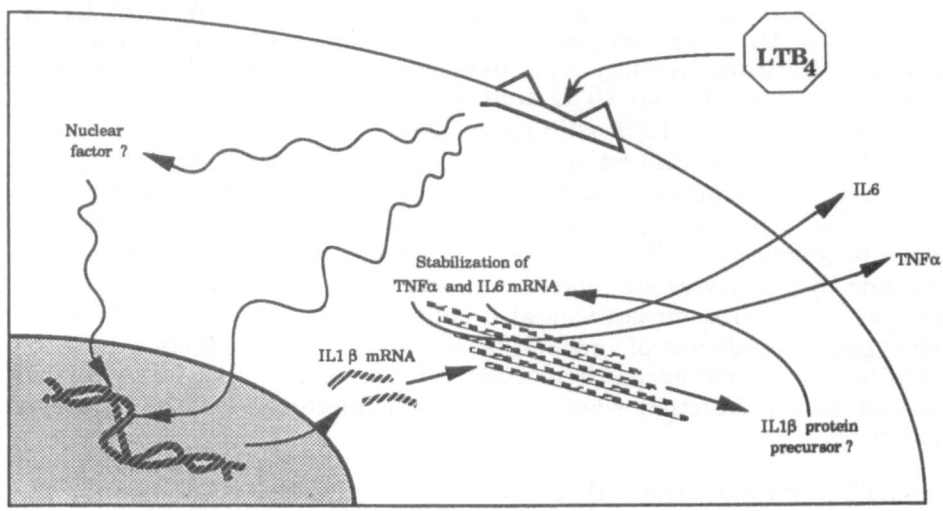

Figure 1.   Proposed molecular regulation of IL1β, IL 6 and TNFα gene expression by LTB₄.

induced augmentation in thymocyte responses to monocyte supernatants is due to augmented IL 6 contents in the presence of baseline IL 1 production.

Since regulation of IL 6 gene expression by LTB₄ appears to be post-transcriptional (no augmentation of transcripts on nuclear run ons), it is possible that LTB₄ augments IL 6 production indirectly by inducing IL 1β mRNA transcription with secondary stimulation of IL 6, and TNFα, production via stabilization of their mRNAs (Figure 1).

C.   TNF-α

Tumor necrosis factor (TNFα) is a potent mediator of cytotoxic activity exerted by monocytes and macrophages, and possibly natural cytotoxic (NC) lymphocytes. LTB₄ markedly enhances cytotoxicity mediated by NC cells and monocyte-mediated cytotoxicity is also augmented by LTB₄ . To further define the underlying mechanism(s), we studied the effect of LTB₄ on TNFα production by human monocytes, as measured by lysis of L929 target cells[44]. Addition of 10^{-14} - 10^{-8}M LTB₄ to monocyte cultures enhanced TNF activity of their 24-h supernatants by an average of 50% with maximal enhancement at 10^{-10} - 10^{-8}M LTB₄. Higher concentrations of LTB₄ were inhibitory. Kinetic studies indicated a maximal effect during the first 8 h of incubation. Addition of indomethacin to the cultures augmented TNF production, while addition of the lipoxygenase inhibitor nordihydroguaiaretic acid diminished TNF production[45]. Regulation of TNFα production by LTB₄ appears to be post-transcriptional as no nuclear transcription of TNFα mRNA could be induced by LTB₄ (unpublished observation). Our data suggest that LTB₄ may affect cytotoxic activity of human monocytes by augmenting their production of the cytolytic cytokine TNFα.

Alveolar macrophages (AM) can play a crucial role in the pathogenesis of pulmonary disease via their ability to produce potent inflammatory and fibrogenic mediators. We found that rat AM cultured with 1 to 100 μg/ml of silica particles or asbestos fibers produced TNFα and LTB_4 in a concentration-dependent fashion whereas latex beads, an inert phagocytic stimulus, failed to induce significant augmentation of either TNFα or LTB_4[17]. In a time course study, AM stimulated for 2 hr with silica or asbestos produced an increased amount of LTB_4 which preceded the rise in TNF activity detected 7 h and 24 h after culture initiation. In view of those results, we next examined the role of LTB_4 in mineral dust-induced TNFα production. The lipoxygenase inhibitors nordihydroguaiaretic acid (NDGA) and AA861 used at 1 to 50 μg/ml reduced in a concentration-dependent fashion asbestos- or silica-stimulated TNFα release[17]. On the other hand, "reconstitutive" experiments in which we added exogenous LTB_4 (10^{-14} - 10^{-8} M) to AM treated with lipoxygenase inhibitors showed partial restoration of TNFα production induced by chrysotile or silica with peak effect at 10^{-10}M LTB_4. These studies demonstrated that AM incubated in the presence of chrysotile A or silica could produce both LTB_4 and TNFα that endogenous lipoxygenase metabolites as well as exogenous LTB_4 could act to amplify TNFα production[17]. These observations suggest a common mechanism by which asbestos and silica may modulate the production of inflammatory and fibrogenic cytokines.

D.   IL 2 and IFNγ

When lymphokine production was studied following interaction of LTB_4 with T cell subsets, it was found that production of both IL 2 and IFNγ by CD4+ T cells was enhanced by nanomolar concentrations of LTB_4, while their production by CD8+ cells was impaired[46,47] (see also section II A.). This was observed with human cells with a variety of stimuli including lectins, phorbol ester plus ionomycin, or poly I:C. In the mouse, LTB_4 (as well as other leukotrienes and arachidonic acid itself) could replace helper cells[48] or IL 2[49] in IFNγ production.

III.   PAF AND CYTOKINE PRODUCTION

That PAF can activate monocyte or macrophage functions has been known for some time, as shown by Hartung[50]. More recently, Ho et al[51] observed that PAF could modulate the expression of c-fos and c-myc oncogenes in human monocytes, indicating that PAF can initiate activation or differentiation of monocytes at the gene level. Studies on the early PAF-induced events of GTPase stimulation, adenylate cyclase inhibition, phosphoinositide metabolism, calcium mobilization and protein phosphorylation[52-57] have lead to suggestions that there is more than one class of PAF receptors and that PAF interacts with a GTP-binding (guanine nucleotide regulatory) protein which differs from the classical N_s and N_i proteins by its insensitivity to cholera and pertussis toxins. Elucidation of the molecular events leading to modulation of the various monocyte-macrophage functions is presently the object of numerous studies, in particular as it relates to cytokine synthesis and release.

A.   IL 1

When exogenous PAF is added to monocytes or macrophages, several patterns of responsiveness in terms of IL 1 production can be seen, as reported from different laboratories. Mouse peritoneal macrophages, elicited either with thioglycollate or starch, respond in a varied and irregular manner to PAF (unpublished observations). In contrast, rat splenic macrophages present a biphasic response[58]: low concentrations of PAF (10^{-12} - 10^{-10}M) significantly enhance IL 1 production while higher concentrations (10^{-8} - 10^{-6}M) are inhibitory. Continuous low rate administration of PAF in vivo, in rats, via osmotic mini-pumps also revealed a biphasic effect on subsequent ex vivo IL 1 production by rat splenic monocytes[59]: cells from rats having received 28 μg of PAF during 7 days showed impaired IL 1 (both released and cell-associated) production in response to LPS stimulation. In contrast, cells from rats having received 1-9 μg of PAF showed augmented IL 1 production. These effects were abolished in animals having received a concomitant treatment with the PAF receptor antagonist BN52021, given orally.

IL 1 production by human monocytes is also modulated by PAF. Barrett et al.[60] initially reported an enhanced production of IL 1 when human monocytes were cultured in the presence of graded concentrations of PAF. Even in the absence of any other stimulus, such as LPS, (R)PAF and PAF analogs PR 1501 and PR 1502, but not (S)PAF, were capable of inducing IL 1 production. Salem et al.[61] also showed that PAF could increase IL 1 production by human monocytes stimulated with muramyl dipeptide, with higher concentrations of PAF inducing the highest responses.

Studies in our laboratory also indicate a positive effect of PAF on IL 1 production by human monocytes stimulated with muramyl dipeptide or LPS[62]. The production of IL 1 by human monocytes was analyzed following their stimulation with muramyl dipeptide (MDP; 1 μg/ml), in the absence or presence of graded concentrations of PAF. Significantly enhanced production of both IL 1 α and β was observed at two concentration ranges of PAF: a major enhancement was observed at 10^{-8} - 10^{-6}M and this was blocked by the PAF antagonist BN 52021 (10^{-4}M). A second enhancement was observed at 10^{-15} - 10^{-14}M PAF, which was not blocked by BN 52021. Monocytes isolated either by adherence or counterflow elutriation had similar responses to PAF. The biologically inactive precursor-metabolite, lyso-PAF, had no effect on cytokine production. PAF was shown to augment the production of both bioactive TNF and IL 1 and immunoreactive TNFα and IL 1α and β. Fractionation of monocytes on a discontinuous Percoll gradient yielded a denser subpopulation which responded preferentially to higher PAF concentrations, while the less dense subpopulation responded to both concentration ranges. These data indicate that PAF can modulate monocyte functions as related to cytokine production, and may thus contribute to amplification of inflammatory reactions and regulation of immune responses by interacting with subsets of human monocytes.

PAF can also synergize with interferon-gamma (IFN) or tumor necrosis factor (TNF) to induce higher levels of IL 1 production by human monocytes[63]. This may be part of an amplification loop since cytokines (IL 1, TNF, IFN) stimulate a biphasic synthesis of PAF by human monocytes[64]. We have recently shown that PAF can prime (following an 18 h preincubation) monocytes to produce augmented quantities of IL 1. This may be analogous to the observations in polymorphonuclear (PMN) leukocytes where priming with PAF augments their superoxide generation, elastase release and lysis of endothelial cells in response to phorbol ester or formyl-methionyl-phenylalamine (FMLP)[65]. TNFα can also prime PMN for enhanced PAF-induced superoxide production[66] and, again, an amplification loop may be operative since endotoxin can prime PMN for augmented PAF production in response to FMLP[67]

B. IL 6

Interleukin 6 (IL 6) is a cytokine produced by numerous cell types which has been found to affect many immune and inflammatory processes. Its production can be positively or negatively regulated by several cytokines, and our studies were undertaken to assess its possible regulation by PAF. When rat alveolar macrophages were cultured for 24 h with graded concentrations of PAF, a 3- to 6-fold increase in IL 6 production was observed, with peak activity at 10^{-10}M PAF. Human umbilical vein endothelial cells showed an even greater response to PAF with a 10- to 100-fold increase in IL 6 production, an effect comparable to that observed with synergistic TNF and IFN-γ treatment. Human monocytes responded to PAF in a biphasic manner in terms of IL 6, as they did in terms of IL 1 or TNF, with peak activities at nanomolar and femtomolar ranges[68]. Since PAF can markedly augment IL 1 and TNF production by human and rat mononuclear phagocytes, the present findings may indicate a potential amplification mechanisms for IL 6 production during inflammatory processes involving PAF.

C. TNFα

Another major cytokine produced by monocytes and macrophages and regulated by PAF is TNFα.

When rat alveolar macrophages (AM) were cultured with PAF alone, no change in TNFα production was observed. However, the concomitant addition of PAF and muramyl dipeptide (MDP) to AM cultures markedly enhanced (2-3 fold) TNFα production in a concentration-dependent fashion with peak effect at 10^{-10}M PAF[69,70]. This enhancement occurred when MDP and PAF were present together at the initiation of the 24-h culture. Stimulation of TNFα production by PAF was blocked by specific PAF receptor antagonists, BN 52020, BN 52021 and WEB 2086 with IC_{50} of respectively 1 x 10^{-6}M, 2.0 x 10^{-7}M, and 8.0 x 10^{-8}M. Additionally, the stereoisomer of PAF, [S]PAF, and the biologically inactive precursor/metabolite of PAF, lyso PAF, used at 10^{-10} - 10^{-12}M failed to induce significant enhancement in TNFα production. In parallel, addition of PAF to AM triggered LTB_4 release in a concentration-dependent manner. Inhibition of 5-lipoxygenase by NDGA or AA-861 blocked the PAF-induced augmentation of both TNFα and LTB_4 production. Thus, the findings of a concentration-dependent LTB_4 production by AM stimulated with PAF and the concomitant inhibition of PAF-induced TNFα production as well as LTB_4 generation by lipoxygenase inhibitors are consistent with the possibility that the action of PAF on AM can be mediated by the generation of endogenous lipoxygenase metabolites, probably LTB_4, which will further act as a second messenger to induce TNF production.

The ability of lipoxygenase products, mainly leukotriene B_4, to participate in immunoregulatory processes and to modulate cytokine production has been described in section II.

Human monocytes-macrophages can also be stimulated by PAF to produce augmented quantities of TNFα[62,69,71]. TNFα production, moreover, is enhanced in a bimodal manner at two concentration ranges of PAF, 10^{-15}-10^{-13}M and 10^{-9} - 10^{-7}M in LPS- or MDP-treated monocytes. These responses are partially associated with monocyte subsets presenting different characteristics in terms of size, density, phenotypic markers and $[Ca^{2+}]_i$ mobilization responses. TNF production parallelled IL 1 production as described in section IIIA.

We used the human promyelocytic leukemia cell line HL-60, at various times during differentiation with 1,25-dihydroxyvitamin D$_3$ (1,25(OH)$_2$D$_3$) towards the monocyte lineage in order to study the relation of cell differentiation to responsiveness to PAF in terms of cytokine production[72]. TNF production was induced by pretreatment with IFN-γ for 24 h and treatment with muramyl dipeptide. Although detectable TNF was produced by 4 day-differentiated cells, no effect was seen with PAF (10^{-16} - 10^{-6}M) at this or earlier stages. In contrast, 5 day-differentiated cells had a comparable baseline production of TNF, but responded with a 2.5 -fold increase to PAF with a single peak, maximal at 10^{-8}M. Moreover, 6 day- or 7 day-differentiated HL-60 cells showed a further increase in TNF production in response to PAF, and the response was bimodal, similar to the less dense subset of monocytes, with peaks at 10^{-14} and 10^{-7}M PAF. In parallel, undifferentiated HL-60 failed to respond to PAF in terms of $[Ca^{2+}]_i$ mobilization. The earliest responsiveness to PAF (10^{-7}M) was observed by 4 days of treatment with 1,25(OH)$_2$D$_3$ and by day 7, the response to PAF became bimodal (10^{-14} and 10^{-7}M). These results indicate that myeloid cells acquire, during maturation towards the monocyte lineage, a progressive responsiveness to PAF in terms of $[Ca^{2+}]_i$ mobilization and enhanced cytokine production and suggest that the heterogeneity in responses to PAF observed in normal monocytes may be related to their stage of differentiation or maturation (Figure 2).

Furthermore, PAF or IL 1 can prime monocytes to respond to LPS with enhanced production of TNFα during a subsequent culture with IL 1 or PAF, respectively[73]. These findings suggest again that a cascade of inflammatory signals may have much greater effects than simultaneous or individual activities.

Additional studies were undertaken to evaluate the possibility that PAF could modulate NK activity by enhancing the production of cytotoxic cytokines. To do so, large granular lymphocytes (LGL) were incubated with various target cell lines (K562, Clone I, MA-160, SK-MEL 109), in the presence or absence of PAF. Supernatants were collected and assayed for cytotoxicity in a 20 h ^{51}Cr-release assay against WEHI 164 and U-937 target cells which are sensitive to cytotoxic cytokines. The results showed that LGL incubated with several types of target cells, in the presence of PAF, released significantly increased amounts of cytotoxic factors[74]. The maximal release was obtained after 8 to 10 h of incubation. Mannose-6-PO$_4$ (10mM) which inhibits natural killer cytotoxic factor (NKCF) was not able to block supernatant cytotoxicity against U 937 target cells, while anti-TNF antibody

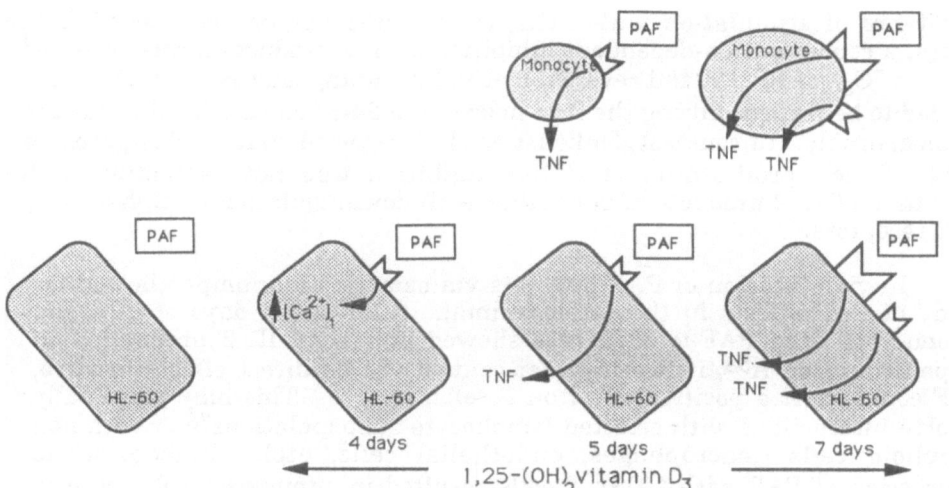

Figure 2.   Differentiaton-dependent responsiveness of myeloid cells to PAF.

completely abolished supernatant activity. These findings suggest that picomolar concentrations of PAF enhance the release of cytotoxic factor(s), mainly TNF, by LGL following interaction with target cells.

During inflammation, blood lymphocytes cross the wall of blood vessels to infiltrate the tissues. Endothelial cells, which line the vessels, can thus modulate lymphocyte functions; particularly, some products of endothelial cells, such as PAF, can inhibit the proliferation of lymphocytes. We have therefore exposed human endothelial cells (HEC), isolated from the umbilical vein, for 1 h to PAF, the PAF antagonist BN52021 (10^{-4}M), IL 1 or TNF and the cyclooxygenase inhibitor indomethacin (10^{-6}M), alone or in combination. The HEC were then treated with mitomycin C, washed and incubated 72 h with lymphocytes and concanavalin A. We observed that the preincubation of HEC with IL 1 or TNF alone or with indomethacin induced a significant suppression of lymphocyte proliferation of 20 and 25%, respectively[75]. Concomitant use of BN52021 completely prevented this suppressor activity exerted by HEC, suggesting that endogenous PAF from HEC was mediating the suppressive effect. Depletion of monocytes abrogated the effect. Sequential incubation of HEC with TNF or IL 1 followed by addition of monocytes before addition to lymphocytes resulted in greatest (> 40%) suppression. Preincubation of HEC with exogenous PAF (10^{-12} - 10^{-8}M) also induced suppressor cell activity and this was blocked by BN52021. These results suggest that IL 1 and indomethacin together stimulate the production of PAF by HEC, which in turn suppresses lymphocyte proliferation. This effect of PAF is blocked by BN52021 which can thus restore normal lymphocyte proliferation. Immunoregulation by endothelial cell-derived PAF and its prevention by a PAF antagonist may have important bearing in the numerous disease states where vascular inflammation is a prominent component.

D.   IL 2

Because lymphocyte proliferation involves, in its early stages, the production of IL 2, we measured the production of this lymphokine by PBML at 24 hrs of stimulation with PHA, in the presence or absence of PAF. Again, a concentration-dependent inhibition of IL 2 production was observed with an IC_{50} of $10^{-11}M$ and reversal by the PAF antagonist BN52021[76]. PAF needed to be present during the first hour of the 24-hr culture for the effect to be measurable. In contrast, Dulioust et al.[77] reported that PAF suppressed $CD4^+$ T cell proliferation, but the inhibition was not associated with inhibition of IL 2 production, but rather with downregulation of high-affinity IL 2 receptors.

In vivo infusion of PAF into rats via osmotic minipumps allowed for study of PAF effects in the whole animal. After seven days of infusion, splenocytes from PAF-treated rats showed enhanced IL 2 production in response to Con A[58], indicating that contrary to its direct effects in vitro, PAF could exert a positive effect on T cells, in vivo. This may potentially involve interactions with selected lymphocyte subpopulations[78] or with non lymphoid cells (macrophages, endothelial cells, etc). Indeed, while interaction of PAF with $CD8^+$ T cells resulted in suppressor cell activity, interaction with $CD4^+$ cells induced helper cell function[79].

E.   GM-CSF and IFNγ

Preliminary data from our laboratory indicate that human PBML exposed in culture, in the presence of PHA, to graded concentrations of PAF produced markedly augmented quantities of both GM-CSF and IFNγ. GM-CSF production was enhanced by a factor of 2, with peak activity at 10^{-14} to $10^{-11}M$ PAF. On the other hand, IFNγ production was augmented by a factor of 20 to 30, with peak activity at 10^{-12} to $10^{-8}M$ PAF.

IV.   CONCLUSIONS

An increasing body of evidence suggests that LTB_4 may be an important modulator of many immune cell functions, either as an internal messenger within some cells, or as an external signal, linking inflammatory cells to lymphocytes.

LTB_4 and, to some extent, other leukotrienes, are known protagonists of numerous inflammatory processes. As reviewed in this chapter, they can also mediate varied and powerful effects on many cells involved in immunological reactions.

PAF also appears to exert potent immunoregulatory functions at several levels of the immune response. Some studies, such as those on NK cell activity and macrophage production of TNF suggest that these cells may bear more than one class of receptors for PAF, with different patterns of inhibition by PAF antagonists and distinct coupling to GTP-binding proteins. Further studies on binding and molecular definition of PAF receptors on cells of the immune system will be needed to unravel this potentially important mechanism of modulation of immune function. The numerous ongoing studies should establish even more clearly the interactions between inflammation and immune responses, thus opening new venues for research and eventual therapeutic intervention.

V. REFERENCES

1. P. Borgeat, and B. Samuelsson, Metabolism of arachidonic acid in polymorphonuclear leukocytes: unstable intermediate in formation of dihydro acids. Proc. Natl. Acad. Sci. U.S.A. 76:3213 (1979).

2. P. Sirois, and P. Borgeat, Leukotrienes: a new approach to the biochemistry of hypersensitivity. Surv. Immunol. Res. 1:279 (1982).

3. P. Borgeat, and B. Samuelsson, Transformation of arachidonic acid by rabbit polymorphonuclear leukocytes: formation of a novel dihydro-eicosa-tetraenoic acid. J. Biol. Chem. 254:2643 (1979).

4. A.W. Ford-Hutchinson, M.A. Bray, M.V. Doig, M.E. Shipley, and M.J.H. Smith, Leukotriene B, a potent chemokinetic and aggregating substance released from polymorphonuclear leucocytes. Nature 286:264 (1980).

5. J. Palmblad, C.L. Malmsten, A.M. Uden, K.O. Radmar, L. Engstedt, and B. Samuelsson, Leukotriene B_4 is a potent stereospecific stimulator of neutrophil chemotaxis and adherence. Blood 58:658 (1981).

6. H.J. Showell, P.H. Naccache, P. Borgeat, S. Picard, P. Valerand, E.L. Becker, and R.I. Sha'afi, Characterization of the secretory activity of LTB_4 toward rabbit neutrophils. J. Immunol. 128:811 (1982).

7. D.A. Bass, M.J. Thomas, E.J. Goetzl, E.R. DeChatelet, and C.E. McCall, Lipoxygenase-derived products of arachidonic acid mediate stimulation of hexose uptake in human polymorphonuclear leukocytes. Biochem. Biophys. Res. Commun. 100:1 (1981).

8. M.A. Bray, A.W. Ford-Hutchinson, and M.J.H. Smith, Leukotriene B_4: an inflammatory mediator in vivo. Prostaglandins 22:213 (1981).

9. T.F.P. Molski, P.H. Naccache, P. Borgeat, and R.I. Sha'afi, Similarities in the mechanisms by which formylmethionyl-leucyl-phenylalanine, arachidonic acid and leukotriene B_4 increase calcium and sodium influxes in rabbit neutrophils. Biochem. Biophys. Res. Commun. 103:227 (1981).

10. D.P. Lew, J.-M. Dayer, C.B. Wollheim, and T. Pozzan, Effect of leukotriene B_4 and arachidonic acid on cytosolic-free calcium in human neutrophils. FEBS Lett. 166:44 (1984).

11. D.W. Goldman, L.A. Gifford, D.M. Olson, and E.J. Goetzl, Transduction by leukotriene B_4 receptors of increases in cytosolic calcium in human polymorphonuclear leukocytes. J. Immunol. 135:525 (1985).

12. T. Andersson, W. Schlegel, A. Monod, K.-H. Krause, O. Stendahl, and D.P.Lew, Leukotriene B_4 stimulation of phagocytes results in the

formation of inositol 1, 4, 5-trisphosphate. Biochem. J., 240:333 (1986).

13. S. Mong, G. Chi-Rosso, J. Miller, K. Hoffman, K.A. Raggaitis, P. Bender, and S.T. Crooke, Leukotriene B_4 induces formation of inositol phosphates in rat peritoneal polymorphonuclear leukocytes. Mol. Pharmacol. 30:235 (1986).

14. A.O. Fels, N.A. Pawlowski, E.B. Cramer, T.K. King, A.Z. Cohen, and W.A. Scott, Human alveolar macrophages produce leukotriene B_4. Proc. Natl. Acad. Sci. USA 79:7866 (1982).

15. P.A.J. Hewricks, M.E. VanDertol, F. Engels, F.P. Nijkamp and J. Verhoef, Human polymorphonuclear leukocytes release leukotriene B_4 during phagocytosis of staphylococus aureus. Inflammation 10:37 (1986).

16. N.R. Ferreri, W.C. Howland and H.L. Spiegelberg, Release of leukotrienes C_4 and B_4 and prostaglandin E_2 from human monocytes stimulated with aggregated IgG, IgA and IgE. J. Immunol. 136:4188 (1986).

17. Dubois, C., Bissonnette, E., Rola-Pleszczynski, M.: Asbestos fibers and silica particles stimulate rat alveolar macrophages to release TNF; autoregulatory role of leukotriene B_4. Am. Rev. Resp. Dis. , 139:1257-1264 (1989).

18. J. Maclouf, B. Fruteau de Laclos, and P. Borgeat, Stimulation of leukotriene biosynthesis in human blood leukocytes by platelet-derived 12-hydroperoxy-icosatetraenoic acid. Proc. Natl. Acad. Sci. USA, 79: 6042 (1982).

19. F.H. Chilton, J.T. O'Flaherty, C.E. Walsh, M.J. Thomas, R.L. Wykle, L.R. DeChatelet, and B.M. Waite, Stimulation of the lipoxygenase pathway in polymorphonuclear leukocytes by 1-0-alkyl-2-0-acetyl-SN-glycero-3-phosphocholine. J. Biol. Chem. 257:5402 (1982).

20. A.H. Lin, D.R. Morton, and R.R. Gorman, Acetyl glyceryl ether phosphorylcholine stimulates leukotriene B_4 synthesis in human polymorphonuclear leukocytes. J. Clin. Invest., 70:1058 (1982).

21. M.E. Goldyne, G.F. Burrish, P. Poubelle, and P. Borgeat, Arachidonic acid metabolism among human mononuclear leukocytes. Lipoxygenase related pathways. J. Biol. Chem. 259:8815 (1984).

22. P. Poubelle, P. Borgeat, and M. Rola-Pleszczynski, Assessment of leukotriene B_4 synthesis in human lymphocytes using high performance liquid chromatography and radioimmunoassay methods. J. Immunol. 139:1273 (1987).

23. M.E. Goldyne, and L. Rea, Stimulated T cell and natural killer (NK) cell lines fail to synthesize leukotriene B_4. Prostaglandins 34:783 (1987).

24. E.J. Goetzl, Selective feed-back inhibition of the 5-lipoxygenation of arachidonic acid in human T-lymphocytes. Biochem. Biophys. Res. Commun., 1901:344 (1981).

25. D. Atluru, E.A. Lianos, J.S. Goodwin, Arachidonic acid inhibits 5-lipoxygenase in human T cells. Biochem. Biophys. Res. Commun. 135:670 (1986).

26. B. Odlander, P.-J. Jakobsson, A. Rosen, and H.-E. Claesson, Human B and T lymphocytes convert leukotriene A_4 into leukotriene B_4. Biochem. Biophys. Res. Commun. 153:203 (1988).

27. J.L. Ambrus, C.H. Jurgensen, N.L. Witzel, R.A. Lewis, J.L. Butler, and A.S. Fauci, Leukotriene C_4 produced by a human T-T hybridoma suppresses Ig production by human lymphocytes. J. Immunol. 140:2382 (1988).

28. E.M. Davidson, S.A. Rae, and M.J.H. Smith, Leukotriene B_4 in synovial fluid. J. Pharm. Pharmacol. 34:410 (1982).

29. S.D. Brain, R.D.R. Camp, P.M. Dowd, A.K. Black, P.M. Woolard, A.I. Mallet, and M.W. Greaves, Psoriasis and leukotriene B_4. Lancet 2:762 (1982).

30. Y. Kikawa, Y. Shigematsu, and M. Sudo, Leukotriene B_4 and 20-OH-LTB_4 in purulent peritoneal exudates demonstrated by GC-MS. Prostagl. Leukotr. Med. 23:85 (1986).

31. P. Sharon, and W.F. Stenson, Production of leukotrienes by colonic mucosa from patients with inflammatory bowel disease. Gastroenterol. 84:1306 (1983).

32. R.D. Zipser, C.C. Nast, M. Lee, H.W. Kao, and R. Duke, In vivo production of leukotriene B_4 and leukotriene C_4 in rabbit colitis. Gastroenterol. 92:33 (1987).

33. F.H. Chilton, J.M. Ellis, S.C. Olson, and R.L. Wykle. 1-0-alkyl-2-arachidonyl-sn-glycero-3-phosphocholine. A common source of platelet-activating factor and arachidonate in human polymorphonuclear leukocytes. J. Biol. Chem. 259:12014 (1984).

34. P. Braquet, L. Touqui, T.Y. Shen, and B.B. Vargaftig. Perspectives in platelet activating factor research. Pharmacol. Review 39:97 (1987).

35. E. Jouvin-Marche, E. Ninio, G. Beauvain, M. Tence, P. Niaudet and J. Benveniste. Biosynthesis of PAF-acether (platelet-activating factor). VII. Precursors of PAF-acether and acetyl-transferase activity in human leukocytes. J. Immunol. 133:892 (1984).

36. F. Bussolino, R. Foa, F. Malavasi, M.L. Ferrando, and G. Camussi. Release of platelet-activating factor (PAF)-like material from human lymphoid cell lines. Exp. Haematol. 12:688 (1984).

37. F. Malavasi, C. Tetta, A. Funaro, G. Bellone, E. Ferrero, and F. Caligaris-Cappio. F_c receptor triggering induces expression of surface activation antigens and release of platelet-activating factor in large granular lymphocytes. Proc. Natl. Acad. Sci. (USA), 83:2443 (1986).

38. M. Rola-Pleszczynski, Immunoregulation by leukotrienes and other lipoxygenase metabolites. Immunol. Today 6:302 (1985).

39. P. Braquet, M. Rola-Pleszczynski. Platelet-activating factor and cellular immune responses. Immunol. Today, 8:345 (1987).

40. C.A. Dinarello, I. Bishai, L.J. Rosenwasser and F. Coceani. The influence of lipoxygenase inhibitors on the in vitro production of human leukocytic pyrogen and lymphocyte activating factor (interleukin 1). Int. J. Immunopharmacol. 1:43 (1984).

41. M. Rola-Pleszczynski and I. Lemaire. Leukotrienes augment interleukin 1 production by human monocytes. J. Immunol. 135:3958 (1985).

42. M. Rola-Pleszczynski, L. Bouvrette, D. Gingras and M. Girard. Identification of interferon-γ as the lymphokine that mediates leukotriene B_4-induced immunoregulation. J. Immunol. 139:513 (1987).

43. P. Poubelle, J. Stankova, J. Grassi, M. Rola-Pleszczynski. Leukotriene B_4 up-regulates IL-6 rather than IL-1 synthesis in human monocytes. Agents and Actions, in press, 1991.

44. L. Gagnon, L. Fillion, C. Dubois, and M. Rola-Pleszczynski. Leukotrienes and macrophage activation: augmented cytotoxic activity and enhanced interleukin 1, tumor necrosis factor and hydrogen peroxide production. Agents and Actions, 26:141 (1989).

45. L. Gagnon, L.G. Filion, M. Rola-Pleszczynski. Enhanced production of tumor necrosis factor (TNF)-α by human monocytes exposed to leukotriene B4. Int. J. Immunopathol. Pharmacol. 2:155 (1989).

46. M. Rola-Pleszczynski, P.-A. Chavaillaz, and I. Lemaire. Stimulation of interleukin 2 and interferon-γ production by leukotriene B_4 in human lymphocyte cultures. Prostagl. Leukotr. Med. 23:207 (1986).

47. M. Rola-Pleszczynski, L. Gagnon, and P.-A. Chavaillaz. Immune regulation by leukotriene B_4. in: "Biology of the leukotrienes.", R. Levi and R.D. Krell, eds., Ann. N.Y. Acad. Sci., 524:218 (1988).

48. H.M. Johnson, and B.A. Torres. Leukotrienes, positive signals for regulation of interferon- production. J. Immunol., 132:413 (1984).

49. W.L. Farrar, and J.L. Humes. The role of arachidonic acid metabolism in the activities of interleukin 1 and 2. J. Immunol. 135:1153 (1985).

50. H.-P. Hartung. Acetyl glyceryl ether phosphoryl-choline (platelet-activating factor) mediates heightened metabolic activity in macrophages. Studies on PGE_1, TXA_2 and O_2 production, spreading and the influence of calmodulin inhibitor W-7. FEBS Lett. 160:209 (1983).

51. Y.-S. Ho, W.M.F. Lee, and R. Snyderman. Chemoattractant-induced activation of c-fos gene expression in human monocytes. J. Exp. Med. 165:1524 (1987).

52. H. Homma, D.J. Hanahan. Attenuation of platelet-activating factor (PAF)-induced stimulation of rabbit platelet GTPase by phorbol ester, dibutyryl cAMP, and desensitisation: concomitant effects on PAF receptor binding characteristics. Arch. Biochem. Biophys. 262:32 (1988).

53. M. Bachelet, M.J.P. Adolfs, J. Masliah, G. Bereziat, B.B. Vargaftig, and I.L. Bonta. Interaction between PAF-acether and drugs that stimulate cyclic AMP in guinea-pig alveolar macrophages. Eur. J. Pharmacol. 149:73 (1988).

54. G. Barzaghi, and S. Mong. Platelet-activating factor (PAF) stimulates a phospholipase C(PLC) in differentiated human monocytic leukemia (U-937) cells, resulted in phosphoinositide (PI) hydrolysis and intracellular calcium mobilization. Prostaglandins 35:819 (1988).

55. S. Hopple, R.Meurer, J. Westwick, and D.E. MacIntyre. PAF-induced Ca^{2+} flux and formation of inositol tris- and tetrakis-phosphates in U937 cells. FASEB J 2:A415 (1988).

56. V. Prpic, R.J. Uhing, J.E. Weiel, L. Jakoi, G. Gawdi, B. Herman, and D.O. Adams. Biochemical and functional responses stimulated by platelet-activating factor in murine peritoneal macrophages. J. Cell. Biol. 107:363 (1988).

57. F. Bussolino, F. Turrini, E. Fischer, D. Alessi, M.D. Kazatchkine, and P. Arese. PAF enhances erythrophagocytic activity of human monocytes by the protein kinase C dependent phosphorylation of C36 receptor (CR1). Role of PAF receptor antagonists. Prostaglandins 35:803 (1988).

58. B. Pignol, S. Hénane, J.-M. Mencia-Huerta, M. Rola-Pleszczynski, and P. Braquet. Effect of platelet-activating factor (PAF-acether) and its specific receptor antagonist, BN 52021, on interleukin 1 (IL 1) release and synthesis by rat spleen adherent monocytes. Prostaglandins, 33:931 (1987).

59. B. Pignol, S. Hénane, B. Sorlin, B. M. Rola-Pleszczynski, J.-M. Mencia-Huerta, P Braquet. Effect of long-term treatment with platelet-activating factor on IL 1 and IL 2 production by rat spleen cells. J. Immunol. 145:980 (1990).

60. M.L. Barrett, G.P. Lewis, S. Ward, and J. Westwick. Platelet-

activating factor induces interleukin 1 production from human adherent macrophages. Br. J. Pharmacol. 90:113P (1987).

61. P. Salem, S. Derickx, A. Dulioust, E. Vivier Y. Denizot, C. Damais, C. Dinarello, Y. Thomas. Immunoregulatory functions of paf-acether. Enhancement of IL 1 production by muramyl dipeptide-stimulated monocytes. J. Immunol. 144:1338 (1990).

62. P. Poubelle, D., Gingras, C., Demers, C., Dubois, D., Harbour, and M.Rola-Pleszczynski, M. Platelet activating factor (PAF-acether) enhances the concomitant production of tumor necrosis factor alpha and interleukin 1 by subsets of human monocytes. Immunol. in press, 1991.

63. R. Barthelson, F. Valone. Interaction of platelet-activating factor with interferon-γ in the stimulation of interleukin-1 production by human monocytes. J. Allergy Clin. Immunol. 86:193 (1990).

64. F.H. Valone, and L.B. Epstein. Biphasic platelet-activating factor (PAF) synthesis by human monocytes stimulated with interleukin 1 beta (IL 1β), tumor necrosis factor (TNF) or IFN-γ. J. Immunol. 141:3945 (1988).

65. G.M.Vercellotti, H.Q Yin, K.S. Gustabson, R.D. Nelson, and H.S. Jacob. Platelet-activating factor primes neutrophils responses to agonists: role in promoting neutrophil-mediated endothelial damage. Blood 71:1100 (1988).

66. M. Paubert-Braquet, M.-O. Lonchampt, P. Klotz, and J. Guilbaud. Tumor necrosis factor (TNF) primes platelet-activating factor (PAF)-induced superoxide generation by human neutrophils (PMN): consequences in promoting PMN-mediated endothelial cell (EC) damages. Prostaglandins 35:803 (1988).

67. G.S. Worthen, J.F. Seccombe, K.L. Clay, L.A. Guthrie, and R.B. Jr Johnston. The priming of neutrophils by lipopolysaccharide for production of intracellular platelet-activating factor. J. Immunol. 140:3553 (1988).

68. M. Rola-Pleszczynski, L., Bouvrette, M., Thivierge, C., Lacasse. Platelet-activating factor enhances interleukin 6 production by monocytes, alveolar macrophages and endothelial cells. FASEB J. 4:A1713 (1990).

69. M. Rola-Pleszczynski, J. Bossé, E. Bissonnette and C. Dubois. PAF-acether enhances the production of tumor necrosis factor by human and rodent lymphocytes and macrophages. Prostaglandins 35:802 (1988).

70. C. Dubois, E. Bissonnette, M. Rola-Pleszczynski, M. Platelet-activating factor (PAF) stimulates tumor necrosis factor production by alveolar macrophages: prevention by PAF receptor antagonists and lipoxygenase inhibitors· J. Immunol. 143:964 (1989).

71. Bonavida, Braquet, P. Effect of platelet-activating factor (PAF) on

monocyte activation and production of tumor necrosis factor (TNF). *Int. Arch. Allergy Appl. Immunol.* (1988).

72. M. Rola-Pleszczynski, J., Stankova. Differentiation-dependent modulation of TNF production by PAF in human HL-60 myeloid leukemia cells. Submitted for publication, *J. Immunol.* (1990).

73. M. Rola-Pleszczynski. Priming of human monocytes with PAF augments their production of tumor necrosis factor. *J. Lipid Mediators* 2:S77 (1990).

74. J. Bossé, S. Turcotte, and M. Rola-Pleszczynski. Platelet activating factor (PAF) enhances the production of cytotoxic cytokines during natural cell-mediated cytotoxicity. *FASEB J.* 2:A415 (1988).

75. C. Lacasse, and M. Rola-Plesczcynski. Immune regulation by PAF II. Mediation of suppression by cytokine-stimulated endothelial cells. *J. Leuk. Biol.*, in press, 1990.

76. M. Rola-Pleszczynski, B. Pignol, C. Pouliot, and P. Braquet, Inhibition of human lymphocyte proliferation and interleukin 2 production by platelet-activating factor (PAF-acether): Reversal by a specific antagonist: BN52021. *Biochem. Biophys. Res. Commun.*, 142:754 (1987).

77. A. Dulioust , V. Duprez, C. Pittou, P. Salem, A. Hemar, J. Benveniste, Y. THomas,. Immunoregulatory functions of PAF-acether. Down-regulation of CD4+ T cells high affinity IL 2 receptor expression. *J. Immunol.* 144:3123 (1990).

78. M. Rola-Pleszczynski, C. Pouliot, S. Turcotte, B. Pignol, P. Braquet and L. Bouvrette.: Immune regulation by platelet-activating factor. I. Induction of suppressor cell activity in human monocytes and CD8+ T cells and of helper cell activity in CD4+T cells. *J. Immunol.* 140:3547 (1988).

Studies by the author presented in this chapter were supported by grants from the National Cancer Institute and the Medical Research Council of Canada. The author is a Research Scholar of the Fonds de la Recherche en Santé du Québec.

Correspondence should be addressed at:
Immunology Division
Faculty of Medicine
3001, 12th Avenue North
SHERBROOKE (QC)
Canada J1H 5N4

PLATELET-ACTIVATING FACTOR (PAF) - A PUTATIVE MEDIATOR

IN INFLAMMATORY TISSUE INJURY

Tian-Li Yue, Reuven Rabinovici and Giora Feuerstein

Department of Pharmacology, SmithKline Beecham
Pharmaceuticals
709 Swedeland Road, King of Prussia, PA, USA 19406-0939

Platelet-activating factor (PAF) is a potent lipid autacoid identified as 1-0-alkyl-2-acetyl-sn-glycero-3-phosphocholine (figure 1). Originally isolated from immunoglobulin E-stimulated basophils as a potent inducer of platelet aggregation[1], it soon became apparent that in addition to activation of platelets, PAF exerts diverse biological actions including stimulation of neutrophils, monocytes, and macrophages, increased vascular permeability, marked hemodynamic effects, and others[2]. It is now clear that PAF synthesis can be stimulated by various stimuli[3] in many different cell types, some of which also release PAF into extracellular milieu. PAF actions are achieved at concentrations as low as 10^{-12} M in some systems and almost always by 10^{-9} M, suggesting a role as an intercellular messenger[4]. These observations have suggested that PAF plays a role both in normal physiological events and a variety of pathological responses. This chapter will emphasize the effects of PAF relevant to inflammatory responses and tissue injury. Some essential biochemical background of PAF will also be outlined.

BIOCHEMISTRY OF PAF

The molecular structure of PAF

Several features of PAF are required for its biological activity[2]. The first is the ether linkage at the sn-1 position of the glycerol backbone. Conversion of the ether bond results in much reduced activity. The potency of PAF is also influenced by the length of the alkyl chain at the sn-1 position. PAF carrying a 16:0 alkyl chain is more active than PAF carrrying an 18:0 alkyl chain. The second unique feature of PAF molecule is that instead of the fatty acid at the sn-2 position of glycerol there is an acetate. Other short chain organic acids can substitute the acetate but the biological activity is reduced as the chain is lengthened. Most importantly, hydrolisis of the acetate results in a lyso-PAF form which lacks biological activity. The polar head group of PAF is phosphocoline, which is required for optimal activity. Lastly, only the natural stereoisomer of PAF (the R-form), but not the S-isomer, exhibits biological activity.

Cell-Cell Interactions in the Release of Inflammatory Mediatiors
Edited by P. Y-K Wong and C.N. Serhan, Plenum Press, New York, 1991

Biosynthesis and catabolism of PAF[3]

PAF is not stored in cellular organelles, but is enzymatically synthesized in response to a variety of stimuli by either a remodeling or a de novo pathway (figure 1). The first step of the remodeling pathway is the activation of a phospholipase A_2 (PLA_2), which catalyzes the hydolysis of the sn-2 fatty acyl residue to yield an intermediate, 1-0-alkyl-sn-glycero-3-phosphocholine (lyso-PAF) and a free fatty acid such as arachidonic acid (AA). The lyso-PAF is then converted to PAF by the addition of acetate, a reaction catalyzed by a specific acetyl-coenzyme A: lyso-PAF

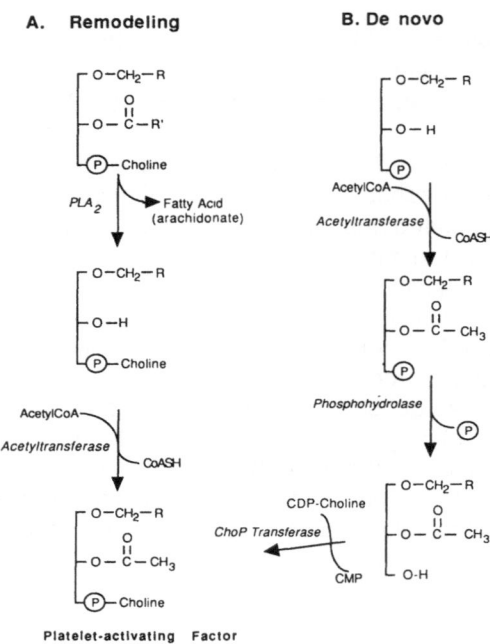

Fig. 1 Pathways for biosynthesis of PAF

acetyltransferase. Both PLA_2 and acetyltransferase may be regulated by protein kinase C (PKC). Moreover, the PLA_2 step is also affected by the Ca^{++} concentration in the cellular cytoplasm. Therefore, PAF synthesis can be initiated by any manipulation that results in marked influx of Ca^{++}, including treatment with calcium ionophores or permeabilization of the cellular membrane in response to bacterial toxins. These latter manipulations may reflect the potential for PAF synthesis under pathological conditions. Since many of the same stimuli that activate PAF synthesis also stimulate the release of AA from phospholipids and its conversion to prostaglandin (PG) and leukotriene (LT), this phenomenon may play an important role in mediating inflammatory responses[5].

Alternately, PAF can be produced by a de novo route beginning with 1-0-alkyl-2-lyso-sn-glycero-3-phosphate (figure 1), which can be acetylated by an acetyl-CoA:acetyl-transferase to form the corresponding 1-0-alkyl-2-acetyl-sn-glycerol, which is then converted to PAF by a unique CDP-choline:1-alkyl-2-acetyl-sn-glycerol-choline-phosphotransferase. It has been reported that some cells (e.g. endothelial cells) do not produce PAF via the de novo route[4]. However, this pathway is present in some cell types and may serve to make a small amount of PAF constitutively in tissues where PAF might serve a basal physiological role. Our own observations in rat cerebellar granule cell suggest a relative importance of the de novo pathway over the remodeling pathway[6], which is consistent with the observation in rat brain, where the acetyltransferase activity is very low or undetectable[7].

The primary route of PAF degradation to inactive product is catalyzed by a cytosolic acetylhydrolase that hydrolyzes the acetate moiety to produce lyso-PAF; lyso-PAF however is rapidly re-acylated by a microsomal transacylase to form alkyl-acyl-GPC. The PAF acetylhydrolase is highly specific for phospholipids with short acyl chains at the sn-2 position and this enzymatic activity is found in plasma and in a variety of cells and tissues[4].

Although there are presently no specific inhibitors for PAF synthesis or metabolism, some anti-inflammatory peptides such as antiflammins, which were recently reported to inhibit PAF synthesis in neutrophils, macrophages and endothelial cells. The mechanism of this inhibition is due to inhibiting PLA_2 and acetyltransferase[8]. Inhibition of PAF release by dexamethasone in endotoxemia is also due to inhibition of PLA_2, thereby reducing the supply of lyso-PAF for PAF synthesis.

PAF receptor and cellular signal transduction

PAF exerts its effects via a cellular receptor[9]. Although PAF receptor has not been purified as yet, it has been characterized to some degree in crude systems. The existence of a specific PAF receptor was first suggested by the demonstration that only the naturally occurring stereoisomer stimulated various PAF responses. Additional corroborating evidence include: very low concentrations of PAF (nM-pM) are required to trigger biological effects; the binding sites and physiological responses show saturability and specificity; specific desensitization takes place after tissue exsposure to PAF; and the blockade of PAF actions by highly specific PAF antagonists. High affinity receptors for PAF have been found in platelets, neutrophils, macrophages, monocytes and a variety of tissues such as brain and lung.

An increasing body of evidence has suggested that the biological responses of PAF may be mediated via a receptor-signal transduction mechanism[9] (Figure 2). When PAF binds to the receptor, it activates a G-protein that, in turn, can activate either a phosphoinositide-specific phospholipase C or inhibit adenylate cyclase. Inositolphosphate hydrolysis mobilizes intracellular Ca^{++} and activates protein kinase C, which could act independently or simultaneously to promote many of the physiologic and pathophysiologic effects of PAF. While the majority of studies have been confined to the platelets[9], we recently observed PAF receptor-mediated signal transduction in neurohybrid cells. In brief, PAF induces intracellular Ca^{++} release and phosphoinositide hydrolysis that are regulated by protein kinase C,[10] through a G-protein which is partially sensitive to pertussis toxin. (Figure 2)

CELLULAR ORIGINS OF PAF RELEVANT TO INFLAMMATORY RESPONSES

PAF can be synthesized by a wide variety of cells and tissues, and the list of cells capable of producing PAF continues to grow. Table 1 shows some cell types which synthesize PAF and are likely to be involved in the development of inflammatory responses. Most of these cells secrete PAF to the exterior of the cell. However, almost all of the newly-synthesized PAF in endothelial cell remained associated with the cell[4]. Likewise, it was recently reported that the majority of newly synthesized PAF in human neutrophils is retained intracellularly, mainly in the membranous fractions but some of the newly synthesized PAF is also found in fractions corresponding to specific granules[11]. Most studies indicate that cells do not synthesize significant amounts of PAF at the resting (unstimulated) stage. A comprehensive review regarding the cellular origins of PAF has been presented by Bratton and Henson[12].

PAF AS A MODULATOR FOR RELEASE OF PROSTAGLANDIN (PG), LEUKOTRIENE (LT) AND CYTOKINE [5,13]

The primary evidence supporting a role for PAF in inflammatory tissue injury is derived from pharmacological studies. PAF produces vasoconstriction or vasodilation, and enhances vascular permeability and extravasation of plasma proteins and edema formation[2]. PAF recruits and activates platelets, neutrophils, and monocytes, and enhances the migration of inflammatory cells into the tissue. Furthermore, PAF plays an important role in mediating the adhesive interaction between circulating leukocytes and microvascular endothelin[14].

A growing body of evidence has demonstrated that PAF modulates the release of PG, LT and cytokines (table 2), many of which are known as potent inflammatory mediators and have effect on PAF production and action[28,29]. Therefore, the modulatory effect of PAF and the interaction of PAF with these mediators may be an important mechanism for PAF induced-inflammatory tissue injury, particularly for prolonged inflammation[42]. The interaction of PAF and monokines in endotoxemia and septic shock will be discussed in the following chapter in this book.

ROLES OF PAF IN DISEASE PROCESSES RELEVANT TO INFLAMMATORY RESPONSES

Renal immune-mediated inflammatory reactions

PAF has been implicated as a potential mediator in glomerular inflammatory reactions[30]. It has been reported that PAF can be produced by isolated kidneys, isolated glomeruli, and medullary cells[31,32]. Several experimental models of renal immune injury, including hyperacute kidney allograft rejection, acute serum sickness and nephrotoxic nephritis, including human studies have clearly indicated that renal damage is associated with increased intraglomerular formation of PAF. Direct evidence in support for a pathological effect of PAF on the glomerulus were

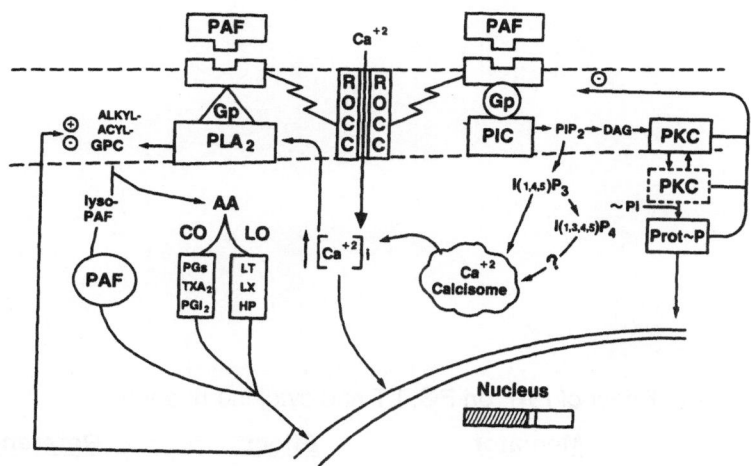

Figure 2. Schematic representation of PAF receptor-mediated
cellular signal transduction

Abbreviations: Gp, G protein; PLA$_2$, phospholipase A$_2$; PIC,
phosphoinositidase C; PIP$_2$, phosphatidylinositol 4, 5-bisphosphate; DAG,
diacylglycerol; PKC, protein kinase C; I(1,4,5)P$_3$, Inositol 1,4,5 trisphosphate; I
(1,3,4,5)P$_4$, Inositol 1,3,4,5 tetrakis phosphate; AA, Arachidonic Acid; CO,
cyclooxygenase; LO, lipoxygenase; LT, leukotriene; LX, lipoxin; PG,
prostaglandin; TX, thromboxane.

Table 1. Cell origins of PAF relevant to inflammatory responses

Cell Type	Activator
Neutrophil	A23187, zymosan, FMLP, PMA, IL-1, TNF, IgG, C5a, LTB$_4$, 5-HETE
Monocyte	A23187, zymosan, PMA, IL-1, TNF, IFN-γ, endotoxin
Macrophage	A23187, antigen, PMA, zymosan
Basophil	A23187, neutrophil cationic protein
Eosinophil	A23187, FMLP, C5a
Platelet	A23187, thrombin
Endothelial Cell	A23187, IL-1, LTC$_4$, LTD$_4$, thrombin, TNF, angiotensin II, LPS, vasopressin
Fibroblast	A23187

Abbreviations: A23187, calcium ionophore; FMLP, N'-formyl-methionyl-
leucyl-phenylanine; PMA, phorbol-12-myristate-13-acetate; LPS,
lipopolysaccharide; TNF, tumor necrosis factor; IL-1, interleukin-1; IFN,
interferon; IgG, immunoglobulin G; LTB$_4$, leukotriene B$_4$; LTC$_4$, leukotriene
C$_4$; LTD$_4$, leukotriene D$_4$; 5-HETE, 5-hydroxyeicosatetraenoic acid.

Table 2 Effect of PAF on PG, LT and cytokine release

Cell or Tissue	Mediator	Effect*	Reference
Neutrophil	LTB_4	+	(15)
Macrophage	TNF, IL-1	+	(16)
Monocyte	IL-1,	+ or -**	(17)
	TNF-α,	+	(18,19,20)
	IL-1 messenger RNA	+	(21)
Lymphocyte	IL-2, TNF	+	(22)
Lung	TXA_2, PGI_2	+	(23)
	LTC_4, LTD_4	+	(24)
Gastro-intestinal tract	LTC_4	+	(25)
Heart	LTC_4, LTD_4	+	(26)
Spleen	IL-1	+	(27)

* "+": increase; "-": decrease.

** (1) PAF stimulates IL-1 release at 100fM to 1pM and 100 pM to 1 nM, and inhibits IL-1 release at 1 or 10pM. (2) PAF and Interferon-γ have synergistic effect to enhance IL-1 release.

obtained in studies where PAF was perfused in the renal circulation. The following changes have been observed in the glomeruli: accumulation of platelets and neutrophils in the capillary lumen; endothelial cell lesions consisting of swelling and increase in number and size of the fenestrae; loss of negative charge of the glomerular capillary wall coinciding with the localization along the capillary wall of cationic proteins from platelets and neutrophils[33]. In isolated rat kidney preparations perfused with a cell-free medium, the addition of PAF to the perfusate caused a progressive increase in urinary protein which can be prevented by PAF antagonist. It has been suggested that in vivo two concomitant mechanisms are likely to be responsible for PAF-induced proteinuria. First, PAF increases glomerular permeability to proteins through the release of cationic proteins from platelets and neutrophils infiltrating the glomerular tuft, resulting in the loss of fixed anionic charges. The seond mechanism is a direct effect of PAF on the glomerular capillary wall which involves changes in the size-selective properties of glomerular membrane[30].

Gastrointestinal (GI) tract injury

It has been reported that PAF production and release by GI mucosa from duodenal ulcer patients and colonic mucosa from ulcerative colitis patients are significantly higher than controls[34,35]. Moreover, the PAF levels return to normal after 4 weeks of treatment with cimetidine which resulted in ulcer healing[34]. PAF induces long-lasting reduction in flow rates in an isolated perfused rat stomach model[36], and causes inflammatory changes in the gastrointestinal tract which can be inhibited by pretreatment with PAF antagonists[37,39]. Further studies demonstrated that PAF stimulates GI tissues to produce peptidoleukotrienes which may play an important role in PAF-induced damage since pretreatment with lipoxygenase inhibitors or corticosteroids significantly ameliorated PAF-induced gastric damage and haemoconcentration[38]. It has been recently reported that PAF infusion into the hepatic artery in anesthetized cats in which the gallbladder was continuously perfused caused a significant gallbladder inflammation with the increase in the histologic score of inflammatory cells and tissue lysosomal enzyme activities; these effects of PAF were attenuated by indomethacin, suggesting the involvement of prostanoid-mediated mechanism[40]. In rabbits, local injection of PAF results in acute pancreatitis evidenced by accumulation of leukocytes and platelets in the small arteries just 60 sec following PAF administration. These changes were all inhibited by pretreatment with PAF antagonists[41].

Pulmonary injury

Most studies of PAF-induced pulmonary injury are related to PAF-induced hypersensitivity. Thus far, the accumulated data suggest that PAF has an important function in the complex interactions between the humoral and cellular events in inflammation[42]. When IgG-sensitized rabbits were challenged with antigen inhalation, the animals showed respiratory distress and release of PAF into the circulation indicating that PAF was responsible at least in part for the lung lesion, consisting of interstitial edema and accumulation and infiltration of inflammatory cells[43]. Intravenous injection of PAF initiates acute and reversible lung tissue injury. Within seconds after PAF injection, activated platelets accumulate locally within the lumen arterioles and alveolar capillaries, along with sequestered neutrophils. The neutrophils appear to be adherent to the endothelium or are closely associated with platelet aggregates. An increased number of activated

monocytes in direct contact with the alveolar endothelium are also observed[44].
Injection of PAF into the pulmonary artery of the perfused rat lung induces a dose-dependent vasoconstriction and edema. The intrabronchial instillation of PAF in rabbits causes a dose-dependent inflammatory lung reaction characterized by the accumulation of predominately, sometimes degranulated, macrophages in the alveolar space, and of neutrophils and platelets in the alveolar capillary wall epithelium and endothelium. These initial responses are followed by infiltration of the alveolar septa by inflammatory cells and, in a later stage, by pulmonary fibrosis. Therefore, it seems that PAF acts locally as a mediator of intercellular communication[45]. When PAF is infused into isolated unsensitized guinea pig lungs perfused with a cell-free solution, increased production of TXA_2 and PGI_2 occurs; these effects are suppressed by various PAF antagonists. Moreover, intravenous or intratracheal administration of PAF also leads to the margination of eosinophils in lung tissue. The PAF-induced recruitment of inflammatory cells in lungs is an important mechanism for inducing pulmonary injury[46].

Ischemia/reperfusion-induced injury in heart and CNS

Reperfusion of ischemic tissues is associated with a variety of microvascular alterations similar to those observed in PAF-mediated inflammatory reactions, and it makes PAF a potential candidate in mediating this injury[47]. PAF production or release is increased in the ischemic isolated rabbit heart[48], gerbil brain[49] and rabbit spinal cord[50]. It was demonstrated in cat mesenteric venules subjected to 60 min of ischemia that PAF plays an important role in mediating the adhesive interaction between circulating leukocytes and microvascular endothelium and promotes the leukocyte extravasation induced by ischemia/reperfusion[51]. The PAF antagonist, CV3988, prevents the increase vascular permeability observed in the ischemic reperfused rat heart[52]. It was recently reported that the PAF antagonist, SDZ 63-675, significantly reduced hematologic and hemodynamic alterations as well as the size of necrotic area in rabbits subjected to 40 min of coronary occlusion and reperfusion. The accumulation of platelets in the central ischemic area of the heart and in the lungs are also markedly reduced [47]. Similarly, it has been demonstrated in multiple studies using the PAF antagonists that the consequences of brain injury such as edema and inflammatory cell accumulation was reduced or prevented[53].

Inflammatory injury in cornea and skin

PAF has been found to accumulate rapidly in the cornea as part of the inflammatory response to injury associated with an increase in the synthesis of lipoxygenase products of AA. The PAF antagonist, BN52021, inhibited PAF activity and formation of lipoxygenase products when applied to the inflamed cornea[54]. Inflammatory responses in human skin following intradermal injection of PAF have been demonstrated[42], suggesting that PAF may be involved in the pathogenesis of inflammatory skin disorders such as psoriasis[55] and cold urticaria[56]. In this regard, PAF has been identified in psoriatic scales and in fluid collected in chambers placed over psoriatic lesions. Moreover, a PAF-like lipid has been demonstrated in the blood of cold urticaria patients following challenge.

CONCLUDING REMARKS

PAF is released in vitro as well as in vivo from inflammatory cells, and the synthesis of PAF can be triggered by exogenous and endogenous stimuli, which are inducers of inflammatory responses. PAF exerted diverse biological effects relevant to inflammation. PAF recruits and activates various inflammatory cells, and enhance the migration of these cells into the tissue, and causes release of inflammatory mediators. Data has been accumulated to strongly suggest that PAF may be an important inflammatory mediator and has a function particularly in the regulation of cell-cell interaction and the communication between cells and tissues. PAF antagonists have shown some protective effects against PAF-induced inflammatory reactions; however, the data concerning the therapeutic effect of these antagonists in acute inflammatory reaction are still scarce. Moreover, the evidence implicating PAF in chronic inflammatory reactions is still dubious. Therefore, further studies are needed to better understand the precise role of PAF in tissue inflammatory reactions.

REFERENCES

1. J. Benveniste, P.M. Henson, and C.G. Cochrane, Leukocyte-dependent histamine release from rabbit platelet: The role of IgE basophils and a platelet activating factor, J Exp Med 136:1356 (1972).
2. P. Braquet, L. Touqui, T.Y. Shen, and B.B. Vargaftig, Perspectives in platelet-activating factor research, Pharmacol Rev 39:97 (1987).
3. F. Snyder, Platelet-activating factor and related acetylated lipids as potent biologically active cellular mediators, Am J Physiol 259:C697 (1990).
4. S.M. Prescott, G.A. Zimmerman, and T.M. McIntyre, Platelet-activating factor, J Biol Chem 265:17381 (1990).
5. P.V. Peplow, and D.P. Mikhailidis, Platelet-activating factor (PAF) and its relation to prostaglandins and leukotrienes and other aspects of arachidonate metabolism, Prostaglandins Leukotrienes Essential Fatty Acids 41:71 (1990).
6. T.L. Yue, P.G. Lysko, and G. Feuerstein, Production of platelet-activating factor from rat cerebellar granule cells in culture, J. Neurochem 54:1809 (1990).
7. R.L. Wykle, B. Malone, and F. Snyder, Enzymatic synthesis of 1-0-alkyl-2-acetyl-sn-glycero-3-phosphocholine, a hypotensive and platelet-aggregating lipid, J Biol Chem 255:10256 (1980).
8. G. Camussi, C. Tetta, F. Bussolino, and C. Baglion, Antiinflammatory peptides (antiflammins) inhibit synthesis of platelet-activating factor neutrophil aggregation and chemotaxis, and intradermal inflammatory reactions, J Exp Med 171:913 (1990).
9. S.B. Hwang, Specific receptors of platelet-activating factor, receptor heterogeneity, and signal transduction mechanisms, J Lip Mediators 2:123 (1990).
10. T.L. Yue, M.M. Gleason, J. Hallenbeck, and G. Feuerstein, Characterization of platelet-activating factor-induced elevation of cytosolic free-calcium level in neurohybrid NCB-20 cells, Neurosci (1991). (In press).
11. D.W.H. Riches, S.K. Young, J.F. Seccombe, J.E. Henson, K.L. Clay, and P.M. Henson, The subcellular distribution of platelet-activating factor in stimulated human neutrophils, J Immunol 145:3062 (1990).
12. D. Bratton and P.M. Henson, Cellular origins of PAF, in: "Platelet-activating factor and human disease," P.J. Baarnes, C.P. Page and P.M. Henson, eds., Blackwell, Oxford, London (1989).
13. P. Braquet, M.P. Braquet, R.H. Bourgain, F. Bussolino, and D. Hosford, PAF cytokine auto-generated feedback networks in microvascular immune injury, J Lip Mediators 1:75 (1989).
14. S.M. Prescott, T.M. McIntyre, and G.A. Zimmerman, The role of platelet-activating factor in endothelial cells, Thromb Haemost 64:99 (1990).
15. A.H. Lin, D.R. Morton, and R.R. Gorman, Acetyl-glyceryl-ester-phosphorylcholine stimulates leukotriene B_4 synthesis in human polymorphonuclear leukocytes, J Clin Invest 70:1058 (1982).

16. C. Dubois, E. Bissonnette, and M. Rola-Pleszczynski, Platelet-activating factor (PAF) enhances tumor necrosis factor production by alvaolar macrophages, J Immunol 143:964 (1989).

17. R. Barthelson and F. Valone, Interaction of platelet-activating factor with interferon-γ in the stimulation of interleukin-1 production by human monocytes, J Allergy Cin Immunol 86:193 (1990).

18. J. Rose, R. Barthelson, R. Debs, and F. Valone, Selective monocyte activation by a 2-0-methoxy analog of platelet-activating factor, Clin Res 37:418A (1989).

19. B. Bonavida, J.M. Mencia-Huerta, and P. Braquet, Effects of platelet-activating factor on peripheral blood monocytes: induction and priming for TNF secretion, J Lipid Mediator 2 (suppl):S65 (1990).

20. M. Rola-Pleszczynski, Priming of human monocytes with PAF augments their production of tumor necrosis factor, J Lipid Mediator 2 (suppl):S77 (1990).

21. R. Barthelson and F. Valone, Platelet-activating factor (PAF)-induced interleukin-1 beta (IL-1) messenger RNA, Clin Res 38:111A (1990).

22. M. Rola-Pleszczynski, B. Pignol, C. Pouliot, and P. Braquet, Inhibition of human lymphocyte proliferation and interleukin-2 production by platelet-activating factor (PAF-acether): reversed by a specific antagonist: BN52021, Biochem Biophys Res Commun 142:754 (1987).

23. M. Pretolani, J. Lefort, C. Dumarey and B.B. Vargaftig, Role of lipoxygenase metabolites for the hyperresponsiveness to PAF-acether of lungs from actively sensitized guinea-pigs, J Pharmacol Exp Ther 248:355 (1989).

24. N.F. Voelkel, S. Worthen, J.T. Reeves, P.M. Henson, and R.C. Murphy, Nonimmunological production of leukotrienes induced by platelet-activating factor, Science 218:286 (1982).

25. W. Hsueh, F. Gonzalez-Crussi, and J.L. Arroyave, Release of leukotriene C_4 by isolated, perfused rat small intestine in response to platelet-activating factor, J Clin Invest 78:108 (1986).

26. P.J. Piper, and A.G. Stewart, Coronary vasoconstriction in the rat, isolated perfused heart induced by platelet-activating factor is mediated by leukotriene C_4, J Br Pharmacol 88:595 (1986).

27. B. Pignol, S. Henane, J.M. Mencia-Huerta , and P. Braquet, Effect of long term treatment with platelet-activating factor on cytokine production by rat spleen cells, Int Arch Appl Immunol 88:161 (1989).

28. M. Paubert-Braquet, D. Hosford, P. Klotz, J. Guibaud, and P. Braquet, Tumor necrosis factor alpha 'primes' the platelet-activating factor-induced superoxide production by human neutrophils: possible involvement of G proteins, J Lipid Mediator 2 (suppl):S1 (1990).

29. F. Bussolino, G. Camussi, C. Tetta, G. Barbarino, A. Bosia, and C. Baglioni, Selected cytokines promote the synthesis of platelet-activating factor in vascular endothelial cells: comparison between tumor necrosis factor alpha and beta and interleukin-1, J Lipid Mediator 2 (suppl):S15 (1990).

30. N. Perico, and G. Remuzzi, Role of platelet-activating factor in renal immune injury and proteinuria, Am J Nephrol 10 (suppl 1):98 (1990).

31. D. Schlondorff, P. Goldwasser, R. Neuwirth, A. Satriano, and K.L. Clay, Production of platelet-activating factor in glomeruli and cultured glomerular mesangial cells, Am J Phyaiol 250:F1123 (1986).

32. E. Pirotzky, J. Bidault, C. Burtin, M.C. Gubler, and J. Benveniste, Release of platelet-activating factor, slow reacting substance and vasoactive amines from isolated rat kidneys, Kidney Int 25:404 (1984).

33. G. Camussi, C. Tetta, R. Coda, G. Segoloni, and A. Vercellone, Platelet-activating factor-induced lose of glomerular anionic charges, Kidney Int 25:73 (1984).

34. Z. Ackerman, F. Karmeli, M. Ligumsky, and D. Rachmilewitz, Enhanced gastric and duodenal platelet-activating factor and leukotriene generation in duodenal ulcer patients, Scand J Gastroenterol 25:925 (1990).

35. D. Rachmilewitz, F. Karmeli, and R. Eliakim, Platelet-activating factor - a possible mediator in the pathogenesis of ulcerative colitis, Scand J Gastroenterol 25 (suppl 172):19 (1990).

36. A. Dembinska-Kiec, B.A. Peskar, M.K. Muller, and M.A. Peskar, The effects of platelet-activating factor on flow rate and eicosanoid release in the isolated perfused rat gastric vascular bed, Prostaglandins 37:69 (1989).

37. J.L. Wallace, and B.J.R. Whittle, Profile of gastrointestinal damage induced by platelet-activating factor, Prostaglandins 32:137 (1986).

38. J.L. Wallace, C.M. Hogaboam, and G.W. McKnight, Platelet-activating factor mediates gastric damage induced by hemorrhagic shock, Am J Physiol 259:G140 (1990).

39. M.S. Caplan, X.M. Sun, and W. Hsueh, Hypoxia causes ischemic bowel necrosis in rats: the role of platelet-activating factor (PAF-acether), Gastroenterol 99:979 (1990).

232

40. D. Kaminski, C.H. Andrus, and D. German, The role of prostanoids in the production of acute acalculous cholecystitis by platelet-activating factor, Ann Surg 212:455 (1990).
41. G. Emanuelli, G. Montrucchio, and E. Grain, Experimental acute pancreatitis induced by platelet-activating factor in rabbits, Am J Pathol 134:314 (1989).
42. G. Camussi, and J.R. Brentjens, The role of platelet-activating factor in inflammation, in: "Platelet-activating factor and related lipid mediators," F. Snyder, ed., Plenum, New York (1987).
43. G. Camussi, and J.R. Brentjens, Inhalation of horseradish peroxidase by rabbits with specific IgE antibodies results in release into the circulation of platelet-activating factor and in lung lesions, Clin Immunol Immunopathol 34:332 (1985).
44. J. Mcmanus, and R.N. Pinckard, Kinetics of acetyl glycerol ether phosphorylcholine (AGEPC)-induced acute lung alterations in the rabbit, Am J Pathol 121:55 (1985)
45. G. Camussi, I. Pawlowski, C. Tetta, C. Roffinello, M. Alberton, J. Brentjens, and S. Andres, Acute lung inflammation induced in the rabbit by local instillation of 1-0-octadecyl-2-acetyl-sn-glyceryl-3-phosphorylcholine or of native platelet-activating factor, Am J Pathol 112:78 (1983).
46. J.M. Mencia-Huerta, D. Hosford, and P. Braquet, Acute and long-term pulmonary effects of platelet-activating factor, Clin Exp Allergy 19:125 (1989).
47. G. Montrucchio, G. Alloatti, F. Mariano, R. Paulis, and G. Camussi, Role of platelet-activating factor in the reperfusion injury of rabbit ischemic heart, Am J Pathol 137:71 (1990).
48. G. Montrucchio, G. Alloatti, C. Tetta, R. DeLuca, R.N. Saunders, G. Emanuelli, and G. Camussi, Release of platelet-activating factor (PAF) from ischemic-reperfused rabbit heart, Am J Physiol 2456:H1236 (1989).
49. P. Braquet, D. Hosford, B. Spinnewyn, D. Dunerger, N.G. Bazan, and E. Pirotzky, The therapeutic potential of Ginkgolides and other PAF antagonists in cerebral ischemia, in: "Platelet-activating factor (PAF) and diseases," K. Saito and D.J. Hanahan, eds., Int Med Publisher, Tokyo (1989).
50. P.J. Lindsberg, T.L. Yue, K.U. Frerichs, J.M. Hallenbeck, and G. Feuerstein, Evidence for platelet-activating factor as a novel mediator in experimental stroke in rabbits, Stroke 21:1452 (1990).
51. P. Kubes, G. Ibbotson, J. Russell, J.L. Wallace, and D.N. Granger, Role of platelet-activating factor in ischemia/reperfusion-induced leukocyte adherence, Am J Physiol 259:G300 (1990).
52. G.L. Stahl, Z. Terashita, and A. Lefer, Role of platelet-activating factor in propagation of cardiac damage during myocardial ischemia, J Pharmacol Exp Ther 244:898 (1988).
53. G. Feuerstein, T.L. Yue, and P.G. Lysko, Platelet-activating factor a putative mediator in central nervous system injury?, Stroke 21 (suppl III):90 (1990).
54. H.E.P. Bazan, S.T.K. Reddy, J.M. Woodland, and N.G. Bazan, The accumulation of platelet-activating factor in the injured cornea may be interrelated with the synthesis of lipoxygenase products, Biochem Biophys Res Commun 149:915 (1987).
55. A.I. Mallet, and F.M. Cunningham, Structural identification of platelet-activating factor in psoriatic scale, Biochem Biophys Res Commun 126:192 (1985).
56. K.E. Grandel, R.S. Farr, A.A. Wanderer, T.C. Eisenstaat, and S.I. Wasserman, Association of platelet-activating factor with primary acruired cold urticaria, N Engl J Med 313:405 (1985).

TRANSCELLULAR METABOLISM OF LEUKOTRIENES IN THE LUNG

Timothy D. Bigby

Department of Medicine
University of California, San Diego and
Veterans Administration Medical Center, San Diego

INTRODUCTION

Leukotrienes are a family of oxygenated metabolites derived from the 5-lipoxygenase pathway of arachidonic acid metabolism via the unstable epoxide intermediate 5(S)5,6-oxido-7,9-trans-11,14-cis-eicosatetraenoic acid or leukotriene A_4 (LTA_4). Leukotrienes have potent effects on inflammatory cell chemotaxis, adherence, and activation [1]. They also have effects on non-inflammatory cells and tissues thus increasing vascular permeability [1], inducing smooth muscle contraction [1], modulating T- and B-cell function [2], augmenting natural killer activity [3], and inducing fibroblast chemotaxis [4]. In humans, the distribution of the 5-lipoxygenase enzyme appears to be limited to inflammatory cells [1]. Within these cells, the 5-lipoxygenase enzyme is stimulated by millimolar calcium concentrations and ATP [5], associating with the cell membrane via the 5-lipoxygenase-activating protein [6]. The enzyme then inserts molecular oxygen at the C_5 position of unesterified arachidonic acid and, in turn, further converts the resultant hydroperoxide, 5(S)-hydroperoxyeicosatetraenoic acid (5-HPETE), into LTA_4. This epoxide is the pivotal intermediate in the 5-lipoxygenase pathway and can be metabolized within inflammatory cells to 5(S),12(R)-dihydroxy-6,14-cis,8,10-trans-eicosatetraenoic acid (leukotriene B_4, LTB_4) by LTA_4 hydrolase or to 5(S)-hydroxy-6(R)-S-glutathionyl-7,9-trans-11,14-cis-eicosatetraenoic acid (leukotriene C_4, LTC_4) by glutathione-S-transferase [1].

Cell-Cell Interactions in the Release of Inflammatory Mediators
Edited by P. Y-K Wong and C.N. Serhan, Plenum Press, New York, 1991

Other fates are possible for LTA_4. For example, the epoxide bond of LTA_4 is acid labile [7] and is extremely unstable in aqueous media [8]. Many investigators have assumed, therefore, that the principal fate of excess LTA_4 generated by 5-lipoxygenase is non-enzymatic hydrolysis into biologically inactive stereoisomers of LTB_4, 5(S),12(R)-dihydroxy-6,8,10-trans-14-cis-eicosatetraenoic acid (Δ^6-trans-LTB_4) and 5(S),12(S)-dihydroxy-6,8,10-trans-14-cis-eicosatetraenoic acid (Δ^6-trans-12-epi-LTB_4) or the two stereoisomers of 5,6-dihydroxyeicosatetraenoic acid (5,6-diHETEs) [9]. LTA_4 can, however, be released intact by inflammatory cells [10] and can serve as an intermediate for further metabolism to the biologically active compounds LTB_4 [11-13] or LTC_4 [14-16]. Non-inflammatory cells, devoid of 5-lipoxygenase, contain enzymes capable of metabolizing LTA_4, reflecting a wider distribution for these enzymes than the 5-lipoxygenase. For example, glutathione-S-transferase capable of converting LTA_4 to LTC_4 has been found in human liver [14], human endothelial cells [15], and human platelets [16]. Likewise, LTA_4 epoxide hydrolase activity has been found in human plasma [11], erythrocytes [12], liver [13], and whole lung [17]. Lipoxygenase metabolism is, therefore, possible via exchange of unstable intermediates from inflammatory cells to non-inflammatory cells. This metabolic fate of LTA_4 is a previously unrecognized potential source of lipoxygenase products derived from cell to cell interactions. The synthesis of leukotrienes via such cell to cell interactions has been termed transcellular metabolism [11,18].

Work performed by other investigators has demonstrated that neutrophils are capable of interacting with non-inflammatory cells and donating LTA_4 for further metabolism. Feinmark and Cannon demonstrated that porcine aortic endothelial cells were capable of converting LTA_4 to LTC_4 and that the neutrophil was capable of providing the LTA_4 to the endothelial cell [15]. Subsequent investigations by Maclouf and Murphy demonstrated that human neutrophils could provide LTA_4 to human platelets for further conversion to LTC_4 [16]. Other examples have been reported.

Our laboratory has focused on the 5-lipoxygenase pathway of

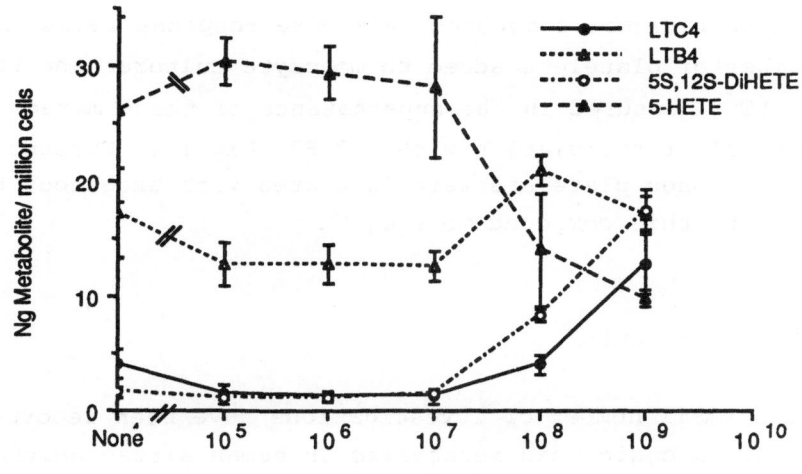

Figure 1. Platelet Dose-Response Curve. $2.5-3 \times 10^6$ monocytes were incubated with up to 10^9 platelets in log increments. The cells were stimulated with 1 μM A23187 for 15 min at 37°C. Products were identified and quantified by RP-HPLC, SP-HPLC, and UV spectroscopy. Data are reported as ng product/ 10^6 monocytes. (•) LTC$_4$, (O) 5S, 12S-diHETE , (Δ) LTB$_4$, (Δ) 5-HETE. Data represent the mean ±sd for 3 separate experiments.

mononuclear phagocytes. In early studies, we questioned whether mononuclear phagocytes could also participate in transcellular metabolism [19]. These studies demonstrate a complex transcellular metabolic interaction involving multiple aspects of lipoxygenase pathways. In short, monocytes provide arachidonic acid, 5-HETE, and LTA$_4$ which are, in turn, avidly taken up and further metabolized by platelets to 12-HETE, 5S,12S-diHETE, and LTC$_4$, respectively. Monocytes do not appear to take up lipoxygenase products from platelets (i.e.- 12-HETE) and monocyte 5-lipoxygenase activity is not significantly stimulated by 12-HPETE. These findings differ substantially from observations of neutrophil interactions with platelets [20,21] reflecting a unidirectional instead of a bidirectional interaction as is the case for the platelet and neutrophil. Release of LTA$_4$ by monocytes and the uptake of this compound by platelets with conversion to LTC$_4$ was suggested by a dose response relationship to the number of platelets added to monocyte cultures and the amount of LTC$_4$ measured in the supernatants of these mixed incubations after stimulation with A23187 (Fig 1). Further work revealed that when platelets were incubated with exogenous LTA$_4$, they converted this compound to LTC$_4$ [19].

AIRWAY EPITHELIAL CELLS

Only a small number of investigations have been reported regarding arachidonic acid metabolism in human airway epithelia, but some features of the pathway in this cell have been established. For example, these cells contain significant quantities of esterified arachidonic acid which is present in membrane phospholipids, principally in the phosphatidylcholine (3.8%) and the phosphatidylethanolamine (13.4%) fractions [22]. As shown by radioimmunoassay, human tracheal epithelial cells can metabolize this arachidonic acid via the cyclooxygenase pathway to a variety of prostaglandins [23]. With respect to lipoxygenase activity, dog airway epithelial cells contain a 5-lipoxygenase pathway capable of generating 5-HETE and LTB$_4$ [24]. Initial work suggested that human airway epithelial cells might also have a 5-lipoxygenase pathway [25]. Subsequent investigations have, however, established that this cell does not have a 5-lipoxygenase pathway, but, instead, has a potent 15-lipoxygenase pathway [26].

Products of the 5-lipoxygenase pathway have significant effects in the lung and are thought to play an important role in a variety of lung diseases [27,28]. Inflammation in the lung is prominent in many of these diseases and, specifically, in airway diseases [29]. Substantial evidence indicates that leukotrienes are important mediators in this inflammatory response in the airway and that leukotrienes also have direct effects on airway function [27]. We therefore designed studies to determine whether human airway epithelial cells could metabolize exogenous LTA$_4$ into lipoxygenase products with biological effects. We also questioned whether inflammatory cells could serve as a source of this intermediate [30].

LTA$_4$ was prepared by alkaline hydrolysis of its methyl ester [31]. For most experiments, transformed human tracheal epithelial cells were utilized. This cell line has been designated 9/HTEo⁻ and has been previously characterized [32]. These cells were grown to confluence on 100 mm tissue culture plastic dishes coated with collagen, fibronectin, and BSA in minimal essential media with 10% FCS. Human airway epithelial cells were also obtained from post-mortem tracheal specimens (i.e.-patients dying without known airway pathology), acutely disaggregated [33], placed in primary culture [33], and studied at confluence. Human mononuclear cells were prepared by density gradient centrifugation [19] and resuspended at 0.8×10^6 cells/ml in HBSS prior to study.

To identify products, 1 to 2×10^7 human tracheal epithelial cells were incubated with 3 μM LTA$_4$ in HBSS with 0.35% BSA, pH 7.4 for 30 min at 37°C. A dose-response curve was constructed to LTA$_4$ by incubating 9/HTEo⁻ cells with 0.09 to 10 μM LTA$_4$ for 30 min at 37° C. The time course of metabolite generation was assessed by incubating 9/HTEo⁻ cells with 3.0 μM LTA$_4$ for up to 4h. These cells were also incubated with LTA$_4$ in the absence of calcium or BSA. To determine the role of cell stimulation, 9/HTEo⁻ cells were also challenged with the calcium ionophore A23187 in the presence or absence of added LTA$_4$. To determine if hydrolase activity could be inactivated by substrate, 9/HTEo- cells were incubated with 10 μM LTA$_4$ for 30 min, the supernatant collected and the cells were re-challenged with 10 μM LTA$_4$, twice. All three supernatants were analyzed for metabolic products. To evaluate the role of inflammatory cells as a source of LTA$_4$, 2.5 ml of mononuclear cell suspension was added to confluent epithelial cell cultures containing 1 to 2×10^7 cells. The cells were then co-incubated for 30 min and then stimulated with 1 μM A23187 for 30 min at 37° C. These incubations were compared to control incubations of mononuclear cells without epithelial cells which were also stimulated with A23187.

Immediately after experimental incubations, samples were collected, placed on ice, and extracted using organic solvents. Extracts were analyzed by reversed phase-HPLC, straight-phase HPLC, ultraviolet spectroscopy, and gas chromatography-mass spectrometry [30].

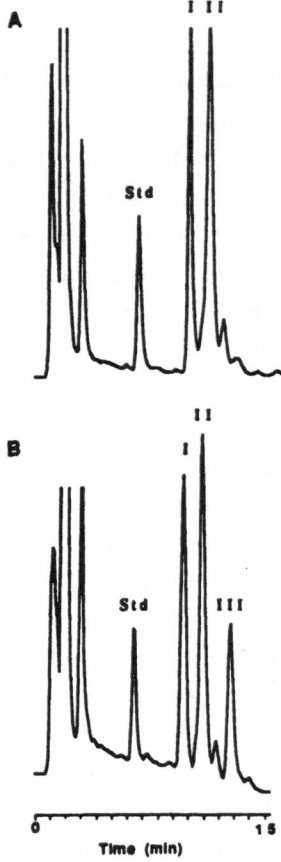

Figure 2. A) Reversed phase-HPLC chromatogram of an extract of
HBSS incubated at 37° C with 3 μM LTA$_4$ for 30 min
without cells (control). B) Reversed phase-HPLC
chromatogram of a supernatant extract from 9/HTEo⁻
cells (1 x 10^7) incubated with 3 μM LTA$_4$ for 30 min.
Both samples monitored at 270 nm. Peak I, Peak II,
and Peak III corresponded to retention time of
authentic Δ^6-trans-LTB$_4$, Δ^6-trans-12-epi-LTB$_4$, and
LTB$_4$, respectively. Monitoring at 280 nm did not
reveal evidence of LTC$_4$ or LTD$_4$.

Results

    Reversed phase-HPLC analysis of extracts of media alone
incubated with 3.0 μM LTA$_4$ in the absence of cells for 30 min
revealed two peaks (Peak I and II) (Fig 2a) . When extracts of
media from transformed and non-transformed tracheal epithelial
cells incubated with 3.0 μM LTA$_4$ for 30 min were analyzed they
consistently revealed three peaks of absorbance (Fig 2b).   The

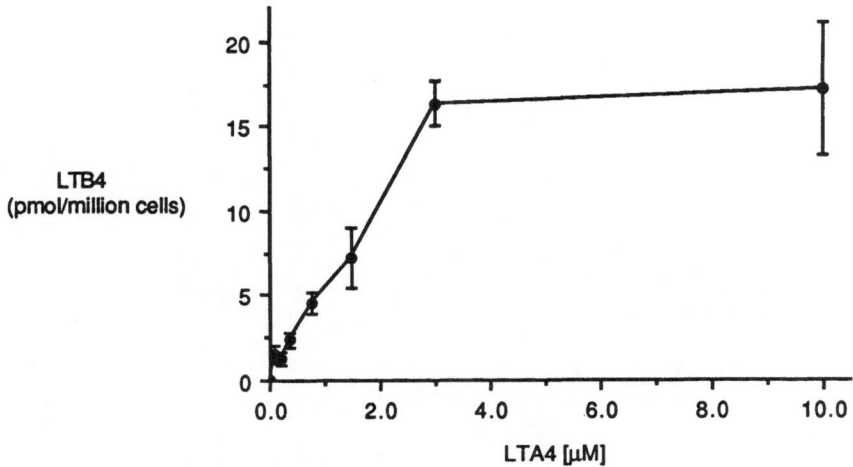

Figure 3. Dose-Response Curve. $1-2 \times 10^7$ 9/HTEo⁻ cells were incubated with up to 10 μM LTA$_4$ for 30 min and then extracted and analyzed as above. Maximum LTB$_4$ release was observed at or above 3 μM LTA$_4$. Similar results were obtained with primary cultures of non-transformed human tracheal epithelial cells.

retention times of peaks I and II corresponded exactly to the retention times of Δ^6-trans-LTB$_4$ (9.9 min) and Δ^6-trans-12-epi-LTB$_4$ (10.9 min), respectively. The presence of these two peaks was presumed to represent non-enzymatic hydrolysis of LTA$_4$ as previously shown [7]. Peak III eluted exactly at the retention time of authentic LTB$_4$ (12.7 min). No peaks corresponding to the retention times of LTC$_4$ (7.7 min) or LTD$_4$ (19.7 min) were found in either sample. The ultraviolet absorption spectrum of peak III was characteristic of authentic LTB$_4$ with maximal absorption at 270 nm with shoulders at 260 and 281 nm [9]. The methyl ester of peak III eluted as a single peak on straight phase-HPLC and corresponded exactly to the retention time of methylated authentic LTB$_4$ (8.0 min). Gas chromatography-mass spectrometry analysis of the trimethylsilyl ether, methyl ester derivative of the compound contained in peak III was identical to that of the derivatized authentic standard, LTB$_4$, and previously published spectra [9].

When 9/HTEo⁻ cells were incubated with increasing quantities of freshly prepared LTA$_4$, from 0.09 to 10 μM for 30 min at 37° C

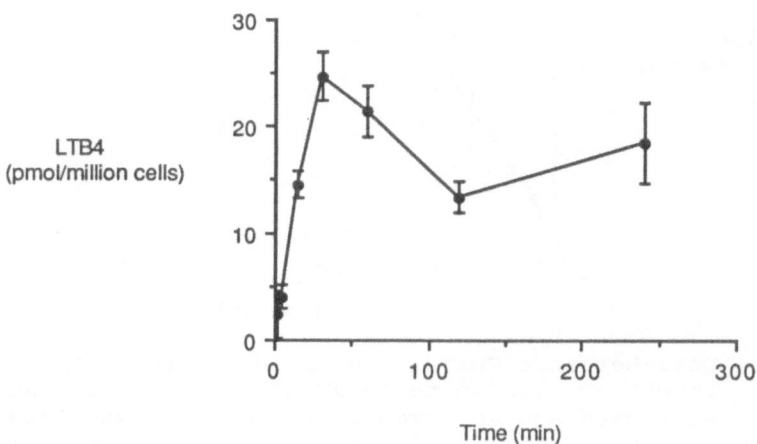

Figure 4.   Time Course.   9/HTEo⁻ cells were incubated with 3 μM
            LTA₄ cells for up to 4h.   Supernatants were then
            extracted and analyzed as above.   Maximum LTB₄ release
            occured by 30 min.

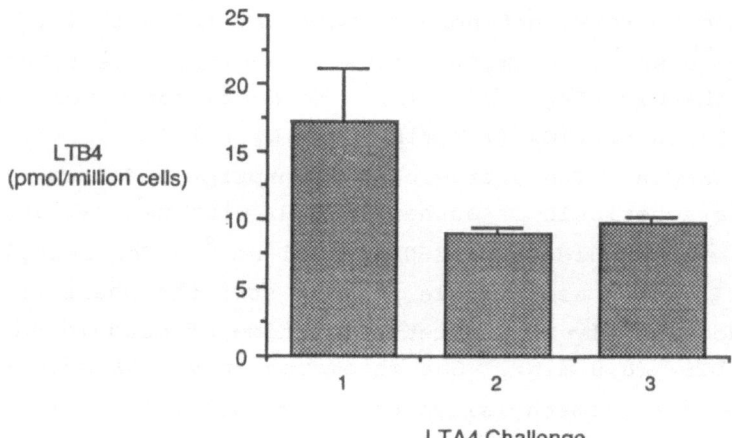

Figure 5.   Repeated Substrate Challenge.   9/HTEo⁻ cells were
            incubated with 10 μM LTA₄ for 30 min at 37° C.   The
            supernatant was then collected and the cells were
            re-challenged with 10 μM LTA₄, twice.   All three
            samples were examined for metabolic products.

maximum product (16.4 ± 1.3 pmol/10^6 cells)(mean $\pm$ sd) was detected at concentrations of LTA_4 at or above 3 μM (Fig 3). When the time-course of LTB_4 generation and release from 9/HTEo⁻ cells was examined, maximal LTB_4 release was observed at 30 min with a subsequent decrease in the amount of LTB_4 detectable in the media (Fig 4). Similar quantities of LTB_4 were detected when primary cultures of human tracheal epithelial cells were incubated with LTA_4 for 30 min (21.5 ± 8.9 pmol/10^6 cells). When 9/HTEo⁻ cells were incubated in media without added BSA, only 45.7 % (7.5 ± 1.1 pmol/10^6 cells) of the amount of LTB_4 formed in the presence of BSA was detected. However, no significant difference in the quantity of LTB_4 released was observed in the absence of calcium in the media and no increase was observed when the cells were stimulated with A23187 in the presence of LTA_4 (data not shown). Likewise, no 5-lipoxygenase products were detected when cells were stimulated with A23187 in the absence of LTA_4 (data not shown), as has been previously reported [26]. When 9/HTEo⁻ cells were challenged three times with 10 μM LTA_4, the cells released LTB_4 with each successive challenge, although there was a decrease in product after the first challenge (Fig 5). When 9/HTEo⁻ cells were incubated with mononuclear cells for 30 min and then challenged with 1 μM A23187 and compared to mononuclear cells challenged with A23187 in the absence of epithelial cells, significantly greater quantities of LTB_4 were detected in supernatants (5.4 ± 0.5 vs 3.7 ± 0.2 pmol/10^6 mononuclear cells, respectively).

DISCUSSION

The described study demonstrated that human airway epithelial cells have the capacity to convert LTA_4, an unstable epoxide, to LTB_4 [30]. Because previous investigations have established that the generation of biologically active LTB_4 from LTA_4 requires a stereospecific hydrolase enzyme [7], we presume that our findings reflect the presence of a LTA_4 hydrolase in airway epithelial cells. Preliminary investigations examining this activity in a cell-free fraction of airway epithelial cells further substantiates its enzymatic nature [34]. We have not, however, found evidence of conversion of LTA_4 to LTC_4 and, thus, we have found no evidence for a glutathione-S-transferase activity specific for LTA_4 in airway epithelial cells. We have also demonstrated greater LTB4 generation when inflammatory cells are incubated with epithelial cells which suggests that inflammatory cells are capable of providing LTA4 as a substrate for conversion by the epithelial cell to LTB4.

The importance of transcellular metabolism of lipoxygenase intermediates is not known either with respect to the cell biology of inflammation or the biology of the organism, as a whole. Nevertheless, substantial evidence indicates that transcellular metabolism can occur. Initial reports regarding the conversion of exogenous LTA4 by intact neutrophils [35], isolation of intact LTA4 from stimulated neutrophils [35], stabilizing effects of albumin on LTA4 [8], and evidence that transcellular metabolism occurs with other unstable intermediates [36] all suggest that LTA4 can be released intact by inflammatory cells and metabolized by another cell. Subsequent studies have utilized several techniques including the property of "suicide inactivation" of LTA4 hydrolase [37], alcohol trapping studies of LTA4 released from neutrophils [10], and radiolabeled peptide adducts [15,16] to establish extracellular transfer of LTA4. Likewise, the microenvironment of cells may facilitate transfer of this labile intermediate through the presence of extracellular albumin [8] or, possibly, the co-adherence of cells interacting in the exchange.

The human neutrophil LTA4 hydrolase has been purified to homogeneity [38], cloned [39], and expressed [40]. The properties of this enzyme are that it has a M_r of 69,140 kd, consists of 610 amino acids, is cytosolic, has a K_m of 20-30 μM, a pH optimum of approximately 8, and is monomeric. Other LTA4 hydrolases have been purified from whole human lung parenchyma [17] and erythrocytes [41]. The hydrolase purified from whole human lung appears to be the same as the neutrophil-derived enzyme as suggested by its n-terminal amino acid sequence, molecular weight, and similar biochemical characteristics. The cell of origin of this enzyme in the lung parenchyma is unknown; however, this enzyme could be a feature common to all cells in the lung, a feature of a specific non-inflammatory cell, or be derived solely from resident inflammatory cells in the lung such as macrophages. The LTA4 hydrolase present in erythrocytes may be different based on preliminary purification data indicating a lower molecular weight of 54,000 kd [41]. A feature shared by LTA_4 hydrolases purified from a variety of sources, including neutrophils [40], whole lung [17], and erthrocytes [41], is inactivation by exposure to substrate ("suicide inactivation") presumably by formation of a covalent link between enzyme and substrate [42]. These enzymes have also been shown to have rapid initial velocities in all cases whether in a whole cell or an enzyme preparation [40,41]. Recent reports indicate that human neutrophil LTA_4 hydrolase is a zinc containing enzyme with remarkable homology to the aminopeptidase-N family of proteases [43]. Further studies have established that this enzyme does indeed have peptidase activity, albeit less than its LTA_4 hydrolase activity [44]. The biological significance of these findings is unknown.

In investigations in our own laboratory we have localized LTA_4 hydrolase activity to the human airway epithelial cell. We have not found suicide inactivation of this activity up to a substrate concentration of 10 μM in the intact cell and have found a slower time course than those previously described for LTA_4 hydrolase activity in intact neutrophils or erythrocytes. These preliminary differences may reflect differences in assay conditions and require confirmation with a purified LTA_4 hydrolase from human airway epithelial cells. These findings do, however, raise the question of whether the human airway epithelium has an LTA_4 hydrolase which is distinct from previously described enzymes. Complete characterization through purification to homogeneity and protein sequencing will be necessary to answer this question.

The function and significance of an intermediate enzyme present in the epithelium involved in the generation of the potent chemotactic lipid, LTB_4, is not clear. A number of functions are, however, possible. For example, LTB_4 formed in the epithelium may serve an intracellular function in the epithelium or may serve to amplify the amount of LTB_4 formed in the airway from inflammatory cell 5-lipoxygenase activity. Hydrolase activity in the airway epithelium may induce directed chemotaxis of inflammatory cells to the epithelium. More importantly, if this enzyme activity is not suicide-inactivated and has slower kinetics than that described for the neutrophil enzyme, the airway epithelium could provide a sustained source of

LTB_4. Determination of the true function of this enzyme activity will require further investigation.

TRANSCELLULAR METABOLISM ELSEWHERE IN THE LUNG

Further reports directly investigating transcellular metabolism in the lung have not yet been published. However, transcellular synthesis of leukotrienes in the lung is likely to be a frequent and important event. For example, human lung fibroblasts have been shown to have LTA_4 hydrolase activity and that this activity is enhanced by SV40 transformation [45]. The distribution of LTA_4 hydrolase enzymatic activity [46] and immunoreactivity [47] in animal lungs is extensive. Further, the finding that human lymphocytes have LTA_4 hydrolase activity and that their function is modulated by LTB_4 [2] suggest that this may well occur in the lung. Although the pulmonary endothelium, smooth muscle cells, and alveolar epithelium have not been examined in humans to determine if they are capable of metabolizing LTA_4, these are logical experiments. Nevertheless, the ultimate importance of transcellular metabolism in the lung will need to be determined by in vivo experiments.

ACKNOWLEDGMENTS

This work was supported in part by grants from the National Institutes of Health (HL-01888) and the Veterans Adminstration.

REFERENCES

1. Samuelsson, B. Leukotrienes: Mediators of immediate hypersensitivity reactions and inflammation. Science. 220: 568-575, 1983.
2. Samuelsson, B. and H.-E. Claesson. Leukotriene B_4: Biosynthesis and role in lymphocytes. Advances in Prostaglandin, Thromboxane, and Leukotriene Research. 20: 1-13, 1990.

3. Rola-Pleszczynski, M., L. Gagnon and P. Sirois. Leukotriene B_4 augments human natural cytotoxic cell activity. Biochem Biophys Res Commun. 113: 531-537, 1983.

4. Mensing, H. and B. M. Czarnetzki. Leukotriene B_4 induces in vitro fibroblast chemotaxis. J Investig Derm. 82: 9-12, 1984.

5. Rouzer, C. A. and B. Samuelsson. On the nature of the 5-lipoxygenase reaction in human leukocytes: Enzyme purification and requirement for multiple stimulatory factors. Proc. Natl. Acad. Sci. USA. 82: 6040-6044, 1985.

6. Miller, D. K., J. W. Gillard, P. J. Vickers, et al. Identification and isolation of a membrane protein necessary for leukotriene production. Nature. 343: 278-281, 1990.

7. Radmark, O., C. Malmsten and B. Samuelsson. Leukotriene A: Stereochemistry and enzymatic conversion to leukotriene B. Biochem. Biophys. Res. Commun. 1980(3): 954-961, 1980.

8. Fitzpatrick, F. A., D. R. Morton and M. A. Wynalda. Albumin stabilizes leukotriene A4. J. Biol. Chem. 257(9): 4680-4683, 1982.

9. Borgeat, P. and B. Samuelsson. Metabolism of arachidonic acid in polymorphonuclear leukocytes. Structural analysis of novel hydroxylated compounds. J. Biol. Chem. 254(16): 7865-7869, 1979.

10. Dahinden, C. A., R. M. Clancy, M. Gross, J. M. Chiller and T. E. Hugli. Leukotriene C_4 production by murine mast cells: Evidence of a role for extracellular leukotriene A4. Proc. Natl. Acad. Sci. USA. 82: 6632-6636, 1985.

11. Fitzpatrick, F., J. Haeggstom, E. Granstrom and B. Samuelsson. Metabolism of leukotriene A_4 by an enzyme in blood plasma: A posible leukotactic mechanism. Proc. Natl. Acad. Sci. USA. 80: 5425-5429, 1983.

12. Fitzpatrick, F., W. Liggett, J. McGee, et al. Metabolism of leukotriene A_4 by human erthrocytes. A novel cellular source of leukotriene B4. J. Biol. Chem. 259(18): 11403-11407, 1984.

13. Haeggstrom, J., O. Radmark and F. A. Fitzpatrick. Leukotriene A_4-hydrolase activity in guinea pig and human liver. Biochim. Biophys. Acta. 835: 378-384, 1985.

14. Bach, M. K., J. R. Brashler, R. E. Peck and D. R. Morton. Leukotriene C synthetase, a special glutathione S-transferase: Properties of the enzyme and inhibitor studies with special reference to the mode of action of U-60,257, a

selective inhibitor of leukotriene synthesis. J. Allergy
Clin. Immunol. 74: 353-357, 1984.

15. Feinmark, S. J. and P. J. Cannon. Endothelial cell
    leukotriene C_4 synthesis results from intercellular transfer
    of leukotriene A_4 synthesized by polymorphonuclear
    leukocytes. J. Biol. Chem. 261(35): 16466-16472, 1986.

16. Maclouf, J. A. and R. C. Murphy. Transcellular metabolism of
    neutrophil-derived leukotriene A_4 by human platelets. J.
    Biol. Chem. 263(1): 174-181, 1988.

17. Ohishi, N., T. Izumi, M. Minami, et al. Leukotriene A_4
    hydrolase in the human lung. J. Biol. Chem. 262(21): 10200-
    10205, 1987.

18. Marcus, A. J. "Eicosanoids: Transcellular metabolism."
    Inflammation: Basic Principles and Clinical Correlates.
    Gallin, Goldstein and Snyderman ed. 1988 Raven Press, Ltd.
    New York.

19. Bigby, T. D. and N. Meslier. Transcellular lipoxygenase
    metabolism between monocytes and platelets. J. Immunol. 143:
    1948-1954, 1989.

20. Marcus, A. J., M. J. Broekman, L. B. Safier, H. L. Ullman N.
    Islam, C. N. Serhan, L. E. Rutherford, H. M. Korchak, and G.
    Weissmann. Formation of leukotrienes and other hydroxy acids
    during platelet-neutrophil interactions in vitro. Biochem.
    Biophys. Res. Commun. 109(1): 130-137, 1982.

21. Maclouf, J., B. Fruteau de Laclos and P. Borgeat. Stimulation
    of leukotriene biosynthesis in human blood leukocytes by
    platelet-derived 12-hydroperoxy-icosatetraenoic acid. Proc.
    Natl. Acad. Sci. USA. 79: 6042-6046, 1982.

22. Holtzman, M. J., D. Grunberger and J. A. Hunter. Phospholipid
    fatty acid composition of pulmonary airway epithelial cells:
    potential substrates for oxygenation. Biochim. Biophys. Acta.
    877: 459-464, 1986.

23. Widdicombe, J. H., I. F. Ueki, D. Emery, et al. Release of
    cyclooxygenase products from primary cultures of tracheal
    epithelia of dog and human. Am J Physiol. 257: L361-365,
    1989.

24. Holtzman, M. J., H. Aizawa, J. A. Nadel and E. J. Goetzl.
    Selective generation of leukotrienes by tracheal epithelial
    cells from dogs. Biochem. Biophys. Res. Commun. 114: 1071-,
    1983.

25. Holtzman, M. J., W. E. Finkbeiner, E. J. Goetzl and J. A. Nadel. Selective generation of leukotriene B_4 and 15-hydroxy-eicosatetraenoic acid by human tracheal epithelial cells. Fed Proc. 43: 829A, 1984.

26. Hunter, J. A., W. E. Finkbeiner, J. A. Nadel, E. J. Goetzl and M. J. Holtzamn. Predominant generation of 15-lipoxygenase metabolites of arachidonic acid by epithelial cells from human trachea. Proc. Natl. Acad. Sci. USA. 82: 4633-4637, 1985.

27. Drazen, J. M. and K. F. Austen. State of Art. Leukotrienes and airway responses. Am. Rev. Respir. Dis. 136: 985-998, 1987.

28. Henderson, W. R. Eicosanoids and lung inflammation. Am. Rev. Respir. Dis. 135: 1176-1185, 1987.

29. Bigby, T. D. and J. A. Nadel. "Asthma." Inflammation: Basic Principles and Clinical Correlates. Gallin, Goldstein and Snyderman ed. 1988 Raven Press, Ltd. New York.

30. Bigby, T. D., D. M. Lee, N. Meslier and D. C. Gruenert. Leukotriene A_4 hydrolase activity in human airway epithelial cells. Biochem Biophys Res Commun. 164(1): 1-7, 1989.

31. Maycock, A. L., M. S. Anderson, D. M. DeSousa and F. A. Kuehl Jr. Leukotriene A_4: Preparation and Enzymatic Conversion in a Cell-free Ssytem to Leukotriene B_4. J. Biol. Chem. 257(23): 13911-13914, 1982.

32. Gruenert, D. C., C. B. Basbaum, M. J. Welsh, et al. Characterization of human tracheal epithelial cells transformed by an origin-defective simian virus 40. Proc. Natl. Acad. Sci. USA. 85: 5951-5955, 1988.

33. Coleman, D. L., I. K. Tuet and J. H. Widdicombe. Electrical properties of dog tracheal epithelial cells grown in monolayer culture. Am. J. Physiol. 246(C355-C359): 1984.

34. Bigby, T. D., D. M. Lee, M. J. Banda, et al. Human airway epithelial cell leukotriene A_4 hydrolase. J Cell Biochem. Suppl 14C: 330, 1990.

35. Radmark, O., C. Malmsten, B. Samuelsson, et al. Leukotriene A: Isolation from human polymorphonuclear leukocytes. J. Biol. Chem. 255(24): 11828-11831, 1980.

36. Marcus, A. J., B. B. Weksler, E. A. Jaffe and M. J. Broekman. Synthesis of prostacyclin from platelet-derived endoperoxides by cultured human endothelial cells. J. Clin. Investig. 66: 979-986, 1980.

37. McGee, J. G. and F. A. Fitzpatrick. Erythrocyte-neutrophil interactions: Formation of leukotriene B_4 by transcellular biosynthesis. Proc. Natl. Acad. Sci. USA. 83: 1349-1353, 1986.

38. Radmark, O., T. Shimizu, H. Jornvall and B. Samuelsson. Leukotriene A_4 hydrolase in human leukocytes. Purification and properties. J. Biol. Chem. 259(20): 12339-12345, 1984.

39. Funk, C. D., O. Radmark, J. Y. Fu, et al. Molecular cloning and amino acid sequence of leukotriene A_4 hydrolase. Proc. Natl. Acad. Sci. USA. 84: 6677-6681, 1987.

40. Minami, M., Y. Minami, Y. Emori, et al. Expression of human leukotriene A_4 hydrolase cDNA in Escherichia coli. FEBS Letters. 229(2): 279-282, 1988.

41. McGee, J. and F. Fitzpatrick. Enzymatic hydration of leukotriene A_4. Purification and characterization of a novel epoxide hydrolase from human erythrocytes. J. Biol. Chem. 260(23): 12832-12837, 1985.

42. Evans, J. F., D. J. Nathaniel, R. J. Zamboni and A. W. Ford-Hutchinson. Leukotriene A_3. A poor substrate but a potent inhibitor of rat and human neutrophil leukotriene A_4 hydrolase. J Biol Chem. 260: 10966-10970, 1985.

43. Toh, H., M. Minami and T. Shimizu. Molecular evolution and zinc ion binding motif of leukotriene A_4 hydrolase. Biochem Biophys Res Commun. 171: 216-221, 1990.

44. Haeggstrom, J. Z. Leukotriene A_4 hydrolase: an epoxide hydrolase with peptidase activity. Biochem Biopys Res Commun. 173: 431-437, 1990.

45. Medina, J. F., C. Barrios, C. D. Funk, et al. Human fibroblasts show expression of the leukotriene A_4 hydrolase gene, which is increased after simian virus 40 transformation. Eur J Biochem. 191: 27-31, 1990.

46. Medina, J. F., J. Haeggstrom, M. Kumlin and O. Radmark. Leukotriene A_4: metabolism in different rat tissues. Biochim Biophys Acta. 961: 203-212, 1988.

47. Ohishi, N., M. Minami, J. Kobayashi, et al. Immunological quantitation and immunohistochemical localization of leukotriene A_4 hydrolase in guinea pig tissues. J Biol Chem. 265: 7520-7525, 1990.

THE ROLE OF LIPID MEDIATORS IN OXYGEN-INDUCED LUNG INJURY

Ronald B. Holtzman, Joseph R. Hageman

Evanston Hospital and Northwestern University Medical School
2650 Ridge Avenue; Evanston, IL. 60201

Oxygen therapy has been incorporated in the treatment of critically ill patients of all ages since the 1930's. Although often life saving in this context, prolonged exposure to high concentrations of oxygen is toxic to all cells, particularly those comprising the alveolar-capillary unit of the lung. Pulmonary oxygen toxicity has been implicated in the pathogenesis of the adult respiratory distress syndrome, and among premature neonates oxygen injury is believed to play a central role in the development of bronchopulmonary dysplasia and retinopathy of prematurity.

Although the potentially damaging effects of oxygen were described in the 1700's by Scheele, Priestley and Lavoisier, the pathogenic mechanisms of oxygen injury remain incompletely understood. In 1954 Gerschman and colleagues noted a similarity in the histologic appearance of pulmonary oxygen toxicity and that induced by ionizing radiation.[1] The free radical theory of oxygen toxicity proposed a common intermediate relating these two phenomena. Oxygen is normally reduced in the cell by combining with two electrons to form water. Under conditions of prolonged exposure to high concentrations of oxygen, oxygen is additionally metabolized to partially reduced intermediates including the superoxide anion, singlet oxygen, hydroxyl radical, and hydrogen peroxide. These radical intermediates are highly reactive, and damage the cell by peroxidizing double bonds present in membrane lipids, inducing breaks in and cross-linking of DNA fragments, and altering protein structure.[2] These intermediates are presumably damaging when produced in excess of the cell's capacity to detoxify their actions through interaction with host antioxidant defenses.

Pulmonary oxygen toxicity has been documented in humans and a number of animal species, and has been extensively studied utilizing both in vivo and in vitro models.[3] Hageman et al. exposed adult rabbits to >95% O_2 for up to 65 hours.[4] Compared to normoxic controls (Figure 1A), oxygen-exposed rabbits demonstrated scattered areas of separation of the pulmonary capillary endothelial cell cytoplasm from the basement membrane at 48 hours (Figure 1B). Following 65 hours of O_2 exposure alveolar septae were significantly thickened by interstitial edema, increased capillary permeability was demonstrated by the recovery of proteinaceous bronchoalveolar lavage fluid, and an influx of neutrophils was noted (Figure 1C). Alveolar type I epithelial cells were extensively damaged, with swelling and vacuolization of the cytoplasm and separation from the basal lamina, while capillary endothelial cells were widely necrotic (Figure 1D). Necrosis of endothelial and type I epithelial cells, interstitial and alveolar edema, and the influx of polymorphonuclear leukocytes comprise the central features of acute oxygen injury to the lung, and has been reproduced in a number of animal species.[5]

Cell-Cell Interactions in the Release of Inflammatory Mediators
Edited by P. Y-K Wong and C.N. Serhan, Plenum Press, New York, 1991

251

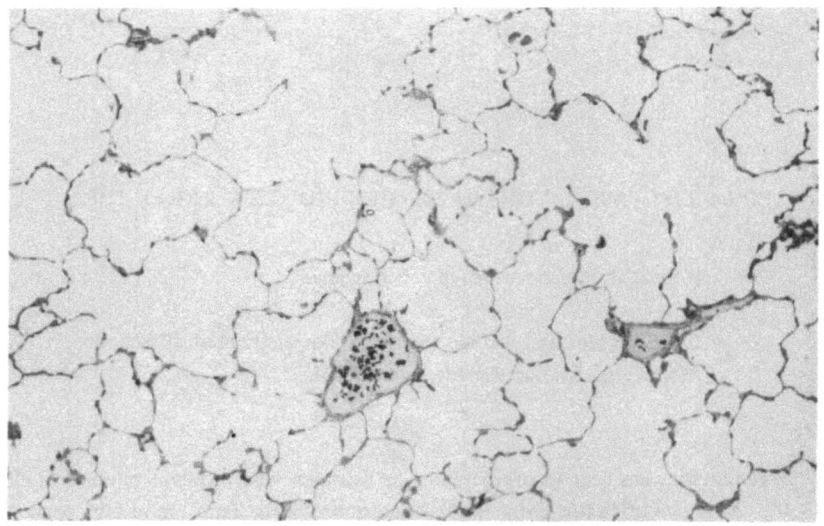

Figure 1A. Light microscopy of a rabbit lung exposed to air for 65 hours. As in the other control groups, the alveolar walls are thin, with occasional nucleated cells. The alveolar spaces contain rare macrophages. There is perivascular or interstitial edema. Methylene blue stain: original magnification x 100.

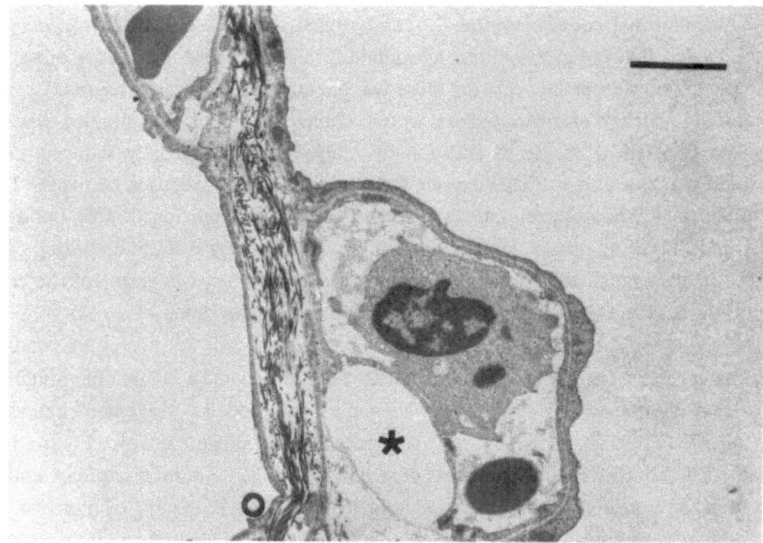

Figure 1B. Electron microscopy of a rabbit lung exposed to > 95% O_2 for 48 hours. The Type I epithelial cell is normal. The endothelial cell cytoplasm is also normal. However, there is one area where the endothelial cell cytoplasm is separated from its basement membrane, resulting in bleb formation (asterisk). Original magnification x 10375. Marker - 2μ.

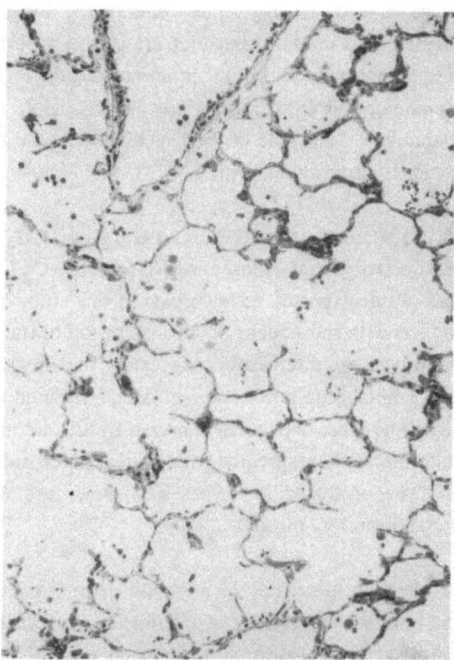

Figure 1C. **Figures 1C and 1D:** Morphologic lung changes in a rabbit lung exposed to > 95% O_2 for 65 hours. Figure 1C: Light microscopy. The alveolar walls are thickened by interstitial edema. Perivascular edema is also present. The alveolar spaces contain proteinaceous exudate and increased number of AMs and PMNs. Methylene blue stain: original magnification x 100.

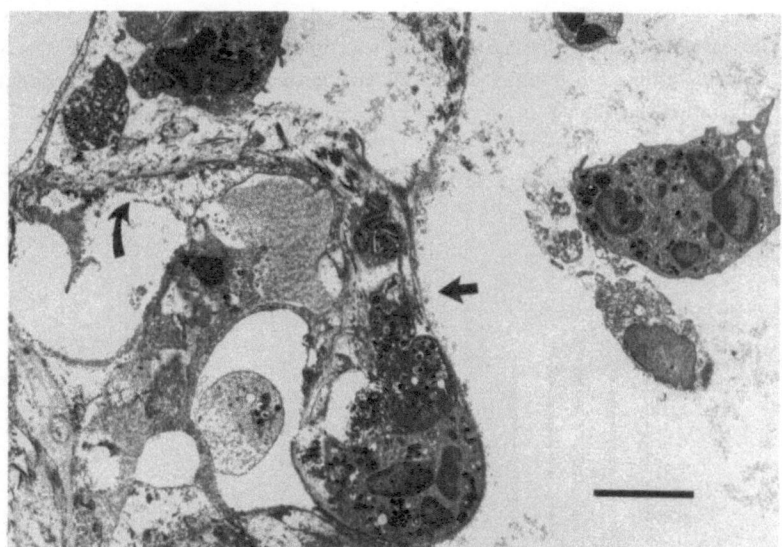

Figure 1D. Electron microscopy: The Type I epithelium is severely damaged; only a bare basement membrane lines a large area of the alveolus (straight arrow). Endothelial cells are also severely damaged (curved arrow). PMNs are present in the interstitial space. Original magnification x 5190. Marker = 4μ. Figures 1A, 1B, 1C, 1D reproduced with permission from <u>Prostaglandins, Leukotrienes and Medicine</u>, (1986;25:105-122).

253

Although the notion that reactive O_2 intermediates generated in hyperoxic exposure directly damage lung cells is widely accepted, it is unlikely that they are centrally participant in the cell-cell communication resulting in inflammation. Bioactive lipid mediators, including metabolites of arachidonic acid, have been documented to modulate various aspects of the inflammatory response in a number of experimental paradigms. The role of lipid mediators in the inflammatory response of acute pulmonary oxygen toxicity is currently unclear.

De Nucci et al. found no increases in TXB2, 6-oxo-PGF$_{1\alpha}$ or LTB$_4$ in pulmonary venous effluent from isolated perfused guinea pig lungs exposed in vivo to 100% O_2 for up to 96 hours when compared to normoxic controls.[6] Prior exposure to hyperoxia in this study did result in significant increases in cyclooxygenase products with subsequent in vitro perfusion of the pulmonary artery with bradykinin. This effect was blocked by dexamethasone, suggesting that oxygen exposure modulates a priming effect on cyclooxygenase metabolite production, possibly mediated through phospholipase A$_2$. Sporn et al. reported a dose-and time-dependent increase in arachidonate release and TXB$_2$ production in cultured rat alveolar macrophages treated with hydrogen peroxide.[7] This study suggests that other oxidants, including those produced in oxygen exposure, have the capacity to induce arachidonic acid release and metabolism by lung cells.

Hageman et al. reported elevated quantities of metabolites of prostaglandin E$_2$ and PGI$_2$, determined by radioimmunoassay, in bronchoalveolar lavage fluid of adult rabbits following 40 hours of exposure to >95% O_2 compared to normoxic time-matched controls.[4] These elevations in eicosanoid levels preceded the appearance of overt microscopic evidence of lung injury. Thromboxane A$_2$ metabolite, as well as PGE$_2$ and 6-keto-PGF$_{1\alpha}$ levels were additionally elevated at 65 hours of exposure, and corresponded to widespread oxygen-induced injury. Radioimmunoassay of cultured bronchoalveolar lavage cell supernatants obtained from in vivo-exposed hyperoxic vs. normoxic rabbits, predominantly alveolar macrophages in early injury, and equal populations of macrophages and neutrophils at 65 hours, demonstrated significant elevations in PGE$_2$, PGI$_2$, and TXA$_2$ metabolites at 40 hours of O_2 exposure. Thromboxane B$_2$ was also elevated at 40 hours, while PGE$_2$ was increased at 65 hours. Plasma levels of these metabolites was not different between hyperoxic and normoxic groups, suggesting that the cellular source of these mediators is resident and recruited lung cells rather than blood.

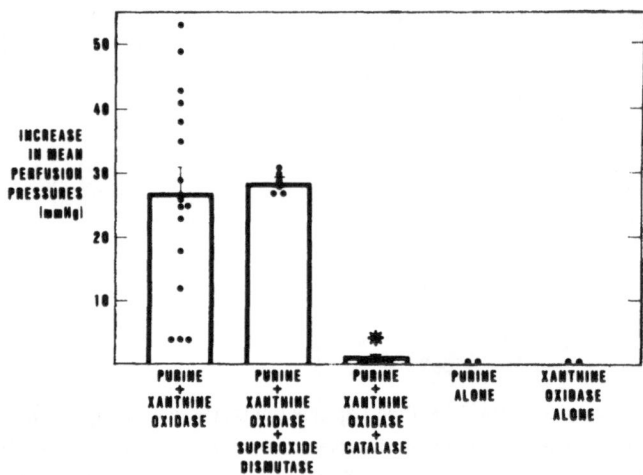

Figure 2

Although oxidants may stimulate the production of cyclooxygenase metabolites by the lung, dependent upon the species, fluid selected for analysis, and assay technique utilized, it is unclear whether these products mediate the pathophysiology of acute pulmonary oxygen toxicity. Tate at al. demonstrated that infusion of purine and xanthine oxidase in the pulmonary artery of isolated perfused adult rabbit lungs resulted in an increase in pulmonary artery perfusion pressure associated with a 30-fold increase in TXB_2 in venous effluent (Figure 2).[8] This effect was inhibited by prior administration of the cyclooxygenase inhibitors aspirin and indomethacin, and by the antioxidant catalase. Administration of the organic peroxide tert-butyl-hydroperoxide by Gurtner et al. resulted in pulmonary artery vasoconstriction and increased TXB_2 production in isolated perfused rabbit lungs.[9] This effect was blocked by indomethacin, and was strongly correlated with an elevation in the ratio of TXB_2:6-keto-$PGF_{1\alpha}$ (Figure 3), suggesting that an imbalance between vasoconstricting and vasodilating eicosanoids mediated this effect. Hageman et al. reported that infusion of PGE_1, a potent vasodilator and inhibitor of neutrophil chemotaxis, attenuated alveolar protein leakage and polymorphonuclear leukocyte influx in bronchoalveolar lavage fluid of oxygen-exposed adult rabbits.[10] However oxygen-induced lung injury remained extensive in the treated animals. Jones et al. found that exposure to 87-90% O_2 for 7 days in rats resulted in a striking loss of cross-sectional area of intra-acinar pulmonary arteries and pulmonary hypertension.[11] This was associated with a reduction in arterial external diameter and increased medial wall thickness. This study represents the sole report of oxygen-induced pulmonary hypertension in vivo, however vasoactive and cell proliferative mediators were not measured.

Hageman et al. infused the isotonic fat emulsion Intralipid, which contains a high concentration of prostaglandin precursors, into the external jugular vein of adult rabbits.[12] Intralipid infusion resulted in no deterioration in gas exchange as determined by arterial and venous blood gas measurements, and no alterations in plasma prostaglandin levels. Fat emulsion administration to rabbits with lung injury, induced by prior administration of oleic acid, resulted in a 16 torr decrement in PaO_2 and an increase in $PaCO_2$ of 6 torr. These alterations corresponded to an increase in plasma PGE_2 and 6-keto-$PGF_{1\alpha}$, and were blocked by prior administration of indomethacin. This study again suggests that oxidants may have a priming effect on cyclooxygenase metabolite production in the lung, and that these metabolites may in part mediate changes in vascular tone resulting in the gas exchange abnormalities seen in oxygen lung injury.

The leukotrienes derived from the 5-lipoxygenase pathway have been implicated in a variety of inflammatory processes. Leukotriene B_4 binds with high affinity and stereospecificity to receptors on human neutrophils.[13] Aspects of post-receptor signal transduction are currently unclear, however LTB_4 has been shown to be a potent chemokinetic and chemoactivating substance for neutrophils.[14]

Figure 2. Purine (2 mM) plus xanthine oxidase(0.02 U/ml) caused increase in mean perfusion pressure in isolated lungs. All perfusion pressure data are reported as the maximum pressure increases prior to any lung weight gains. Superoxide dismutase (100 μg/ml, n = 4, 10 μ/ml, n = 1; 1 μg/ml, n = 1) did not alter the xanthine oxidase-induced pressor responses. Catalase (100 μg/mo, n = 4) or purified catalase (50 μg/ml, n = 2) inhibited the pressor responses. Each symbol represents one experiment. Asterisk indicates significant difference (P <0.05) from purine plus xanthine oxidase group.

Figure 2 reproduced with permission from Journal of Clinical Investigation (1984;74:608-613), by copyright permission of the American Society of Clinical Investigation.

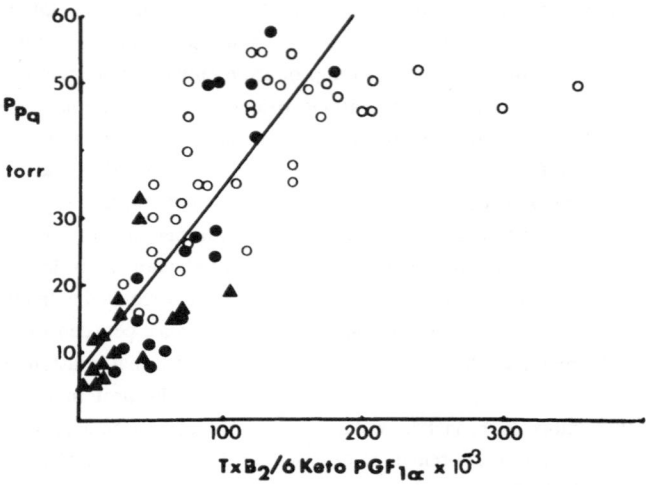

Figure 3. Pulmonary arterial pressure is plotted against ratio of thromboxane B_2 (TxB_2) to prostacyclin (6-ketoprostaglandin $F_{1\alpha}$, 6-keto-$PGF_{1\alpha}$) metabolites. At ratios less than 200×10^{-3} there is a very strong correlation between the 2 variables. From this relationship we conclude that magnitude of pressor response is modulated by relative release of these 2 mediators.

Figure 3 reproduced with permission from Journal of Applied Physiology (1983:55:949-954) by copyright permission of The American Physiological Society.

The sulfidopeptide leukotrienes LTC_4, LTD_4, and LTE_4 have been found to induce bronchial and vascular constriction, and increase microvascular permeability.[15,16]

Jackson et al. reported increased production of immunoreactive LTB_4 from 95% oxygen-exposed bovine pulmonary artery endothelial cell supernatants subsequently treated with the calcium ionophore A23187, compared to stimulated cells grown in 5% CO_2.[17] Exposure of adult rats to 97% O_2 for 60 hours by Taniguchi et al. resulted in significant increases of immunoreactive LTB_4 in bronchoalveolar lavage fluid compared to normoxic controls (Figure 4A).[18] A strong correlation between quantities of LTB_4 and influx of neutrophils was demonstrated. Administration of the 5-lipoxygenase inhibitor AA861 inhibited the hyperoxia-induced increases in LTB_4 as well as neutrophil influx (Figure 4B). Martin et al. found that LTB_4 was the predominant 5-lipoxygenase product from human alveolar macrophages stimulated in vitro with A23187.[19] In contrast, stimulated human peripheral blood neutrophils produced comparatively little LTB_4. These authors further showed that synthetic LTB_4 was strongly chemotactic for human neutrophils and weakly chemoa actant for alveolar macrophages in vitro.

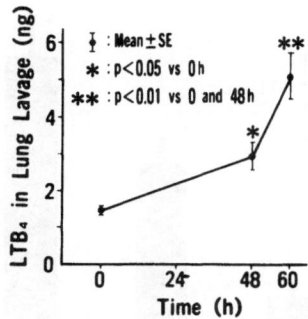

Figure 4A. Effect of exposure of rats to hyperoxia on the level of LTB_4 in their lung lavages (mean ± SE). The level of LTB_4 after 60 h of O_2 exposure was significantly higher than after zero or 48 h (p < 0.01). Each point represents 8 determinations.

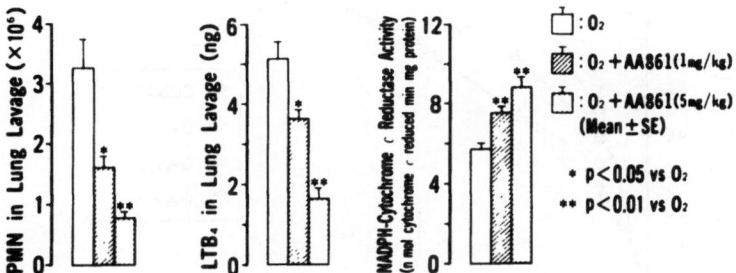

Figure 4B. Effects of AA861 on the number of PMN, the level of LTB_4 in lung lavages, and the specific activity of lung microsomal NADPH-cytochrome c reductase in rats exposed to hyperoxia for 60 h. AA861 (5 mg/kg) reduced the number of PMN and the level of LTB_4 in lung lavages in rats exposed to hyperoxia for 60 h, respectively (p <0.01). AA861 also inhibited the reduction of the activity of NADPH-cytochrome c reductase in rats exposed to hyperoxia (p<0.01). Administration of AA861 (1 mg/kg) showed a similar tendency; however, the effect was weak, i.e., the effects of AA861 were observed in a dose-dependent fashion. Data are presented as the mean ± SE of 8 determinations.

Figures 4A and 4B reproduced with permission by <u>American Review of Respiratory Disease</u> (1986;133:805-808) by copyright permission from the American Lung Association.

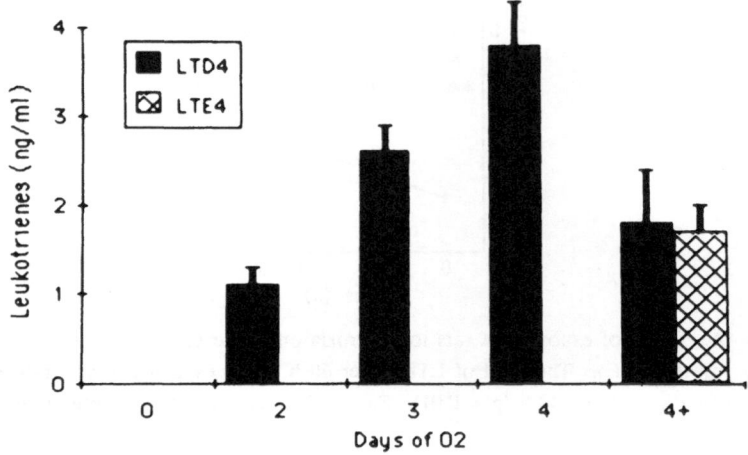

Figure 5A. Leukotriene D_4 (LTD_4) and LTE_4 concentrations in bronchoalveolar lavage fluid. Values are mean ± SE; n = 5-10 mice each time.

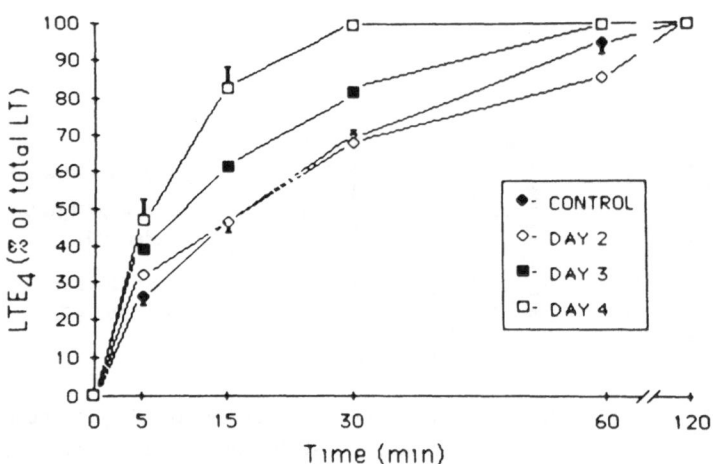

Figure 5B. Bioconversion of leukotriene D_4 (LTD_4) in bronchoalveolar (BAL) fluid. Incubation time on x-axis. Total sulfidopeptide leukotrienes recovered from BAL fluid after incubation (%), identified as LTE_4, on y-axis. Values are means; SE shown only for control (*day 0*) and *day 4* data. Number of experiments performed at each incubation time for each exposure condition was as follows: control = 3 - 10; *day 2 = 2; day 3 = 2; day 4 = 3*. *Day 4* values are significantly different from control values at 5)P < 0.05), 15 (P <0.01), and 30 min (P <0.001). Figures 5A and 5B reproduced with permission from Journal of Applied Physiology (1988; 64(3):944-951) by copyright permission from the American Physiological Society.

Smith et al. studied the appearance and bioconversion of peptide leukotrienes in hyperoxic lung injury.[20] Compared to normoxic controls, adult mice exposed to 100% O_2 had no increases in LTB_4 and LTC_4 in bronchoalveolar lavage fluid. The appearance of LTD_4 was noted in lavage fluid after 48 hours of O_2 exposure, reaching their highest levels on day 4 (Figure 5A). With more prolonged oxygen exposure, when lung damage was severe, LTD_4 levels declined, concomitant with the appearance of LTE_4. Addition of synthetic LTD_4 in vitro to bronchoalveolar lavage cell supernatants exposed to hyperoxia in vivo resulted in a greater production of LTE_4 compared to normoxic cells (Figure 5B). Synthetic LTC_4 was not converted to other leukotriene metabolites. This effect was correlated with the duration of oxygen exposure and quantities of protein in lavage fluid, and was blocked by the addition of the dipeptidase inhibitor L-cysteine. This study demonstrates that LTD_4 is the predominant 5-lipoxygenase product elaborated in the lungs of oxygen injured mice, and that this metabolite is rapidly bioconverted to LTE_4 by a dipeptidase. Production of LTD_4 may have in part accounted for the increased microvascular permeability.

Stenmark et al. documented increased leukotriene-like activity in tracheal aspirates from oxygen- and ventilator-dependent newborns with respiratory distress and chronic lung disease.[21] The previous studies demonstrate that peptide leukotrienes are produced in pulmonary oxygen toxicity, with species-specific variability determining the particular metabolite elaborated.

Although 5-lipoxygenase products may be elaborated in hyperoxic lung damage, some authors have suggested that their appearance is an epiphenomenon of cellular injury, rather than a participant in the pathophysiology of the disease. Kertesz et al. administered the specific LTD_4/LTE_4 receptor antagonist ICI 198,615 to oxygen-exposed young rabbits.[22] This compound had no protective effect on lung neutrophil influx, protein leak, or pulmonary edema compared with placebo-treated O_2-exposed controls. In contrast, Burghuber et al. reported that administration of a leukotriene receptor antagonist (FPL 55712), a leukotriene synthesis blocker (U60,257), or a combined lipoxygenase and cyclooxygenase inhibitor (BW 755C) inhibited the hydrogen peroxide-mediated increase in alveolar edema in isolated perfused adult rat lungs.[23]

Polymorphonuclear leukocyte influx into an oxygen-damaged lung has been documented in a number of animal species, and frequently correlates with mortality from oxygen toxicity. Neutrophils have the capacity to amplify lung injury in the context of hyperoxia through their production of oxygen radicals, myeloperoxidase, elastase, and additional arachidonic acid metabolites. Whether they are non-specifically recruited to sites of lung injury, or attracted by specific chemotactic signals elaborated in hyperoxia is currently unclear.

Suttorp and Simon reported that phorbol myristate acetate- and opsonized zymosan-stimulated rat peritoneal neutrophils damaged rat type II alveolar epithelial cells, rat lung fibroblasts, and calf pulmonary artery endothelial cells quantitated by [51]-chromium release.[24] Neutrophil-mediated cytotoxicity was greater in cell lines previously exposed to sustained hyperoxia. Pre-treatment of cells with the antioxidants superoxide dismutase or catalase protected against cellular injury, suggesting that superoxide anion and hydrogen peroxide mediated the neutrophil-induced cytotoxicity.

Exposure of rats to >95% O_2 by Fox et al. resulted in a significant influx of neutrophils into bronchoalveolar lavage fluid by 66 hours.[25] Mortality in hyperoxia was temporally associated with neutrophil influx in 71% of the subjects. Lavage fluid from 66 hour O_2-exposed rats in this study demonstrated a 10-fold greater chemoattractant activity for human and rat neutrophils in vitro compared to normoxic lavage controls. Granulocyte depletion by administration of nitrogen mustard to rabbits attenuated hyperoxia-induced accumulation of lung neutrophils and reduced edematous lung injury.[26] Raj and Bland found similar reduction in lung neutrophils following nitrogen mustard treatment in O_2-exposed rabbits, but no decrement in lung water content nor improvement in survival.[27]

The cellular sources of chemotactic signals for polymorphonuclear leukocytes recruited to oxygen-injured lungs, and the nature of these signals is incompletely understood. Increasing evidence implicates the lung macrophage in the control of alveolar-interstitial cellular traffic.[28] Harada et al. found that rabbit alveolar macrophage exposure to 95% O_2 in culture resulted in significant ultra-structural damage and release of lactate dehydrogenase (Figure 6).[29] Hyperoxia also stimulated these macrophages to release chemotaxins for human peripheral blood neutrophils. This chemotactic signal was similar in molecular weight to that obtained from lung lavages of rabbits exposed to hyperoxia in vivo. Chauncey et al. reported that calcium ionophore induced rat alveolar type II epithelial cells to release ^{14}C-arachidonic acid.[30] The labeled arachidonate was subsequently metabolized by co-cultured alveolar macrophages to LTB_4 and 12-HETE. Hageman et al. found that administration of dexamethasone or indomethacin to O_2-exposed adult rabbits failed to suppress lung production of LTB_4 or neutrophil influx, and did not attenuate the development of lung injury.[31] Smith found a significant recruitment of neutrophils to oxygen-damaged mouse lungs despite no elevations in LTB_4, suggesting that additional chemotactic signals are elaborated in hyperoxia.[20]

While it is clear that prolonged exposure to hyperoxia results in significant lung injury, a number of studies have documented relative oxygen tolerance among term newborn animals compared to adult members of the same species.[3] Mechanisms affording this tolerance are incompletely understood. Frank and Groseclose documented increases in the lung antioxidant enzymes superoxide dismutase, catalase, and glutathione peroxidase occurring in late gestation in the fetal rabbit.[32] Yam, Frank, and Roberts reported increased quantities of these antioxidants in oxygen-tolerant newborn rats following hyperoxic exposure, compared to normoxic controls.[33] In contrast, levels of these enzymes decreased from their baseline values in O_2-injured adult rats. These studies suggest that relative oxygen tolerance in the newborn derives from their lung's capacity to induce antioxidant enzyme production in response to hyperoxia. Holtzman et al. confirmed that 65 hours of exposure to >95% O_2 resulted in no biochemical or structural evidence of lung injury in rabbits when exposure was initiated on day one of life.[34] An equal interval of oxygen exposure initiated on day seven of life resulted in significant lung injury. Lung antioxidants were lower in the older animals compared to newborns, and lung injury occurred in the seven day-old rabbits despite induction in antioxidants with hyperoxic exposure. This study suggests that a minimum quantity of antioxidants may be required to support oxygen tolerance, rather than their induction with oxygen exposure.

Holtzman et al. examined the role of lipid mediators in newborn tolerance to hyperoxia.[35] Exposure of adult rabbits to >95% O_2 produced significantly greater quantities of protein and neutrophils in broncholaveolar lavage fluid, and overt histologic evidence of lung injury compared to oxygen-exposed newborn rabbits. Oxygen-injured adult rabbits had significantly higher lavage LTB_4 and LTC_4 than did newborns (Figure 7A), while the prostacyclin metabolite 6-keto-$PGF_{1\alpha}$ was higher both in normoxic and hyperoxic newborn groups (Figure 7B). Developmental changes in lung prostacyclin and other eicosanoids was recently examined by Holtzman et al.[34] Highest levels of PGI_2 metabolites in rabbits raised under normoxic conditions were found on the third day of life, and demonstrated a developmental decrement to day 10 (Figure 8). Thromboxane B_2 levels were lower on day 10 than day one, while LTB_4, though somewhat higher on day one, was not statistically different during this time period.

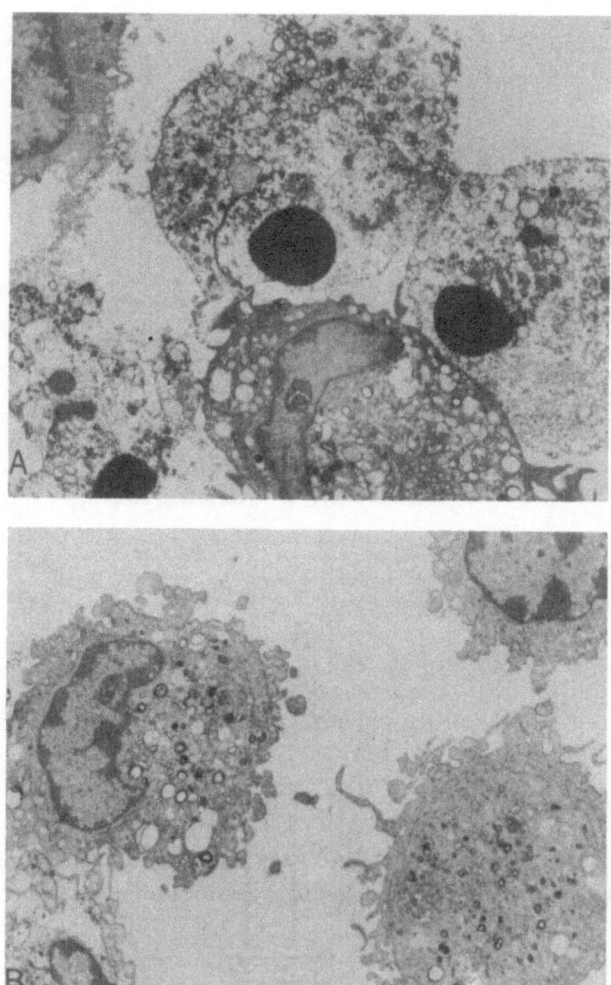

Figure 6. AM exposed to hyperoxia for 48 hr (A) had evidence of cell death and lysis while AM exposed to normoxia were normal (B). These electron micrographs are representative of chánges found in five separate experiments. (x 4,500.)

Figure 6 reproduced with permission from Journal of Leukocyte Biology with copyright permission from John Wiley Publishing. 1984;35:373-383.

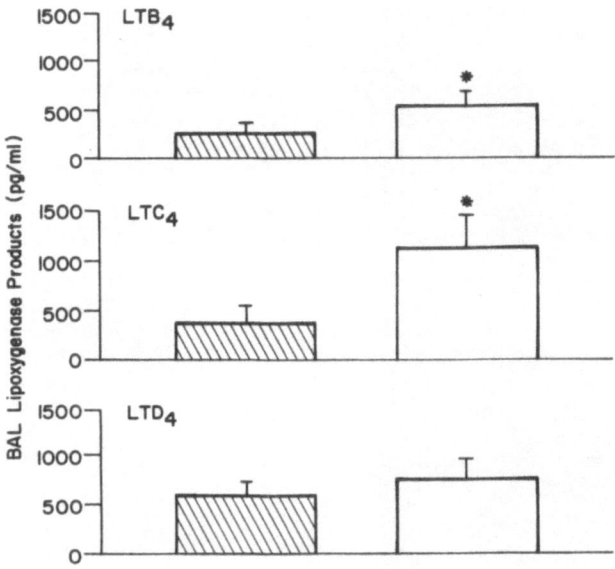

Figure 7A. 5-lipoxygenase products in BAL fluid from hyperoxid newborn and adult rabbits (mean group values ± SEM). (*=p<.05; hyperoxic newborn vs. adult).

Newborn-Hyperoxic (n=9)
Adult-Hyperoxic (n=15)

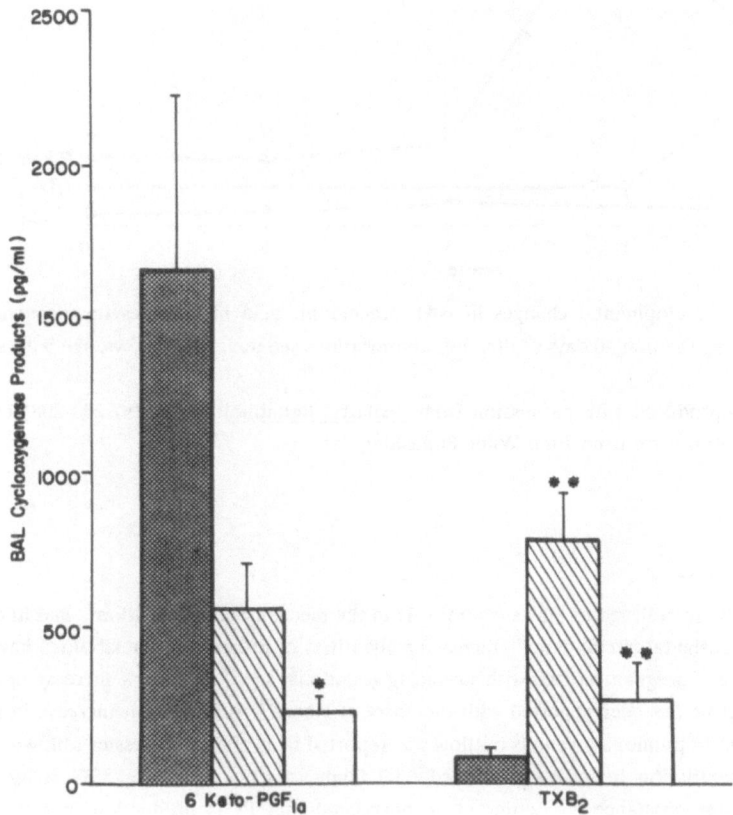

Figure 7B. Cyclooxygenase products in BAL fluid from normoxic newborn and hyperoxic newborn and adult rabbits (mean group values ± SEM). (*=p<.05, **=p<.01; normoxic vs. hyperoxic newborn, and hyperoxic adult vs. hyperoxic newborn). Figures 7A and 7B reproduced with permission from <u>Prostaglandins</u> (1989;37:481-491).

Newborn-Normoxic (n=7)
Newborn-Hyperoxic (n=9)
Adult-Hyperoxic (n=15)

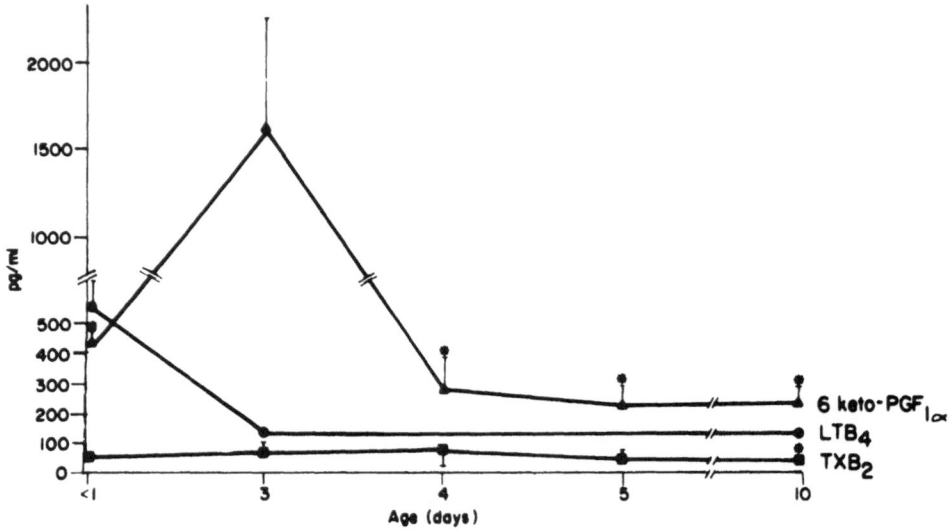

Figure 8. Developmental changes in BAL arachidonic acid metabolites in normoxic rabbits during the first 10 days of life. For abbreviations see text. *P <0.05 vs. age 3 days.

Figure 8 reproduced with permission from Pediatric Pulmonology (1989;7:200-208) by copyright permission from John Wiley Publishing.

Eicosanoids are believed to play a central role in the mechanics of parturition[36] and in control of the fetal and perinatal circulation.[37] Increasing quantities of prostacyclin metabolites have been found in the urine of pregnant women with advancing gestational age,[38] and a 50% increase in 6-keto-$PGF_{1\alpha}$ in fetal urine has been reported with the onset of labor.[39] A significant increase in prostacyclin metabolites in pulmonary venous outflow was reported by Leffler and Hessler following onset of mechanical ventilation in term exteriorized fetal lambs, associated with a 58% reduction of pulmonary vascular resistance.[40] Leffler et al. postulated that PGI_2 production in response to ventilation is related to distention or stress of pulmonary tissue.[41] These studies suggest that increased quantities of prostacyclin in the perinatal period play an important role in successful fetal circulatory and respiratory adaptation to extrauterine life.

Prostacyclin may additionally afford relative oxygen tolerance to newborn animals. The protective actions of prostacyclin may derive from its modulating effect on neutrophil release of lysosomal enzymes and oxygen radicals.[42] Prostacyclin increases intracellular cyclic adenosine monophosphate via stimulation of adenylate cyclase,[43] which inhibits the production of arachidonic acid metabolites by down-modulating activity of protein kinases.[44] Kuehl et al. found that prostacyclin inhibited cellular release of LTB_4 and counteracted the effects of leukotrienes C_4 and D_4.[45] The loss of oxygen tolerance in older rabbits previously described may have resulted, in part, from a maturational decrement in prostacyclin.

Holtzman et al. treated oxygen-exposed young rabbits with the stable carbaprostacyclin derivative iloprost.[46] Compared to O_2-exposed controls, iloprost-treated rabbits had a significantly smaller percentage of neutrophils in bronchoalveolar lavage fluid at 65 and 84 hours of exposure, and greater survival in hyperoxia at 96 hours. However iloprost failed to diminish other biochemical parameters of lung injury.

Dietary effects on lung lipid composition, eicosanoid production, and tolerance to hyperoxia have also been studied. Sosenko et al. administered a diet high in either n-3 polyunsaturated fatty acids (PUFA), n-6 PUFA, or a standard chow to female rats prior to and during gestation.[47] Compared to rat pups born to mothers fed a standard chow, high n-3 PUFA offspring had increased eicosapentaenoic and docosahexaenoic acid, and decreased linoleic and arachidonic acid in their lung lipids, and decreased lung levels of PGE_2, $PGF_{2\alpha}$, and TXB_2. High n-6 PUFA-fed offspring had the opposite profile. Irrespective of their lung lipid composition and eicosanoid profiles, both the n-3 and n-6 PUFA-fed offspring had superior survival in >95% O_2 than standard chow pups. These authors speculate that improved oxygen tolerance conferred by high dietary PUFA is related to improved scavenging of oxygen free radicals rather than eicosanoid production.

In summary, prolonged exposure to high concentrations of oxygen results in significant lung injury in both humans and a number of animal species. Oxygen-induced lung injury is comprised of necrosis of the pulmonary capillary endothelial and type I alveolar epithelial cell, interstitial and alveolar edema, influx of neutrophils and, in some models, pulmonary hypertension. Arachidonic acid release is stimulated by oxidants, and metabolites of both the cyclooxygenase and 5-lipoxygenase pathways have been documented with oxygen exposure both in vivo and in vitro. Alveolar macrophages and type II alveolar epithelial cells secrete bioactive lipid mediators in response to oxidants, while macrophages may play a central role in modulating cell-cell communication. Pharmacologic inhibition of arachidonate release or subsequent metabolism has a variable protective effect against hyperoxia. Newborn animals are relatively oxygen tolerant compared to adults, and exhibit a distinct pattern of eicosanoid production in response to oxygen challenge.

Pulmonary oxygen toxicity remains a challenging clinical problem in the treatment of patients of all ages. The study of the role of lipid mediators has produced greater insight into the pathophysiology and pathogenesis of this disease. However, many fundamental issues remain unresolved. Our understanding of the basic mechanisms of oxygen toxicity is additionally limited by fundamental differences in response to oxygen exposure, and exposure to other oxidants. Results obtained from in vitro experiments may not be strictly analogous to the intact organism. Although the production of lipid mediators in pulmonary oxygen toxicity has been established by ample experimental evidence, because of the diversity of mediators elaborated in this context it seems unlikely that inhibition of their actions will completely ameliorate oxygen-induced lung injury.

The authors wish to express their gratitude to Drs. Carl E. Hunt and Lewis J. Smith, Ms. Luba Adler and Ms. Jeanne Zemaitis for their collaboration in many of the studies cited, and to Ms. Esther Werre for expert preparation of the manuscript. This work was supported in part by grants from the Dee and Moody Research Fund of Evanston Hospital, and from the Dean's Intramural Research Fund of Northwestern University Medical School.

References

1. R. Gerschman, D.L. Gilbert, S.W. Nye, P. Dwyer, and W.O. Fenn. Oxygen poisoning and x-irradiation: a mechanism in common., Science., 119:623 (1954).
2. L. Frank and D. Massaro. Oxygen toxicity., Am. J. Med., 69:117 (1980).
3. J.M. Clark and C.J. Lambertson. Pulmonary oxygen toxicity: a review., Pharmacol. Rev., 23:37 (1971).

4. J.R. Hageman, S. Babler, S.C. Lee, M. Cobb, L.M. Pachman, L.J. Smith, and C.E. Hunt. The early involvement of pulmonary prostaglandins in hyperoxic lung injury., <u>Prostaglandins, Leukotrienes and Medicine.</u>, 25:105 (1986).

5. J.D. Crapo. Morphologic changes in pulmonary oxygen toxicity., <u>Ann. Rev. Physiol.</u>, 48:721 (1986).

6. G. De Nucci, R. Astbury, N. Read, J.A. Salmon, and S. Moncada. Release of eicosanoids from isolated lungs of guinea-pigs exposed to pure oxygen: effect of dexamethasone., <u>Eur. J. Pharmacol.</u>, 126:11 (1986).

7. P.H.S. Sporn, M. Peters-Golden, and R.H. Simon. Hydrogen-peroxide-induced arachidonic acid metabolism in the rat alveolar macrophage., <u>Am. Rev. Respir. Dis.</u>, 137:49 (1988).

8. R.M. Tate, H.G. Morris, W.R. Schroeder, and J.E. Repine. Oxygen metabolites stimulate thromboxane production and vasoconstriction in isolated saline-perfused rabbit lungs., <u>J. Clin. Invest.</u>, 74:608 (1984).

9. G.H. Gurtner, A. Knoblauch, P.L. Smith, H. Sies, and N. F. Adkinson. Oxidant- and lipid-induced pulmonary vasoconstriction mediated by arachidonic acid metabolites., <u>J. Appl. Physiol.</u>, 55:949 (1983).

10. J.R. Hageman, S.E. Lee, J. Zemaitis, L.J. Smith, and C.E. Hunt. Prostaglandin E_1 infusion fails to prevent hyperoxic lung injury in adult rabbits., <u>Crit. Care Med.</u>, 17:339 (1989).

11. R. Jones, W.M. Zapol, and L. Reid. Pulmonary artery remodeling and pulmonary hypertension after exposure to hyperoxia for 7 days: a morphometric and hemodynamic study., <u>Am. J. Pathol.</u>, 117:273 (1984).

12. J.R. Hageman, K. McCulloch, P. Gora, E. Olsen, L. Pachman, and C.E.Hunt. Intralipid alterations in pulmonary prostaglandin metabolism and gas exchange., <u>Crit. Care Med.</u>, 11:794 (1983).

13. D.W. Goldman and E.J. Goetzl. Specific binding of leukotriene B_4 to receptors on human polymorphonuclear leukocytes., <u>J. Immunol.</u>, 129:1600 (1982).

14. A.W. Ford-Hutchinson, M.A. Bray, M.V. Doig, M.E. Shipley, and M.J. Smith. Leukotriene B_4, a potent chemokinetic and aggregating substance released from polymorphonuclear leukocytes., <u>Nature</u>, 286:264 (1980).

15. S.E. Dahlen, P. Bjork, P. Hedqvist, K.E. Arfors, S. Hamarstrom, J. A. Lingren, and B. Samuelsson. Leukotrienes promote plasma leakage and leukocyte adhesion in postcapillary venules--in vitro effects with relevance to the acute inflammatory response., <u>Proc. Natl. Acad. Sci. USA.</u> 78:3887 (1981).

16. S.E. Dahlen, P. Hedqvist, S. Hammarstrom, and B. Samuelsson. Leukotrienes are potent constrictors of human bronchi., <u>Nature</u>, 288:484 (1980).

17. R.M. Jackson, D.B. Chandler, and J.D. Fulmer. Production of arachidonic acid metabolites by endothelial cells in hyperoxia., <u>J. Appl. Physiol.</u> 61:584 (1986).

18. H. Taniguchi, F. Taki, K. Tagaki, T. Satake, S. Sugiyama, and T. Ozawa. The role of leukotriene B_4 in the genesis of oxygen toxicity in the lung., <u>Am. Rev. Respir. Dis.</u>, 133:805 (1986).

19. T.R. Martin, L.C. Altman, R.K. Albert, and W.R. Henderson. Leukotriene B_4 production by the human alveolar macrophage: a potential mechanism for amplifying inflammation in the lung., <u>Am. Rev. Respir. Dis.</u>, 129:108 (1984).

20. L.J. Smith, M. Shamsuddin, J. Anderson, and W. Hsueh. Hyperoxic lung damage in mice: appearance and bioconversion of peptide leukotrienes., <u>J. Appl. Physiol.</u>, 64:944 (1988).

21. K.R. Stenmark, M. Wyzaguirre, L. Remigio, J. Seccombe, and P.M. Henson. Recovery of platelet-activating factor and leukotrienes from infants with severe bronchopulmonary dysplasia: clinical improvement with cromolyn treatment., <u>Am. Rev. Respir. Dis.</u>, 131:A236 (1985).

22. N.J. Kertesz, R.B. Holtzman, L. Adler, and J.R. Hageman. Evaluation of a leukotriene (LT) receptor antagonist in prevention of hyperoxic lung injury., <u>Pediatr. Res.</u>,25:316A (1989).

23. O.C. Burghuber, R.J. Strife, J. Zirrolli, P.M. Henson, J.E. Henson, M.M. Mathias, J.T. Reeves, R.C. Murphy, and N.F. Voelkel. Leukotriene inhibitors attenuate rat lung injury induced by hydrogen peroxide., Am. Rev. Respir. Dis., 131:778 (1985).

24. N. Suttorp and L.M. Simon. Lung cell oxidant injury: enhancement of polymorphonuclear leukocyte-mediated cytotoxicity in lung cells exposed to sustained in vitro hyperoxia., J. Clin. Invest., 70:342 (1982).

25. R.B. Fox, J.R. Hoidal, D.M. Brown, and J.E. Repine. Pulmonary inflammation due to oxygen toxicity: involvement of chemotactic factors and polymorphonuclear leukocytes., Am. Rev. Respir. Dis., 123:521 (1981).

26. D.M. Shasby, R.B. Fox, R.N. Harada, and J.E.Repine. Reduction of the edema of acute hyper-oxic lung injury by granulocyte depletion., J. Appl. Physiol., 52:1237 (1982).

27. J.U. Raj and R.D. Bland. Neutrophil depletion does not prevent oxygen-induced lung injury in rabbits., Chest, 83:20S (1983).

28. C.F. Nathan, H.W. Murray, and Z.A. Cohn. The macrophage as an effector cell., N. Engl. J. Med., 303:622 (1980).

29. R.N. Harada, A.E. Vatter, and J.E.Repine. Macrophage effector function in pulmonary oxygen toxicity: hyperoxia damages and stimulates alveolar macrophages to make and release chemo-taxins for polymorphonuclear leukocytes., J. Leukocyte Biol., 35:373 (1984).

30. J.B. Chauncey, R.H. Simon, and M. Peters-Golden. Rat alveolar macrophages synthezize leuko-triene B_4 and 12-hydroxyeicosatetraenoic acid from alveolar epithelial cell-derived arachidonic acid., Am. Rev. Respir. Dis., 138:928 (1988).

31. J.R. Hageman, J. Zemaitia, R.B. Holtzman, S.E. Lee, L.J. Smith, and C.E. Hunt. Failure of non-selective inhibition of arachidonic acid metabolism to ameliorate hyperoxic lung injury., Prostaglandins, Leukotrienes, and Essential Fatty Acids, 32:145 (1988).

32. L. Frank and E.E. Groseclose. Preparation for birth into an O_2-rich environment: the antioxi-dant enzymes in the developing rabbit lung., Pediatr. Res., 18:240 (1984).

33. J. Yam, L. Frank, and R.J. Roberts. Oxygen toxicity: comparison of lung biochemical responses in neonatal and adult rats., Pediatr. Res., 12:115 (1978).

34. R.B. Holtzman, L. Adler, L.J. Smith, M. Shamsuddin, C.E. Hunt, and J.R. Hageman. Loss of oxygen tolerance in newborn rabbits: relationship to changes in eicosanoid and antioxidant levels., Pediatr. Pulmonol., 7:200 (1989).

35. R.B. Holtzman, J. Zemaitis, L. Adler, L.J. Smith, C.E. Hunt, and J.R. Hageman. Role of eicasanoids in relative oxygen tolerance of newborn rabbits., Prostaglandins, 37:481 (1989).

36. M.D. Mitchell Prostaglandins during pregnancy and the perinatal period., J. Reprod. Fertil., 62:305 (1981).

37. S. Cassin. Arachidonic acid metabolites and the pulmonary circulation of the fetus and new-born., In: S.M. MacLeod, A.B. Okey, and S.P. Speilberg (eds.). "Developmental Pharmacology." New York: Alan R. Liss, Inc., pp.227-250 (1983).

38. R.P. Goodman, A.P. Killam, A.R. Brash, and R.A. Branch. Prostacyclin production during pregnancy: comparison of production during normal pregnancy and pregnancy complicated by hypertension., Am. J. Obstet. Gynecol., 142:817 (1982).

39. M.L. Casey, S. Cutrer, and M.D. Mitchell. Origin of prostanoids in human amniotic fluid: The fetal kidney as a source of amniotic fluid prostanoids., Am. J. Obstet. Gynecol., 147:547 (1983).

40. C.W. Leffler and J.R. Hessler. Perinatal pulmonary prostaglandin production., Am. J. Physiol., 241:H756 (1981).

41. C.W. Leffler, J.R. Hessler, and R.S. Green. Mechanism of stimulation of pulmonary prostacyclin synthesis at birth., Prostaglandins, 28:877 (1984).

42. R.H. Demling. Role of prostaglandins in acute pulmonary microvascular injury., Ann. N.Y. Acad. Sci., 384:517 (1982).

43. J.E. Tateson, S. Moncada, and J.R. Vane. Effects of prostacyclin (PGX) on cyclic AMP concentrations in human platelets., Prostaglandins, 13:389 (1977).

44. M.L. Steer and E.W. Salzman. Cyclic nucleotides in hemostasis and thrombosis., Adv. Cyclic Nucleotide Res.., 12:71 (1980).

45. F.A. Kuehl, H.W. Dougherty, and E.A. Ham. Interaction between prostaglandins and leukotrienes., Biochem. Pharmacol., 33:1 (1984).

46. R.B. Holtzman, L. Adler, M. Shamsuddin, and J.R. Hageman. Iloprost infusion decreases mortality in fourteen day old rabbits exposed to hyperoxia for 96 hours., Pediatr. Res., 27:307A (1990).

47. I.R.S. Sosenko, S.M. Innis, and L. Frank. Menhaden fish oil, n-3 polyunsaturated fatty acids, and protection of newborn rats from oxygen toxicity., Pediatr. Res., 25:399 (1989).

INTERACTIONS BETWEEN MACROPHAGES AND GRANULOCYTES IN BRONCHIAL ASTHMA

James R. Wilkinson, Stephen J. Lane, Tak H. Lee

Department of Allergy and Allied Respiratory
Disorders, U.M.D.S., Guy's Hospital, London SE1 9RT
U.K.

INTRODUCTION

Histological examination of the airways in patients with bronchial asthma demonstrates a number of characteristic features. There is plugging of the bronchial lumen with secretions and cellular debris, the epithelium shows evidence of hyperplasia and metaplasia with the shedding of epithelial cells, the epithelial basement membrane is thickened, and there is marked oedema of the sub-mucosa.

Although infiltration by eosinophil granulocytes (EOS) is a feature of asthmatic airways, there is also an increase in the number of neutrophils (PMN) and of mononuclear cells, including T-lymphocytes and mononuclear phagocytes (1). In order to elucidate the cellular biology of airways inflammation, it is important to understand the interactions between the different inflammatory cells which are recruited into the airways, as well as the biology of each individual cell type.

Lung macrophages are well equipped to play an important role in the initiation and continuation of the inflammatory response seen in asthma by virtue of their anatomic distribution, their availability to exposure to inhaled allergens, their ability to be activated by IgE-mediated stimuli, and their ability to generate a wide range of molecules with pro-inflammatory effects.

Both in normal and in asthmatic individuals alveolar macrophages are the predominant inflammatory cells resident in the airways, comprising over 90% of the cell population (2). Thus they would be one of the first inflammatory cells to be exposed to any inhaled allergen.

Activation of Macrophages and Monocytes by IgE-dependent Stimuli in Asthma

In the early 1970's, Capron et al (3) discovered evidence for the existence of Fcϵ receptors on macrophages by implicating the role of IgE antibodies in the macrophage-dependent cytotoxic killing of Schistosoma mansoni larvae (3). The FcϵR expressed on the surface of macrophages is now believed to be similar, if not identical to that found on T and B lymphocytes and on platelets. This receptor, referred to as FcϵR2, differs in both structure and function from the FcϵR expressed by mast cells and

Cell-Cell Interactions in the Release of Inflammatory Mediators
Edited by P. Y-K Wong and C.N. Serhan, Plenum Press, New York, 1991

269

basophils, which is referred to as $Fc_\epsilon R1$. The major functional difference is that the $Fc_\epsilon R2$ on macrophages is a low affinity receptor. The estimated K_a for monomeric IgE binding to rat macrophages (4) and to U937 cells (5) is approximately 10^7 M^{-1} which is 200-fold lower than that of IgE binding to the high affinity receptor on mast cells (6). There are approximately $4-5x10^4$ low affinity receptors on rat macrophages (4) and $5-9x10^4$ receptors on U937 cells (5). This suggests that the low affinity macrophage $Fc_\epsilon R2$ may be activated preferentially by IgE immune complexes since these offer multiple binding sites and therefore higher affinities.

Between 5 and 10% of peripheral blood mononuclear cells (MNC) and lung macrophages from normal non-atopic humans express $Fc_\epsilon R2$ (6-8). This proportion is increased in atopic subjects compared with normal individuals (9), with up to 20% of alveolar macrophages expressing $Fc_\epsilon R2$ in mild atopic asthmatic subjects (7). However, in patients with more severe asthma who have been treated with corticosteroids, the proportion of peripheral blood monocytes bearing $Fc_\epsilon R2$ is substantially lower (9), indicating that the enhanced expression of $Fc_\epsilon R2$ found in the monocytes and macrophages of patients with asthma may be modulated by corticosteroid treatment.

Speigelberg (6) also noted that peripheral blood monocytes obtained from severely atopic asthmatic subjects induced significantly more ^{51}Cr release from IgE coated ^{51}Cr labeled erythrocytes than did monocytes from either non-atopic subjects or mild atopic subjects, indicating that monocytes are able to phagocytose and lyse IgE coated target cells, and that IgE appears to be an effective opsonin for monocytes.

Stimulation of macrophages by IgE-antigen complexes (10) induces the subsequent release of lysosomal enzymes and generation of the superoxide anion (O_2-). Macrophage products of oxygen metabolism such as O_2- have pro-inflammatory effects and may contribute to the airway inflammation and bronchial hyper-reactivity seen in patients with asthma. Furthermore, stimulation of normal rat alveolar macrophages by monoclonal IgE and its specific antigen has been demonstrated to induce the generation and release of leukotriene (LT) LTB_4 (11), a potent neutrophil (12, 13) and monocyte (14) chemotaxin, and LTC_4 (15), which constricts human airways, increases vascular permeability (16-18), and enhances bronchial responsiveness in asthmatic subjects (19). Several laboratories have demonstrated that mononuclear phagocytes may play an important role in IgE-mediated disease. Ferreri et al (20) challenged human periphertal blood MNC in vitro with chemically aggregated IgE and found that these cells release small amounts of LTB_4, LTC_4, and prostaglandin (PG) E_2. Fuller et al (21) observed the release of LTB_4, PGF_{2a}, thromboxane (Tx) B_2 and beta-glucuronidase, following stimulation with anti-IgE, by macrophages obtained from patients suffering from a wide variety of lung diseases. In macrophages cultured in vitro, the respiratory burst is increased in macrophages derived from asthmatic subjects compared to that in those derived from normal subjects (22). The number of peripheral blood monocytes forming rosettes with complement-coated sheep erythrocytes is increased in asthmatic subjects following allergen bronchoprovocation, but there is no increase seen after histamine-induced bronchoconstriction (23).

Recent in vivo studies add further support to the view that macrophages may play a role in the pathological mechanisms involved in asthma. Analysis of the intra-cellular and extra-cellular concentrations of beta-glucuronidase in bronchoalveolar lavage (BAL) samples from patients with asthma following local bronchial challenge with allergen reveals increased amounts of beta glucuronidase in the BAL fluid whereas the macrophage intracellular levels of beta-glucuronidase are reduced compared with BAL samples obtained from non-atopic control subjects (24, 25). This suggests that macrophage secretory processes may be activated by allergens in vivo in a similar fashion to the IgE-mediated activation of macrophages and monocytes observed in vitro. Finally, Metzger et al (26) have demonstrated that the total numbers of mononuclear phagocytes present in BAL fluid are increased at 48 hours and at 96 hours after bronchial challenge with allergen. Furthermore, Metzger et al (26) found that the numbers of macrophages in the BAL fluid, staining positively for peroxidase, increased significantly 48 hours after allergen challenge, thus suggesting that a population of monocytes had been recruited into the lung from the vascular compartment after allergen exposure.

MNC derived from patients with asthma exhibit a number of other functional abnormalities. They show increased complement receptor expression compared with MNC from normal non-atopic subjects, and also a greater enhancement of complement receptor expression following stimulation with casein (27). This enhanced complement receptor expression reverts to normal levels following treatment with prednisolone in asthmatic subjects who respond clinically to treatment with corticosteroids (CS asthmatics), but in a minority who fail to improve clinically on treatment with corticosteroids (CR asthmatics), the level of complement receptor expression remains elevated following treatment with prednisolone. More recently, it has been demonstrated that the formation of colonies by phytohaemagglutinin-stimulated mixed MNC cultures is substantially inhibited by the presence in vitro of methyl-prednisolone at a concentration of 10^{-8} M in CS asthmatics, but not in CR asthmatics (28). Crossover studies, incubating MNC from CR subjects with lymphocytes from CS subjects and vice versa indicated that, in this system, the monocytes rather than lymphocytes were responsible for the corticosteroid resistance in vitro. However, other recent evidence has also implicated functional abnormalities of lymphocytes in corticosteroid resistance (29, 30).

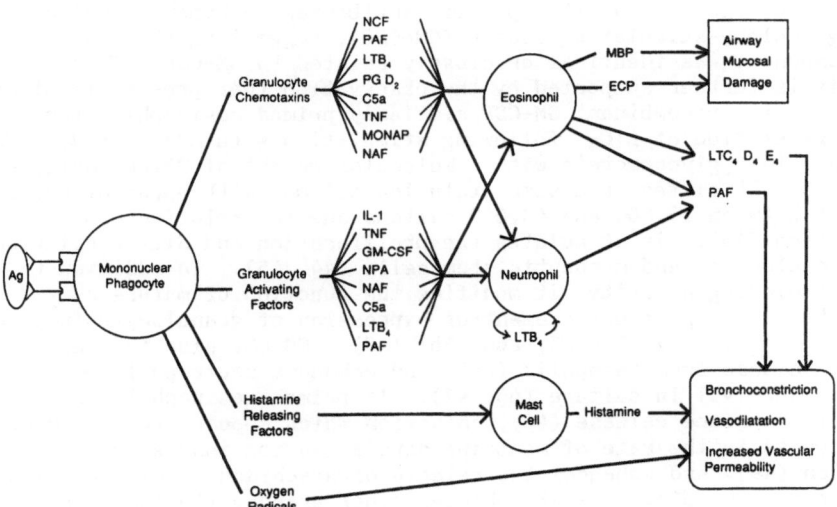

Fig. 1. The generation of granulocyte activating inflammatory mediators, cytokines and products of oxidative metabolism by macrophages. (C5a = complement component 5a; ECP=eosinophil cationic protein; GM-CSF= granulocyte macrophage colony stimulating factor; IL-1-interleukin 1; LT=leukotriene; MBP=major basic protein; MONAP=monocyte derived neutrophil activating peptide; NCF=neutrophil chemotactic factor; NAF=neutrophil activating factor; NPA=neutrophil priming activity; PG=prostaglandin; PAF=platelet activating factor; TNF=tumour necrosis factor.

Interactions between Macrophages, Monocytes and Granulocytes

Macrophages and monocytes may contribute to the airways inflammation seen in asthmatic subjects not only through the generation of pro-inflammatory products, but also through the release of factors which exert modulatory effects on the functions of other inflammatory cells (Fig. 1).

Macrophages are able to generate and release a variety of chemotactic factors. These include LTB_4 (31-33), platelet activating factor (PAF) (34), tumor necrosis factor (TNF)(35), the complement fragment C5a (36), platelet-derived growth factor (37) and a 10,000 dalton molecule that has yet to be completely characterized (38).

Both alveolar macrophages and peripheral blood monocytes also secrete a number of pro-inflammatory cytokines with neutrophil and eosinophil-activating properties. Pennington et al (39, 40) isolated a 6000 dalton glycoprotein of pI 7.6 generated by macrophages, which enhanced the bactericidal activity of neutrophils. In addition the partially purified neutrophil activating factor enhanced neutrophil superoxide anion release upon subsequent stimulation either by bacterial phagocytosis or by phorbol myristate acetate (PMA), compared with control neutrophils stimulated in an identical manner.

More recently, Howell et al have studied the interactions between alveolar macrophages and eosinophils in asthmatic subjects (41). They found that eosinophils incubated with supernatants from cultures of alveolar macrophages obtained by BAL from asthmatic patients, generated increased amounts of LTC_4 when subsequently stimulated by the calcium ionophore A23187. Alveolar macrophage supernatants derived from normal donors had no enhancing effects on eosinophil LTC_4 generation when compared with culture medium alone. This eosinophil priming activity could be neutralized by incubation with specific antibodies to human granulocyte-macrophage colony-stimulating factor (GM-CSF), suggesting that the major active component was identical or closely related to GM-CSF. This hypothesis is further supported by the observation that pre-treatment of eosinophils with recombinant GM-CSF similarly primed eosinophils for enhanced generation of LTC_4 following stimulation with A23187 (42). GM-CSF is an acidic glycoprotein with a molecular weight of 22,000 daltons and a pI of 4.5. It elutes from size exclusion columns with apparent molecular weights of between 15,000 and 40,000 daltons due to variations in its glycosylation (43). It stimulates the proliferation and differentiation of normal granulocytes and monocytic stem cells (44, 45). In addition to its colony stimulating activity, it modifies the function of mature granulocytes leading to enhancement of expression of granulocyte functional antigens 1 and 2, Mo 1, Leu M5, and C3bi (45). GM-CSF also induces histamine release from basophils (46), and enhances neutrophil and eosinophil survival in culture (45, 47). It primes neutrophils for enhanced leukotriene release (48), an action which appears to be mediated by an increase in the rate of membrane depolarisation induced by subsequent stimulation (49), and enhances the release of arachidonic acid from membrane phospholipids (50, 51). GM-CSF enhances other granulocyte functions in-cluding phagocytosis of zymosan (45), possibly by its effects on the expression of C3b, cytotoxic killing and superoxide anion generation (51). In addition it activates macrophages to become tumoricidal (52), and to release PGE_2 (53), and it stimulates the secretion of other cytokines including IL-1, TNF, G-CSF and M-CSF.

Thus, the presence of GM-CSF in the lung may pre-condition eosinophils for enhanced pro-inflammatory functions upon subsequent stimulation by other factors, and either alone, or in conjunction with other cytokines, may lead to eosinophil colony formation from bone marrow progenitors, so playing a potentially important role in the amplification of eosinophil numbers and activation which is characteristic of asthmatic airways.

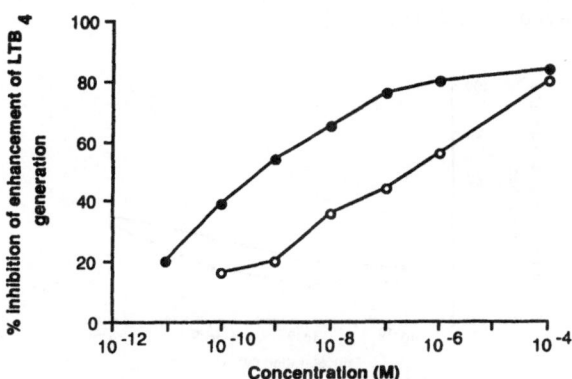

Fig. 2. The effects of different concentrations of hydrocortisone (O) and dexamethasone (●) on the % inhibition of the enhancement of LTB$_4$ generation in one representative experiment. Peripheral blood monocytes were isolated from CS asthmatic patients and incubated for 24 hours in the presence of increasing concentrations of corticosteroids. The monocyte-derived supernatant was then tested for its capacity to enhance LTB$_4$ generation by human neutrophils stimulated by the calcium ionophore A23187. There was a dose-dependent inhibition of the generation of the enhancing activity by each of the corticosteroids.

Similarly to mature macrophages, monocytes generate a number of granulocyte-activating cytokines including interleukin-1 (IL-1), TNF, colony-stimulating factors (CSFs) and histamine releasing factors (HRFs) (54-57). They have also been found to secrete a range of heat stable factors of heterogeneous size and charge which prime eosinophils and neutrophils for enhanced generation of leukotrienes when subsequently stimulated by the calcium ionophore A23187 (58), and further factors, including eosinophil cytotoxicity enhancing factor (ECEF) and eosinophil activating factor (EAF), which prime eosinophils for enhanced cytotoxicity against schistosomula (59-61).

Recently, Wilkinson et al demonstrated enhanced generation of a low molecular weight granulocyte activating factor by cultured monocytes derived from asthmatic subjects. A23187-activated neutrophils primed by the 24 hour culture supernatants of monocytes from asthmatic subjects generated 3-fold more LTB$_4$ than neutrophils primed by monocyte supernatants derived from normal donors (62). Incubation of monocytes from CS asthmatic subjects with hydrocortisone and dexamethazone significantly inhibited the production of enhancing activity (Fig. 2), whereas in CR subjects, there was failure of suppression of enhancing activity (Fig. 3). hydrocortisone at concentrations up to 10^{-4} M failed to suppress the enhancing activity (62). Characterization of the predominant factor revealed it to be of molecular weight 3000 daltons, with a pI of 7.1. The activity was inhibited by a mean of 95% by heating to 60°C for 60 minutes and was sensitive to digestion by pronase with a mean 76% inhibition of activity. The purified cytokine was found to enhance generation of LTB$_4$ and LTC$_4$ by neutrophils and eosinophils respectively, when subsequently stimulated by either A23187 or unopsonized zymosan (63). Its effects were not limited to modulation of lipoxygenase metabolism. Pre-incubation with the granulocyte priming factor also enhanced the generation of superoxide anion by neutrophils subsequently stimulated by PMA (62), and generation of PAF by eosinophils stimulated by A23187 (63).

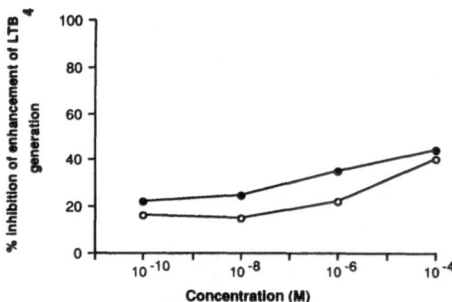

Fig. 3. The effects of different concentrations of hydrocortisone (O) and dexamethasone (●) on the % inhibition of the enhancement of LTB$_4$ generation. Peripheral blood monocytes were isolated from CR asthmatic patients. Hydrocortisone and dexamethasone were both very weak in inhibiting the generation of the enhancing activity and did not reach 50% inhibition at 10^{-4}M corticosteroid.

Interactions Between Macrophages and Basophils

Macrophages (57), in common with lymphocytes (64) and platelets (65) produce a number of histamine releasing factors (HRFs). HRFs are an incompletely characterized group of molecules with molecular weights of between 15 and 30 kDa. Although they are of similar molecular weight, they are distinct from other cytokines such as IL-1, IL-2, IL-3, IL-4 and GM-CSF, which may in themselves cause histamine release. HRFs can cause histamine release from basophils and mast cells via an interaction with cell surface-bound IgE (66), the subsequent histamine release having the same kinetics and the same characteristics of temperature dependence and calcium dependence as those seen following stimulation with anti-IgE antibodies (67). The response to HRFs may be abolished by removal of the cell surface IgE (68).

There is now a considerable body of evidence implicating interactions between HRFs and basophils in the mechanisms of allergic inflammation. Following allergen challenge to the upper airways, there is a marked influx of basophils, their numbers increasing approximately 8-fold (69). This influx coincides temporally with the development of the late phase response.

Furthermore, after both upper and lower airways challenge with allergen, the late phase response is accompanied by the appearance of histamine and other mediators which are thought to be at least partly basophil-derived (70, 71). Basophil infiltration has been observed in chronic allergic rhinitis (72) and also in dermal reactions to chronic allergen exposure (73).

HRFs have also been detected during the late phase response following allergen exposure. Warner et al (67) demonstrated the presence of HRF in blister fluid during the late phase response following dermal exposure to antigen in allergic subjects, but not after similar challenge in normal subjects, or in control blisters not challenged. Similarly, Bascomb et al have demonstrated HRF in nasal lavage fluid following nasal antigen challenge (69). It is interesting that although HRF generation itself does not correlate with the presence of symptoms in allergic subjects, the ability of basophils to respond to HRFs correlates with symptoms (74). Only basophils derived from atopic subjects are able to respond to HRF. This has led Lichtenstein (66) to propose that in atopic subjects the acquisition of a specific subset of IgE which interacts with HRF (termed IgE$^+$) correlates with the clinical severity of the disease.

CONCLUSIONS

The causes of the airways narrowing found in asthma are complex and likely to be multifactorial. The histological features are characterized by both acute and chronic inflammatory changes, with severel different inflammatory cell types participating. There is now a substantial and increasing body of evidence from both in vitro and in vivo studies indicating recruitment of neutrophils, eosinophils and of mononuclear phagocytes into the airways in asthma. These cells are activated after controlled antigen challenge and in naturally occuring asthma.

Macrophages and monocytes are also well equipped to play a pivotal role in the amplification of the inflammatory response in asthma by virtue of their distribution in the airways and by their ability to be activated by IgE-mediated stimuli to generate a variety of pro-inflammatory mediators and cytokines. There is increasing evidence of complex and extensive interactions between these and the other inflammatory cells seen in asthmatic airways inflammation. The recognition of these interactions may lead to a better understanding of the mechanisms by which the airways inflammation is initiated and amplified in asthma.

REFERENCES

1. M.S. Dunnill, G.R. Massarella, and J.A. Anderson. A comparison of the quantitative anatomy of the bronchi in normal subjects in status asthmaticus in chronic bronchitis and in emphysema. Thorax, 24: 176, (1969).
2. M.S. Balter, W.L. Eschenbacher, and M. Peters-Golden. Arachidonic acid metabolism in cultured alveolar mascrophages from normal, atopic and asthmatic subjects. Am. Rev. Respir. Dis. 138: 1134, (1988).
3. A. Capron, J.P. Dessaint, R. Rousseau, M. Capron, and H. Bazin. Specific IgE antibodies in immune adherence of normal macrophages to Schistosoma mansoni schistosomules. Nature. 253: 474, (1975).
4. D.S. Findbloom, and H. Metzger. Binding of immunoglobulin E to the receptor on rat peritoneal macrophages. J. Immunol. 129: 2004, (1981).
5. C.L. Anderson, and H.L. Spiegelberg. Macrophage receptors for IgE: binding of IgE to specific IgE receptors on a human macrophage cell line U937. J. Immunol. 126: 2470, (1981).
6. H.L. Spiegelberg. Structure and function of Fc receptors for IgE on lymphocytes, monocytes and macrophages. Adv. Immunol. 35: 61, (1984).
7. M. Joseph, A.B. Tonnel, G. Torpier, A. Capron, B. Arnoux, and J. Benveniste. Involvement of immunoglobulin E in the secretory processes of alveolar macrophages from asthmatic patients. J. Clin. Invest. 71: 221, (1983).
8. F.M. Melewicz, L.E. Kline, A.B. Cohen, and H.L. Speigelberg. Characterisation of IgE receptors for IgE on human alveolar macrophages. Clin. Exp. Immunol. 49: 364 (1982).

9. F.M. Melewicz, R.S. Zeiger, M.H. Mellon, D. O'Connor, and H.L. Speigelberg. Increased peripheral blood monocytes with Fc receptors for IgE in patients with severe allergic disorders. J. Immunol. 126: 1592 (1981).

10. J.P. Dessaint, A. Capron, M. Joseph, and H. Bazin. Cytophilic binding of IgE to the macrophage. II. Immunologic release of lysosomal enzyme from macrophages by IgE and anti-IgE in the rat. Cell. Immunol. 46: 24 (1946).

11. J.A. Rankin. IgE immune complexes induce LTB_4 release from rat alveolar macrophages. Ann. Inst. Pasteur Immunol. 137: 364 (1986).

12. E.J. Goetzl, and W.C. Pickett. Novel structural determinants of the human neutrophil chemotactic activity of leukotriene B_4. J. Exp. Med. 153:482 (1981).

13. L. Nagy, T.H. Lee, E.J. Goetzl, W.C. Pickett, and A.B. Kay. Complement receptor enhancement and chemotaxis of human neutrophils and eosinophils by leukotrienes and other lipoxygenase products. Clin. Exp. Immunol. 47: 541 (1982).

14. R.M. Palmer, R.J. Stepney, G.A. Higgs, and K.E. Eakin. Chemokinetic activity of arachidonic acid lipoxygenase products on leukocytes of different species. Prostaglandins. 20:411 (1980).

15. J.A. Rankin, M. Hitchcock, W. W. Merrill, M.K. Bach, J.R. Brasher, and P.W. Askenase. IgE-dependent release of leukotriene C_4 from alveolar macrophages. Nature. 279: 329 (1982).

16. J.M. Drazen, K.F. Austen, R.A. Lewis, D.A. Clark, G. Goto, A. Marfat, and E.J. Corey. Comparative airway and vascular activities of leukotrienes C and D in vivo and in vitro. Proc. Natl. Acad. Sci. USA. 77: 4354 (1980).

17. M.J. Peck, P.J. Piper, and T.J. Williams. The effect of leukotrienes C_4 and D_4 on the microvasculature of guinea-pig skin. Prostaglandins. 21: 315 (1981).

18. S.E. Dahlen, J. Bjark, P. Hedqvist, K.E. Arfas, S. Hammarstrom, J.-A. Lindgren, and B. Samuelsson. Leukotrienes promote plasma leakage and leukocyte adhesion in post capillary venules. In vivo effects with relevance to acute inflammatory response. Proc. Natl. Acad Sci. USA. 78: 3887 (1981).

19. J.P. Arm, B.W. Spur, and T.H. Lee. The effects of inhaled leukotriene E_4 on the airways hyperresponsiveness to histamine in asthmatic and normal subjects. J. Allergy Clin. Immunol. 82,654 (1988).

20. N.R. Ferreri, W.C. Howland, and H.L. Spiegelberg. Release of leukotrienes C_4 and B_4 and prostaglandin E_2 from human monocytes stimulated with IgG, IgA, and IgE. J. Immunol. 136: 4188 (1986).

21. R.W. Fuller, P.K. Morris, R.D. Sykes, I.M. Varndell, D.M. Kemeny, P.J. Cole, C.T. Dollery, and J. MacDermot. Immunoglobulin E-dependent stimulation of human alveolar macrophages: significance in type I hypersensitivity. Clin. Exp. Immunol. 65: 416 (1986).

22. M. Cluzel, M. Damon, and P. Chanez. Enhanced alveolar cell luminol-dependent chemiluminescence in asthma. J. Allergy Clin. Immunol. 80:195 (1987).

23. M. Carroll, S.R. Durham, G. Walsh, and A.B. Kay. Activation of neutrophils and monocytes after allergen and histamine-induced broncho-constriction. J. Allergy Clin. Immunol. 75: 290 (1985).

24. A.B. Tonnel, P.H. Gosset, M. Joseph, E. Fournier, and A. Capron. Stimulation of alveolar macrophages in asthmatic patients after local provocation test. Lancet. 1: 1406 (1983).

25. J.J. Murray, A.B. Tonnel, A.R. Brash, L.J. Roberts II, P. Gosset, R. Workman, A. Capron, and J.A. Oates. Release of prostaglandin D_2 into human airways during acute allergen challenge. N. Eng. J. Med. 315: 800 (1986).

26. W.J. Metzger, D. Zavala, H.B. Richerson, P. Moseley, P. Imamota, M. Mouiek, K. Sjoerdsma, and G.W. Hunninghake. Local allergen challenge and bronchoalveolar lavage of allergic asthmatic lungs. Am. Rev. Respir. Dis. 135: 433 (1987).

27. A.B. Kay, P. Diaz, J. Carmichael, and I.W.B. Grant. Corticosteroid resistant chronic asthma and monocyte complement receptors. Clin. Exp. Immunol. 44: 576 (1981).

28. M.C. Poznansky, A.C.H. Gordon, J.G. Douglas, A.S. Krajewski, A.H. Wyllie, and A.W.B.Grant. Resistance to methyl prednisolone in cultures of blood mononuclear cells from glucocorticoid-resistant asthmatic patients. Clin. Sci. 67: 639 (1984).

29. K.B. Walker, J.M. Potter, and A.K. House. Interleukin-2 synthesis in the presence of steroids: a model of steroid resistance. Clin. Exp. Immunol. 68: 162 (1987).

30. C.J. Corrigan, P. Brown, N. Barnes, A.J. Frew, J.-J. Tsai, M.B. Allen, S. Ray, G.K. Crompton, A.G. Leitch, and A. B. Kay. Corticosteroid resistance in chronic asthma: correlation of clinical response with in vitro sensitivity of lymphocytes to anti-inflammatory drugs. Thorax. (abstract) 44: 319 (1989).

31. A.O.S. Fels, N.A. Pawlowski, E.B. Cramer, Z.A. Cohn, and W.A. Scott. Human alveolar macrophages produce leukotriene B_4. Proc. Natl. Acad. Sci. USA. 79: 7866 (1982).

32. T.R. Martin, G. Raugi, T. Merritt, and W.R. Henderson. Relative contribution of leukotriene B_4 to the neutrophil chemotactic activity produced by the resident human alveolar macrophage. J. Clin. Invest. 8:1114 (1987).

33. M. Laviolette, R. Coulombe, S. Picard, P. Braquet, and P. Borgeat. Decreased leukotriene B_4 synthesis in smokers' alveolar macrophages in vitro. J. Clin. Invest. 77: 54 (1986). ·

34. B. Arnoux, D. Duval, and J. Benveniste. Release of platelet activating factor (PAF-acether) from alveolar macrophages by the calcium ionophore A23187 and phagocytosis. Eur. J. Clin. Invest. 10: 437 (1980).

35. B. Beutler, and A. Cerami. Cachectin: More than a tumour necrosis factor. N. Eng. J. Med. 316: 379 (1987).

36. C.F. Nathan. Secretory products of macrophages. J. Clin. Invest. 79: 319 (1987).

37. Y. Martinet, W.N.Rom, G.R. Grotendorst, G.R. Martin, and R.G. Crystal. Exaggerated spontaneous release of platelet-derived growth factor by alveolar macrophages from patients with idiopathic pulmonary fibrosis. N. Eng. J. Med. 317: 202 (1987).

38. W.W. Merrill, G.P. Naegel, R.A. Mathey, and H.Y. Reynolds. Alveolar macrophage-derived chemotactic factor: Kinetics of in vitro production and partial characterisation. J. Clin. Invest. 65: 268 (1980).

39. J.E. Pennington, T.H. Rossing, L.W. Boerth, and T.H. Lee. Isolation and partial characterization of a human alveolar macrophage-derived neutrophil activating factor. J. Clin. Invest. 75: 1230 (1985).

40. J.E. Pennington, T.H. Rossing, and L.W. Boerth. The effect of human alveolar macrophages on the bactericidal capacity of neutrophils. J. Infect. Dis. 148: 101 (1983).

41. C.J. Howell, L.-L. Pujol, A.E.G. Crea, R. Davidson, A.J.H. Gearing, Ph. Godard, and T.H. Lee. Identification of an alveolar macrophage derived activity in bronchial asthma which enhances LTC_4 generation by human eosinophils stimulated by ionophore (A23187) as granulocyte macrophage colony stimulating factor (GM-CSF). Am. Rev. Respir. Dis. 140:1340 (1989).

42. C.J. Howell, and T.H. Lee. GM-CSF primes eosinophils for enhanced eicosanoid mediator generation. J. Allergy Clin. Immunol. (in press).

43. S.C. Clark, and R. Kamen. The human hematopoietic colony stimulating factors. Science. 236: 1229 (1987).
44. C.N. Abboud, J.R. Brennan, G.H. Barlow, and L.A. Lichtman. Hydrophobic adsorption chromatography of colony-stimulating activities and erythroid enhancing activity from the human monocyte-like cell line GCT. Blood. 58: 1158 (1981).
45. A.F. Lopez, D.J. Williamson, J.R. Gamble, C.G. Begley, J.M. Harlan, S.J. Klebanoff, A. Waltersdorph, G. Wang, S.C. Clark, and M.A. Vadas. Recombinant human granulocyte-macrophage colony stimulating factor stimulates in vitro mature human neutrophiland eosinophil function, surface receptor expression and survival. J. Clin. Invest. 78:1220 (1986).
46. M. Haak-Frendscho, N. Arai, K.-I. Arai, M.L. Baeza, A. Finn, and A.P. Kaplan AP. Human recombinasnt granulocyte-macrophage colony stimulating factor and interleukin-3 cause basophil histamine release. J. Clin. Invest. 82: 17 (1988).
47. W.F. Owen Jnr, M.E. Rothenberg, D.R. Silberstein, J.C. Gasson, R.L. Stevens, K.F. Austen, and R.J. Soberman. Regulation of human eosinophil viability, density and function by granulocyte/macrophage colony-stimulating factor in the presence of 3T3 fibroblasts. J. Exp. Med. 166:129 (1987).
48. C.A. Dahinden, J. Zingg, F.E. Maly, and A.J. de Weck. Leukotriene production in human neutrophils primed by recombinant human granulocyte-macrophage colony stimulating factor and stimulated with the complement component C5a and FMLP as second signals. J. Exp. Med. 167: 1281 (1988).
49. R. Sullivan, J.D. Griffin, E.R. Simons, A.I. Schafer, T. Meshulam, J.P. Fredette, A.K. Maas, A.S. Gadenne, J.L. Laevitt, and Melnick. Effects of recombinant human granulocyte and macrophage colony-stimulating factors on signal transduction pathways in human granulocytes. J. Immunol. 139:3422 (1987).
50. J.F. Di Persio, P. Billing, R. Williams, and J.C. Gasson. Human granulocyte-macrophage colony-stimulating factor and other cytokines prime human neutrophils for enhanced arachidonic acid release and leukotriene B$_4$ synthesis. J. Immunol. 140: 4315 (1988).
51. A. Kapp, G. Zeck-Kapp, M. Donner, and T.A. Luger. Human granulocyte-macrophage colony stimulating factor: an effective direct activator of human polymorphonuclear neutrophilic granulocytes. J. Invest. Dermatol. 91: 49 (1988).
52. K.H. Grabstein, D.L. Urdal, R.J. Tushinski, D.Y. Mochizuki, V.L. Price, M.A. Cantrell, S. Gillis, and P.J. Conlon. Induction of macrophage tumoricidal activity by granulocyte-macrophage colony stimulating factor. Science. 232: 506 (1986).
53. W.W. Hancock, M.E. Pleau, and L. Kobzik. Recombinant granulocyte-macrophage colony-stimulating factor down-regulates expression of IL-2 receptor on human mononuclear phagocytes by induction of prostaglandin Eur. J. Immunol. 140: 3021 (1988).
54. P.A. Chervenick, and A.F. LoBuglio. Human blood monocytes: stimulators of granulocytes and mononuclear colony formation in vitro. Science. 178: 164 (1972).
55. J.A. Elias, A.D. Schreiber, K. Gostilo, P. Chien, M.D. Rossman, P.J. Lammie, and R.P. Daniele. Differential interleukin-1 elaboration by un-fractionated and density fractionated human alveolar macrophages and blood monocytes: Relationship to cell maturity. J. Immunol. 135: 3198 (1985).
56. M. Berger, E.M. Wetzler, and S.R. Wallis. Tumour necrosis factor is the major monocyte product that increases complement receptor expression on mature human neutrophils. Blood. 71: 151 (1988).

57. M.C. Liu, D. Proud, L.M. Lichtenstein, D.W. MacGlashan Jr, R.P. Schleimer, L.F. Adkinson Jr, A. Kagey-Sobotka, E.S. Schulman, and M. Plaut. Human lung macrophage-derived histamine-releasing activity is due to an IgE-binding factor(s). J. Immunol. 136: 2588 (1986).

58. A.J. Dessein, T.H. Lee, P. Elsas, J. Ravalese III, D. Silberstein, J.R. David, K.F. Austen, and R.A. Lewis RA. Enhancement by monokines of leukotriene generation by human eosinophils and neutrophils stimulated with calcium ionophore A23187. J. Immunol. 136: 3829 (1986).

59. A.J. Dessein, M. Vadas, N.A. Nicola, D. Metcalf, and J.R. David. Enhancement of human blood eosinophil cytotoxicity by semi-purified eosinophil colony-stimulating factor(s). J. Exp. Med. 156: 90 (1982).

60. H.L. Lenzi, A.D. Mendis, and A.J. Dessain. Activation of human eosinophils by monokines and lympphokines: source and biochemical characteristics of the eosinophil cytotoxicity-enhancing activity produced by blood mononuclear cells. Cell Immunol. 94: 333 (1985).

61. P. Elsas, T.H. Lee, H.L. Lenzi, and A.J. Dessein. Monocytes activate eosinophils for enhanced helminthotoxicity and increased generation of leukotriene C_4. Ann. Inst. Pasteur Immunol. 138: 97 (1987).

62. J.R.W. Wilkinson, A.E.G. Crea, T.J.H. Clark, and T.H. Lee. Identification and characterization of a monocyte-derived neutrophil activating peptide in corticosteroid resistant bronchial asthma. J. Clin. Invest. 84: 1930 (1989).

63. J.R. Wilkinson, A.E. Crea, T.J. Clark, and T.H. Lee. Enhanced generation of a novel cytokine by monocytes in corticosteroid-resistant asthma. Am. Rev. Respir. Dis. 141: A876 (Abstract) (1990).

64. D.O. Thueson, L.S. Speck, M.A. Lett-Brown, and J.A. Grant. Histamine-releasing activity (HRA). I. Production by mitogen or antigen-stimulated human mononuclear cells. J. Immunol. 123: 626 (1979).

65. M.A. Orchard, A. Kagey-Sobotka, N.F. Adkinson Jr, et al. Platelet augmentation of IgE-dependent histamine release from human basophils and mast cells. Int. Arch. Allergy Appl. Immunol. 74: 26035 (1984).

66. L.M. Lichtenstein. Histamine-releasing factors and IgE heterogeneity. J. Allergy Clin. Immunol. 81: 814 (1988).

67. J.A. Warner, M.M. Pienowski, M. Plaut, and P.S. Norman. Identification of histamine-releasing factor(s) in the late phase of cutaneous IgE-mediated reactions. J. Immunol. 136: 2583 (1986).

68. S.M. MacDonald, L.M. Lichtenstein, D. Proud, and M. Plaut. Studies of IgE-dependent histamine releasing factors: heterogeneity of IgE. J. Immunol. 139: 506 (1987).

69. R. Bascom, M. Wachs, R.M. Naclerio, U. Pipkorn, S.J. Galli, and L.M. Lichtenstein. Basophil influx occurs after nasal antigen challenge: effects of topical corticosteroid pretreatment. J. Allergy Clin. Immunol. 81: 580 (1988).

70. R.M. Naclerio, D. Proud, A.G. Togias, N.F. Adkinson Jr, D.A. Meyers, A. Kagey-Sobotka, M. Plaut, and P.S. Norman. Inflammatory mediators in late antigen-induced rhinitis. N. Eng J. Med. 313: 65 (1985).

71. L. Nagy, T.H. Lee, and A.B. Kay. Neutrophil chemotactic activity in antigen-induced late asthmatic reactions. N. Eng. J. Med. 306: 497- (1982).

72. M. Okuda, S. Kawabori, and H. Otsaka. Electron microscopic study of basophilic cells in the nose. Arch. Otolaryngol. 221: 215 (1978).

73. E.B. Mitchell, J. Crow, G. Williams, and T.A. Platts-Mills. Increase in skin mast cells following chronic house dust mite exposure. Br. J. Dermatol. 114: 65 (1986).

74. R.H. Fisher, A. Kagey-Sobotka, D. Proud, M.A. Orchard, and L.M. Lichtenstein. Platelet-basophil interactions: clinical correlates. J. Allergy Clin. Immunol. 79: 196 (1987).

METABOLISM OF GRANULOCYTE-DERIVED LEUKOTRIENE A_4 IN HUMAN PLATELETS AND RESPIRATORY TISSUE: TRANSCELLULAR FORMATION OF LIPOXINS AND LEUKOTRIENES

Charlotte Edenius, Leif Stenke, Susanne Tornhamre, Katarina Heidvall, Inger Forsberg, Barbro Näsman-Glaser, and Jan Åke Lindgren

Department of Physiological Chemistry
Karolinska Institutet
Box 60 400
S-104 01 Stockholm
Sweden

Leukotriene (LT)A_4, the unstable intracellular intermediate in leukotriene biosynthesis, may be released to the extracellular space by activated leukocytes (1). As a consequence, the metabolism of LTA_4 is not restricted to cells with 5-lipoxygenase activity, but can also be exerted by other cell types equipped with LTA_4-metabolizing enzymes. Thus, LTA_4, released by 5-lipoxygenase expressing cells, may be converted to LTB_4 by surrounding erythrocytes (2), endothelial cells (3) or lymphocytes (4), all possessing LTA_4 hydrolase activity. Similarly, mast cells (1), endothelial cells (3, 5) and smooth muscle cells (6) have been demonstrated to convert LTA_4 to cysteinyl-containing leukotrienes. The present chapter describes some of our recent data regarding the metabolism of synthetic or granulocyte-derived LTA_4 in human platelets and respiratory tissue leading to formation of cysteinyl-containing leukotrienes and lipoxins (LX).

Formation of leukotrienes C_4, D_4 and E_4 in human platelets

Addition of autologous platelets to human granulocyte suspensions (platelet/granulocyte ratio 100:1), resulted in a 2-3-fold increase in ionophore A23187-induced LTC_4 formation (Fig 1; ref. 7). In contrast, LTB_4 levels decreased to 12-72% of those obtained in control incubations with granulocytes only.

The mechanism leading to increased LTC_4 synthesis in the presence of platelets was investigated in experiments using suspensions of granulocytes and [^{35}S]cysteine-prelabeled platelets (7). Isolation of radiolabeled LTC_4 from these incubations suggested transcellular LTC_4 synthesis, via platelet dependent conjugation of glutathione to granulocyte-derived LTA_4. This mechanism of formation was further indicated by the finding that human platelet suspensions efficiently converted exogenous LTA_4 to leukotrienes C_4, D_4 and E_4, with

Cell-Cell Interactions in the Release of Inflammatory Mediators
Edited by P. Y-K Wong and C.N. Serhan, Plenum Press, New York, 1991

281

maximal levels of LTC$_4$ after 5 min. Thereafter, a pronounced catabolism of LTC$_4$ to LTD$_4$ and LTE$_4$ was observed, with the latter compound as major metabolite after 30 min. In agreement, platelet suspensions efficiently metabolized exogenously added LTC$_4$. This metabolic capacity may be of physiological importance for the removal of LTC$_4$ from the circulation.

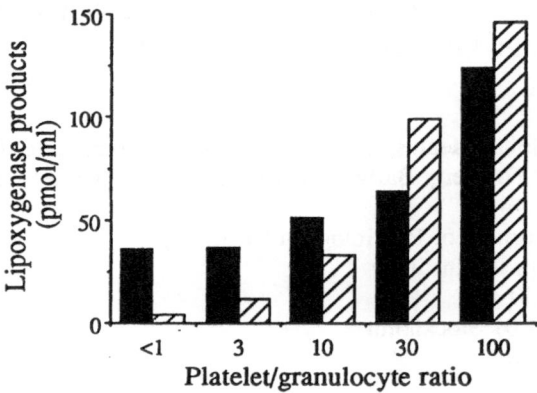

Figure 1. Effect of platelet concentration on LTC$_4$ (filled bars) and LXA$_4$ (hatched bars) formation in mixed platelet/granulocyte suspensions incubated with ionophore A23187.

In experiments with subcellular platelet preparations only the particulate fraction (100.000 x g pellet) converted LTA$_4$ to LTC$_4$ (7). In contrast, GSH S-transferase activity, measured as conjugation of GSH to 1-chloro-2,4-dinitrobenzene, was almost exclusively recovered in the soluble fraction. The results suggested the presence of a specific LTC$_4$ synthase (8) in human platelets.

Platelet dependent lipoxin formation

A platelet-concentration dependent formation of LXA$_4$, LXB$_4$, 6(S)-LXA$_4$ and the all-*trans* isomers of LXA$_4$ and B$_4$ was observed in mixed platelet/granulocyte suspensions after ionophore A23187 stimulation (Fig 1; ref. 9). Neither cell type alone produced significant amounts of lipoxins after ionophore stimulation. In contrast, pure platelet suspensions converted exogenous LTA$_4$ to a similar profile of lipoxin isomers, although LXB$_4$ was not formed. The isolation of 6(S)-LXA$_4$ together with the all-*trans*-LXB$_4$ isomers indicated a platelet-dependent lipoxygenation of LTA$_4$ at C-15, leading to production of the putative intermediate in lipoxin synthesis, 5(6)-epoxy-15-hydroxyeicosatetraenoic acid (15-OH-LTA$_4$) (10), which was then further hydrolyzed to the trihydroxylated lipoxins. The occurrence of 15-OH-LTA$_4$ was also suggested by short time incubations of pure platelets or granulocytes with LTA$_4$ or 15-HETE, respectively (11). In these experiments, instant conversion of the tetraene-epoxide to methoxy derivatives was elicited by the addition of large volumes of warm acidified methanol. Analysis by RP-HPLC demonstrated the presence of five less polar, novel peaks with characteristic UV-absorbance of conjugated tetraenes, in agreement with formation of methoxy-lipoxins. The products appeared with identical elution times irrespective of incubated cell type. These peaks were also observed in material from trapping experiments performed with ionophore A23187-stimulated suspensions of mixed platelets and granulocytes. In contrast, the less polar lipoxin derivatives totally disappeared when the incubations were terminated by addition of acid.

An alternative pathway for platelet-dependent lipoxin synthesis was indicated by the finding that these cells converted the four positional isomers of 5,6-dihydroxy-7,9-*trans*-11,14-*cis*-eicosatetraenoic acid (5,6-diHETE) to lipoxins and lipoxin-like compounds (12). In particular, 5(S),6(R)-diHETE was transformed to LXA$_4$, while 6(S)-LXA$_4$ was produced from the 5(S),6(S)-isomer. These two 5,6-diHETEs are formed by nonenzymatic hydrolysis of LTA$_4$. In addition, the epoxide can be enzymatically transformed to 5(S),6(R)-diHETE by an epoxide hydrolase in e.g. the liver (13).

Several findings obtained with normal platelets indicated that the lipoxin production was mediated by the platelet 12-lipoxygenase (11): i) Conversion of exogenous LTA$_4$ to lipoxins and metabolism of arachidonic acid to 12-HETE in intact platelets were inhibited with almost identical ID$_{50}$s (7.0 μM and 8.2 μM, respectively) by 15-HETE, which is known as a lipoxygenase inhibitor (14, 15). ii) The transformation of LTA$_4$ to lipoxins by platelet sonicates was dose-dependently inhibited by arachidonic acid, suggesting substrate competition for the same enzyme. Lipoxin synthesis was not significantly inhibited by addition of exogenous 12-HETE, excluding a possible inhibitory action of this product on lipoxin synthesis. iii) The high speed (100,000 x g) pellet and supernatant from platelet sonicates were incubated with either LTA$_4$ or arachidonic acid and indomethacin, respectively. Conversion of LTA$_4$ to lipoxins, including LXA$_4$ and 6(S)LXA$_4$, was observed both in the cytosolic and the particulate fraction. Thus, 54 % of the total LXA$_4$-producing capacity was recovered in the cytosol and 46 % in the membrane fraction. The formation of 12-HETE from arachidonic acid displayed a similar subcellular distribution with 69 and 31 % of the total capacity detected in the soluble and particulate fraction, respectively. In agreement, a dual subcellular distribution of the platelet 12-lipoxygenase has earlier been described (16). Synthesis of 15-HETE was not observed in any of the subcellular fractions.

Involvement of platelet 12-lipoxygenase in lipoxin formation was further indicated by results obtained with 12-lipoxygenase deficient platelets from patients with myeloproliferative disorders (17). These platelets demonstrated a significantly decreased capacity to convert LTA$_4$ (4 μM) to lipoxins (183.7 $\pm$ 42.2 as compared to 438.7 $\pm$ 62.8 pmol LXA$_4$ per 10^9 platelets in healthy controls; p=0.0003). Six of the patients were in blastic crisis of chronic myelogenous leukemia (CML). Interestingly, platelets from each of these individuals produced only <7% of mean control LXA$_4$ levels (cfr. Fig 2). These six CML patients also demonstrated a dramatically decreased capacity to convert arachidonic acid to 12-HETE (2.5 $\pm$ 1.1 as compared to 81.5 $\pm$ 13.1 nmol 12-HETE per 10^9 platelets in controls; p=0.0006). A correlation between the disease stage of CML and the ability to form lipoxins was further indicated by longitudinal studies (17). Thus, when one of the patients in blastic crisis of CML converted back to a chronic phase a temporary restoration of lipoxin/12-HETE production was

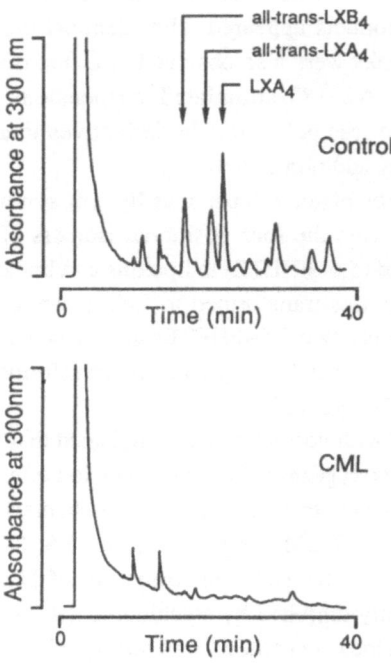

Figure 2. Comparison of lipoxin formation from LTA$_4$ in normal and 12-lipoxygenase deficient leukemic platelets (CML).

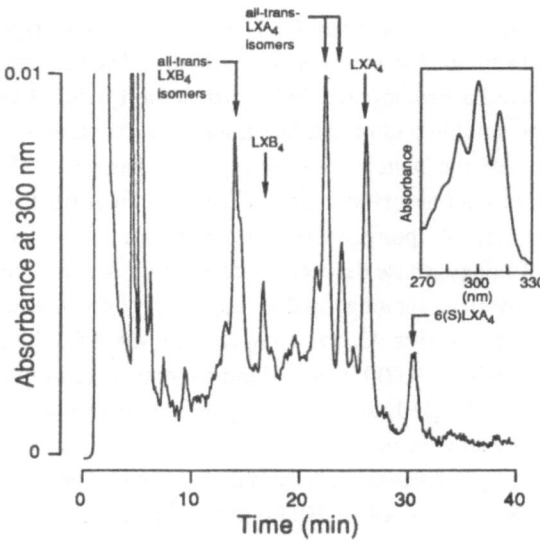

Figure 3. HPLC-chromatography demonstrating lipoxin formation by a polyp tissue/granulocyte mixture incubated with ionophore A23187. **Inset:** UV-spectrum of biosynthetic LXA$_4$.

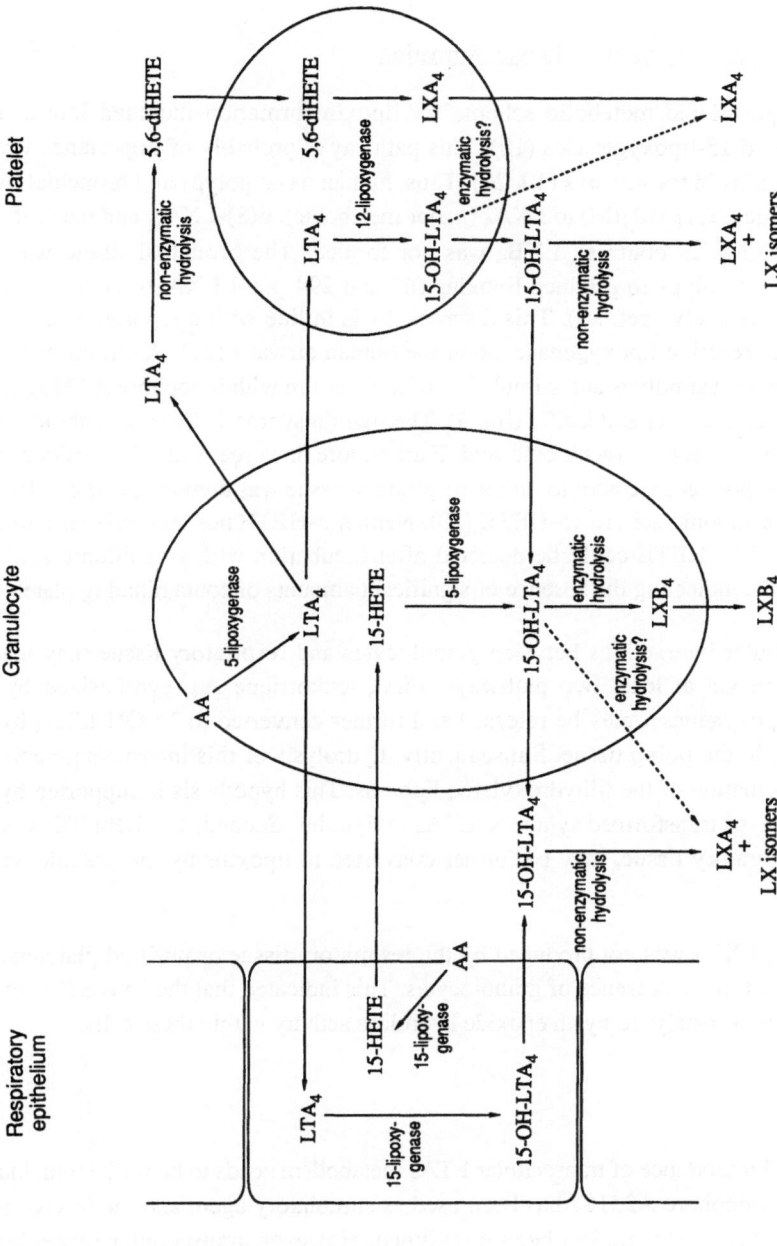

Figure 4. Hypothetical scheme of transcellular lipoxin formation.

285

observed, while retransformation into a second blastic crisis again was paralleled by an almost total inability to produce lipoxins/12-HETE. The prostagladin endoperoxide synthase activity was ordinarily intact in platelets from patients with myeloproliferative disorders. Taken together, these findings suggest that essentially abolished 12-lipoxygenase activity may be a general phenomenon in blastic crisis of CML.

Involvement of respiratory tissue in lipoxin formation

The originally postulated metabolic scheme for lipoxin formation included interactions between the 5- and 15-lipoxygenases (18). This pathway is probably of importance for the formation of lipoxins in the airways (19, 20). Thus, human nasal polyps and bronchial tissue converted authentic LTA_4)10 μM) to LXA_4 (major metabolite), 6(S)-LXA_4, and the all-*trans*-lipoxin isomers (20). In contrast, LXB_4 was not formed. The bronchial tissue was less efficient than nasal polyps to produce lipoxins (63 and 294 pmol LXA_4/g of bronchial or polyp tissue, respectively, ref. 20). This discrepancy is in line with a reported gradual decrease of immunoreactive lipoxygenase along the human airways (21). Addition of granulocytes to chopped nasal polyps and stimulation of the mixture with ionophore A23187 led to the formation of both LXA_4 and LXB_4 (fig. 3). The lipoxin synthesis increased about 3-fold in the presence of exogenous arachidonic acid. Furthermore, in agreement with earlier reports (22, 23) high 15-lipoxygenase activity in the respiratory tissue was demonstrated by efficient conversion of arachidonic acid to 15-HETE (20). Neither 5-HETE nor leukotrienes and only minor amounts of 12-HETE could be detected after incubation with arachidonic acid and ionophore A23187, indicating the absence of significant amounts of contaminating platelets or leukocytes.

The transcellular interactions between granulocytes and respiratory tissue may lead to lipoxin formation via at least two pathways. First, leukotriene A_4, synthesized by the granulocyte 5-lipoxygenase, may be released and further converted to 15-OH-LTA_4 by the 15-lipoxygenase in the polyp tissue. Subsequently, hydrolysis of this intermediate epoxide would lead to formation of the trihydroxylated lipoxins. This hypothesis is supported by the fact that nasal polyps transformed synthetic LTA_4 to lipoxins. Second, 15-H(P)ETE, synthesized by the respiratory tissue, may be further converted to lipoxins by the granulocyte 5-lipoxygenase.

Interestingly, LXB_4 was not produced by the respiratory tissue or purified platelets, but exclusively formed in the presence of granulocytes. This indicated that the conversion of 15-OH-LTA_4 to LXB_4 is catalyzed by an epoxide hydrolase activity within these cells.

Conclusion

The physiological importance of transcellular LTA_4 metabolism needs to be further elucidated. In most studies, ionophore A23187 has been used as stimulatory agent and the in vivo relevance of the findings has therefore been questioned. However, transcellular formation of leukotrienes (24) and lipoxins (25) may also occur after stimulation with receptor-mediated agonists, suggesting a (patho)-physiological role for transcellular LTA_4 metabolism. Platelet dependent modulation of LTA_4 metabolism can be of importance in allergy and inflammation, conditions where accumulation and activation of platelets have been reported (26). Thus, increased amounts of platelets in e.g. the asthmatic lung may direct the conversion of leukocyte-derived LTA_4 towards elevated formation of the potent bronchoconstrictors, leukotrienes C_4, D_4 and E_4. Furthermore, transcellular formation of lipoxins, due to interactions between activated leukocytes and respiratory epithelium or accumulated platelets, may also occur in allergic and non-allergic inflammation (Fig 4). The importance of lipoxin synthesis under such conditions is still unclear, since lipoxins are inflammatory agonists (18), but also counteract leukotriene-induced biological effects (27-31).

Acknowledgements

This project was supported by the Swedish Heart Lung Foundation, The Swedish Medical Research Council, The Swedish Cancer Society, King Gustaf V 80-years Fund, The Swedish Medical Society, Magnus Bergvalls Foundation and Karolinska Institutet´s Research Funds.

References

1. Dahinden, C. A., Clancy, R. M., Gross, M., Chiller, J. M. and Hugli, T. E. (1985) *Proc. Natl. Acad. Sci. USA* 82, 6632-6636.
2. McGee, J. E. and Fitzpatrick, F. A. (1986) *Proc. Natl. Acad. Sci. USA* 83, 1349-1353.
3. Claesson, H.-E. and Haeggström, J. (1988) *Eur. J. Biochem.* 173, 93-100.
4. Odlander, B., Jakobsson, P-J., Rosén, A. and Claesson, H.-E. (1988) *Biochem. Biophys. Res. Commun.* 153, 203-208.
5. Feinmark, S. J. and Cannon, P. J. (1986) *J. Biol. Chem.* 261, 16466-16472.
6. Feinmark, S. J. and Cannon, P. J. (1987) *Biochim. Biophys. Acta* 922, 125-135.
7. Edenius, C., Heidvall, K. and Lindgren, J. Å. (1988) *Eur. J. Biochem.* 178, 81-86.
8. Bach, M. K., Brashler, J. R. and Morton, Jr. D. R. (1984) *Arch. Biochem. Biophys.* 230, 455-465.
9. Edenius, C., Haeggström, J. and Lindgren, J. Å. (1988) *Biochem. Biophys. Res. Commun.* 157, 801-807.
10. Puustinen, T., Webber, S. E., Nicolaou, K. C., Haeggström, J., Serhan, C. N. and Samuelsson, B. (1986) *FEBS Lett* 207, 127-132.
11. Edenius, C., Stenke, L. and Lindgren, J. Å. (1990) *Eur. J. Biochem.* In press.
12. Tornhamre, S., Edenius, C., Barbedette, A., Lellouch, J. P. and Lindgren, J. Å. (1991) *FEBS Lett.* Submitted.
13. Haeggström, J., Wetterholm, A., Hamberg, M., Meijer, J., Zipkin, R. and Rådmark, O. (1988) *Biochim. Biophys. Acta* 958, 469-476.
14. Vanderhoek, J. Y., Bryant, R. W. and Bailey, J. M. (1980) *J. Biol. Chem.* 255, 5996-5998.
15. Vanderhoek, J. Y., Bryant, R. W. and Bailey, J. M. (1980) *J. Biol. Chem.* 255, 10064-10066.
16. Lagarde, M., Croset, M., Authi, K. S. and Crawford, N. (1984) *Biochem. J.* 222, 495-500.
17. Stenke, L., Edenius, C., Samuelsson, J. and Lindgren, J. Å. (1991) *Blood* Submitted
18. Samuelsson, B., Dahlén, S.-E., Lindgren, J. Å., Rouzer, C. A. and Serhan, C. N. (1987) *Science* 237, 1171-1176.
19. Lee, T. H., Crea, A. E. G., Gant, V., Spur, B. W., Marron, B. E., Nicolaou, K. C., Reardon, E., Brezinski, M. and Serhan, C. N. (1990) *Am. Rev. Respir. Dis.* 141, 1453-1458.
20. Edenius, C., Kumlin, M., Björck, T., Änggård, A. and Lindgren, J. Å. (1990) *FEBS Lett.* 272, 25-28.
21. Shannon, V. R., Hansbrough, J. R., Takahashi, Y., Ueda, N., Yamamoto, S. and Holtzman, M. J. (1990) in: *Adv in Prostaglandin, Thromboxane and Leukotriene Research*, Vol. 21 (Samuelsson, B., Paoletti, R. and Ramwell, P., eds.), Raven Press, New York. pp 37-40.
22. Henke, D., Danilowicz, R. M., Curtis, J. F., Boucher, R. C. and Eling, T. E. (1988) *Arch. Biochem. Biophys.* 267, 426-436.

23. Kumlin, M., Hamberg, M., Granström, E., Björck, T., Dahlén, B., Matsuda, H., Zetterström, O. and Dahlén, S.-E. (1990) *Arch. Biochem. Biophys.* 282, 254-262.
24. Maclouf, J., Murphy, R. C. and Henson, P. M. (1990) *Blood* 76, 1838-1844.
25. Serhan, C. N. and Sheppard, K.-A. (1990) *J. Clin. Invest.* 85, 772-780.
26. Page, C. P. (1989) *Immunopharmacology* 17, 51-59.
27. Dahlén, S.-E., Franzén, L., Raud, J., Serhan, C. N., Westlund, P., Wikström, E., Björk, T., Matsuda, H., Webber, S. E., Veale, C. A., Puustinen, T. and Samuelsson, B. (1988) *Adv. Exp. Med. Biol.* 229, 107-130.
28. Hedqvist, P., Raud, J., Palmertz, U., Haeggström, J., Nicolaou, K. C. and Dahlén, S.-E. (1989) *Acta Physiol. Scand.* 137, 571-572.
29. Badr, K. F., DeBoer, D. K., Schwartzberg, M. and Serhan, C. N. (1989) *Proc. Natl. Acad. Sci. USA* 86, 3438-3442.
30. Lee, T. H., Horton, C.E., Kyan-Aung, U., Haskard, D., Crea, A. E. and Spur, B. W. (1989) *Clin. Sci.* 77, 195-203.
31. Brady, H. R., Persson, U., Ballerman, B. J., Brenner, B. M. and Serhan C. N. (1990) *Am. J. Physiol.* 259, F809-F815.

METABOLISM OF ARACHIDONIC ACID BY ISOLATED LUNG CELLS

AND TRANSCELLULAR BIOSYNTHESIS OF THROMBOXANES

Karim Maghni[1], Chantal Robidoux[1], Johanne Laporte[1], Annie Hallée[1], Johanne Carrier[1], Pierre Borgeat[2] and Pierre Sirois[1]

[1] Department of Pharmacology
Faculty of Medicine
University of Sherbrooke
Sherbrooke (P.Q.) Canada J1H 5N4
[2] Inflammation et Immunologie-Rhumatologie
CHUL, Quebec, Canada G1V 4G2

INTRODUCTION

Harkavy[1] was the first to report that an alcohol-soluble extract of sputum from allergic asthmatic patients contained an agent which provoked spasms of cat and rabbit intestines in vitro. Few years later, von Euler[2] and Goldblatt[3] showed that extracts from human prostate gland and seminal vesicles decreased the blood pressure and stimulated the smooth muscles of the uterus. The compounds were designated as prostaglandins[2]. A new area of research on arachidonic acid metabolism started by the characterization of prostaglandins E and F by Bergström and Sjövall[4,5]. The final structures of PGE_1, $PGF_{1\alpha}$ and $PGF_{1\beta}$ were elucidated later[6]. The thromboxanes were first described as the rabbit aorta contracting substance (RCS) by Piper and Vane[7]. Hamberg et al.[8] showed that RCS corresponded to the unstable thromboxane A_2. Another line of investigations focused on the nature of substances causing a slowly developing and long-lasting contraction of guinea pig jejunum in vitro[9]. This substance called slow reacting substance (SRS) was shown to be released during the antigen-antibody reaction in guinea pig anaphylaxis and later designated as slow reacting substances of anaphylaxis (SRS-A). SRS-A were purified and characterized for the first time by Morris et al.[10]. The first product of the SRS-A has been identified as leukotriene C by Murphy et al.[11] and Morris et al.[12]. The two other components of SRS-A were demonstrated to be the leukotriene D_4 (LTD_4) formed through cleavage of the γ-glutamyl moiety of the gluthathione side chain of LTC_4[13,15] and the leukotriene E_4 (LTE_4) formed through the cleavage of the glycine residue from the peptide chain of LTD_4[14,16]. Leukotriene B_4 (LTB_4) was originally isolated by Borgeat and Samuelsson, in incubates of rabbit PMNL[17] or human PMNL[18] stimulated with arachidonic acid and calcium ionophore. The biological activities of eicosanoids have been extensively studied in various physiological and pathophysiological conditions. The putative role of eicosanoids and especially leukotrienes in asthma has been mainly illustrated by: 1) the bronchoconstrictor response of healthy and asthmatic

Cell-Cell Interactions in the Release of Inflammatory Mediators
Edited by P. Y-K Wong and C.N. Serhan, Plenum Press, New York, 1991

subjects to peptido-leukotrienes inhalation[19]; 2) the hyperreactive response to inhaled leukotrienes in asthmatics[20]; and 3) the inhibition of allergen-induced bronchoconstriction in atopic asthmatics by leukotriene antagonist[21]. Leukotriene B_4 causes chemokinesis, chemotaxis, aggregation, degranulation with release of lysosomal enzymes and generation of superoxide anion by neutrophils[22]. LTB_4 also possesses marked myotropic activity on the guinea pig lung strip[23] which is mediated by release of cyclooxygenase products, most likely TxA_2[24,25]. With regards to the proinflammatory effects and the myotropic activity of LTB_4, it could be postulated that LTB_4 is a mediator of both the inflammatory and bronchoconstrictor components of bronchial asthma. The role of eicosanoids in asthma is supported by experimental evidence but the cell(s) responsible for their synthesis remain(s) unknown. Mast cells were shown to play a major role in the early event of mediator release. The activation of mast cells by the IgE-dependent cross-linking of membrane Fc receptors leads to different events such as arachidonic acid mobilization and eicosanoid release[26]. Several other kind of cells are able to generate eicosanoids and thus, could contribute to the pathogenesis of asthma. Eosinophils are believed to play a major role in asthma[27] in damaging bronchial epithelium and generating LTC_4. It has been shown that some eicosanoids may be formed through cellular interactions as described for erythrocyte-neutrophil[28], endothelial-neutrophil[29] and platelet-neutrophil[30] interactions. These interactions could explain some inflammatory processes involved in bronchial asthma. The purpose of the present study was to isolate and purify different cell types of guinea pig lungs and to investigate arachidonic acid metabolism. Transcellular biosynthesis of TxA_2 was also investigated between platelets and alveolar cells.

METHODS

Isolation and purification of lung cells

Enzymatic digestion of guinea pig lungs and cell elutriation were performed as previously described[31] and type II pneumocytes cells[32], Clara cells[33] and Kurloff cells[34] were isolated and purified. Briefly, Dunkin-Hartley guinea pigs (350-400 g) purchased from Charles River Lab. were injected with heparin (5000 units) and killed by cervical dislocation. The isolated lungs were perfused via the pulmonary artery at a flow rate of 10 ml/min (37°C) with calcium- and magnesium-free Krebs solution containing 2.5 % bovine serum albumin to eliminate residual red cells. The lungs were then perfused via the trachea with calcium- and magnesium-free Krebs solution containing 0.005 % protease type XXIV and 1 mM EDTA (pH = 7.3, 37°C) for 15 min. The cells were readily dispersed into Krebs solution supplemented with 10 mM Hepes (pH 7.4), 2.4 mM $MgSO_4.7H_2O$, 50 Kunitz units DNAse and 0.5 % bovine serum albumin. After filtration of the cell suspension through nylon bolting clothes (450 and 110 nm size), they were injected into a Beckman JEGB rotor equiped with a standard elutriation chamber, centrifuged in a Beckman J2-21M centrifuge and collected in 50 ml centrifuge tubes. Elutriation parameters were: constant pressure, temperature (22°C), centrifugation speed (2000 RPM) and volume of aliquots (150 ml); variable flow of buffer 9, 12, 15, 18, 22, 25, 28 and 30 ml/min corresponding to fractions F_1, F_2, F_3, F_4, F_5, F_6, F_7 and F_8, respectively. The cells were injected into the elutriation chamber at an initial flow rate of 9 ml/min and three aliquots (50 ml) were collected at each flow rate.

Type II pneumocytes were purified from fraction F_3 according to the technique of Dobbs et al.[35] as modified by Maghni et al.[32]. This method involves adherence of cells bearing Fc fragment receptors on 60 mm Petri dishes coated with guinea pig IgG

(500 μg/ml in 50 mM Tris-base, pM 9.5; 3 hours, 22°C, washed 5 times with PBS and once with RPMI). The cells from elutriation fraction F_3 were added to each IgG-coated dish (25 x 10^6 cells/dish) and allowed to adhere for 20 min at 37°C. The supernatants were collected and the dishes were gently washed 3 times and pelleted by centrifugation.

Clara cells were purified from fractions 6 and 7 of two separate lungs. Fractions were centrifuged and mixed together. A second elutriation was performed and the following conditions were used: variable rotor speed (2000 to 1200 RPM), flows of buffer (15, 22 and 28 ml/min) and constant temperature (22°C). The second and third fractions were collected and centrifuged. The macrophages and endothelial cells were eliminated by differential adherence on Petri dishes, for 4 h at 37°C. The supernatant containing the non-adherent cells were collected and centrifuged. Clara cells were purified by centrifugation on a discontinuous Percoll gradient (1.044, 1.049 and 1.054 densities) at 1200 RPM for 30 min. The Clara cells were obtained from the lower interface 1.049 - 1.054 of the Percoll gradient.

Foà-Kurloff cells, commonly called Kurloff cells, are a small percentage of total lung cells in normal guinea pigs. Their number can be increased by intraperitoneal injection of 1 mg/week for three weeks of 17-β-estradiol-3 benzoate (Sigma Chem., st. Louis, USA). One week after the third injection, animals were used. Kurloff cells were purified from fraction 8 as previously described[34]. Pulmonary Kurloff cells were further purified by centrifugation on Percoll continuous density gradient (360 g, 30 min, 20°C). The gradient was built by centrifuging a solution of 50 % Percoll 15 min at 20,000 g. A high density (1.100 g/ml) pulmonary Kurloff cell population was collected from the bottom of the Percoll gradient.

Alveolar macrophages and eosinophils were purified from bronchoalveolar lavage (BAL), as described previously[36]. Briefly, guinea pigs were killed and exsanguinated. BAL was done by cannulation of the trachea with phosphate buffered saline (PBS) at 37°C. The lavage fluid was combined and centrifuged. Red cell lysis was achieved by osmotic shock. After centrifugation, BAL cells were resuspended in Hank's balanced salt solution (HBSS) with Ca^{2+} and Mg^{2+} free. Alveolar macrophages were purified by differential adherence on Petri dishes for 60 min at 37°C. The supernatant containing the non-adherent cells was collected and centrifuged. The cell suspension was purified by centrifugation on discontinuous Percoll gradients (20 to 70 % Percoll). After centrifugation (360 g, 30 min, 20°C), alveolar eosinophils were obtained at the interface 60-70 % Percoll.

Platelets were obtained from guinea pig blood, as previously described[36]. Briefly, guinea pigs were anesthetized with an intraperitoneal injection of sodium pentobarbital (10 mg/kg body weight). Guinea pig blood was collected in syringes with sodium citrate solution (3.15 g/100 ml; pH 7.4). Blood was centrifuged at 270 g for 10 min. The plasma was collected and recentrifuged. The plasma was collected, 77 mM EDTA was added (9:1, v/v), and the mixture was centrifuged at 540 g for 10 min. The supernatant was discarded and the cell pellet was resuspended at appropriate dilutions.

Cell counts and identification

The cells were counted and viability was assessed using the Trypan blue exclusion test at each purification step. Differential cell counts were made after periodic acid-Schiff (PAS), Papanicolaou technique without acid alcohol, nitroblue tetrazolium, Wright-Giemsa stainings to identify the Kurloff cells, the type II pneumocytes, the Clara cells, the macrophages and eosinophils, respectively. Electron microscopy was also performed using a procedure described previously[31]. Ultrathin sections of cell pellets were stained with 6 % lead citrate (pH 10) and observed under an EM Philips electron microscope. Cells were counted in each square of grids.

Arachidonic acid metabolism

The incubations were performed in HBSS containing 2 mM Ca^{2+} and 0.5 mM Mg^{2+}. The cells were preincubated at 37°C for 5 min. The formation of cyclooxygenase products was studied by stimulation of each type of lung cells ($1x10^6$ cells/ml) with 10 µM of exogenous arachidonic acid (Sigma Chem., St. Louis, USA) in the absence or presence of 2 µM calcium ionophore A23187 (Calbiochem, La Jolla, USA) during 10 min at 37°C. For the Kurloff cells, the incubation was also performed with 30 and 100 µM arachidonic acid. Incubation were stopped by centrifugation and supernatants were collected and stored at -80°C until eicosanoid assays.

Cell cooperation was studied by incubating fixed concentration of platelets (1 x 10^8 cells/ml) with various ratios of alveolar macrophages or eosinophils: no alveolar cells, 100:1, 100:2 and 100:6 (platelets:cells). Alveolar macrophages or eosinophils (2 x 10^6 cells/ml) were incubated with various ratios of platelets: no platelet, 2:100, 1:100 and 1:200 (cells:platelets). The cells were incubated with 5 µM calcium ionophore for 30 min at 37°C. Incubations were stopped as described above.

Arachidonic metabolism via the 5-lipoxygenase pathway was studied by incubating Kurloff cells (2 x 10^6 cells/ml), type II pneumocytes (1-5 x 10^6 cells/ml), Clara cells (4 x 10^6 cells/ml), macrophages (3 x 10^6 cells/ml) or eosinophils (2 x 10^6 cells/ml) with 10 µM arachidonic acid and 2 µM calcium ionophore A23187 or with 2.5 µM leukotriene A_4, (generously supplied by Dr. Jean-Marie Ferland, Bio-Mega Inc., Montreal) for 10 min at 37°C. Incubations were stopped by addition of one volume of methanol containing 12.5 ng each of PGB_2 and 19-OH-PGB_2 as internal standards and stored at -80°C until assays.

Eicosanoid analyses

Lipoxygenase products were analyzed by reverse phase high performance liquid chromatography (RP-HPLC) using an on-line extraction procedure as described elsewhere[37]. Briefly, denatured incubation mixtures were diluted with water and methanol to obtain a final volume of 2 ml containing 25 % methanol. Samples were centrifuged to remove precipitated material and injected directly without further treatment. The column used was the Resolve Radial Pak C_{18} 5µm particule (5 x 100 mm cartridge) from Waters Millipore. Identification of the various lipoxygenase products was assessed on the basis of their migration with authentic standard and specificity of UV absorbance at either 229 or 280 nm. The detection limit was 0.5 ng for the leukotrienes and 2 ng for the HETE. Quantitation was done by comparison of product peak heights with peak heights of known amounts of the corresponding lipoxygenase products after correction of the recovery by using the internal standard PGB_2.

Prostaglandin E_2, thromboxane B_2 (the stable metabolite of thromboxane A_2) and 6-keto-prostaglandin $F_{1\alpha}$ (the stable metabolite of prostacyclin) were measured in cell supernatants by enzyme immunoassays (EIA) as described by Pradelles et al.[38]. The lower limite of detection for prostaglandins was around 0.4 pg/ml; the lower limit of detection for thromboxane B_2 was 4 pg/ml.

Statistical Analysis

Data were analyzed with the Student T test and a one-way analysis of variance. A P value less than 0.05 was considered significant.

RESULTS

Cell purification

The enzymatic digestion and mechanical dispersion of guinea pig lungs yielded approximately 650×10^6 cells, with a viability of 95 %. Type II pneumocytes were purified from elutriation fraction 3 which contained about 60×10^6 cells with 30 % Type II pneumocytes as determined with Papanicolaou staining and electron microscopy. After incubation of cell fraction in Petri dishes coated with guinea pig IgG, the percentage of type II pneumocytes in supernatant was 92 % (viability over 85 %), as shown in Table 1. The contamining cells (below 8 %) have not been identified yet; electron microscopy of these cells did not provide characteristic features useful for their identification. Clara cells were purified from elutriation fractions 6 and 7 which contained 17 and 25 % of Clara cells, respectively (as shown in Table 1). After the second elutriation of the pooled fractions, Clara cells were enriched to 43 %. The discontinuous Percoll gradient has permitted to obtain at the lower interface (1.049 - 1.054) a cell fraction (2.6×10^6 cells) composed of 72 or 83 % Clara cells, depending upon whether the cell counts were done by electron microscopy or optical microscopy using nitroblue tetrazolium staining. The contaminating cells were endothelial cells (13 %), type II pneumocytes (3.7 %), eosinophils (1.3 %), mast cells (0.5 %),

Table 1.   Cell number and purity in the main steps of purification of type II pneumocytes, Clara cells and Foà-Kurloff cells.

	Cell number (x10^6)			Cell type (%)[a]		
	T	C	FK	T	C	FK
Total lung cells	122	21	178[b]	17	3	25[b]
Fraction 3	60	-	6	30	-	9.5
Purified type II pneumocytes	15	-	-	92	-	-
Fraction 6	0.8	3.4	1.8	4	17	9
Fraction 7	0.3	3.8	1.5	2	25	10
Purified Clara cells	0.1	1.9	-	3.7	72	-
Fraction 8	0.4	3.2	90[b]	0.8	7.0	75[b]
Purified FK cells	-	-	50[b]	-	-	99.3[b]

T: Type II pneumocyte   C: Clara cell   FK: Foà-Kurloff cell
a: Percentages were determined by electron microscopy
b: data provided from estradiol-treated animal

Table 2.

Table 2. Purification of alveolar macrophages and eosinophils from BAL.

	BAL cells (x10⁶)	(%)	Purified macrophages (x10⁶)	(%)	Purified eosinophils (x10⁶)	(%)
Macrophages	21	88	15	99	0.03	2
Eosinophils	2	9	-	-	1.1	96

Data provided from one guinea pig.

macrophages (0.5 %) and non-identified cells (9.0 %). Kurloff cells were purified from estradiol-treated animals. After centrifugal elutriation, fraction 8 contained approximately 90 x 10⁶ cells with 75 % Kurloff cells. As shown in Table 1, centrifugation on continuous Percoll gradient has allowed to obtain a highly pure ($\approx$ 100 %) population of Kurloff cells (50 x10⁶ cells) with a 99 % viability. This Kurloff cells population has a density of 1.100 g/ml as assessed by calibration of the Percoll with density marker beads. Alveolar macrophages and eosinophils were purified from BAL. As shown in Table 2, the purity of alveolar macrophages obtained after plating was above 99 %. After plating, non-adherent cells consisted of 60 % eosinophils. These cells were purified on discontinuous Percoll gradient and collected at the 60-70 % interface. Purified eosinophils (1.1 x 10⁶ cells) were contaminated by macrophages and lymphocytes (3.1 %) and neutrophils (1 %).

Identification of lung cell populations

As shown in Figure 1A, the guinea pig isolated type II pneumocytes have an oval shape and are covered by short microvilli on their surface. Type II pneumocytes contain variable amount of lamellar bodies. The residual lamellar body is made of phospholipids, the principal material in the pulmonary surfactant. The cytoplasm contains some mitochondria, ribosomes and rough endoplasmic reticulum. The guinea pig Clara cells (Fig. 1B) shows some characteristic structural features such as secretory granules, smooth endoplasmic reticulum and two types of mitochondria. There are a few types of Clara cells that can be classified according to structural features including the electron density of the secretory granules and the abundance of smooth endoplasmic reticulum (not shown here). The electron micrograph of an isolated pulmonary Kurloff cell (Fig. 1C) shows an ellipsoid nucleus bordered by an envelope, a single large homogenous electron-dense inclusion body, a Golgi apparatus with microvesicles of variable density and an abundant endoplasmic reticulum. The periphery of the inclusion exhibits myelin figures located in close proximity of the inclusion membrane and sometimes appearing to be within the matrix of the inclusion body. Isolated guinea pig alveolar macrophages (Fig. 1D) are covered by several pseudopodia on their surface. The nucleus is folded several times and the cytoplasm contains a small number of mitochondria, endoplasmic reticulum and a few Golgi apparatus. Alveolar macrophages have various amounts of lysosome-like structures and residual bodies in the cytoplasm. The characteristic features of the guinea pig eosinophil (Fig. 1E) are defined by the presence of circular or elleptical shapes of granules with a central crystal. The nucleus is sometimes divided in 2 lobes depending of the section. The cytoplasm contains a small amount of mitochondria, glycogen and endoplasmic reticulum and the cells are covered by some pseudopodia.

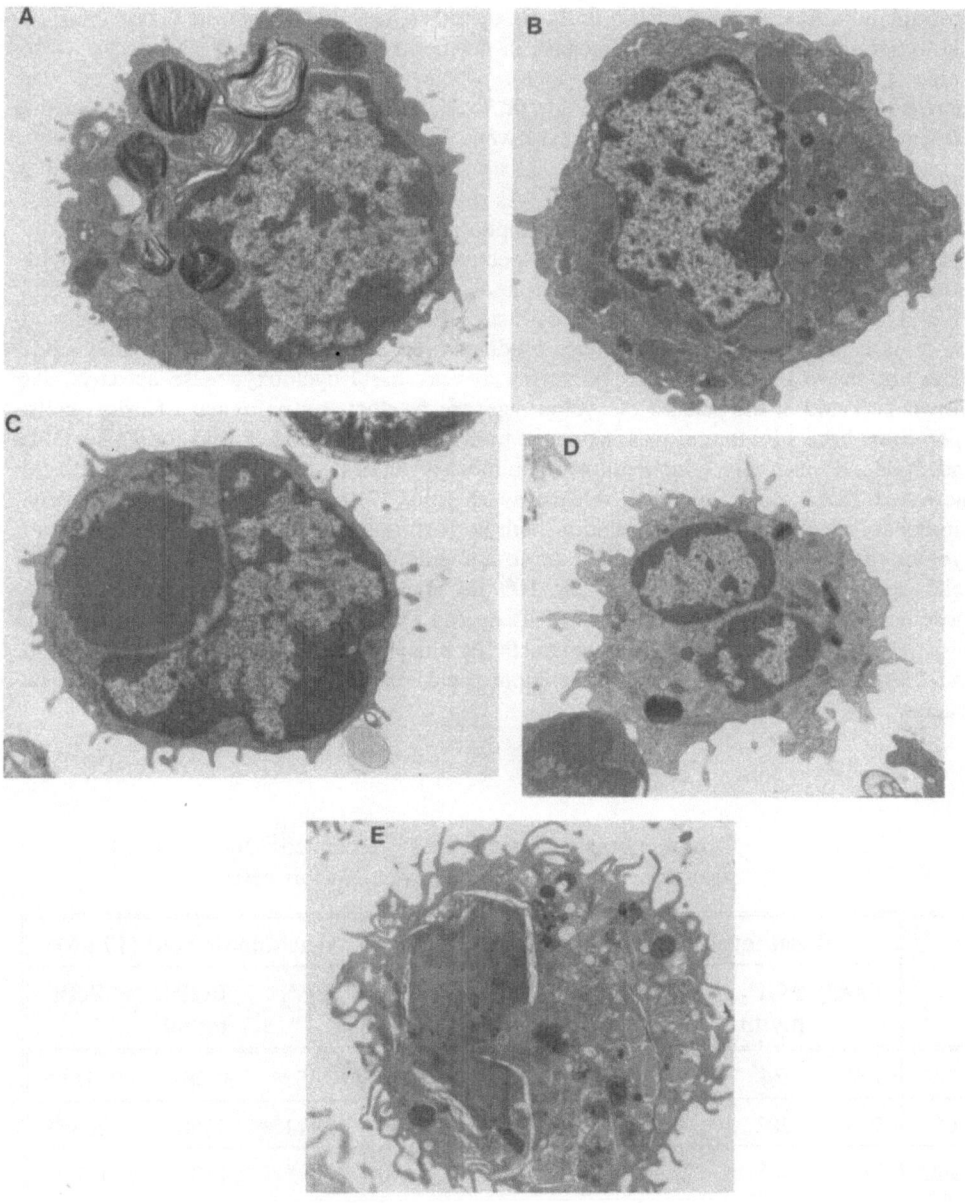

Figure 1. Electron microscopy of guinea pig type II pneumocytes (A), Clara cells (B), Foà-Kurloff cells (C) alveolar eosinophils (D) and alveolar macrophages (E).

Arachidonic acid metabolism

Cyclooxygenase product formation was studied by incubating the cell populations with 2 µM ionophore A23187 or 10 µM arachidonic acid. Three cyclooxygenase products were determined: thromboxane A_2 (TxA_2, measured as its stable metabolite TxB_2), prostaglandin E_2 and prostacyclin (PGI_2, measured as its stable metabolite 6-keto-$PGF_{1\alpha}$). The basal activity of the cyclooxygenase in purified type II pneumocytes was shown in Table 3. The release of cyclooxygenase products upon ionophore stimulation was increased by 4 fold, 2 fold and 3 fold for TxA_2, PGI_2 and PGE_2, respectively. In the presence of arachidonic acid the releases went up to 51, 50 and 212 fold respectively. No additive effect was observed when cells were stimulated with both arachidonic acid and calcium ionophore (data not shown). As shown in Table 3, the release of TxA_2, PGI_2 and PGE_2 by purified Clara cells, upon ionophore stimulation was increased 5 fold, 7 fold and 7 fold for TxA_2, PGI_2 and PGE_2, respectively. The incubation of Clara cells with arachidonic acid increased the release of TxA_2, PGI_2 and PGE_2 by 42 fold, 19 fold and 32 fold, respectively. Combination of both stimuli did not produce an additive effect and the release profiles of cyclooxygenase products were similar to arachidonic acid alone (data not shown). Kurloff cells possessed a weak basal cyclooxygenase activity. As shown in Table 3, the release of cyclooxygenase products by pulmonary Kurloff cells, upon ionophore stimulation was increased by 3 fold, 3 fold and 2.5 fold for TxA_2, PGI_2 and PGE_2, respectively. Incubation of pulmonary Kurloff cells with arachidonic acid increased TxA_2, PGI_2 and PGE_2 release by 25 fold, 27 fold and 78 fold, respectively. Similar to other cell types, stimulation with both stimuli did not enhance the release of cyclooxygenase products, in comparison to arachidonic acid alone. Pulmonary Kurloff cells were also stimulated with 30 and 100 µM arachidonic acid (data not shown). In these conditions, the release profiles of TxA_2 and PGI_2 were similar to the stimulation with 10 µM arachidonic acid. However, PGE_2 biosynthesis was increased by 750 fold and 1500 fold for 30 and 100 µM arachidonic acid, respectively (in comparison to basal levels).

Table 3.   Cyclooxygenase metabolites produced by type II pneumocytes, Clara cells and Foà-Kurloff cells.

	Basal level			Ionophore (2 µM)			Arachidonic acid (10 µM)		
	TxA_2^a	PGI_2^b pg/ml	PGE_2	TxA_2^a	PGI_2^b pg/ml	PGE_2	TxA_2^a	PGI_2^b pg/ml	PGE_2
T^c	122	74	10	464**	153*	32*	6255**	3708**	2116**
C^d	309	283	65	1508*	1978**	476*	12911**	5350*	2035*
FK^d	11.4	4.5	2.0	36.2*	12.0*	5.0	290**	120**	157**

T: Type II pneumocyte   C: Clara cell   FK: Foà-Kurloff cell
a: TxA_2 was measured as its stable metabolite TxB_2
b: PGI_2 was measured as its stable metabolite 6-keto-$PGF_{1\alpha}$
c: Stimulation of 2×10^6 cells/ml
d: Stimulation of 1×10^6 cells/ml
*: $p < 0.05$; **: $p < 0.01$

Table 4.   Cyclooxygenase product formation by guinea pig alveolar eosinophils in response to various stimuli.

	Control	PMA (5×10^{-8} M)	PAF (5×10^{-7} M)	fMLP (2.5×10^{-7} M)
TxA^a_2	147	4430**	1282**	1500**
PGE^a_2	165	163	151	118
PGI^a_2	68	144*	94	135*

a: eicosanoid release was expressed in pg/10^6 cells
*: $p < 0.05$;   **: $p < 0.01$

The prostaglandin production by guinea pig alveolar eosinophils was studied by incubating these cells with phorbol myristate acetate (PMA), platelet-activating factor (PAF) and formyl-methionyl-leucyl-phenylalanine (fMLP). As shown in Table 4, PMA (5×10^{-8} M), PAF (5×10^{-7} M) and fMLP (2.5×10^{-7} M) increased by 30 fold, 9 fold and 10 fold, the production of TxA_2, respectively. The production of PGE_2 after stimulation of the cells by PMA, PAF and fMLP did not increase significantly. PMA and fMLP enhanced PGI_2 production by approximately 2 fold, but PAF stimulation did not induce a significant increase.

The 5-lipoxygenase activity was also investigated in these purified guinea pig lung cells. In order to demonstrate 5-lipoxygenase activity, each cell population was stimulated with arachidonic acid in the absence or presence of calcium ionophore. As summarized in Table 5, type II pneumocytes, Clara cells and pulmonary Kurloff cells did not seem to express 5-lipoxygenase activity as determined by RP-HPLC (data not shown).

Table 5.   5-lipoxygenase and LTA_4 hydrolase activities in purified guinea pig lung cells.

	5-lipoxygenase activity	LTA_4 epoxide hydrolase activity	other products
Type II pneumocyte	-	+	LTC_4, LTD_4 in small amount
Clara cell	-	+	LTC_4, LTD_4 in small amount
Kurloff cell	-	+	-
Eosinophil	LTB_4	+	LTB_4 metabolites 5-HETE
Macrophage	LTB_4	+	5-HETE

Positive and negative activities were mentioned as (+) and (-), respectively.

297

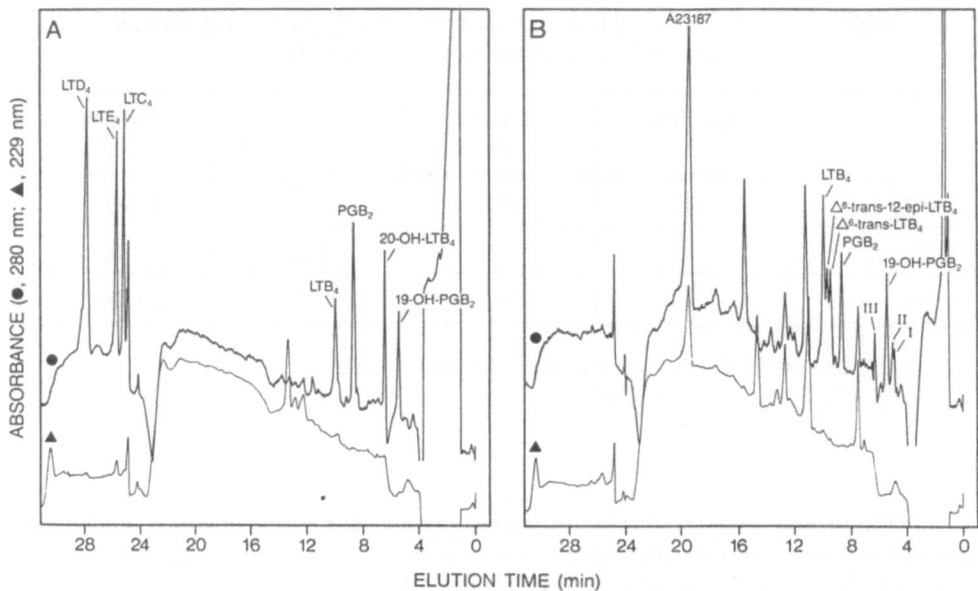

Figure 2. RP-HPLC chromatograms of arachidonic acid metabolites released by guinea pig alveolar eosinophils stimulated with 10 μM arachidonic acid and 2 μM ionophore (B). Chromatogram A shows the separation and retention times of several lipoxygenase products and the two internal standards PGB$_2$ and 19-OH-PGB$_2$.

However, alveolar macrophages and eosinophils were able to convert exogenous arachidonic acid into LTB$_4$ whereas no peptido-leukotrienes were observed, as determined by RP-HPLC (data not shown). Guinea pig alveolar eosinophils showed a peculiar profile of lipoxygenase products. As shown in Figure 2, stimulation of eosinophils with arachidonic acid and ionophore A23187 produced three new compounds I, II and III. These compounds appeared to be LTB$_4$ metabolites since addition of exogenous LTB$_4$ to eosinophil suspensions leads to a time-dependent disappearance of LTB$_4$ and to the concomitant formation of compounds I, II and III (data not shown). The UV spectra of the three compounds clearly indicated that all compounds carried a UV chronophore identical to that of LTB$_4$ (data not shown) and thus, confirmed their biochemical origin. Type II pneumocytes, Clara cells, and pulmonary Kurloff cells were also incubated with exogenous leukotriene A$_4$ (LTA$_4$) and it was observed that LTB$_4$ was formed by the three cell types, as determined by RP-HPLC (data not shown). A little amount of leukotriene C$_4$ and D$_4$ was found in type II pneumocytes and Clara cells incubates when the cell number was increased. The synthesis of these peptido-leukotrienes could be associated to a weak gluthatione-S-transferase activity, or more likely to contaminating cells.

Transcellular metabolism of arachidonic acid

Transcellular metabolism of arachidonic acid has been studied by measuring the synthesis of TxA$_2$ by cells in co-cultures. Interactions were studied between blood platelets and alveolar macrophages or eosinophils. As shown in Table 6, incubations of alveolar macrophages (2 x 10^6 cells/ml) with various ratios of platelets (macrophages: platelets) 2:100, 1:100 and 1:200 increased TxA$_2$ synthesis by 32, 35 and 61 %,

Table 6.   Synthesis of TxA$_2$ following incubation of a fixed concentration of alveolar cells and various concentrations of platelets.

	C	P1	2:100	P2	1:100	P4	1:200
Macrophages	4.58	1.37	24.55	4.25	11.94	31.93	58.71
Eosinophils	1.46	5.34	28.27	10.77	37.67	72.14	95.22

TxA$_2$ was measured in ng/10^6 cells.
C: ionophore-stimulated alveolar cells alone (2 x 10^6 cells/ml).
P1, P2 and P4 correspond to 1, 2 and 4 x 10^8 platelets/ml, respectively, stimulated with ionophore.
2:100, 1:100 and 1:200 correspond to cell ratios (cells:platelets).

Table 7.   Synthesis of TxA$_2$ following incubation of a fixed concentration of platelets with various concentrations of alveolar cells.

	P	C1	100:1	C2	100:2	C6	100:6
Macrophages	4.25	2.62	10.85	4.58	11.95	34.31	59.86
Eosinophils	10.77	0.58	25.92	1.46	37.67	5.87	61.82

TxA$_2$ was measured in ng/10^6 cells.
P: ionophore-stimulated platelets alone (1 x 10^8 cells/ml).
C1, C2 and C6 correspond to 1, 2 and 6 x 10^6 alveolar cells/ml, respectively, stimulated with ionophore.
100:1, 100:2 and 100:6 correspond to cell ratios (platelets:cells).

respectively, in comparison to the sum of TxA$_2$ release by each stimulated cell population alone.   When platelets (1 x 10^8 cells/ml) were incubated with various ratios of macrophages (platelets:macrophages) 100:1, 100:2 and 100:6, TxA$_2$ synthesis was enhanced by 58, 35 and 55 %, respectively, in comparison to the sum of separate stimulated cells (Table 7). In similar experiments, incubations of alveolar eosinophils (2 x 10^6 cells/ml) with various ratios of platelets (eosinophils:platelets) 2:100, 1:100 and 1:200 increased TxA$_2$ synthesis by 315, 208 and 29 % respectively, in comparison to the sum of stimulated cell alone (Table 6). Incubations of platelets (1 x 10^8 cells/ml) with various ratios of eosinophils (platelets:eosinophils) 100:1, 100:2 and 100:6 increased TxA$_2$ synthesis by 305, 208 and 272 % respectively, in comparison to the sum of separate stimulated cells.

DISCUSSION

The enzymatic digestion of one guinea pig lung with low concentration of protease have allowed to obtain a large number of cells (650 x 10^6 cells) with a good viability (75 %). After centrifugal elutriation, eight fractions were collected which contained cell populations enriched in type II pneumocytes (Fraction 3), Clara cells (Fractions 6 and 7) and Kurloff cells (Fraction 8). Type II pneumocytes were collected at a flow of 18 ml/min which is similar to the flow used for the purification of type II pneumocytes of other species[39,40]. The guinea pig type II pneumocytes were purified according to the technique previously described by Dobbs et al.[35]. Differential adherence allowed to eliminate cells with Fc fragment receptors and to further purify up to 15 x 10^6 type II pneumocytes with a purity of about 92 %. The contaminating cells (below 8 %) have not been identified yet; electron microscopic studies of these cells did not provide characteristic features useful for their identification. The purity of type II pneumocytes was also measured using the alkaline phosphatase biochemical stain (data not shown) a specific marker of alveolar type II cell[41]. Clara cells were purified from fractions 6 and 7 of centrifugal elutriation. After a second elutriation followed by elimination of contaminating adherent cells by adherence, and centrifugation on discontinuous Percoll gradients, about 2 x 10^6 Clara cells were obtained with an approximative purity of 72 %. The cell preparation contained small percentages of endothelial cells, type II pneumocytes and macrophages. The purity of cell preparation was comparable to that reported by several groups in other mammal species[42-44]. Kurloff cells are mononuclear cells found exclusively in guinea pig and characterized by a large inclusion body identified as a mucoprotein-sulphated mucopolysaccharide complex[45] synthetized by the Kurloff cell itself[46]. Kurloff cells were found in the bone marrow, in the peripheral blood, the vascular compartment of the splenic red pulp, the septal capillaries of the lungs and the sinusoid capillaries of the liver[47]. The number of Kurloff cells is increased in female guinea pig during pregnancy, and in both sexes, by estrogen administration[47]. Kurloff cells possess a cytotoxic activity against natural killer-sensitive target cell lines[48-50]. The organ distribution of Kurloff cells strongly correlates with the distribution of natural killer (NK) activity in the guinea pig as well as in human[51]. More recently, Sewell et al.[52] demonstrated that the immunophenotype of Kurloff cells were analogous to human NK cells. With regards to other similarities between Kurloff cells and NK cell of different mammal species, we postulate that the enigmatic Kurloff cell correspond to an important NK cell line in guinea pigs. The pulmonary Kurloff cells were purified from estradiol-treated animals from fraction 8 of the centrifugal elutriation and purified on continuous gradient centrifugation. A high-density (1.100 g/ml) Kurloff cell population (about 50 x 10^6 cells) were obtained with a purity close to 100 %. Kurloff cells have previously been isolated from the spleen of estrogen-treated animals[53]. Alveolar macrophages and eosinophils were purified from guinea pig BAL fluid. Our results showed that BAL cells were composed of 9 % eosinophils. In humans, the BAL cell populations of normal subjects contained less than 1 % eosinophils[54] whereas in asthmatic patients there is a substantial increase in the percentage of eosinophils[55]. Alveolar macrophages were easily isolated by adherence with a purity of 99 % and a viability of 98 %. Alveolar eosinophil were purified on discontinuous Percoll gradient centrifugation and recovered at the 60-70 % interface of the Percoll gradient; about 1 x 10^6 cells were recovered per animal and the purity was 96 %.

The arachidonic acid metabolism were assessed in these highly-purified cell populations. The cyclooxygenase pathway was evaluated by estimating the release of TxA$_2$, the PGI$_2$ and the PGE$_2$. Unstimulated type II pneumocytes, Clara cells and Kurloff cells released variable amounts of the three cyclooxygenase products. The calcium ionophore A23187 stimulated the release of TxA_2 by type II pneumocytes and Clara cells and to a smaller extent by Kurloff cells. PGE_2 and PGI_2 were found to be released in appreciable amount by Clara cells. However, the calcium ionophore stimulation produced a small increase of PGI_2 and PGE_2 synthesis by type II pneumocytes and Kurloff cells. Incubation with arachidonic acid (10 μM) stimulates the release of three cyclooxygenase products by type II pneumocytes, Clara cells and Kurloff cells. TxA_2 was the major product released by the three cell types in comparison to prostaglandins. Pulmonary Kurloff cells were also incubated with 30 and 100 μM arachidonic acid. A significant increase of PGE_2 synthesis was observed but no significant augmentation of TxA_2 and PGI_2 productions. These results could be explained by the possible saturation of thromboxane and prostacyclin synthetases by high concentrations of the free endoperoxide substrate. Arachidonic acid metabolism have also been studied by a number of investigators in rat type II pneumocytes. These studies[56-59] showed that PGE_2 and/or PGI_2 were the major arachidonate metabolites synthesized whereas TxA_2 was released in smaller amount. These studies showed that for a similar cell type, various profiles of arachidonic acid metabolism are found in different mammal species. Arachidonate metabolites have been reported by Xu et al.[58] in rat Clara cells. It was found that rat Clara cells produced mostly PGI_2. In contrast, Scott et al.[60] showed that sonicated Clara cells released TxB_2, PGD_2, PGE_2, $PGF_{2\alpha}$, HHT and 12-HETE. At the present time, there is no report on arachidonic acid metabolism by Kurloff cells. The release of arachidonate metabolites by guinea pig type II pneumocytes, Clara cells and Kurloff cells could have significant physiological consequences. Cyclooxygenase metabolites have been demonstrated to modulate surfactant secretion in different species[61,62]. These three cell types could play a role in the secretion of surfactant by generating PGE_2 that could in turn stimulate type II pneumocytes to produce surfactant. PGE_2 and PGI_2 were described as bronchodilatators whereas TxA_2 has potent bronchoconstrictor activity[63]. The formation of these three cyclooxygenase products in vivo by the different pulmonary cells could contribute to the regulation of smooth muscle tonus. The release of potential lipid mediators by the non-inflammatory type II pneumocytes and Clara cells is not clear but suggest that these non-inflammatory cell could be involved in inflammatory processes. The ability of Kurloff cells to produce large amount of PGE_2 and to express NK cell activity seems paradoxical because of its immunosuppressive activity. In fact, it has been showed that PGE_2 induced both a stimulatory or an inhibitory effect on NK cells activity, depending on the dose used and the presence of PGE_2 in the course of the response[64-66]. Release of PGE_2 by Kurloff cells could be associated to an auto-regulatory process in which NK activity would be stimulated or repressed depending on the biological stimuli. Purified lung eosinophils produced TxA_2 predominantly and small amount of PGI_2 but produced no PGE_2 after stimulation with PMA, fMLP and PAF. However, small amounts of cyclooxygenase metabolites are released by human eosinophils[67] reflecting mammal species differences. The release of inflammatory mediators by guinea pig eosinophils support their role not only in asthma, but in various inflammatory processes.

Our results also showed that type II pneumocytes, Clara cells and Kurloff cells did not seem to express 5-lipoxygenase activity. However, these three pulmonary cells seem to possess the LTA_4 epoxide hydrolase since they converted exogenous LTA_4 into LTB_4. The identification of LTB_4 was confirmed by on-line UV spectroscopic analysis of the compound using a high resolution photodiode array detector permitting to discriminate between LTB_4 and its various geometrical isomers (data not shown). The inability of guinea pig type II pneumocytes, Clara cells and Kurloff cells to express 5-lipoxygenase activity could either be explained by the absence of the 5-lipoxygenase or the absence of the 5-lipoxygenase activating protein (FLAP)[68,69] necessary for leukotriene biosynthesis. However, the presence of LTA_4 epoxide hydrolase may suggest that type II pneumocytes,

Clara cells and Kurloff cells could participate in the formation of LTB_4 in the lungs following a process involving cell-cell interactions. This transfer of LTA_4 could occur from another cell type containing active 5-lipoxygenase, as described between platelets and neutrophils[30] or endothelial cells and neutrophils[70]. Such a process of transcellular metabolism between these three pulmonary cells and others guinea pig lung cells could influence a number of immune responses such as chemotaxis and chemokinesis[71], stimulation of suppressor lymphocytes[72] or augmentation of NK cells activity[73] that are stimulated by LTB_4. Alveolar macrophages and eosinophils both produced LTB_4 and 5-HETE but the peptidoleukotrienes were not formed. These results could not be explained by their catabolism in peptidoleukotriene sulfoxides (data not shown). In the eosinophil, LTB_4 appeared to be metabolized into three compounds (I, II and III) which carried a UV chromophore identical to that of LTB_4 but did not correspond to ω-oxidation products (data not shown). The 5-lipoxygenase pathway in guinea pig eosinophils strikingly contrast with that of human eosinophils which almost exclusively produced leukotriene C_4[74].

The transcellular formation of cyclooxygenase products was studied in guinea pig blood platelets and alveolar macrophages or alveolar eosinophils. The stimulation of a fixed amount of platelets with various concentrations of macrophages or eosinophils with calcium ionophore A23187 have produced a significant increase of TxA_2 synthesis compared to the sum of TxA_2 release by individual cell type. Similar results were obtained when a fixed amount of alveolar macrophages or eosinophils were incubated with various concentrations of platelets in presence of calcium ionophore. The increase of TxA_2 production through the transcellular metabolism of arachidonic acid in guinea pig cells could be explained by: 1) the release of large amounts of endoperoxides which in turn may be used by the other cell types[75,76] or 2) the release of eicosanoid or an unidentified compound which could act as a stimulus in eicosanoid production[77]. These two cases correspond to type IA and type III cell-cell interactions according to the classification of cell-cell interactions involving eicosanoid precursors and intermediates proposed by Marcus[78]. Studies in progress should better define the mechanism of platelet-alveolar cell interactions and investigate transcellular metabolism of arachidonic acid in non-immune vs immune cell interactions.

ACKNOWLEDGEMENTS

The authors would like to thank Ms. Solange Cloutier and Mr. Serge Picard for technical assistance, Ms. Carmen Labrecque for secretarial assistance, and the Medical Research Council of Canada for support. P.B. and P.S. are in receipt of a "Fonds de la Recherche en Santé" scholarship and a MRC Scientist Award, respectively. C.R. and J.C. are in receipt of "Fonds de la Recherche en Santé" studentships and A.H. is supported by the George Phénix Foundation.

REFERENCES

1. J. Harkavy, Spasm-producing substance in the sputum of patients with bronchial asthma, Arch. Int. Med., 45: 641 (1930).
2. U.S. Von Euler, On the specific vaso-dilating and plain muscle stimulating substances from accessory genital glands in man and certain animals (prostaglandin and vesiglandin), J. Physiol., 88: 213 (1936).

3. M.W. Goldblatt, Properties of human seminal plasma, J. Physiol. (London), 84: 208 (1935).

4. S. Bergström, and J. Sjövall, The isolation of prostaglandin E from sheep prostate glands, Acta Chem. Scand., 14: 1701 (1960).

5. S. Bergström, and J. Sjövall, The isolation of prostaglandin F from sheep prostate glands, Acta Chem. Scand., 14: 1693 (1960).

6. S. Bergström, R. Ryhage, B. Samuelsson, and J. Sjövall, Prostaglandins and related factors. The structures of prostaglandins E_1, $F_{1\alpha}$ and $F_{1\beta}$, J. Biol. Chem., 238: 3555 (1963).

7. P.J. Piper, and J.R. Vane, Release of additional factors in anaphylaxis and its antagonism by anti-inflammatory drugs, Nature (London), 223: 29 (1969).

8. M. Hamberg, J. Svensson and B. Samuelsson, Thromboxanes: A new group of biologically active compounds derived from prostaglandin endoperoxides, Proc. Natl. Acad. Sci. USA, 71: 345 (1975).

9. W. Feldberg, and C.H. Kellaway, Liberation of histamine and formation of lysocithin-like substances by cobra venom, J. Physiol., 94: 187 (1938).

10. H.R. Morris, G.W. Taylor, P.J. Piper, P. Sirois, and J.R. Tippins, Slow-reacting substance of anaphylaxis: purification and characterization, FEBS Lett., 87: 203 (1978).

11. R.C. Murphy, S. Hammarström, and B. Samuelsson, Leukotriene C: a slow-reacting substance from murine mastocytoma cells, Proc. Natl. Acad. Sci. USA, 76: 4275 (1979).

12. H.R. Morris, G.W. Taylor, P.J. Piper, M.N. Samhoun, and J.R. Tippins, Slow reacting substances (SRS_s): the structure identification of SRS_s from rat basophilic leukemia (RBL-1) cells, Prostaglandins, 19: 185 (1980).

13. L. Örning, S. Harmmarström, and B. Samuelsson, Leukotriene D_4: A slow reacting substance from rat basophilic leukemia cells, Proc. Natl. Acad. Sci. USA, 77: 2014 (1980).

14. C.W. Parker, S.F. Falkenhein, and M.M. Huber, Sequential conversion of the gluthationyl side chain of slow reacting substance (SRS) to cysteinyl-glycine and cysteine in rat basophilic leukemia cells stimulated with A23187, Prostaglandins, 20: 863 (1980).

15. H.R. Morris, G.W. Taylor, P.J. Piper, and J.R. Tippins, Structure of slow-reacting substance of anaphylaxis from guinea-pig lung, Nature, 285: 104 (1980).

16. L. Örning, Bernström, K., and S. Harmmarström, Formation of leukotrienes E_3, E_4 and E_5 in rat basophilic leukemia cells. Eur. J. Biochem., 120: 41 (1981).

17. P. Borgeat, and B. Samuelsson, Transformation of arachidonic acid by rabbit polymorphonuclear leukocytes. Formation of a novel dihydroxy acid, Proc. Natl. Acad. Sci. USA, 76: 3213 (1979).

18. P. Borgeat, and B. Samuelsson, Arachidonic acid metabolism in polymorphonuclear leukocytes: Effects of ionophore A23187, Proc. Natl. Acad. Sci. USA, 76: 2148 (1979).

19. J.M. Drazen, Inhalation challenge with sulfidopeptide leukotrienes in human subjects, Chest, 89: 414 (1986).

20. M. Griffin, J.W. Weiss, A.G. Leitch, E.R.J. McFadden, E.J. Corey, K.F. Austen and J.M. Drazen, Effects of leukotriene D on the airways in asthma, N. Engl. J. Med., 308: 436 (1983).

21. S.E. Dahlén, B. Dahlén, E. Eliasson, H. Johansson, T. Björck, M. Kumlin, K. Boo, J. Whitney, S. Binks, B. King, R. Stark, and O. Zetterström, Inhibition of allergic bronchoconstriction in asthmatics by the leukotriene-antagonist ICI-204, 219, Adv. Prostaglandin Thromboxane Leukotriene Res., 21: 461 (1990).

22. A.W. Ford-Hutchinson, Leukotriene B_4 in inflammation, Immunology, 10: 1 (1990).

23. P. Sirois, P. Borgeat, A. Jeanson, S. Roy, and G. Girard, The action of leukotriene B_4 (LTB$_4$) on the lung, Prostaglandins Med., 5: 429 (1980).

24. P. Sirois, S. Roy, P. Borgeat, S. Picard, and P. Vallerand, Evidence for a mediator role of thromboxane A_2 in the mytropic action of leukotriene B_4 on the guinea pig lung, Prostaglandins Med., 8: 157 (1982).

25. P.J. Piper, and M.N. Samhoun, Stimulation of arachidonic acid metabolism and generation of thromboxane A_2 by leukotrienes B_4, C_4 and D_4 in guinea-pig lung in vitro, Br. J. Pharmacol., 77: 267 (1982).

26. T. Ishisaka, and K. Ishisaka, Activation of mast cells for mediator release through IgE receptors, Prog. Allergy, 34: 188 (1984).

27. E. Frigas, and G.J. Gleich, The eosinophil and the pathophysiology of asthma, J. Allergy Clin. Immunol., 77: 527 (1986).

28. J. McGee, and F.A. Fitzpatrick, Erythrocyte-neutrophil interactions: Formation of leukotriene B_4 by transcellular biosynthesis, Proc. Natl. Acad. Sci. USA, 83: 1349 (1986).

29. H.E. Claesson, and J. Haeggström, Human endothelial cells stimulate leukotriene synthesis and convert granulocyte released leukotriene A_4 into leukotriene B_4, C_4, D_4 and E_4, Eur. J. Biochem., 173: 93 (1988).

30. A.J. Marcus, M.J. Brockman, .B. Safier, H.L. Ullman, K.J. Islam, C.N. Serhan, L.E. Rutherford, H.M. Korchak, and G. Weissman, Formation of leukotrienes and other hydroxy acids during platelet-neutrophils interaction in vivo, Biochem. Biophys. Res. Commun., 109: 130 (1982).

31. J.P. Pelé, C. Robidoux, and P. Sirois, Guinea pig lung cells. Method of isolation and partial purification, identification, ultrastructure and cell count, Inflammation, 13: 103 (1989).

32. K. Maghni, C. Robidoux, J. Laporte, A. Hallée, and P. Sirois, Release of prostaglandins and thromboxanes by guinea pig isolated type II pneumocytes, Prostaglandins, 40: 217 (1990).

33. J. Laporte, A. Hallée, K. Maghni, C. Robidoux, P. Borgeat, and P. Sirois, Metabolism of arachidonic acid by guinea pig Clara cells, Prostaglandins, 41: 263, (1991).

34. K. Maghni, C. Robidoux, J. Laporte, A. Hallée, P. Borgeat, and P. Sirois, P., Purification of natural killer-like Kurloff cells and arachidonic acid metabolism, (submitted).

35. L.G. Dobbs, R. Gonzalez, and M.C. Williams, An improved method for isolating type II cells in high yield and purity, Am. Rev. Respir. Dis., 134: 141, (1986).

36. K. Hirata, K. Maghni, P. Borgeat, and P. Sirois, Guinea pig alveolar eosinophils and macrophages produce leukotriene B_4 but no peptido-leukotriene, J. Immunol., 144: 1880, (1990).

37. P. Borgeat, S. Picard, P. Vallerand, S. Bourgoin, A. Odeimat, P. Sirois, and P.E. Poubelle, 1990, Automated on-line extraction and profiling of lipoxygenae products of arachidonic acid by high performance liquid chromatography, in: "Methods in Enzymology. Arachidonate related lipid mediators", R.C. Murphy and F. Fitzpatrick eds, Academic Press, New York.

38. P. Pradelles, J. Grassi, and J. Maclouf, Enzyme immunoassay of eicosanoids using acetylcholine esterase as label: An alternative to radioimmunoassay, Anal. Chem., 57: 1170, (1985).

39. V. Castronova, G.S. Jones, and P.R. Miles, Transmembrane potential of isolated rat alveolar type II cells, J. Appl. Physiol., 54: 1511, (1983).

40. T.R. Devereux, and J.R. Fouts, Xenobiotic metabolism by alveolar type II isolated from rabbit lung, Biochem. Pharmacol., 30: 1231 (1981).

41. J.D. Edelson, J.M. Shannon, and R.J. Mason, alkaline phosphatase: a marker of alveolar type II cell differentiation, Am. Rev. Respir. Dis., 138: 1268 (1988).

42. T.R. Devereux, and J.R. Fouts, Isolation and identification of Clara cells from rabbit lung, In vitro, 16: 958 (1980).

43. T.E. Massey, B.A. Geddes, and P.G. Forkert, Isolation of non ciliated bronchiolar epithelial (Clara) cells and alveolar type II cells from mouse lungs, Can. J. Physiol. Pharmacol., 65: 2368 (1987).

44. J.E. Myles, B.A. Geddes, and T.E. Massey, Biotransformation activities in Clara and alveolar type II cells isolated from hamster lungs, Res. Commun. Chem. Path. Pharmacol, 66: 297 (1989).

45. M.F. Dean, and H. Muir, The characterization of a protein-polysaccharide isolated from Kurloff cells of the guinea pig, Biochem. J., 118: 783 (1970).

46. G. Landemore, S.E. Letaïef, J. Bocquet, and J. Izard, Kurloff cell proteoglycans. Evidence de novo synthesis of chondroitin sulfate proteoglycans by purified Kurloff cell, Febs Letters, 209: 299 (1986).

47. P.A. Rewell, The Kurloff cell, Intern. Rev. Cytol., 51: 275 (1977).

48. O. Eremin, R.R.A. Coombs, J. Ashby, and D. Plumb, Natural cytotoxicity in the guinea-pig: the natural killer cell activity of the Kurloff cell. Immunology, 41: 367 (1980).

49. C. Debout, M. Quillec, and J. Izard, Natural killer activity of Kurloff cell: a direct demonstration on purified cell suspensions, Cell. Immunol., 87: 674 (1984).

50. N. Pouliot, K. Maghni, P. Sirois, and M. Rola-Pleszczynski, The cytotoxic activity of the Foà-Kurloff cell, FASEB J., 4: A1892 (1990).

51. A. Attman, and M.J. Rapp, Natural cell-mediated cytotoxicity in guinea pigs: properties and specificity of natural killer cells, J. Immunol., 121: 2244 (1978).

52. H.F. Sewell, L.S. Steward, C.A. McPhee, I.H. Mathie, and A.W. Thomson, Enhanced production and immunophenotypic analysis of Kurloff cells in immunized guinea pigs treated with cyclophosphamide and cyclosporine A: correlation with increased large granular lymphocyte production in rat, Transpl. Proc., 20 (suppl. 2): 196 (1988).

53. G. Landemore, C. Debout, M. Quillec, and J. Izard, Isolation of Kurloff cells by Percoll density-gradient centrifugation. Protein labelling with ^{35}S-methionine of these cells, Biol. Cell., 50: 121 (1984).

54. W.B. Davis, G.A. Fells, X.H. Wum, J.E. Fradek, A. Venet, and R.G. Crystal, Eosinophil-mediated injury possible role for eosinophils in chronic inflammatory disorders of the lower respiratory tract, J. Clin. Invest., 74: 269 (1984).

55. J.G.R.D. Moncky, H.F. Kauffman, P. Venge, G.H. Koeter, HL.M. Jansen, H.J. Sheiter, and K.D. Vries, Bronchoalveolar eosinophilia during allergen-induced late asthmatic reactions, Am. Rev. Respir. Dis., 131: 373 (1985).

56. L. Taylor, P. Polgar, J.A. McAteer, and W.H.J. Douglas, Prostaglandin production by type II alveolar epithelial cells, Biochem. Biophys. Acta, 572: 502 (1979).

57. J.E. Graeber, R.W. Walenga, R.E. Ulane, and M.J. Stuart, Metabolism of ^{14}C-arachidonic acid by type II alveolar epithelial cells in primary culture, Pediat. Res., 16: 350A (1982).

58. G.L. Xu, K. Sivarajak, R. Wu, P. Nettesheim, and T. Eling, Biosynthesis of prostaglandins by isolated and cultured airway epithelial cells, Exp. Lung Res., 10: 101 (1986).

59. J.B. Chauncey, M. Peters-Golden, and R.H. Simon, Arachidonic acid metabolism by rat alveolar epithelial cells, Lab. Invest., 58: 133 (1988).

60. M.R. Van Scott, M.R. McIntire, and D.C. Henke, Arachidonic acid metabolism and regulation of ion transport in rabbit Clara cells, Am. J. Physiol., 259: L213 (1990).

61. A.M. Gilifillan, and S.A. Rooney, Arachidonic acid metabolites stimulate phosphophatidylcholine secretion in primary culture of type II pneumocytes, Biochem. Biophys. Acta, 833: 336 (1985).

62. M.J. Oyrazun, and J.A. Clements, Control of lung surfactant by ventilation, adrenergic mediators, and prostaglandins in the rabbit, Am. Rev. Respir. Dis., 117: 879 (1978).

63. J. Nowak, Eicosanoids and the lungs, Ann. Clin. Res., 16: 269 (1984).

64. M.J. Droller, M.V. Schneider, and P.A. Perlinau, A possible role of prostaglandins in the inhibition of natural and antibody-dependent cell-mediated cytotoxicity against tumor cells, Cell. Immunol., 39: 165 (1978).

65. M.J. Brunda, R.B. Herberman, and M.T. Holden, Inhibition of murine natural killer cell activity by prostaglandins, J. Immunol, 124: 2682 (1980).

66. B. Hacker-Shahin, and W. Droge, Augmentation of cytotoxic responses by prostaglandin E_2, Cell. Immunol., 91: 43 (1985).

67. W.R. Henderson, Eicosanoids and lung inflammation, Am. Rev. Respir. Dis., 135: 1176 (1987).

68. D.K. Miller, J.W. Gillard, P.J. Vickers, S. Sadowski, C. Léveillé, J.A. Mancini, P. Charleson, R.A.F. Dixon, A.W. Ford-Hutchinson, R. Fortin, J.Y. Gauthier, J. Rodkey, R. Rosen, C. Rouzer, I.S. Sigal, C.D. Starder, and J.F. Evans, Identification and isolation of a membrane protein necessary for leukotriene production, Nature, 343: 278 (1990).

69. R.A.F. Dixon, R.E. Diehl, E. Opas, E. Rands, P.J. Vickers, J.F. Evans, J.W. Gillard and D.K. Miller, Requirement of 5-lipoxygenase-activating protein for leukotriene synthesis, Nature, 343: 282, (1990).

70. J.J. Feinmark, and J.J. Cannon, Endothelial cell leukotriene C_4 synthesis results from intercellular transfer of leukotriene A_4 synthesized by polymorphonuclear leukocytes, J. Biol. Chem., 261: 16466 (1986).

71. A.W. Ford-Hutchinson, M.A. Bray, M.V. Doig, M.E. Shipley, and M.J.H. Smith, Leukotriene B, a potent chemokinetic and aggregating substance released from polymorphonuclear leukocytes, Nature, 286: 264 (1980).

72. M. Rola-Pleszczynski, P. Borgeat, and P. Sirois, Leukotriene B_4, induces human suppressor lymphocytes, Biochem. Biophys. Res. Commun., 108: 1531 (1982).

73. M. Rola-Pleszczynski, L. Gagnon, and P. Sirois, Leukotriene B_4 augments human natural cytotoxic cell activity, Biochem. Biophys. Res. Commun., 113: 531 (1983).

74. M. Laviolette, S. Picard, P. Braquet, and P. Borgeat, Comparison of 5- and 15-lipoxygenase activities in blood and alveolar leukocyte preparations from normal subjects and patients with eosinophilia, Prostaglandins Leukotrienes Med., 23: 191 (1986).

75. A.J. Marcus, B.B. Weksler, and E.A. Jaffe, Synthesis of prostacyclin from platelet-derived endoperoxides by cultured human endothelial cells, J. Clin. Invest., 66: 979 (1980).

76. A.I. Schafer, D.D. Crawford, and M.A. Jr. Gimbrone, Unidirectional transfer of prostaglandin endoperoxides between platelets and endothelial cells, J. Clin. Invest., 73: 1105 (1984).

77. B.A. Burall, and E.J. Goetzl, Navigating the sea of eicosanoids (Editorial), West. J. Med., 143: 516 (1985).

78. A.J. Marcus, Transcellular metabolism of eicosanoids, Prog. Hemost. Thrombo., 28: 124 (1986).

THE ROLE OF LEUKOTRIENE A₄ HYDROLASE IN

CELLS AND TISSUES LACKING 5-LIPOXYGENASE

Hans-Erik Claesson, Jesper Z. Haeggström, Björn Odlander, Juan F. Medina, Anders Wetterholm, Per-Johan Jakobsson, Olof Rådmark

Department of Physiological Chemistry II
Karolinska Institutet, Box 60400
S-104 01 Stockholm, Sweden

Leukotriene (LT) A4 hydrolase converts the unstable epoxide intermediate LTA4 into the potent proinflammatory compound LTB4. The formation of LTA4 is catalyzed by the enzyme 5-lipoxygenase and involves the dioxygenation of arachidonic acid with subsequent epoxide formation (1).

Polymorphonuclear leukocytes (PMNL), monocytes / macrophages and mast cells have the enzyme 5-lipoxygenase (2-6). With the possible exception of mast cells, these cells also have the enzyme LTA4 hydrolase, and can thus produce LTB4. Later studies have shown that LTA4 hydrolase is present in a variety of other cells and tissues. Some of these cells apparently lack 5-lipoxygenase activity and thus the ability to synthesize the substrate for LTA4 hydrolase. These seemingly inconsistent findings may be explained by so called "transcellular metabolism", a phenomenon described by several investigators. This term signifies the transfer of a compound, e.g. LTA4, from a donor cell to a recipient cell, where further metabolism may occur.

Below we present some data which further support the hypothesis of a role for transcellular metabolism in leukotriene biosynthesis and metabolism. These involve the export of LTA4 from PMNL and monocytes stimulated with ionophore A23187, and the further metabolism of this intermediate in endothelial cells and lymphocytes. In the subsequent text, the wide distribution of LTA4 hydrolase, and its relevance for transcellular metabolism of LTA4, is described. Finally, the possible implications of some recent findings regarding the stimuli required for release of LTA4 from leukocytes, as well as the structure/function of LTA4 hydrolase, are discussed.

Cell-Cell Interactions in the Release of Inflammatory Mediatiors
Edited by P. Y-K Wong and C.N. Serhan, Plenum Press, New York, 1991

307

Leukotriene A4 hydrolase and leukotriene C4 synthase in human endothelial cells

Human endothelial cells (EC) were isolated from normal umbilical cord veins. Stimulation of EC with the calcium ionophore A23187 did not lead to the formation of leukotrienes, indicating that these cells are devoid of 5-lipoxygenase activity (7,8). However, incubation of EC, that had been cultivated for 10-16 days after isolation, with synthetic LTA4 gave LTB4, LTC4, LTD4 and LTE4. Since PMNL activated with calcium ionophore, release LTA4 (9), we investigated if activated PMNL could supply EC with LTA4. When PMNL were incubated together with EC in the presence of ionophore A23187, a marked increase in the formation of LTB4 (52 %) and cysteinyl-containing leukotrienes (364 %) were observed as compared to the amounts produced by PMNL alone (Fig.1). Leukotrienes D4 and E4 were only detected in the presence of EC. To demonstrate translocation of LTA4 from activated PMNL to EC, cultures of EC were preincubated with [^{35}S]-cysteine to allow for synthesis of [^{35}S]-glutathione, prior to co-incubation with PMNL (stimulated with the calcium ionophore). The cysteinyl-containing leukotrienes that were produced contained [^{35}S]. These results not only demonstrated that transfer of LTA4 from stimulated PMNL to EC occurred, but also that the entire surplus of cysteinyl-containing leukotrienes in PMNL-EC cultures was produced by EC, via metabolism of PMNL-released LTA4. Experiments with EC in which the LTA4 hydrolase had been partially inactivated, indicated that the same mechanism was operative, also regarding the increase in LTB4 synthesis in PMNL/EC cocultures (8). In addition, PMNL-EC interactions resulted in an increased formation of the total amounts of leukotrienes, suggesting a stimulation of the 5-lipoxygenase in PMNL, induced by EC. The transcellular metabolism of PMNL-released LTA4 into LTC4 by endothelial cells has also been demonstrated with porcine endothelial cells (10).

Since monocytes adhere more efficiently than granulocytes to EC, we also investigated the effects of monocyte-EC interactions on leukotriene synthesis. Human monocytes were shown to release LTA4 extracellularly after activation with the calcium ionophore A23187 (11). Incubation of monocytes together with EC, in the presence of the calcium ionophore, lead to an increased formation of LTB4 and cysteinyl-containing leukotrienes (12). The increase in leukotriene formation in monocyte-EC cultures was similar to that found for PMNL-EC cultures.

Subcellular fractionation of EC showed that the formation of LTC4 from LTA4 was only observed in the particulate fraction and was separable from classical S-transferase activity, indicating the presence of a specific LTC4 synthase in EC (12). The identity of LTA4 hydrolase in EC was ascertained by comparison with purified leukocyte LTA4 hydrolase on SDS-PAGE followed by Western blot analysis (12).

Leukotriene A4 hydrolase in lymphocytes

Stimulation of purified B and T lymphocytes as well as monoclonal lymphocytic cells

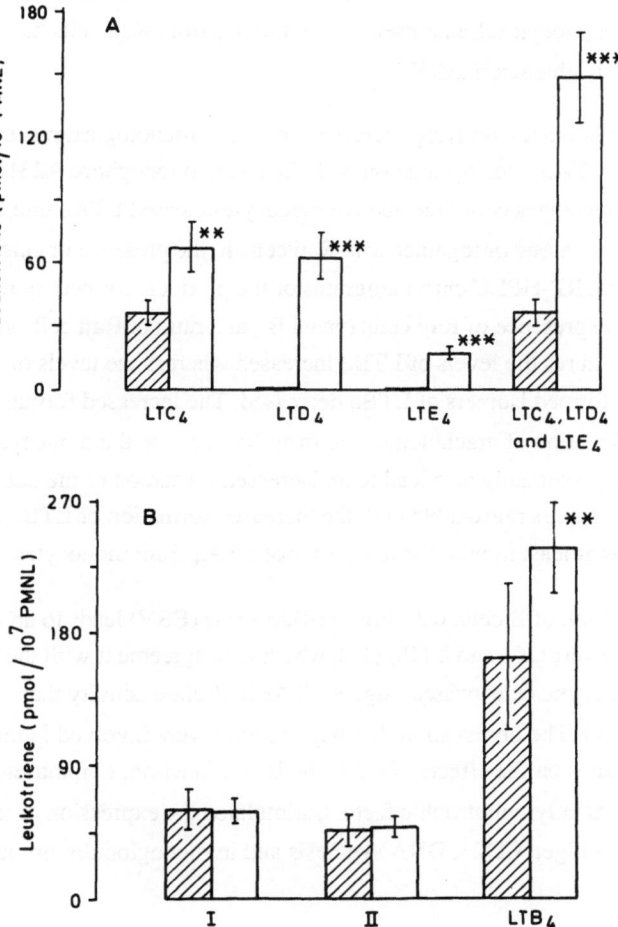

Fig. 1. Effects of PMNL/EC interactions on the formation of (A) cysteinyl-containing leukotrienes and (B) LTB4 and the nonenzymatic isomers of LTB4. PMNL were incubated either alone or in the presence of EC with the ionophore A23187 for 10 min. The products were quantified by HPLC; each bar represents mean±SE, n=11. Shaded bars depict the products formed by PMNL alone and open bars the compounds formed by PMNL/EC cultures. Roman numbers I and II are 6-trans-LTB4 and 12-epi-6-trans-LTB4, respectively.Statistical significance of differences in leukotrienes formation between PMNL and PMNL/EC cultures was estimated by using Student's paired t-test: **0.01 > P > 0.001 ; *** P< 0.001

with the calcium ionophore A23187 did not lead to the synthesis of any detectable amounts of leukotrienes (13-16). However, in incubations of purified lymphocytes (13) or monoclonal lymphocytic cells (13,14) exogenous LTA4 was converted to LTB4. It was found that several monoclonal lymphocytic cells possess higher LTA4 hydrolase activities than normal B and T lymphocytes. Leukotriene A4 hydrolase from Raji cells has recently been purified and partially characterized(17).

Monocytes interact with lymphocytes in certain immunological reactions. Since human monocytes release LTA4 after stimulation with the calcium ionophore A23187 (11), we investigated if lymphocytes could metabolize monocyte-released LTA4 into LTB4. Monocytes were incubated either alone or together with Raji cells in the presence of calcium ionophore. Fig. 2 shows typical RP-HPLC chromatograms of the products formed by monocytes alone (panel A) and in the presence of Raji cells (panel B), at a ratio of Raji cells versus monocytes of 3:1. In mixed cultures the levels of LTB4 increased whereas the levels of the nonenzymatically formed isomers of LTB4 decreased. The increased formation of LTB4 could in part be due to donation of arachidonic acid from Raji cells to the monocytes. However, this mechanism would presumably also lead to an increased formation of the nonenzymatic isomers of LTB4. Thus, it appears reasonable that the increased formation of LTB4 in these coincubations, was at least in part due to transfer of LTA4 from monocytes to Raji cells.

Transformation of B cells with Epstein-Barr virus (EBV) leads to an increased capacity of these cells to convert LTA4 to LTB4 (18), which is in agreement with the earlier finding that monoclonal lymphocytic cells possess higher LTA4 hydrolase activity than normal resting B lymphocytes (13,14). The observation that lymphocytes were involved in the biosynthesis of LTB4 initiated studies on the effects of LTB4 on B cell function. Leukotriene B4 (10^{-10} M), in synergy with certain lymphotrophic factors, stimulated the expression of the activation associated surface antigen CD23, DNA synthesis and immunoglobulin production in B cells (19).

In summary, the results show that lymphocytes, when interacting with ionophore stimulated monocytes, cause an increased formation of LTB4 which might stimulate activation and differentiation of B lymphocytes.

LTA4 hydrolase is a widely distributed enzyme

The first case where LTA4 hydrolase was found outside leukocytes was not in another cell type, but in blood plasma from various species (20). Subsequently, LTA4 hydrolase was also found in erythrocytes (21), which can be efficiently separated from blood leukocytes. This was important for the unambiguous localization of the enzyme to the red cells. Another approach to the problem of defining which cell type harbours LTA4 hydrolase, has been the use of cultured cells. Thus, LTA4 hydrolase was found in cultures of umbilical vein

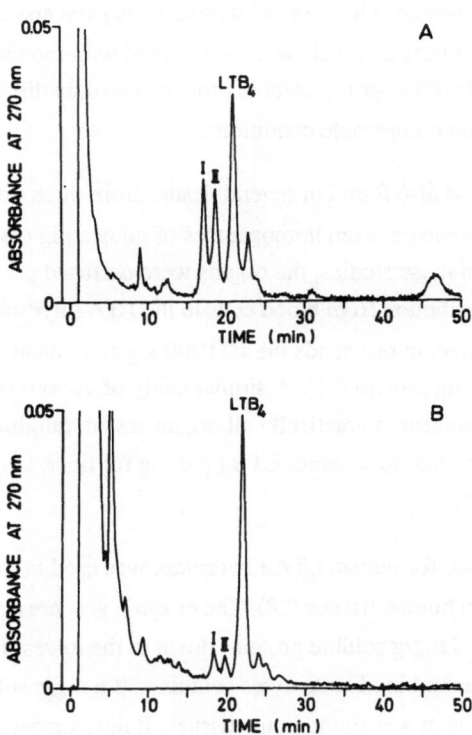

Fig. 2, RP-HPLC chromatograms of the products formed by (A) monocytes (10^7 cells) alone, and (B) monocytes (10^7 cells) plus Raji cells (3×10^7), stimulated with ionophore A23187 (5mM) for 10 min. Peaks designated I and II are 6-trans-LTB4 and 12-epi-6-trans-LTB4, respectively.

endothelial cells (see above), transformed lymphocytes (see above), and lung fibroblasts (22), all of human origin. In the studies on cultured lymphocytes and fibroblasts, the presence of LTA4 hydrolase was demonstrated not only by the conversion of LTA4 to LTB4, also LTA4 hydrolase protein and mRNA were detected in Western and Northern blots, respectively (22,23). The enzyme activities of the transformed lymphocyte cell lines were about 10 -30 pmol LTB4 per 10^6cells, or about 100-300 pmol per mg soluble protein. This can be compared to the activity of human neutrophils (about 15 pmol LTB4 per 10^6cells, 375 pmol LTB4 per mg soluble protein). The LTA4 hydrolase activity of the fibroblasts differed between nontransformed and transformed cells. Normal diploid cells gave about 80 pmol LTB4 per mg of protein. However, fibroblasts that had been transformed with the virus SV-40 had a higher activity, around 240 pmol LTB4 per mg soluble protein. The data discussed above were obtained under different but comparable conditions.

LTA4 hydrolase was also found in several tissues from three different species (guinea pig, rat and human). Supernatants from homogenates of guinea pig organs thus converted LTA4 to LTB4 (24,25). In these studies, the organs were perfused prior to homogenization, in order to minimize the contribution from blood cells to the LTA4 hydrolase activities observed. All organs tested were active, in our hands the 100.000 x g supernatant from guinea pig liver gave 400 pmol LTB4 per mg protein (26). A similar study of various organs from the rat also showed that activity was present in practically all organs tested, ranging from 460 pmol LTB4 per mg soluble protein for lung, to 40 pmol LTB4 per mg for liver. Only for the rat heart, no activity was detectable (27).

A radioimmunoassay for human LTA4 hydrolase was used to measure the LTA4 hydrolase content of seven human tissues (28). The enzyme was present in all tissues with the highest content in lung (1.8 mg/g soluble protein) down to the lowest found for ureter (0.4 mg/g soluble protein). Human blood leukocytes contained 2.6 mg/g soluble protein which is not very much more than what was found in the tissues. It thus appears improbable that all LTA4 hydrolase found for the tissues should stem from contaminating blood (the human tissues could not be perfused). Rather it is reasonable to assume that also resident cells of the human tissues contained LTA4 hydrolase.

Also guinea pig tissues have been studied using immunochemical techniques (29). The quantities of enzyme found (1.6 to 6.5 mg/g) were somewhat higher than those obtained for human tissues (see above). The richest source was the guinea pig intestine (6.5 mg/g) containing almost twice as much hydrolase as guinea pig leukocytes (3.6 mg/g). This could be compared to the human ileum (0.9 mg/g) which thus contained about one third of the amount of hydrolase in human leukocytes (28). Regarding the guinea pig, also immunohistochemical studies were performed, and it could be confirmed that vascular endothelium contains LTA4 hydrolase. In addition, epithelial cells of the airways and the gastrointestinal tract, as well as smooth muscle cells of the bronchi and aorta were identified as novel cell types containing

LTA4 hydrolase. Finally, some cell type of the intestinal nerve plexa was very rich in LTA4 hydrolase.

Taken together, LTA4 hydrolase thus appears to be expressed in many different cell types, i.e. various leukocytes (neutrophils, monocytes/macrophages and lymphocytes), erythrocytes, vascular endothelium, fibroblasts, epithelial cells and smooth muscle cells. Of these cells only neutrophils and monocytes/macrophages are known to express 5-lipoxygenase. The function of LTA4 hydrolase in cells lacking 5-lipoxygenase should thus be to augment the formation of LTB4 around activated leukocytes. Accordingly, this would have to involve transfer of LTA4 between cells, and the diverging distributions of 5- lipoxygenase and LTA4-hydrolase have been taken as evidence for the occurrence of transcellular metabolism of LTA4.

Secretion of LTA4 from PMNL and monocytes

Transcellular metabolism of LTA4 has mostly been demonstrated with the potent ionophore A23187 as stimulus, and could reflect an overflow of LTA4, rather than a physiological process. To get some information on this point we have tried various methods for trapping of extracellular LTA4, released from neutrophils or monocytes. These methods have included trapping with albumin followed by acid methanolysis, trapping with extracellularly added LTA4 hydrolase which converts extracellular LTA4 to LTB4, and trapping with extracellularly added cytosolic epoxide hydrolase which converts extracellular LTA4 to 5,6-dihydroxy- 7,9,11,14-eicosatetraenoic acid (5,6-DHETE). These methods should be able to detect a slow secretion of LTA4 over a longer time period. Using either method we could easily demonstrate the presence of extracellular LTA4 after stimulation with ionophore A23187. However, we have so far not been able to detect extracellular LTA4 after stimulation of human PMNL with fMLP.

Another potential function of LTA4 hydrolase

Recently, a potential zinc binding site was identified in the primary structure of LTA4 hydrolase (30,31). Accordingly, the enzyme was found to contain one zinc atom per enzyme molecule, as determined by atomic absorption spectrometry, and may therefore be classified as a zinc metalloenzyme (32,33). The primary function of zinc in LTA4 hydrolase appeared to be catalytic, as indicated by the pattern of the proposed zinc binding ligands and the intervening sequences (31,32). The predicted zinc binding site of LTA4 hydrolase has a striking similarity to the corresponding primary structures of certain aminopeptidases and neutral proteases typified by thermolysin (31), which cleave peptide bonds. When assayed with synthetic amides, LTA4 hydrolase was also found to possess a peptidase activity (33,34). In view of this finding, i.e. a novel enzymatic activity of the protein, the role of LTA4 hydrolase in biological systems devoid of 5-lipoxygenase activity may have to be reevaluated. Thus it seems possible that LTA4 hydrolase can exert different functions in different cells and tissues, and that this enzyme may operate independent of the availability of LTA4.

Acknowledgement

The work from the authors laboratory was supported by grants from the Swedish Medical Research Council (projects nr. 03X-217, 03X-7135 and 03X-7467), the Swedish Cancer Society (project nr. 2801), and O.E. & Edla Johanssons Foundation.

References

1. Samuelsson, B., Dahlen,S.-E., Lindgren, J.-Å., Rouzer, C.A.,and Serhan, C.N. (1987) Science 237, 1171-1176.
2. Borgeat, P. and Samuelsson, B. (1979) Proc. Natl. Acad. Sci. USA 76, 3213-3217.
3. Jörg, A., Henderson, W.R., Murphy, R.C. and Klebanoff, S.J. (1982) J. Exp. Med. 155, 390-402.
4. Bach, M. K., Brashler, J.R., Hammarström, S. and Samuelsson, B.(1980) J. Immunol. 125, 115-117.
5. Rouzer, C.A., Scott, W.A., Cohn, Z.A., Blackburn, P. and Manning, J.M. (1980) Proc. Natl. Acad. Sci. USA 77, 4928-4932.
6. Razin, E., Mencia-Huerta, J.M., Lewis, R.A., Corey, E.J. and Austen, K.F. (1982) Proc. Natl. Acad. Sci. USA 79, 4665-4667.
7. Claesson, H.-E. and Haeggström, J. (1987) In: Adv Prostaglandin Thromboxane and Leukotriene Res. 17, 115-119.
8. Claesson, H.-E. and Haeggström, J. (1988) Eur. J. Biochem. 173, 93-100.
9. Dahinden, C. A., Clancy, R. M., Gross, M., Chiller, J. M. and Hugli, T. E. (1985) Proc. Natl. Acad. Sci. USA 82, 6632-6636.
10. Feinmark, S. J. and Cannon, P. J. (1986) J. Biol. Chem. 261, 16466-16472.
11. Jakobsson, P.-J., Odlander, O. and Claesson, H.-E. (1991) Eur. J. Biochem. 196, 395-400.
12. Claesson, H.-E., Ax:son Johnson, H., Rådmark, O. and Jakobsson, P.-J. (1990) In: Adv Prostaglandins Thromboxane and Leukotriene Res. 21, 663-666.
13. Odlander, O., Jakobsson, P.-J. and Claesson, H.-E. (1988) Biochem. Biophys. Res. Commun. 153, 203-208.
14. Fu, J.-Y., Medina, J. F., Funk, C. D., Wetterholm, A. and Rådmark, O. (1988) Prostaglandins 36, 241-248.
15. Poubelle, P. A., Borgeat, p. and Rola-Pleszczynski, M. (1987) J. Immunol. 139, 1273-1277.
16. Goldyne, M. and Rea,L. (1987) Prostaglandins 34, 783-795.

17. Odlander, B., Claesson, H.-E., Bergman, T., Rådmark, O., Jörnvall, H. and Haeggström, J. (1991) Arch. Biochem. Biophys. In press.

18. Jakobsson, P.-J., Odlander, B., Yamaoka, K. A., Rosen, A. and Claesson, H.-E. (1990) In: Adv. Prostaglandins Thromboxane and Leukotriene Res. 21, 1005-1012.

19. Yamaoka, K. A., Claesson, H.-E. and Rosen, A. (1989) J. Immunol, 143, 1996-2000.

20. Fitzpatrick, F., Haeggström, J., Granström, E. and Samuelsson, B.(1983) Proc. Natl. Acad. Sci. USA 80, 5425-5429.

21. Fitzpatrick, F, Liggett, W., McGee, J., Bunting, S., Morton, D. and Samuelsson, B. (1984) J. Biol. Chem. 259, 114003-114007.

22. Medina, J.F., Barrios, C., Funk,C.D., Larsson, O., Haeggström, J. and Rådmark, O. (1990) Eur. J. Biochem 191, 27-31.

23. Medina, J.F., Odlander, B., Funk, C.D., Ji-Yi Fu, Claesson, H.-E. and Rådmark, O. (1989) Biochem. Biophys. Res. Commun. 161, 740-745.

24. Haeggström, J., Rådmark, O. and Fitzpatrick, F. (1985) Biochim. Biophys. Acta 835, 378-384.

25. Izumi, T., Shimizu, T., Seyama, Y., Ohishi, N. and Takaku, F. (1986) Biochem. Biophys. Res. Commun. 135, 139-145.

26. Haeggström, J., Bergman, T., Jörnvall, H. and Rådmark, O. (1988) Eur. J. Biochem. 174, 717-724.

27. Medina, J.F., Haeggström, J., Kumlin, M. and Rådmark, O. (1988) Biochim. Biophys. Acta 961, 203-212.

28. Fu, Ji Yi, Haeggström, J., Collins, P. Meijer, J. and Rådmark, O.(1989) Biochim. Biophys. Acta. 1006, 121-126.

29. Ohishi, N., Minami,M., Kobayashi, J., Seyama, Y., Hata, J., Yotsumoto, H., Takaku, F. and Shimizu, T. (1990) J. Biol. Chem.

30. Malfroy, B., Kado-Fong, H., Gros, C., Giros, B., Schwartz, J.-C. and Hellmiss, R. (1989) Biochem. Biophys. Res. Commun. 161, 236-241.

31. Vallee, B.L. and Auld, D.S. (1990) Biochemistry 29, 5647-5659.

32. Haeggström, J.Z., Wetterholm, A., Shapiro, R., Vallee, B.L. and Samuelsson, B. (1990) Biochem. Biophys. Res. Commun. 172, 965-970.

33. Minami, M, Ohishi, N., Mutoh, H., Izumi, T., Bito, H., Wada, H., Seyama, Y., Toh, H. and Shimizu, T. (1990) Biochem. Biophys. Res. Commun. 173, 620-626.

34. Haeggström, J.Z., Wetterholm, A., Vallee, B.L. and Samuelsson, B. (1990) Biochem. Biophys. Res. Commun. 173, 431-437.

KERATINOCYTES CAN REGULATE PROSTAGLANDIN SYNTHESIS

BY FIBROBLASTS: POTENTIAL ROLE FOR INTERLEUKIN 1

Marc E. Goldyne, Kerry L. Blacker, and Mary L. Williams

Departments of Dermatology, Medicine and Pediatrics
Veterans Administration Medical Center and University
of California, San Francisco, CA. 94121

INTRODUCTION

Skin can be considered as a two-compartment tissue consisting of an overlying epidermis and an underlying dermis separated by a selectively permeable basement membrane. Studies on the communication between these skin compartments have shown that the keratinocyte, which is the constitutive cell of the epidermis appears susceptible to modulation by dermally derived influences. In tissue recombinant experiments, Billingham and Silvers (1967) found that regional epidermal histomorphology in guinea pigs (e.g. sole versus ear) was determined by the origin of the dermal component. Melbye and Karasek (1973) and Mackenzie and Fusenig (1983) further found that facilitation of epidermal proliferation may depend on diffusable, dermally-derived factors.

At the same time, it is possible that keratinocyte-derived factors may, in turn, influence the function of fibroblasts that constitute the major cell population of the dermis. We undertook the present studies to determine whether human keratinocytes can modulate the generation of prostaglandins by fibroblasts using an in vitro co-culture system originally developed by Rheinwald and Green (1975) that utilizes a nonproliferating population of 3T3 fibroblasts as a so-called "feeder layer" on which to culture human keratinocytes. Earlier studies from our laboratory (Blacker et al, 1987) have documented that these growth-arrested fibroblasts (treated with mitomycin C and referred to as 3T3M cells) maintain their ability to synthesize prostaglandins E_2 (PGE$_2$) and PGI$_2$ (prostacyclin),the latter prostaglandin being measured as its stable metabolite 6-keto-prostaglandin F$_{1a}$ (6-keto-PGF$_{1a}$).

In the studies to be described, we find that: 1) keratinocytes can enhance 6-keto-PGF$_{1a}$ synthesis by 3T3M cells, 2) 6-keto-PGF$_{1a}$ is generated by fibroblasts but not by keratinocytes and thus becomes a useful marker for keratinocyte stimulation of prostaglandin production by the fibroblasts, 3) interleukin 1 (IL-1), which is a documented keratinocyte-derived cytokine (Sauder et al, 1982; Hauser et al, 1985,1986; Kupper et al, 1986, 1987), may be one of the likely keratinocyte-derived factors that is responsible for the induction of fibroblast prostaglandin synthesis, and 4) IL-1, in the context of inducing prostaglandin synthesis among keratinocytes and fibroblasts may be a unidirectional signal because, while stimulating prostaglandin synthesis among the 3T3M fibroblasts, it fails to do so among human keratinocytes.

METHODS

Materials. All tissue culture media, additives, phosphate buffered saline (PBS) and

Cell-Cell Interactions in the Release of Inflammatory Mediatiors
Edited by P. Y-K Wong and C.N. Serhan, Plenum Press, New York, 1991

317

calcium-, magnesium-free (CMF-PBS) were obtained from the Cell Culture Facility, University of California San Francisco unless otherwise specified. Growth medium for all cell cultures consisted of Dulbecco's modified Eagles Medium (DMEM) supplemented with 10% fetal calf serum (FCS), 1.0 μg/ml glutamine, 0.4 μg/ml hydrocortisone, 1 μM cholera toxin, 20 ng/ml epidermal growth factor , 2.5 μg/ml fungizone, 0.1 KU penicillin, and 0.1 KU/ml streptomycin.

Cell cultures. 3T3 murine fibroblasts were obtained from the laboratories of Dr. Howard Green at Harvard and serially cultivated. Confluent cultures of 3T3 cells were treated with mitomycin C (4 μg/ml) in PBS for 2 hours, washed twice with PBS-CMF and harvested by treatment with 0.1% trypsin in PBS-CMF to detach the cells from the culture dish. The harvested cells were pelleted at 200 x g for 5 minutes, and plated at 1 x 10^4 cells/cm^2 in 3.5 cm^2 tissue culture wells (Falcon Plastics, Cockeysville, MD). The efficiency of the mitomycin C treatment was assessed by monitoring the daily DNA content (see below) of parallel cultures of 3T3M over the total period of experimental co-cultures.

Primary cultures of human keratinocytes were prepared from trypsin and collagenase treated foreskins by a modification of the method of Rheinwald and Green (1975). Foreskins were incubated in a solution of PBS containing 0.5 kU/ml streptomycin and 12.5 μg/ml fungizone for 2 minutes followed by gentle shaking for 1 minute in a 3:1 mixture of the above antibiotic solution with 0.25% hexachlorophene soap. Foreskins were then rinsed 5 times with PBS, scored on the epidermal side with a #15 scalpel blade, placed in a sterile bottle of PBS containing 0.1% trypsin and 0.5% collagenase and stirred for 1 hour. The suspended cells were harvested and fresh trypsin/collagenase solution was added back to the foreskin a total of 4 times with subsequent harvesting of suspended cells. The suspensions were pelleted by centrifugation at 200 x g for 10 minutes and the resulting pellet suspended in growth medium and plated onto 10 cm^2 tissue culture plates (Falcon Plastics) on which 3T3M feeder layers had already been established. Contaminating human dermal fibroblasts were selectively removed from keratinocytes in these primary cultures 4-10 days after plating by washing with 0.01% EDTA. Fresh 3T3M fibroblasts were replaced at this time since they are also removed by the EDTA treatment. Keratinocytes were passed at confluence by first washing with PBS-CMF, harvesting the cells with 0.1% trypsin plus 0.1% EDTA in PBS-CMF, pelleting the cells at 200 x g for 5 minutes, and plating the cells, resuspended in growth medium, at 2.4 x 10^4 cells/cm^2 in 3.5 cm^2 tissue culture wells either with or without 3T3M fibroblast feeder layers. Unless otherwise specified, growth medium was changed every 3 days.

Thin-layer chromatography. To qualitatively profile the prostaglandin generation by keratinocyte or fibroblast cultures, the different cell populations were incubated in suspension in 2 ml of PBS with [1-^{14}C] arachidonic acid ([1-^{14}C]AA - 56 mCi/mmol, New England Nuclear, Boston, MA) and with or without 10^{-6} M indomethacin for 30 minutes in a shaking water bath at 37° C. Incubations were stopped by addition of 2 volumes of methanol. The samples were then centrifuged at 1800 x g for 10 minutes to pellet the cells and any precipitated protein. The resulting supernatants were acidified to pH 3.0 and extracted 3 times with 2 volumes of diethyl ether; the pooled ether fractions were washed with a total of 4 volumes of distilled water. The samples were then evaporated to dryness under nitrogen, immediately resolubilized into ethyl acetate and applied with appropriate standards to silica gel G thin-layer chromatography plates (Redi-plate, Fisher Scientific Co, Pittsburgh, PA). The plates were developed in a solvent system consisting of the equilibrated organic phase of a mixture of ethyl acetate, 2,2,4-trimethyl pentane, acetic acid, and water (55:25:10:50). The locations of the radioactive products were determined using a Berthold LB 2382 linear analyzer linked to an Apple 2E computer. Unlabeled prostaglandin standards, co-chromatographed with the lipid extracts, were visualized using iodine vapor.

Radioimmunoassay. Quantitation of specific prostaglandins was accomplished by radioimmunoassay using appropriate anti-prostaglandin antibodies (Advanced Magnetics, Cambridge, MA) with a technique modified from that of Goldyne et al (1973). Briefly, cell-free supernatants, diluted 1:5 with 0.05 M tris buffer containing 2 μg/ml human gamma globulin (Cohen fraction II, Sigma, St. Louis MO). 100 μl of this diluted sample was incubated for 24 hours at 4°C with the appropriate anti-prostaglandin antibody(100 μl) diluted according to enclosed instructions and 5000 cpm (100 μl) of the corresponding tritiated prostaglandin (New England Nuclear). Separation of free from antibody-bound ligand was accomplished using 2 ml of 0.25% polyethylene glycol (PEG 4000, Fisher Scientific Co) per assay tube followed by centrifugation at 1800 x g for 30 minutes. The supernatant was discarded, the precipitate redissolved in 1 ml of 0.1 N sodium hydroxide and transferred to scintillation vials containing 10 ml of scintillation cocktail. All samples including a set of standards (10 pg to 10 ng) for either PGE_2 or 6-keto-PGF_{1a} were counted in a Beckman LS-100 liquid scintillation counter. All samples were run in duplicate. Values were calculated in ng/μg DNA. The culture medium was also assayed for the possibility of background prostaglandin activity due to serum; any background activity was subtracted from experimental values.

DNA assay. The total DNA content of a dish or well of cultured cells was determined using the method of Labarca and Paigen (1980) that involves treating sonicated cells with the fluorochrome bisbenzamidazole which binds to DNA, and using a Perkin-Elmer Fluorescence Spectrophotometer 204, exposing treated samples to an excitation wavelength of 356 nm and measuring the resultant fluorescence at 458 nm with reference to known quantities of DNA.

IL-1 activity in keratinocyte sonicate supernatants. IL-1 activity in sonicate supernatants obtained from keratinocyte cultures was indirectly assessed by evaluating the ability of a rabbit polyclonal anti-IL-1 antibody (Endogen, Boston MA) to inhibit the supernatant-induced stimulation of PGE_2 generation by the 3T3 M fibroblasts . For these studies, 3T3M fibroblasts were incubated for 72 hours in the presence or absence of 1% sonicate supernatant (v/v) in 5% newborn calf serum (NBCS) with or without 20% (v/v) anti-IL-1 antibody. The sonicate supernatant was obtained by sonicating 1×10^6 human keratinocytes in 1ml of PBS and pelleting the debris. It has been shown that such sonicates contain significant amounts of IL-1 (Hauser et al, 1985; Kupper et al, 1987).

RESULTS

Eicosanoids generated by cultured human keratinocytes (HK) and 3T3 fibroblasts. Figure 1 shows the radiochromatograms resulting from the incubation of [1-^{14}C] AA with HK and with 3T3 or 3T3M fibroblasts. As we previously reported (Blacker et al, 1987), both the 3T3 and 3T3M fibroblasts showed similar metabolism of AA into both 6-keto-PGF_{1a} and PGE_2 eventhough proliferative arrest is achieved in the 3T3 cells with mitomycin C. Of particular note is the absence of a peak co-chromatographing with 6-keto-PGF_{1a} in the HK chromatogram. The addition of indomethacin (10^{-6} M) to similar incubations resulted in the loss of both prostaglandin peaks (data not shown).

Effect of HK on 6-keto-PGF_{1a} generation by 3T3M fibroblasts. Figure 2 graphs the concentrations of 6-keto-PGF_{1a} generated over 24 hour sampling periods by 3T3M fibroblasts alone and in co-culture with HK during 12 days of culture. Combining HK with the 3T3M cells resulted in significantly greater levels of 6-keto-PGF_{1a} from days 6-9 than produced by 3T3M cells alone. Since only the 3T3M are capable of generating 6-keto-PGF_{1a}, the DNA value used to calculate the concentration of this eicosanoid was the mean of the DNA concentrations of the parallel 3T3M cultures over the 12 day incubation period (3.5 ± 0.5 μg DNA/culture dish).

Effect of combining HK with 3T3M fibroblasts on PGE_2 generation. Since both HK and fibroblasts synthesize PGE_2, we evaluated PGE_2 generation by a combined culture of HK and 3T3M cells in comparison to the sum of PGE_2 generation by individual cultures of HK and 3T3M cells. Figure 3 graphs the PGE_2 levels generated over the same 24 hour sampling periods as shown in Figure 2. On days 3 and 7, where statistical analysis was feasible, co-cultures ([3T3M + HK]) contained significantly greater levels of PGE_2 than the sum of levels generated by the individual HK and 3T3M cultures . It is worth noting that during the period over which confluence of the HK occurred (days 9-12), PGE_2 levels in the co-cultures dropped and eventually became significantly less than the sum of the levels in the parallel individual HK and 3T3M cultures suggesting some form of inhibitory activity.

The results of the above experiments raised the question of whether the enhanced eicosanoid generation in co-cultures of HK with 3T3M cells required physical contact between the two different cell populations - in vivo, the basement membrane would physically separate basal keratinocytes from dermal fibroblasts - or whether the enhanced 6-keto-PGF$_{1a}$ generation by the 3T3 fibroblasts could result from some soluble substance(s) released by the HK. To address this question, 3T3M fibroblasts were incubated for 72 hours in conditioned medium obtained from HK cultures. Levels of 6-keto-PGF$_{1a}$ generated in the presence of conditioned medium were compared to those generated in unconditioned medium.

Effect of HK-conditioned medium on 6-keto-PGF$_{1a}$ generation by 3T3M fibroblasts. Table 1 shows that the HK-conditioned medium significantly enhanced generation of 6-keto-PGF$_{1a}$ by the 3T3M fibroblasts. 3T3M-conditioned medium, on the other hand, failed to induce any 6-keto-PGF$_{1a}$ synthesis by the HK in keeping with our finding that HK did not generate labeled 6-keto-PGF$_{1a}$ from labeled arachidonic acid. The level measured (0.7 ± 0.1 ng/μg DNA) when expressed as concentration in the supernatant (3.9 ± 0.2 ng/ml) is virtually equivalent to the level of 6-keto-PGF$_{1a}$ present in the 3T3M-conditioned medium itself (4.3 ng/ml).

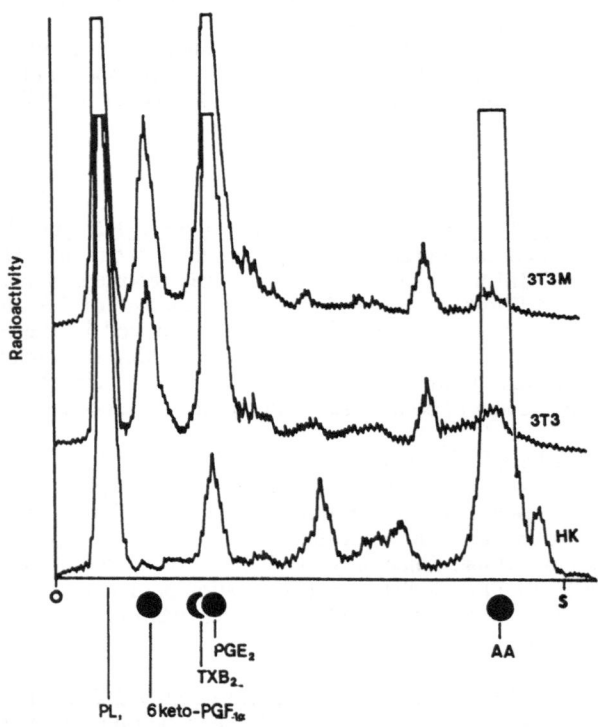

Fig. 1. Thin-layer radiochromatograms of lipid extracts from 30 minute incubations of 3T3M fibroblasts, 3T3 fibroblasts, or human keratinocytes (HK) with [1-^{14}C] arachidonic acid. Tracings are staggered in order to facilitate qualitative comparisons. o = origin; PL = labeled phospholipids, 6-keto-PGF$_{1a}$ and PGE$_2$ as in text; TXB$_2$ = thromboxane B$_2$; AA = arachidonic acid; s = solvent front.

Because of the previously cited studies documenting IL-1 generation by HK and the established ability of IL-1 to enhance fibroblast synthesis of prostaglandins (Zucali et al,1986; Balavoine et al, 1986; Raz et al, 1988) we undertook two sets of studies to determine: 1) if a rabbit polyclonal anti-IL-1 antibody could suppress the induction of prostaglandin generation among 3T3M fibroblasts incubated in medium containing 1% v/v of an HK sonicate supernatant (as a source of IL-1), and 2) if IL-1 had different effects on prostaglandin synthesis by 3T3M fibroblasts in comparison to HK.

Effect of anti-IL-1 antibody on the ability of HK sonicate supernatant to stimulate PGE_2 and 6-keto-PGF_{1a} generation by 3T3M fibroblasts. Figure 4 graphs the ability of HK sonicate supernatant (s.s.-1% v/v) to stimulate PGE_2 and 6-keto-PGF_{1a} generation by 3T3M fibroblasts in the presence or absence of 20% anti-IL-1 antiserum. Over 72 hours, the s.s. significantly enhanced PGE_2 ($p<0.01$) and 6-keto-PGF_{1a} ($p<0.05$, Student's one tailed t test for independent sample means) generation by the 3T3M cells and this enhancement was significantly suppressed by the anti-IL-1 antibody.

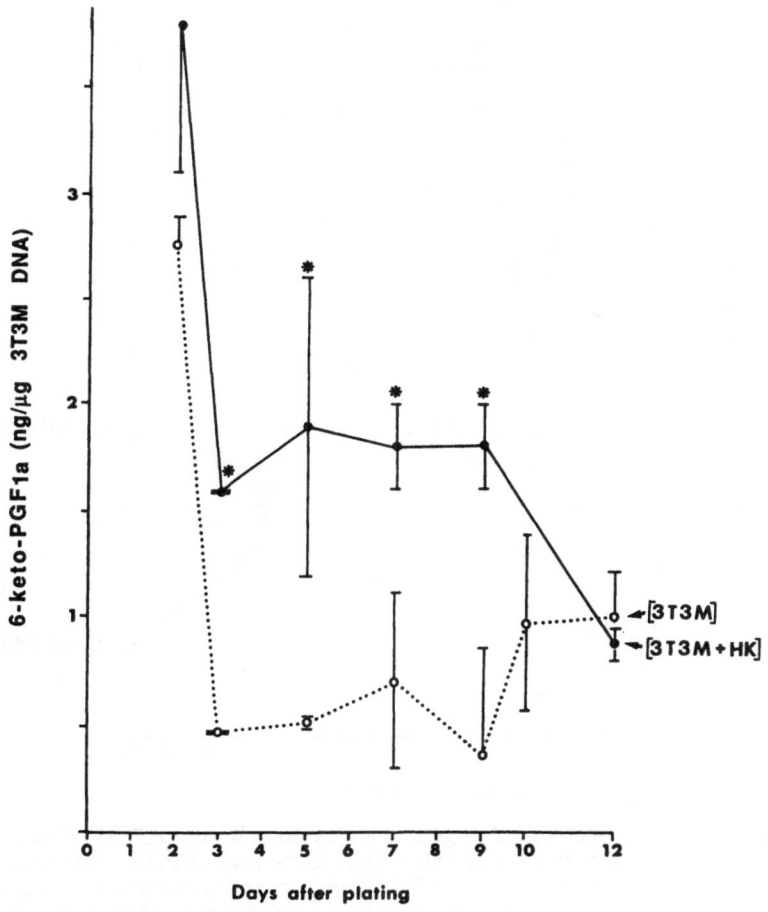

Fig. 2. Comparison of 6-keto-PGF_{1a} generation by cultures of 3T3M fibroblasts alone and by co-cultures of 3T3M fibroblasts with human keratinocytes (HK) during consecutive 24 hour intervals over 12 days of culture. Asterisks signify statistically significant differences between the mean (± S.E.M.) PG levels in the 3T3M cultures and the co-cultures with HK at the times indicated ($p<0.05$, Student's one-tailed t test for independent sample means, n = 2).

Effect of IL-1 on PGE$_2$ generation by 3T3M fibroblasts and by HK. Figure 5 compares the mean levels of PGE$_2$ generated over 72 hours by 3T3M fibroblasts or HK alone or in response to IL-1. For the fibroblasts, 10 half maximum units of human purified IL-1 was used and in the case of the HK, 10 and 100 u/ml of human recombinant IL-1a. We have shown earlier (Goldyne, 1988) that the human purified IL-1 failed to enhance PGE$_2$ synthesis by HK and therefore in this study used the recombinant IL-1a since it represents the active form of IL-1 secreted by HK. IL-1 clearly enhanced PGE$_2$ generation by the 3T3M cells but failed to alter baseline PGE$_2$ generation by HK over a 10-fold concentration range. In other studies (Goldyne and Rea, 1991) we have documented the ability of recombinant human IL-1a to stimulate human fibroblast PGE$_2$ generation.

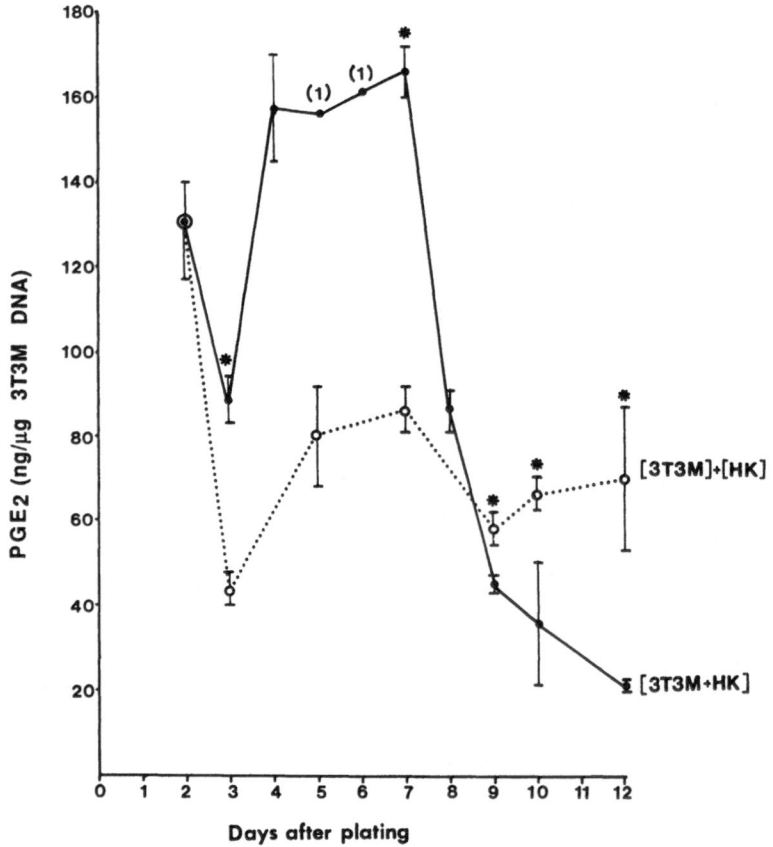

Fig. 3. Comparison of PGE$_2$ generation by co-cultures of 3T3M fibroblasts with human keratinocytes (HK) to the sum of PGE$_2$ generation by separate cultures of 3T3M fibroblasts and of human keratinocytes (HK) during consecutive 24 hour intervals over 12 days of culture. Asterisks signify statistically significant differences between the mean levels (± S.E.M.) in co-cultures versus the mean (± S.E.M.) of the sum of levels in the individual cultures at the times indicated (p<0.05, Student's one-tailed t test for independent sample means, n = 2).

Table 1. Effect of HK conditioned medium on 3T3M production of 6-keto-PGF$_{1a}$.

Cell Type	Medium	[6-keto-PGF$_{1a}$] (mean ± S.E.M.)
-	HK conditioned	0.0 ± 0.0 ng/ml
-	3T3M conditioned	4.3 ng/ml *
3T3M	unconditioned	3.6 ± 0.2 ng/µg DNA
3T3M	HK conditioned	11.2 ± 1.2 ng/µg DNA**
HK	3T3M conditioned	0.7 ± 0.1 ng/µg DNA (3.9 ± 0.2 ng/ml)

* only 1 value

** $p < 0.05$ (Student's t test for independent sample means) when compared to mean value for 3T3M response to unconditioned medium.

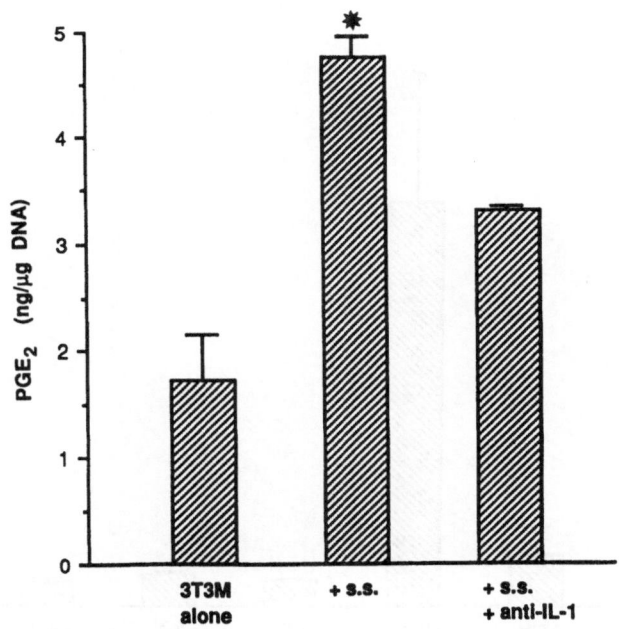

Fig. 4. Effect of human keratinocyte sonicate supernatants (s.s.) on PGE$_2$ generation by 3T3M fibroblasts and the effect of polyclonal anti-human IL-1 antibody on the effect of s.s. on PGE$_2$ generation. Asterisk signifies statistically significant difference between mean (± S.E.M.) value for PGE$_2$ level in response to s.s. and either values for 3T3M alone or in the presence of anti-IL-1 antibody ($p<0.05$, Student's one-tailed t test for independent sample means).

DISCUSSION

The studies summarized in this paper document the ability of HK to modulate prostaglandin synthesis by mitomycin C-treated 3T3 (3T3M) fibroblasts. Co-culture of HK and 3T3M cells, culture of HK-conditioned medium or sonicate supernatants of HK with 3T3M cells all resulted in enhanced prostaglandin synthesis by the 3T3M fibroblasts. Thus, while an effect of cell-cell contact cannot be ruled out, the studies using conditioned supernatants or sonicate supernatants conclusively show the ability of an HK-derived soluble factor or factors to influence fibroblast generation of prostaglandins. At the same time Figure 3 shows that in the co-culture of HK with 3T3M cells, the PGE_2 levels eventually fall significantly below the sum of levels in the individual cultures of these cells; since this suppression occurs at a time coincident with HK reaching confluency, there may be other factors, including cell-cell contact that contribute to the overall modulation of prostaglandin production.

Of significance to the general study of prostaglandin production by skin is the finding that PGI_2 (prostacyclin), as measured by its stable metabolite 6-keto-PGF_{1a}, is a prostaglandin that the 3T3 cells, but not HK, generate (Figure 1 and Table 1). As such, monitoring 6-keto-PGF_{1a} can serve as a marker for 3T3 responses to HK in co-culture studies. Whereas some investigators (Pentland and Needleman, 1986) have claimed 6-keto-PGF_{1a} synthesis by keratinocytes, careful exclusion of contaminating fibroblasts was not considered, and when using first passage keratinocytes as did these authors, it is important to rule out significant fibroblast contamination as a source of prostaglandin activity. And since fibroblasts are clearly responsive to IL-1 and as we have shown, co-cultures of fibroblasts with keratinocytes results in enhanced synthesis of prostaglandins, the need to effectively minimize fibroblast contamination is crucial.

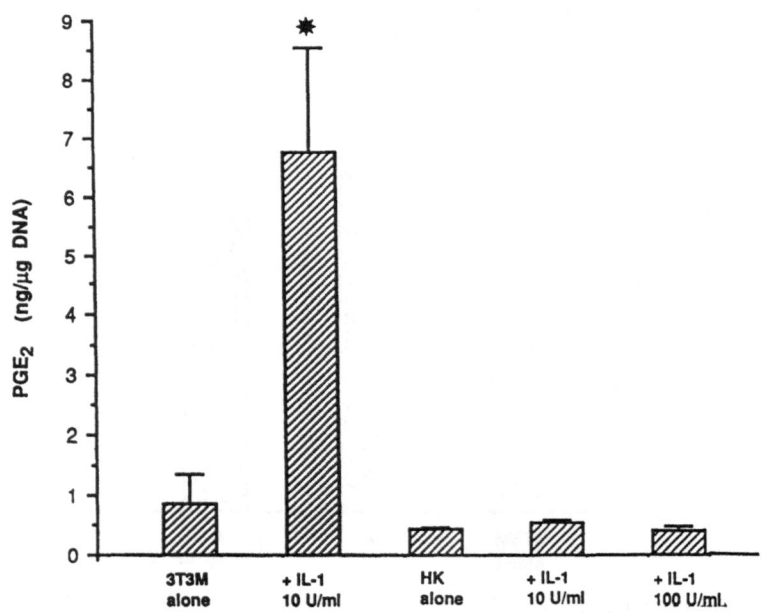

Fig. 5. Effects of human purified IL-1 on PGE_2 generation by 3T3M cells and of human recombinant IL-1a on PGE_2 generation by human keratinocytes (HK) over 72 hours (mean ± S.E.M.). Asterisk signifies statistically significant difference between IL-1-treated and untreated cultures ($p<0.025$, Student's one-tailed t test for independent sample means).

The ability of the anti-IL-1 antibody to suppress the enhancing effect of HK sonicate supernatants on prostaglandin generation by 3T3M fibroblasts provides indirect evidence for IL-1 being at least one keratinocyte-derived cytokine that may be involved in keratinocyte-fibroblast interactions. Further support for this possibility is the demonstration that IL-1 can enhance prostaglandin generation by the 3T3M fibroblasts (Figure 5). Except for the fact that the 3T3M cells studied were proliferation-blocked, it is not unexpected that IL-1 would stimulate prostaglandin generation by 3T3 fibroblasts since Burch et al (1988) have demonstrated the equivalent abilities of both human recombinant IL-1a and ß to stimulate 3T3 fibroblast PGE_2 synthesis.

An important concept suggested by our studies is that, in the specific context of stimulating both cell proliferation and prostaglandin synthesis, IL-1 functions as a unidirectional stimulus from the keratinocyte to the fibroblast.The present studies in addition to data from the literature suggest the following:

1. Keratinocytes, but not dermal fibroblasts, release IL-1. As mentioned in the introduction, studies have clearly identified IL-1 (a and ß) as a secreted product of human keratinocytes (Sauder et al, 1982; Hauser et al, 1985, 1986; Kupper et al, 1986, 1987). In contrast, data regarding fibroblasts suggest that although intracellular IL-1 exists in fibroblasts (Hogquist et al, 1989), it is not actively secreted (Hauser et al, 1985; Libby et al, 1986). And while some investigators have claimed that membrane IL-1 exists on dermal fibroblasts (Kurt-Jones et al, 1987), others have provided convincing evidence that so-called membrane IL-1 is an artifact of preparing the cell membranes for analysis (Minnich-Carruth et al, 1989 and Suttles et al, 1990).

2. IL-1 stimulates fibroblast, but not keratinocyte, proliferation. IL-1 has clearly been established as a mitogen for human dermal fibroblasts (Schmidt et al, 1982; Rupp et al, 1986) albeit this effect may be due to the induction and release of platelet-derived growth factor (PDGF) by the fibroblast (Singh et al, 1988; Raines et al, 1989). Whereas an initial study by Ristow (1986), using tritiated thymidine uptake, claimed IL-1 enhanced murine keratinocyte proliferation, studies by O'Keefe et al (1986), by Morhenn et al (1989) as well as a more recent evaluation by Ristow (1991) show that by itself, IL-1 (a or ß) cannot stimulate HK proliferation. In the studies summarized in Figure 5, we also failed to notice any significant difference over the 72 hour incubation period between the increase in DNA among the untreated HK and those exposed to 10 or 100 U/ml of IL-1a.

3. IL-1 stimulates fibroblast but not keratinocyte prostaglandin production. Similar to the literature on the effect of IL-1 on fibroblast versus keratinocyte proliferation, many studies have uniformly documented the ability of IL-1 to stimulate prostaglandin synthesis by fibroblasts (Hauser et al, 1985, Zucali et al, 1986, Raz et al, 1988; Burch et al, 1988, Goldyne 1989) and the present studies (Figure 5) show that even mitomycin-treated 3T3 fibroblasts will respond to IL-1 with an increase in prostaglandin synthesis. In contrast, our laboratory first reported the inability of IL-1 to enhance prostaglandin production by human keratinocytes at a concentration (3 half-maximal units/ml) that strongly stimulated human dermal fibroblast prostaglandin synthesis (Goldyne, 1988). Subsequently, Pentland and Mahoney (1990) reported the ability of IL-1 (a and ß) to stimulate prostaglandin production by first passage human keratinocytes but at a concentration far above that required to stimulate fibroblasts (100 u/ml); furthermore, 100 u/ml only generated an approximate three-fold increase above the mean unstimulated prostaglandin level and no statistics were provided to verify whether this increase was significant. As shown in figure 5, we failed to observe any stimulation of keratinocyte prostaglandin production using recombinant human IL-1a at concentrations of 10 and 100 u/ml. An explanation for the discrepancy in results may be the fact that the former studies used first passage keratinocytes, thus increasing the chance that the prostaglandin production observed was due to IL-1 stimulation of contaminating fibroblasts. In our studies, second passage keratinocytes were used which had been treated at first passage with 0.1% EDTA to remove contaminating fibroblasts.

In conclusion the above studies document the ability of keratinocytes to stimulate prostaglandin synthesis by fibroblasts and provide indirect evidence for the participation of keratinocyte-derived IL-1 as at least one of the potential signals through which this observed stimulation occurs. Furthermore, these studies are important in demonstrating that the effects of a cytokine need not be uniform for all cells to which it binds as evidenced by the ability of IL-1 to stimulate proliferation and prostaglandin synthesis among fibroblasts but not among keratinocytes from which it is released. Since it has been demonstrated that PGE_2 can enhance keratinocyte proliferation (Furstenberger and Marks 1980; Pentland and Needleman, 1986) a potential feedback loop may exist between the epidermis and dermis wherein keratinocytes, when appropriately stimulated, release IL-1 which passes through the basement membrane of the epidermis, binds to fibroblasts and stimulates PGE_2 production; the PGE_2, in turn, can feed back to the basal keratinocytes and modulate not only their proliferation but possibly IL-1 synthesis itself (Kunkel et al, 1986). In addition the PGE_2 generated may directly affect the fibroblasts themselves (Akahoshi et al, 1988). While such scenarios remain to be confirmed in vivo, the data presented justify further exploration of the role(s) of prostaglandins in keratinocyte-fibroblast interactions.

ACKNOWLEDGMENTS

These studies were supported by Merit Review funding and a Clinical Investigator Award (MG) from the Veterans Administration.

REFERENCES

Akahoshi, T., Oppenheim, J.J., and Matsushima, K., 1988, Interleukin 1 stimulates its own receptor expression on human fibroblasts through the endogenous production of prostaglandins, J. Clin. Invest. 82:1219.

Balavoine, J.-F., de Rochemonteix, B., Williamson, K., Seckinger, P., Cruchaud, A., and Dayer, J.-M., 1986, Prostaglandin E$_2$ and collagenase production by fibroblasts and synovial cells is regulated by urine-derived human interleukin 1 and inhibitor(s), J. Clin. Invest. 78:1120.

Billingham, R., and Silvers, W.K., 1967, Studies on the conservation of epidermal specificities of skin and certain mucosas in adult mammals, J. Exp. Med. 125:429.

Blacker, K.D., Williams, M.L., and Goldyne, M.E., 1987, Mitomycin C-treated 3T3 fibroblasts used as feeder layers for human keratinocyte culture retain the capacity to generate eicosanoids, J. Invest. Dermatol. 89:536.

Burch, R.M., Connor, J.R., and Axelrod, J., 1988, Interleukin 1 amplifies receptor-mediated activation of phospholipase A$_2$ in 3T3 fibroblasts, Proc. Nat. Acad. Sci. 85:6306.

Furstenberger, G., and Marks, F., 1980, Early prostaglandin E synthesis is an obligatory event in the induction of cell proliferation in mouse epidermis in vivo by the phorbol ester TPA, Biochem. Biophys. Res. Commun. 92:749.

Goldyne, M.E., 1988, Human fibroblast and keratinocyte synthesis of eicosanoids in response to interleukin 1: Evidence for fibroblast heterogeneity, In: Endocrine, Metabolic and Immunologic Functions of Keratinocytes, (L.M. Milstone and R.L. Edelson, Eds.), Ann. N.Y. Acad. Sci., New York, 548:108.

Goldyne, M.E., and Rea, L., 1991, CHILD Syndrome: Altered prostaglandin synthesis in involved skin fibroblasts inhibits interleukin 1 alpha-dependent proliferation, submitted for publication.

Goldyne, M.E., Winkelmann, R.K., and Ryan, R.J., 1973, Prostaglandin activity in human cutaneous inflammation: Detection by radioimmunoassay, Prostaglandins 4:737.

Hauser, C., Saurat, J.-H., Jaunin, F., Sizonenko, S., and Dayer, J.M., 1985, Cultured human epidermis cells produce cell-associated interleukin-like prostaglandin E$_2$- and collagenase-stimulating factors, Biochim. Biophys. Acta, 840:350.

Hauser, C., Saurat, J.-H., Schmitt, A., Jaunin, F., and Dayer, J.-M., 1986, Interleukin 1 is present in normal human epidermis, J. Immunol. 136:3317.

Hogquist, K., Katz, I., Chaplin, D., and Strunk, R., 1989, Developmental regulation of IL-1 in human fibroblast lines, J. Cell Biol. 107(6):484a.

Kupper, T.S., Ballard, D.W., Chua, A.O., McGuire, J., Flood, P.M., Horowitz, M.C., Langdon, R., Lightfoot, L., and Gubler, U., 1986, Human keratinocytes contain mRNA indistinguishable from monocyte interleukin 1a and ß mRNA: Keratinocyte epidermal cell derived thymocyte activating factor is identical to interleukin 1. J. Exp. Med. 164:2095.

Kupper, T.S., Chua, A.O., Flood, P., McGuire, J., and Gubler, U., 1987, Interleukin 1 gene expression in cultured human keratinocytes is augmented by ultraviolet irradiation, J. Clin. Invest. 80:430.

Kurt-Jones, E.A., Fiers, W., and Pober, J.S., 1987, Membrane interleukin 1 induction on human endothelial cells and dermal fibroblasts, J. Immunol. 139:2317.

Libby,P., Ordovas, J., M., and Dinarello, C.A., 1986, Regulated expression by human vascular smooth muscle cells of a gene for the inflammatory mediator interleukin 1 (IL-1), Clin. Res. 34(2):321A.

Mackenzie, I.C., and Fusenig, N.E, 1983, Regeneration of organized epithelial structure, J. Invest. Dermatol. 81(suppl):189s.

Melbye, S.W., and Karasek, M.A., 1973, Some characteristics of a factor stimulating skin epithelial cell growth in vitro, Exp. Cell. Res. 79:279.

Minnich-Carruth, L.L., Suttles, J., and Mizel, S.B., 1989, Evidence against the existence of a membrane form of murine IL-1a, J. Immunol. 142:526.

Morhenn, V.B., Wastek, G.J., Cua, A.B., and Mansbridge, J.N., 1989, Effects of recombinant interleukin 1 and interleukin 2 on human keratinocytes. J. Invest. Dermatol. 93:121.

O'Keefe, E.J., Dinarello, C.A., and Chiu, M.J., 1986, Keratinocyte growth-promoting activity from human placenta: Potentiation of thymocyte growth and distinction from IL-1, Clin. Res. 34(2):772A.

Pentland, A.P., and Mahoney, M.G., 1990, Keratinocyte prostaglandin synthesis is enhanced by IL-1, J. Invest. Dermatol. 94:43.

Pentland, A.P., and Needleman, P., 1986, Modulation of keratinocyte proliferation in vitro by endogenous prostaglandin synthesis, J. Clin. Invest. 77:246.

Raines, E.W., Dower, S.K., and Ross, R., 1989, Interleukin 1 mitogenic activity for fibroblasts and smooth muscle cells is due to PGDF-AA, Science 243:393.

Raz, A., Wyche, A., Siegel, N., and Needleman, P., 1988, Regulation of fibroblast cyclooxygenase synthesis by interleukin 1, J. Biol. Chem. 263:3022.

Rheinwald, J.G., and Green, H., 1975, Serial cultivation of strains of human epidermal keratinocytes: The formation of keratinizing colonies from single cells, Cell 6:331.

Ristow, H.J., 1987, a major factor contributing to epidermal proliferation in inflammatory skin diseases appears to be interleukin 1 or a related protein, Proc. Nat. Acad. Sci. U.S.A., 84:1940.

Ristow, H.J., 1990, Interleukin 1 does not stimulate DNA synthesis of cultured human keratinocytes growth-arrested in growth factor-depleted medium, J. Invest. Dermatol. 95:688.

Rupp, E.A., Cameron, P.M., Ranawat, C.S., Schmidt, J.A., and Bayne, E.K., 1986, Specific bioactivities of monocyte-derived interleukin 1a and interleukin 1ß are similar to each other on cultured murine thymocytes and on cultured human connective tissue cells, J. Clin. Invest. 78:836.

Sauder, D.N., Carter, C.S., Katz, S.I., and Oppenheim, J.J., 1982, Epidermal cell production of thymocyte activating factor (ETAF), J. Invest. Dermatol. 79:34.

Schmidt, J.A., Mizel, S.B., Cohen, D., and Green, I, 1982, Interleukin 1, a potential regulator of fibroblast proliferation, J. Immunol. 128:2177.

Singh, J.P., Adams, L.D., and Bonin, P.D., 1988, Mode of fibroblast growth enhancement by interleukin 1, J. Cell Biol. 106:813.

Suttles, J., Carruth, L.M., and Mizel, S.B., 1990, Detection of IL-1a and IL-1ß in the supernatants of paraformaldehyde-treated human monocytes: Evidence against a membrane form of IL-1, J. Immunol. 144:170.

Zucali, J.R., Dinarello, C.A., Oblon, D.J., Gross, M.A., Anderson, L. and Weiner, R.S., 1986, Interleukin 1 stimulates fibroblasts to produce granulocyte-macrophage colony stimulating activity and prostaglandin E_2, J. Clin. Invest. 77:1857.

NEUTROPHIL-EPITHELIAL CELL INTERACTIONS IN THE INTESTINE

James L. Madara, Shirin Nash, Charles Parkos

Division of Gastointestinal Pathology
Department of Pathology
Brigham and Women's Hospital and Harvard Medical School
Boston, MA 02115

INTRODUCTION

Polymorphonuclear leukocytes (PMN) migrate across the intestinal epithelium in a variety of inflammatory diseases afflicting the small intestine and colon. Subsequent collection of PMN in crypt lumens ("crypt abscesses") is a histologic finding used by pathologists to determine disease activity (1, 2). Here we consider the potential impact such PMN-epithelial interactions might have on intestinal epithelial function.

MODEL INTESTINAL EPITHELIUM

The human intestinal epithelial cell line, T84, may be used to grow confluent monolayers with stable high (1500 ohm cm^2) transepithelial electrical resistance (3, 4). Such monolayers consist of polarized columnar epithelial cells which display phenotypic similarities to intestinal crypt epithelia (3, 4). T84 cells also share functional similarities to epithelial cells of the intestinal crypt (5). Namely, when appropriately stimulated (carbachol, cAMP, cholera toxin, adenosine, vasoactive intestinal peptide, heat labile and heat stable E. coli toxins) T84 monolayers respond with electrogenic Cl⁻ secretion (6) – the transport event which underlies secretory diarrhea and which, in natural intestinal epithelia, originates from the crypt. Thus, T84 monolayers are a convenient model to study PMN-intestinal epithelial interactions.

PMN TRANSMIGRATION ACROSS T84 MONOLAYERS

Purified human peripheral blood PMN can be stimulated to cross T84 monolayers in response to transmonolayer gradients of the chemotactic peptide N-formyl methionyl leucyl phenylalanine (FMLP)(7, 8).

Cell-Cell Interactions in the Release of Inflammatory Mediators
Edited by P. Y-K Wong and C.N. Serhan, Plenum Press, New York, 1991

329

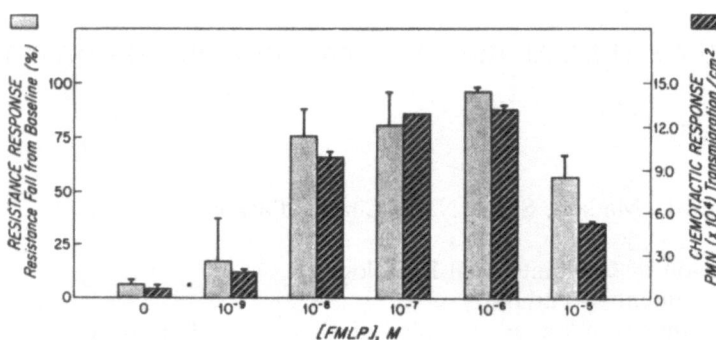

Figure I. Effects of varying transepithelial gradients of the chemotactic peptide FMLP on resistance and chemotactic responses across T84 intestinal epithelial monolayers. Confluent T84 monolayers (7-12 day post-confluency) on ring-mounted Nuclepore filters (5 μM pore) were elevated on glass beads in plastic tissue-culture wells. Monolayers were oriented such that the apical aspect of the cells faced the upper compartment. 10^7 PMN were applied to the apical surface (2 cm^2) and a chemotactic gradient was established across the monolayer with varying doses of FMLP in the basal compartment. Chemotaxis was allowed to progress at 37°C in a 95% air/5% CO_2 atmosphere for 2 h. Resistance response (percent fall in monolayer resistance from its own baseline) and chemotactic response (PMN x 10^4 transmigration/cm^2) are expressed as x±SEM from three to seven experiments for each bar. Maximal resistance and chemotactic responses occurred with chemotactic gradients of 10^{-7} M and 10^{-6} M. We used 10^{-7} M FMLP for subsequent studies. Reproduced with permission from ref. 7.

Table I.   Effect of PMN Transmigration on Transepithelial Resistance and on Mannitol (r = 3.6 Å) and Inulin (r = 11-15 Å) Flux

Monolayers	Baseline R ohms cm^2	Postexperiment R ohms cm^2	Postexperiment mannitol flux μmol/h cm^2	Postexperiment inulin flux μmol/h cm^2
Controls (n=5)	1,257±359	1,238±285[*]	87 X 10^{-4}±36[**]	5 X 10^{-4}±1[***]
Experimental (n=5)	1,280±321	353±154	318 X 10^{-4}±132	14 X 10^{-4}±5

Control vs. experimental: [*]P <0.0005;   [**]P <0.00025;   [***]P <0.01

As shown in Figure 1, the chemotactic response to a 10^{-7} M gradient of FMLP resulted in a large fall in transepithelial resistance. As shown in Table I, the decrease in resistance associated with PMN transmigration resulted in marked increments in monolayer permeability to inert hydrophilic solutes indicating that PMN transmigration perturbed the sieving characteristics of T84 intercellular tight junctions.

The decrease in resistance elicited at a PMN density of $10^{7}/2$ cm^2 in a 10^{-7} M FMLP gradient occurs with a t1/2 of approximately 35 min. (7), and is only slowly reversible. However, after being reversed (a case in which PMN are now underlying the electrically confluent monolayer), application of a FMLP gradient which induces PMN to exit the monolayer again results in a large (>80%) fall in resistance (7) suggesting that the above effects of PMN transmigration on resistance occur regardless of the direction of migration across the monolayer.

The above data suggest that transmigrating PMN transiently disrupt the sieving characteristics of tight junctions – the rate limiting barriers of the major transepithelial permeation pathway (3). Electron microscopic analyses of this event supports this interpretation (Figure 3) by showing that transmigrating PMN impale T84 cell tight junctions. Morphologic assessment of junctional permeation to the macromolecule horseradish peroxidase verifies that transient macromolecular leaks occur at junctions being impaled by PMN (7).

The mucosal preparations from patients with inflammatory bowel disease show decreased epithelial barrier function, as assessed by resistance measurements in Ussing chambers, even in the absence of detectable ulceration (9) and it is possible that PMN transmigration (perhaps in response to the N-formylated peptides naturally present in the intestinal lumen, 10) contributes to such permeation abnormalities. On the basis of intestinal epithelial permeability studies performed on first degree

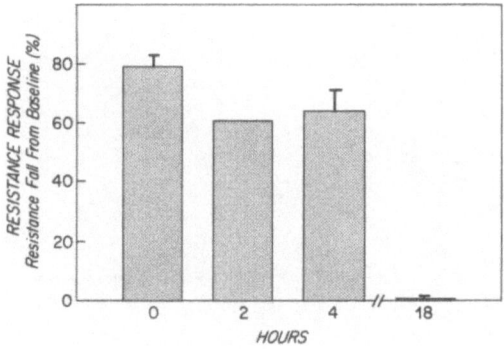

Figure 2. Time course of T84 monolayer resealing after removal of the chemotactic gradient. In these experiments monolayers were exposed to conditions of "Maximal" chemotaxis for 30 min before removing the gradient. Under these conditions resistance responses of ~75% were obtained and did not change for periods up to 4 h after gradient removal. However, by 18 h complete resealing had occurred as evidenced by loss of the resistance response. Resistance response at 0 h represents that obtained at the end of 30 min of chemotaxis. Results represent x±SEM from three experiments for each bar except at 2 h where n = 2. Reproduced with permission from ref. 7.

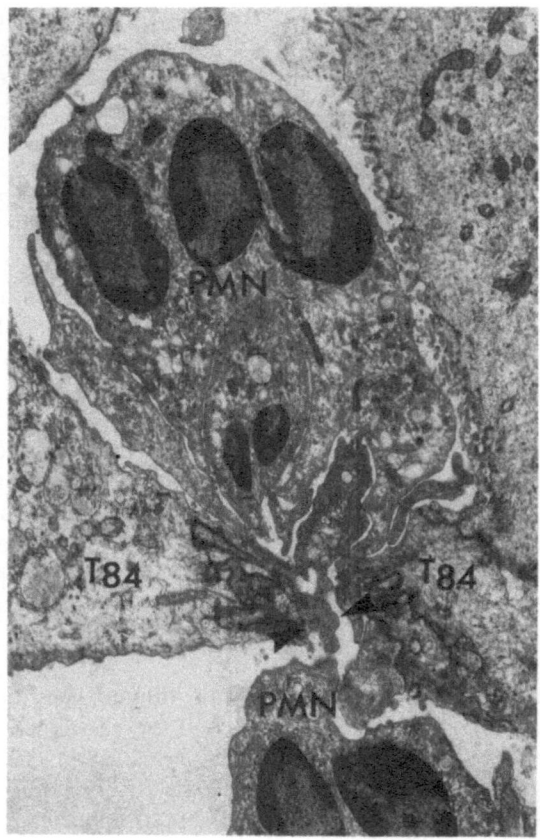

Figure 3. Electron micrograph of PMN indenting the monolayer and passing single file through a junctional impalement site. Transmigration occurs by extension of pseudopodia through the site to epithelial discontinuity (arrowheads). PMN migration is form "mucosal" to "serosal" compartments (~ × 9100) Reproduced with permission from ref. 7.

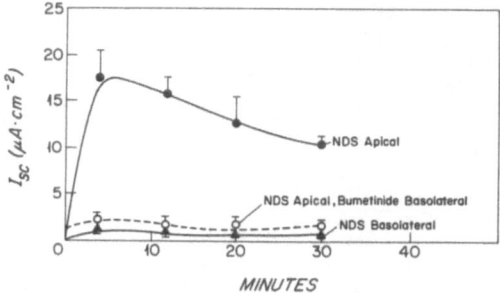

Figure 4. Short circuit current (Isc) response of T84 monolayers to apical addition of cell free supernate obtained from PMN stimulated with FMLP/LPS in a physiological buffer. The neutrophil derived Cl^- secretagogue (NDS) activity in the supernate is not active when added basolaterally. Bumetinide (10^{-4} M, serosal), an inhibitor of the basolateral Na:K:Cl cotransporter necessary for Cl^-secretion (6), inhibits 90% of the NDS response. (all n = 6-16), 16.

relatives of patients with inflammatory bowel disease, it has also been recently suggested (11) that such patients may have, as a primary defect, increased junctional permeability. Such abnormal baseline permeability could promote detection of lumenal chemotactic peptides by subepithelial PMN, thereby promoting PMN transmigration, inducing further enhancement of permeability and thus stimulating a positive feedback loop resulting in clinical disease. It is likely that intercellular adhesion molecules play an important role in such PMN-epithelial interactions, however, currently little is known concerning the mechanisms by which PMN attach to and subsequently migrate over the surface of epithelial cells.

THE CRYPT ABSCESS

Once PMN migrate across intestinal epithelia and collect in crypt lumens as "crypt abscesses", they are bathed in activating agents. Present within this lumen are bacterial products such as LPS (at ~μg/ml concentrations, 12) and n-formylated peptides (at ~ μM concentrations, 10). Since LPS may prime PMN (14) for enhanced respiratory burst (14) and degranulation (15) in response to FMLP, it is likely PMN in crypt abscesses are in markedly altered metabolic states. Given that T84 monolayers express barrier and transepithelial transport characteristic which are comparable to intestinal crypt epithelia (6), this model can also used to probe whether lumenal PMN, as would occur in crypt abscesses, modify crypt epithelial functions. As shown in Figure 4, PMN, when activated by conditions mimicking those in the colonic lumen, release a substance which stimulates a Cl$^-$ secretory short circuit current (Isc) − the transport event defining secretory diarrhea (6). This neutrophil derived secretagogue (NDS) activity passes 500 nominal MW cutoff filters, is heat and acid stable, and is not extracted by the organic solvents hexane, methyl formate or methanol (16). Furthermore, while mM concentrations of H_2O_2 can elicit slight (<5 μAmp·cm^{-2}) Isc responses (16), NDS activity is unaffected by catalase and SOD and measured H_2O_2 in NDS containing supernates is much lower than that required to elicit a Cl$^-$ secretory Isc. In addition, NDS activity is unaffected or only minimally affected by preincubation of PMN with either protease inhibitors, mepacrine and esculatine.

It is possible that the PMN-derived compound(s) associated with NDS activity contribute to the secretory diarrhea characteristically seen in patients with crypt abscesses.

SUMMARY

PMN transmigration across intestinal epithelia and into crypt lumens is a hallmark of active intestinal disease. Models of the transmigratory event indicate that movement of PMN across intercellular tight junctions transiently increases junctional permeability, thus decreasing epithelial barrier function. Once accumulated in the lumen, activated PMN may release unidentified mediators which activate the epithelial transport mechanisms responsible for secretory diarrhea. Further characterization of such epithelial-PMN interactions may allow identification of rational treatment strategies aimed at intervening in the epithelial dysfunction characterizing this type of cell-cell interaction.

Acknowledgements: This work is supported by NIH grant DK33506 (Project 4). Dr. Parkos is recipient of a NFIC Career Development Award.

References

1. Kumar, N. B., T. T. Nostrant, and H. D. Appelman. 1982. The histopathologic spectrum of acute self-limited colitis (acute infectious-type colitis). Am. J. Surg. Path. 6:523-529.

2. Yardley, J. H. 1986. Pathology of idiopathic inflammatory bowel disease and relevance of specific cell findings: an over view. in Recent Developments in the Therapy of Inflammatory Bowel Disease. Proceedings of a symposium. Myerhoff Center for Digestive Disease at Johns Hopkins, Baltimore, MD 3-9.

3. Madara, J. L., and K. Dharmsathaphorn. 1985. Occluding junction structure-function relationships in a cultured epithelial monolayer. J. Cell Biol. 101:2124-2133.

4. Madara, J. L., J. Stafford, K. Dharmsathaphorn, and S. Carlson. 1987. Structural analysis of a human intestinal epithelial cell line. Gastroenterology 92:1133-1145.

5. Dharmsathaphorn, K., J. A. McRoberts, K. G. Mandel, L. D. Tisdale, and H. Masui. 1984. A human colonic tumor cell line that maintians vectorial electrolyte transport. Am. J. Physiol. 246 (Gastrointest. Liver Physiol. 9):G204-G208.

6. Dharmsathaphorn, K., and J. L. Madara. 1990. Established intestinal cell lines as model system for electrolyte transport studies. in Methods in Enzymology: Biomembranes in Biological Transport. Vol. 5. eds, S. Fleisher and B. Fleisher, Academic Press, Orlando, FL.

7. Nash, S., J. Stafford, and J. L. Madara. 1987. Effects of PMN transmigration on the barrier function of cultured intestinal epithelial monolayers. J. Clin. Invest. 80:1104-1113.

8. Nash, S., J. Stafford, and J. L. Madara. 1988. The selective and superoxide independent disruption of intestinal epithelial tight juncitons during leukocyte transmigration. Lab. Invest. 59:531-537.

9. Hawker, P. C., J. S. McKay, and L. A. Turnberg. 1980. Electrolyte transport across colonic mucosa from patients with inflammatory bowel disease. Gastroenterology 79:508-511.

10. Chadwick, V. S., D. M. Mellor, D. B. Myers, A. C. Selden, A. Keshavarzian, M. F. Broom, and C. H. Hobson. 1988. Production of peptides inducing chemo-taxis and lysosomal enzyme release in human neutrophils by intestinal bacteria in vitro and in vivo. Scand. J. Gastro. 23:121-128.

11. Hollander, D. 1989. Is Crohns a tight junction disease? Gut 21:315.

12. vanDeventer, S. J. M., J. W. TenCate, and G. Tytgat. 1988. Intestinal endotoxemia: clinical significance. Gastroenterology 94:825-831.

13. Guthrie, L. A., L. C. McPhail, P. M. Henson, J. R. Johnston. 1984. Priming of neutrophils for enhanced release of oxygen metabolites by bacterial lipo-polysaccharide. J. Exp. Med. 160:1656-1681.

14. Omann, G. M., R. A. Allan, G. M. Bokoch, R. G. Painter, A. E. Traynor, and L. A. Sklar. 1987. Signal transduction and cytoskeletal activation in the neutrophil. Physiological Reviews 67:285-322.

15. Smolen, J. E. 1989. Characteristics and mechanisms of secretion by neutrophils. in The Neutrophil: Cellular Biochemistry and Physiology. M. B. Hallett, editor. CRC, Boca Raton, FL 23-62.

16. Nash, S., C. Parkos, A. Nusrat, C. Delp, and J. L. Madara. 1991. In vitro model of intestinal crypt abscess: a novel neutrophil-derived secretagogue (NDS) activity. J. Clin. Invest., in press.

CELL-CELL INTERACTIONS IN THE REGULATION OF GLOMERULAR INFLAMMATION BY

ARACHIDONATE LIPOXYGENASE PRODUCTS

Kamal F. Badr

Division of Nephrology, Department of Medicine
Vanderbilt University, Nashville, TN, 37232

Introduction

Deposition of antigen-antibody complexes is a common initiating mechanism in a wide variety of human glomerulopathies. While the site of formation of these complexes, the route(s) by which they gain access to the glomerulus, their localization, and the nature of the histopathologic reaction they elicit can vary considerably, the central role of the activated leukocyte in the subsequent pathogenesis is well-established (1). In most immune-mediated glomerulopathies, complement activation in the early phase of injury triggers a polymorphonuclear (PMN) leukocyte infiltrate, and activates resident glomerular macrophages (1-3). This is often followed by macrophage infiltration and proliferation, accompanied, at times, by a proliferative reaction of glomerular mesangial and/or epithelial cells (1). These leukocyte-dependent stages of injury frequently coexist within the same glomerulus, ultimately leading to its sclerosis. Even when an overt inflammatory reaction is absent (as in membranous nephropathy), there is convincing evidence to suggest that the activation of resident macrophages is a central component of the pathogenetic pathways which eventually result in the impairment of glomerular functions (3,4).

Arachidonate Lipoxygenation Products

Leukocyte activation leads to the release of degradative enzymes, the generation of reactive oxygen species, and the biosynthesis of locally acting pro-inflammatory autacoids (1). Among the latter, oxygenated metabolites of arachidonic acid are recognized major products of leukocyte activation and exert potent biological effects on cellular functions . The arachidonate LO family of enzymes catalyze the formation of highly potent biologic mediators in leukocytes and platelets (5,6). The predominant LO pathway in PMNs and macrophages is 5-LO, leading to the formation of LTs and 5-hydroxyeicosatetraenoic acid (5-HETE) (6). In addition, 15-LO activity catalyzes the formation of 15-HETEs in PMNs and macrophages, and, in conjunction with 5-LO, generates the lipoxins (LXs) (6). 12-LO is the major arachidonate LO in the platelet, leading to the formation of 12(S)-HETE.

The sulfidopeptide LTs (LTC4, LTD4, and LTE4) and the non-peptidyl LTB4, elicit potent biological responses: LTC4 and LTD4 contract vascular, pulmonary, and gastrointestinal smooth muscle, and increase vascular permeability to macromolecules (5,6). LTB4 has minimal spasmogenic properties. Its primary target appears to be PMNs, which

Cell-Cell Interactions in the Release of Inflammatory Mediators
Edited by P. Y-K Wong and C.N. Serhan, Plenum Press, New York, 1991

335

express specific high and low affinity receptors for LTB4 (7). Through the former, LTB4 is the most potent chemotactic substance yet described for this cell and also increases PMN aggregation and adhesion to endothelium (6). Through the latter, it acts as a calcium ionophore (8), leading to PMN activation (9-11), stimulation of phosphoinositide turnover (12,13), release of lysosomal enzymes, and an increase in oxidative metabolism (8-13). In turn, activated PMNs are the best studied source of LTB4 (6) where its synthesis is coupled to activation of protein kinase C (14). Interestingly, the dual LO enzyme products, the LXs, inhibit LTB4-induced chemotaxis (15,16), compete for LTD4 receptors on smooth muscle cells (17,18), and are vasodilators (19,20). In these respects, LXs appear to act as couterregulatory influences to the potent pro-inflammatory responses elicited by the LTs.

The potential role of lipoxygenase products in the regulation of glomerular inflammation involves A. The direct actions of lipoxygenase products on the determinants of glomerular functions. B. The interactions of multiple lipoxygenase products on target cells in the glomerulus, and C. The participation of multiple glomerular cell types in the biosynthesis of leukotrienes and lipoxins.

A. The direct actions of lipoxygenase products on the determinants of glomerular functions

In early studies, we observed that systemic administration of LTC4 in the rat led to reductions in renal blood flow (RBF) and glomerular filtration rate (GFR) (21). Since the responses to LTC4 in vascular smooth muscle results from its bioconversion to LTD4 (22), we evaluated the glomerular microcirculatory actions of LTD4 in the intact kidney in vivo, using glomerular micropuncture techniques (23). Intrarenal administration of LTD4 decreased RBF, due to augmentation of post-glomerular arteriolar resistance (Re). Notably, LTD4 administration reduced markedly the glomerular capillary ultrafiltration coefficient (Kf), thereby decreasing GFR (23). In vitro, LTC4 and LTD4 contract glomerular mesangial cells (24,25), a presumed mechanism for the fall in Kf in vivo. We therefore examined the interactions of LTD4 with these cells (26).

Binding of [3H]-LTD4 to mesangial cells in culture was stereoselective, specific, saturable and rapidly reversible. Two binding sites were recognized with dissociation constants and binding site densities at equilibrium of 2.2 and 16.8 nM and 1.1x104 and 3x104 binding sites/cell. LTD4 also induced time and concentration-dependent formation of inositol trisphosphate (IP3) which was maximal at 5 sec, and inhibited by SKF 104353, a specific receptor antagonist. LTD4 also increased intracellular pH, and stimulated [3H]-thymidine incorporation and mitogenesis. These findings were confirmed by others in human mesangial cells, where the Kd for LTD4 binding (12 nM) was remarkably similar to our findings in the rat (27).

We subsequently examined the glomerular actions of lipoxin A4 (LXA4), again in the intact kidney in vivo (19). In dramatic contrast to the vasoconstrictor action of LTD4 on Re, LXA4 elicited selective vasorelaxant responses in pre-glomerular arteriolar resistance (Ra), thereby increasing RBF, transcapillary hydraulic pressure difference (ΔP), and GFR.

B. The interactions of multiple lipoxygenase products on target cells in the glomerulus

To our surprise, and despite its relaxant actions on Ra, administration of LXA4 was consistently associated with a mild, but significant, reduction in Kf (Figure 1). This observation led us to examine the interactions of LXA4 with mesangial cells in culture (17). LXA4 competed for [3H]-LTD4 binding to mesangial cells, its presence prevented LTD4-induced IP3 generation, and its own stimulation of

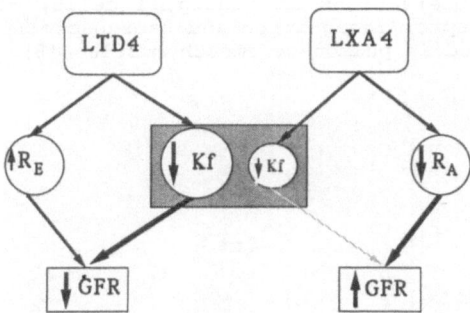

Figure 1. Opposing effects of LTD4 and LXA4 on GFR and renal vascular resistance. Both eicosanoids decrease K_f through activation of a common receptor on glomerular mesangial cells (see text). R_E: efferent resistance, R_A: afferent resistance, K_f: glomerular capillary ultrafiltration coefficient.

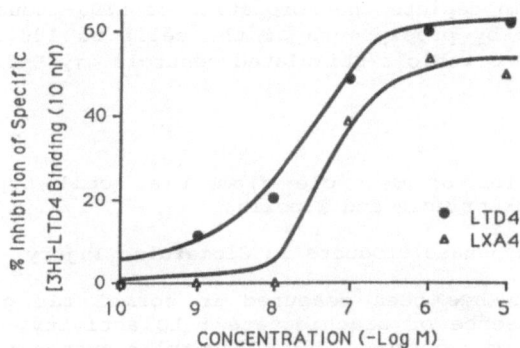

Figure 2. Percent inhibition of [3H]-LTD4 (10 nM) binding to rat glomerular mesangial cells by LTD4 and LXA4. Each point represents the mean of four experiments performed in duplicate. (From Ref. 17).

mesangial cell IP3 was blocked by an LTD4 receptor antagonist (Figures 2 and 3). In vivo, LXA4 antagonized LTD4-induced falls in GFR, but not RPF, implying selective prevention of LTD4-mediated reductions in Kf.

Thus, LTD4 and LXA4 activate a common receptor on rat mesangial cells at which LXA4 provokes partial agonist responses and competitively antagonizes both the cellular and physiological actions of LTD4. The basis for this common interaction is the shared spacial orientation [(S),(R)] of the polar substituents at C5 and C6 in both eicosanoids. Subsequent to our observations in mesangial cells, strikingly similar interactions for receptor recognition and stereospecificity between LTD4 and LXA4 were reported in pulmonary smooth muscle (18).

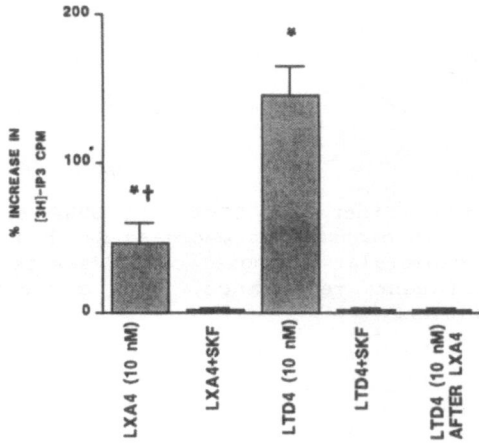

Figure 3. InsP3 formation in rat mesangial cells in response to LXA4 and LTD4 and its complete abolition in the presence of 100-fold concentration of SKF 104353, a specific LTD4 receptor antagonist (SKF). The far right column depicts the abrogation of LTD4-induced stimulation of InsP3 formation by preexposure of the cells to 100 nM LXA4 for 10 min. *, P<0.01 versus vehicle-stimulated controls. †, P<0.05 versus LTD4 alone. (From Ref. 17).

C. The participation of multiple glomerular cell types in the biosynthesis of leukotrienes and lipoxins

Generation of Lipoxygenase Products In Glomerular Injury

LTB4 production has been measured in normal rat glomeruli (28), indicating the presence of arachidonate 5-LO activity, most likely in resident glomerular macrophages (28). Glomerular synthesis of 5- and 12-LO products is enhanced markedly early in the course of several forms of glomerular immune injury, including those characterized by PMN infiltration, such as early nephrotoxic serum (NTS) nephritis (29-32) and anti-thymocyte (a-Thy1.1 Ag)-antibody nephritis (33), or in the absence of PMN involvement, such as in cationic bovine gamma globulin (CBGG)-induced glomerular injury and passive Heymann nephritis (PHN) (31,33,34). In these studies, LTB4 was measured in the supernates of isolated glomeruli either by HPLC/RIA (29,31-34) or by GC/MS (30). The cellular origin of augmented LTB4 synthesis in injured glomeruli remains undefined. Lianos et al. (29) and Schreiner et al. (32) suggest non-leukocyte sources in NTS nephritis, but implicate PMNs in the enhanced LTB4 generation of PHN and a-Thy1.1 Ag-induced nephritides (33). While there is discrepancy as to its cellular origin during glomerular injury, all studies concur as to the transient nature of LTB4 release (29-34):

LTB4 levels are undetectable beyond the first 24 hours of injury in both proliferative (NTS,a-Thy1.1 Ag), as well as non-proliferative (CBGG, PHN) glomerulopathies.

In NTS-induced injury, suppression of LTB4 synthesis accompanies the shift in the glomerular infiltrate from one comprised mainly of PMNs to a macrophage/monocyte infiltrate, localized in the mesangium (29-32). The temporal concordance between the disappearance of PMNs and the suppression of LTB4 generation brings into focus a central theme of this proposal: To what degree is glomerular LTB4 generation PMN-dependent? It is tempting to speculate that PMNs are the only source of LTB4 in early inflammatory injury, and the reduction in LTB4 at 24 hrs results from the absence of PMNs. In studies by Schreiner et al. (32), however, PMN depletion suppressed (from 50- to 10-fold), but did not abolish, LTB4 generation in early NTS nephritis; the same maneuver in studies by Lianos et al. (29) was also ineffectual in totally suppressing LTB4 synthesis. Furthermore, the cellular reaction which follows PMN infiltration, the macrophage-dominated lesion, is a potentially richer source of LTB4 than its predecessor: LT synthetic rates in activated macrophages exceed those in PMNs (5). Why, then, are glomerular LTB4 levels undetectable beyond 24 hours?

Role of Cell-Cell Interactions in the Regulation of Glomerular LTB4 Synthesis

To gain insight into potential cellular sources of LOX products within the glomerulus, as well as mechanisms of biosynthesis, we used recently cloned cDNA probes (35-37) to examine the expression of genes coding for three key enzymes in the biochemical cascades which catalyze the transformation of arachidonic acid to HETEs, lipoxins, and leukotrienes. These are 5-LOX, 15-LOX, and LTA4-OHase. Available probes for these enzymes were all of human origin (35-37). Expression of these genes was examined in cultured cells from glomerular origin in the rat which included glomerular mesangial, endothelial and epithelial cells, as well as human renal microvascular endothelial cells (38,39).

5-LOX is the key initial enzyme in the formation of all LTs and lipoxins. It is abundantly expressed in neutrophils, eosinophils, monocytes, and macrophages. It catalyzes the formation of a key eicosanoid intermediate LTA4, which can be metabolized to form LTs, but also lipoxins, through subsequent catalysis by either 15-LOX or 12-LOX. Northern hybridization analysis failed to reveal the expression of the gene coding for 5-LO or 15-LO in any of the cell lines examined. Western blot analysis, using antibodies directed against rat basophillic leukemia 5-LO and human reticulocyte 15-LO, also failed to demonstrate the presence of these enzymes in all four cell lines. Furthermore, 5-LO gene expression in mesangial cells and human renal microvascular endothelial cells could not be induced by pretreatment of cells with tumor necrosis factor-alpha, γ-interferon, growth factors, interleukins 1/6/8, LTs, LXs, phorbol myristate acetate, A23187, or cAMP (38).

It is therefore unlikely that total synthesis of LTs of LXs can occur in indigenous renal cells in the absence of leukocyte-derived LTA4. We next examined the the expression of LTA4-OHase in our cell lines (39).

Despite the absence of 5-LOX activity in glomerular parenchymal cells, and as shown in Figure 4, the presence of LTA4-OHase gene expression in these cells would allow for the transcellular metabolism of LTA4, generated in, and exported from, adjacent infiltrating and/or resident leukocytes, into biologically active LTB4. Northern blots performed using mesangial cell total RNA, revealed that in contrast to the absence of 5-LOX, LTA4-OHase gene expression can be clearly demonstrated. As positive controls for both LTA4-OHase and 5-LOX, rat PMN RNA was probed for these genes and abundant message expression demonstrated, ascertaining recognition by these human probes, of the rat genes. In addition to rat mesangial cells, human renal microvascular

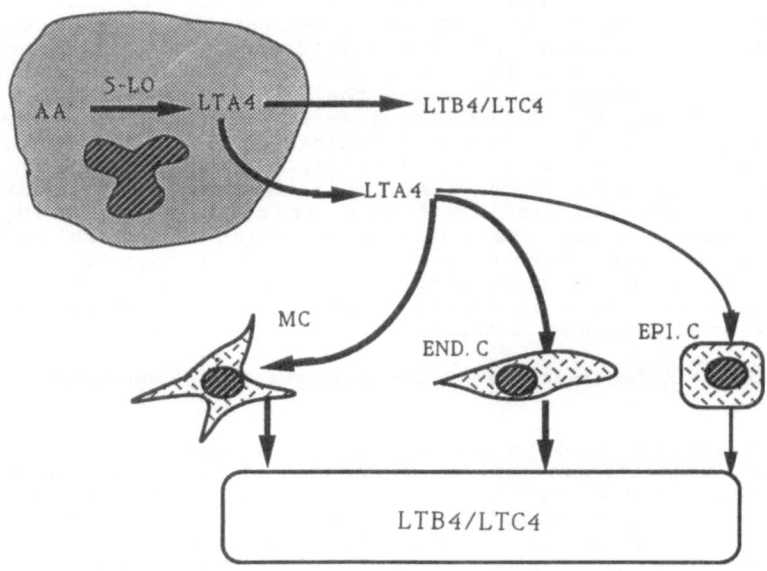

Figure 4. PMNs determine overall intraglomerular LT generation rates during immune injury in two ways: Total synthesis and release of LTB4/LTC4/LTD4 and five lipoxygenase-catalyzed synthesis and extracellular export of the 5,6-epoxy intermediate LTA4 and its subsequent transformation by indigenous glomerular cells to LTB4, a reaction catalyzed by LTA4-hydrolase, and LTC4, a reaction catalyzed by a glutathion-S-transferase. MC: Mesangial cells; END C: Endothelial cells; EPI. C: Epithelial cells; 5-LO: 5-lipoxygenase

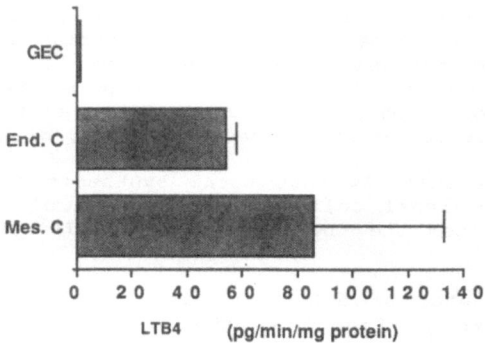

Figure 5. LTB4 generation rate in cultured rat glomerular cells incubated with 1μM LTA4. Abbreviations as in Figure 4.

endothelial cells RNA displayed abundant expression of the LTA4-OHase gene. The same was true of rat glomerular endothelial cells (39).

In contrast to glomerular endothelial and mesangial cells, however, glomerular epithelial cells contained little or no message for this gene, suggesting intraglomerular heterogeneity of its expression in glomerular cells of varying embryonal origins (39).

Incubation of rat glomerular MC and endothelial cells with 1 μM LTA4 free acid at pH 7.4 in the presence of albumin resulted in the formation of LTB4, as measured by HPLC followed by RIA in these two cell lines, and as shown by Figure 5. In keeping with the data from analysis of gene expression which demonstrated absence of LTA4-OHase gene in glomerular epithelial cells, LTB4 biosynthesis was undetectable upon incubation of these cells with LTA4 (Figure 5). These experiments represent the first demonstration of the capacity of endogenous glomerular cells to participate in the synthesis of LTs. Although not shown, both mesangial and endothelial cells generated significant amounts of LTC4 and LTD4 in the same preparation.

Thus, the presence of LTA4-OHase gene expression and catalytic activity in resident glomerular cells provides support for the transcellular metabolism of leukocyte-derived LTA4 as a potentially major source of LTB4 synthesis in the immune injured glomerulus.

Based on these data obtained from analysis of 5-LO and LTA4-hydrolase gene expression in cultured glomerular mesangial, endothelial, and epithelial cells and human renal microvascular endothelial cells, as well as measurements of LTB4 production in vitro, we propose the following: PMNs determine overall intraglomerular LT generation rates during immune injury in two ways: A. Total synthesis and release of LTB4/LTC4/LTD4; B. Five lipoxygenase-catalyzed synthesis and extracellular export of the 5,6-epoxy intermediate LTA4 and its subsequent transformation by indigenous glomerular cells to LTB4, a reaction catalyzed by LTA4-hydrolase, and LTC4, a reaction catalyzed by a glutathion-S-transferase (GST). The ubiquitous cellular expression of GST, including red blood cells and platelets, would provide a likely mechanism for the generation of LTC4 from transcellular transfer of LTA4 within the glomerulus. We propose that the regulation of LTB4 generation is, to a large extent, dependent on the level of LTA4-hydrolase activity in indigenous glomerular cells, rather than in PMNs. We have obtained evidence for the expression, as well as the regulation (40), of the LTA4-hydrolase gene in glomerular mesangial and renal endothelial cells. The suppression of LTA4-hydrolase activity by PMN adhesion (40), and perhaps other factors, would be expected to dramatically reduce intraglomerular LTB4 concentrations. This, in turn, would deprive the glomerular microcirculation of this potent chemoattractant and PMN activator, thereby preventing further influx of PMNs, depleting glomerular LTA4 levels, and suppressing LTB4/LTC4 production. The arrest of PMN infiltration would initiate the cellular shift to the macrophage/monocyte infiltrate which characterizes the chronic phase.

The transcellular metabolism of PMN-derived LTA4 to LTs is well-established (41-45). Cell-cell interactions leading to amplification of LT synthesis include PMN-endothelial cells (41), PMN-red blood cells (42), and PMN-platelets (43-45).

Transcellular metabolism of LTA4 may also generate LXs. The latter has been demonstrated in PMN-platelet interactions (46), and, of particular interest, in the demonstration of LXA4 generation upon incubation of rat mesangial cells with LTA4 (47). In this regard, Serhan and Sheppard have recently provided experimental and theoretical evidence that LX generation during PMN-platelet interactions is catalyzed by the 12-LO enzyme (46) (platelets lack the 15-LO enzyme which catalyzes LX synthesis in PMNs). To the extent that LXs are formed from mesangial cells in vivo, this reaction may also proceed via 12-LO activity. This is based on the demonstration by Baud et al of 12-HETE production from serum-treated-zymosan-exposed mesangial cells in vitro (48), and on our own demonstration of the absence of 15-LO gene expression in these cells. Thus, the metabolic fate of intraglomerular LTA4 likely differs from one stage or form of injury to another, depending on the nature of the cellular infiltrate adjacent to LTA4-generating cells (PMNs, macrophages) and the activities of the enzymes within those cells (LTA4-hydrolase, 12-LO, or 15-LO) for which LTA4 is a substrate.

These data provide a basis for: a) The temporal concordance and tight correlation of PMN infiltration and LT synthesis in early glomerulonephritis. b) The marked, but incomplete, suppression of LT generation in PMN-depleted animals (29,32), since he absence of PMNs would not only deprive the glomerulus of PMN-derived LTs, but also of LTA4 substrate availability for transcellular processing by glomerular cells. PMN depletion, however, would likely not impact on resident macrophage-derived LTA4 or that derived from PMNs not eliminated by "PMN-depleting" maneuvers, such as X-irradiation, thereby leaving residual LTA4 synthetic mechanisms in those cells intact. c) To the extent that glomerular LTB4 generation sustains the influx of PMNs in the early stage of injury, this hypothesis may provide a basis for the mechanism whereby such infiltration is arrested. More importantly, since the biologic consequences of PMN activation are central to the structural and functional components of injury, insight into the mechanisms regulating LTB4 generation may provide a rational basis for the design of interventional strategies aimed at limiting the extent of leukocyte infiltration and activation in the glomerulus.

Pathophysiologic Functional Significance of Lipoxygenase Products In Glomerular Injury

Knowledge of the nature and mechanisms of release of LO products in glomerular injury is of significance only in as much as these eicosanoids are determinants of pathophysiology. The potential influence of LO products on glomerular functions can be divided into three categories:

1. Influences on glomerular hemodynamics by vasoactive LO products We have shown that selective LTD4 receptor blockade results in preservation of Kf and GFR in NTS-induced injury (49). Other investigators have now confirmed these findings, using the same LTD4 receptor antagonist, in a murine model of spontaneous lupus nephritis (50). We have obtained evidence supporting a potential role for LTB4 in amplifying these LTD4-mediated responses (51). The effects of LTs on glomerular dynamics are likely limited to the early stages of injury. In the autologous phase of NTS-induced injury, the major influences on glomerular dynamics appears to be products of the cyclooxygenase (CO) pathway (52) , but the role of LO products, particularly LXs, in the mediation of these changes appears possible, and will be explored further.

2. Influences of LTB4 and, possibly, LXs, on PMN infiltration/adhesion and, hence, the extent of tissue damage due to the sequelae of PMN activation. These would manifest as progression of the histologic lesions due, for example, to the exposure of new antigens and the development of autoimmunity (1), and functionally, as alterations in capillary permeability and proteinuria. Preliminary data indicates that inhibition of LT synthesis in the first three days of injury dramatically reduces proteinuria at two weeks (53), suggesting a role for early LT-mediated structural damage in the chronic impairment of glomerular function.

3. Interactions among LO pathways products in the regulation of eicosanoid synthetic rates and modulation of biological activities. Inverse relationships have been demonstrated in human neutrophils between the biosynthesis of LTB4 and LXA4 (54). As noted above, we have

provided evidence for receptor competition between LTD4 and LXA4 at the level of the mesangial cell, such that the presence of LXA4 prevents LTD4-mediated reductions in Kf in vivo (17). LXA4 also antagonizes LTD4-induced adhesion of PMNs to mesangial cells in vitro (55), and LTB4-evoked inflammatory reactions in the hamster cheek pouch (16), providing further evidence for its counter-inflammatory actions. The potential role of these interactions in modifying glomerular injury remains unexplored. Its investigation requires the capacity to measure LXA4, LXB4 and a newly identified LX in activated neutrophils, 7-cis,11-trans-LXA4 (56), in cortical tissue or in short-term glomerular culture supernates. The recent development of deuterated standards for LXA4 (57) has made such approaches feasible. Measurement of 15- and 12-LO products using similar approaches, as well as elucidation of their biological activities in the kidney, should allow for a more thorough understanding of LO pathways product interactions in the modulation of glomerular injury.

REFERENCES

1. Wilson, C.B and F.J Dixon. 1986. Renal response to immunological injury. In The Kidney. BM Brenner and FC Rector Jr., editors. Saunders, Philadelphia. 800-890.

2. Cochrane, C.G., E. Unanue and F.J. Dixon. 1965. A role of polymorphonuclear leukocytes and complement in nephrotoxic nephritis. J. Exp. Med. 122:99-116.

3. Schreiner, G.F, R.S. Cotran, and E.R. Unanue. 1984. Modulation of Ia and leukocyte common antigen expression in rat glomeruli during the course of glomerulonephritis and aminonucleoside nephrosis. Lab. Invest. 51:524-533.

4. Cook, H.T., J. Smith, J.A. Salmon, and V. Cattell. 1989. Functional characteristics of macrophages in glomerulonephritis in the rat. Am J. Pathol. 134:431-437.

5. Lewis, R.A. and K.F. Austen. 1984. The biologically active leukotrienes. J. Clin. Invest. 73:889-897.

6. Samuelsson, B., S.-E. Dahlén, J. Å. Lindren, C. A. Rouzer, and C. N. Serhan. 1987. Leukotrienes and lipoxins: Structures, biosynthesis and biological effects. Science 237:1171-1176.

7. Goldman, D. W., L. A. Gifford, T. Marotti, C. H. Koo, and E. J. Goetzl. 1987. Molecular and cellular properties of human polymorphonuclear leukocyte receptors for leukotriene B4. Fed. Proc. 46:200-203.

8. Serhan, C. N., J. Fridovich, E. J. Goetzl, P. B. Dunham, and G. Weismann. 1982. Leukotriene B4 and phosphatidic acid are calcium ionophores. J. Biol. Chem. 257:4746-4752.

9. Feinmark, S. J., J. Å. Lindgren, H.-E. Claesson, C. Malmsten, and B. Samuelsson. 1981. Stimulation of human leukocyte degranulation by leukotriene B4 and its w-oxidized metabolites. FEBS letters 136:141-144.

10. Palmblad, J., C. L. Malmsten, A.-M. Udén, O. Rädmark, L. Engstedt, and B. Samuelsson. 1981. Leukotriene B4 is a potent and stereospecific stimulator of neutrophil chemotaxis and adherence. Blood 58:658-661.

11. Naccache, P. H., N. Faucher, S. Therrien, and P. Borgeat. 1988. Calcium mobilization, actin polymerization and right-angle light scatter responses to leukotriene B4, 12(R)- and 12(S)-hydroxyeicosatetraenoic acid in human neutrophils. Life Sciences 42:727-733.

12. Andersson, T., W. Schlegel, A. Monod, K.-H. Krause, O. Stendahl, and D. P. Lew. 1986. Leukotriene B4 stimulation of phagocytes results in the formation of inositol 1,4,5-trisphosphate. A second messenger for Ca2+ mobilization. Biochem. J. 240:333-340.

13. Mong, S., G. Chi-Rosso, J. Miller, K. Hoffman, K. A. Razgaitis, P. Bender, and S. T. Crooke. 1986. Leukotriene B4 induces formation of inositol phosphates in rat peritoneal polymorphonuclear leukocytes. Mol. Pharmacol. 30:235-242.

14. McIntyre, T. M., S. L. Reinhold, S. M. Prescott, and G. A. Zimmerman. 1987. Protein kinase C activity appears to be required for

the synthesis of platelet-activating factor and leukotriene B4 by human neutrophils. J. Biol. Chem. 262:15370-15376.

15. Lee, T,H., Horton, C.E., Kyan-Aung, U., Haskard, D., Crea, A.E.G., and Spur, B.W. 1989 Lipoxin A4 and lipoxin B4 inhibit chemotactic responses of human neutrophils stimulated by leukotriene B4 and N-formyl-L-methionyl-L-leucyl-L-phenylalanine. Clin. Sci. 77:195-203.

16. Hedqvist, P, J. Raud, U. Palmertz, J Haeggstrom, K.C. Nicolau, and S-E Dahlen. 1989. Lipoxin A4 inhibits leukotriene B4-induced inflammation in the hamster cheek pouch. Acta. Physiol. Scand. 137:571-572.

17. Badr, KF, DeBoer, D, Schwartzberg, M, and Serhan, CN. 1989. Lipoxin A4 antagonizes cellular and in vivo actions of leukotriene D4 in rat glomerular mesangial cells: Evidence for competition at a common receptor. Proc. Nat'l. Acad. Science. U.S.A. 86: 3438-3442.

18. Lefer, A.M., G.L Stahl, D. J. Lefer, M.E. Brezinski, K.C. Nicolau, C.A. Veale, Y. Abe, and J. Bryan Smith. 1988. Lipoxins A4 and B4: comparison of icosanoids having bronchoconstrictor and vasodilator actions but lacking lacking platelet aggregatory activity. Proc. Nat'l Acad. Sci, USA. 85:8340-8344.

19. Badr, KF, Serhan, CN, Nicolau, KC, and Samuelsson, B. 1987. The Action of Lipoxin A on Glomerular Microcirculatory Dynamics in the Rat. Biochem. Biophys. Res.Commun. 145:408-414.

20. Dahlen, E-E, J. Raud, C.N. Serhan, J. Bjork, and B Samuelsson. 1987. Biological activities of lipoxin A include lung strip contraction and dilation of arterioles in vivo. Acta. Physiol. Scand. 130:643-647.

21. Badr, KF, C. Baylis, J.M. Pfeffer, M.A. Pfeffer, R.J. Soberman, R.A. Lewis, K.F. Austen, E.J. Corey and B.M. Brenner. 1984. Renal and systemic hemodynamic responses to intravenous infusion of leukotriene C4 in the rat. Circ. Res. 54:492-499.

22. Sun, F.F., Chau, L.Y., and Austen, K.F. 1987. Binding of leukotriene C4 by glutathione transferase: a reassessment of biochemical and functional criteria for leukotriene receptors. Fed, Proc. 46:204-207.

23. Badr, K.F., Brenner, B.M., and Ichikawa, I. 1987. Effects of leukotriene D4 on glomerular dynamics in the rat. Am. J. Physiol. 22:F239-F243.

24. Barnett, R, P. Goldwasser, L.A. Scharschmidt and D. Schlondorff. 1986. Effects of leukotrienes on isolated rat glomeruli and cultured mesangial cells. Am. J. Physiol. 19: F838-F844.

25. Simonson, M.S. and M.J. Dunn. 1986. Leukotriene C4 and D4 contract rat glomerular mesangial cells. Kidney Int. 30:524-531.

26. Badr, KF, Hoover, RL, Mong, S, Ebert, J, Schwartzberg, M., Jacobson, HR, and Harris, RC. 1989. Leukotriene D4 Binding and Signal Transduction in Rat Glomerular Mesangial Cells . Am. J. Physiol. (Renal Fluid and Electrolyte Physiol. 26) F280-F287.

27. Simonson, M. S., Mene, P., Dubyak, G.R., and Dunn, M.J. 1988. Identification and transmembrane signaling of leukotriene D4 receptors in human mesangial cells. Am. J. Physiol. 255:C771-C780.

28. Cattell, V., H. T. Cook, J. Smith, J. A. Salmon, and S. Moncada. 1987. Leukotriene B4 production in normal rat glomeruli. Nephrol. Dial. Transplant 2:154-157.

29. Lianos, E. A. 1988. Synthesis of hydroxyeicosatetraenoic acids and leukotrienes in rat nephrotoxic serum glomerulonephritis: Role of anti-glomerular basement membrane antibody dose, complement, and neutrophiles. J. Clin. Invest. 82:427-435.

30. Fauler, J., A. Wiemeyer, K.-H. Marx, K. Kühn, K. M. Koch, and J. C. Fröhlich. 1989. LTB4 in nephrotoxic serum nephritis in rats. Kidney Int. 36:46-50.

31. Rahman, M.A., M. Nakazawa, S. N. Emancipator, and M. J. Dunn. 1988. Increased leukotriene B4 synthesis in immune injured rat glomeruli. J. Clin. Invest. 81:1945-1952.

32. Schreiner, G. F., B. Rovin, and J. B. Lefkowith. 1989. The antiinflammatory effects of essential fatty acid deficiency in experimental glomerulonephritis: The modulation of macrophage migration and eicosanoid metabolism. J. Immunol. 143:3192-3199.

33. Lianos, E. A., and B. A. Bresnahan. 1990. Origin of leukotrienes and HETE in glomerular epithelial and mesangial cell immune injury. Kidney Int. 37:421. (Abstr.)

34. Lianos, E. A., B. Noble, and B. Hucke. 1989. Glomerular leukotriene synthesis in Heymann nephritis. Kidney Int. 36:998-1002.

35. Funk, C.D., Radmark, O., Fu, J.Y., Matsumoto, T., Jornvall, H. Shimizu, T., and Samuelsson, B. 1987. Molecular cloning and amino acid sequence of leukotriene A4 hydrolase. Proc. Nat'l. Acad. Sci. USA 84:6677-6681.

36. Matsumoto, T., Funk, C.D., Radmark, O., Hoog, J-O, Jornvall, H., and Samuelsson, B. 1988. Molecular cloning and amino acid sequence of human 5-lipoxygenase. Proc. Nat'l Acad. Sci. USA 85:26-30.

37. Sigal, E., C.S. Craik, E. Highland, D. Grunberger, L.L. Costello, R.A.F. Dixon, and J.A. Nadel. 1988. Molecular cloning and primary structure of human 15-lipoxygenase. Biochem. bophys. Res. Commun. 157:457-464.

38. Imai, E, Hoover, RL, Makita, N, Funk, CD and Badr, KF. 1990. Localization and relative abundance of 5-lipoxygenase, 15-lipoxygenase, 12-lipoxygenase and leukotriene A4-hydrolase gene expression in cultured glomerular cells. J. Am. Soc. Nephrol. 1:751.

39. Badr, KF, Frazer, M, Hoover, RL, Imai, E and Funk, CD. 1990. Leukotriene A4 hydrolase gene expression and catalytic activity in cultured glomerular cells: implications for glomerular immune injury. J. Am. Soc. Nephrol. 1:437.

40. Hoover, RL, Imai, E., Makita, N, Funk, CD, and Badr, KF. 1990. Neutrophil adhesion suppresses leukotriene A4 hydrolase gene expression in human renal microvascular endothelial cells: potential mechanism for the arrest of PMN infiltration during inflammation. J. Am. Soc. Nephrol. 1:443.

41. Feinmark, S. J., and P. J. Cannon. 1986. Endothelial cell leukotriene C4 synthesis results from intercellular transfer of leukotriene A4 synthesized by polymorphonuclear leukocytes. J. Biol. Chem. 261:16466-16472.

42. Fitzpatrick, F., Ligget, W., McGee J., Bunting, S., Morton, D., and Samuelsson, B. 1984. Metabolism of leukotriene A4 by human erythrocytes. J. Biol. Chem. 259:11403-11407.

43. Maclouf, J. A., and R. C. Murphy. 1988. Transcellular metabolism of neutrophil-derived leukotriene A4 by human platelets. A potential cellular source of leukotriene C4. J. Biol. Chem. 263:174-181

44. Marcus, A., Broekman, M., Safier, M.L., Ullman, H., and Islam. N. Formation of leukotrienes and other hydroxy acids during platelet-neutrophil interactions in vitro. Biochem. Biophys. Res. Commun. 109:130-137, 1982.

45. Edenius, C, J. Haeggstrom, and J. A. Lindgren. 1988. Transcellular conversion of endogenous archidonic acid to lipoxins in mixed human platelet-granulocyte suspensions. Biochem. Biophys. Res. Commun. 157:801-807.

46. Serhan, C.N. and K-A Sheppard. Lipoxin formation during human neutrophil-platelet interactions. Evidence for the transformation of leukotriene A4 by platelet 12-lipoxygenase in vitro. 1990. J. Clin. Invest. 85:772-780.

47. Garrick, R, Shen, S-Y, Ogunc, S., and Wong, P Y-K. 1989.Transformation of leukotriene A4 to lipoxins by rta kidney mesangial cell. Biochem. Biophys. Res. Commun. 162:626-633.

48. Baud, L., Hagege, J., Sraer, J., Rondeau, E., Perez, J., and Ardaillou, R. 1983. Reactive oxygen production by cultured rat glomerular mesangial cell during phagocytosis is associated with stimulation of lipoxygenase activity. J Exp Med. 158:1836-1852.

49. Badr, KF, Schreiner, GF, Wasserman, M, and Ichikawa, I. 1988. Preservation of the Glomerular Capillary Ultrafiltration Coefficient During Rat Nephrotoxic Serum Nephritis by a Specific Leukotriene D4 Receptor Antagonist. J. Clin. Invest. 81: 1702-1709.

50. Spurney, R.F., P. Ruiz, D.S. Pisetsy, and T. M. Coffman. 1991. Enhanced renal leukotriene production in murine lupus: Role of lipoxygenase metabolites. Kidney. Int'l. 39:95-102.

51. Badr, KF. 1989. Leukotriene D4/leukotriene B4 interactions in the pathophysiology of experimental glomerulonephritis. In Advances in Prostaglandin, Thromboxane, and Leukotriene Research. Vol. 18. Wong, P. K-Y, Samuelsson, B., and Sun, F.F., eds. Raven Press, N.Y. pp 233-236.

52. Takahashi, K., Schreiner, GF, Yamashita, K., Christman, B, Blair, I., and Badr, KF. 1990. Predominant functional roles for thromboxane A2 and prostaglandin E2 during chronic mesangioproliferative glomerulonephritis in the rat. 1990. J. Clin. Invest. 85: 1974-1982.

53. Fischer, D., Takahashi, K, Ebert, J, and Badr, KF. 1990. Limited early therapy with a novel 5-lipoxygenase (5-LO) activating protein (FLAP) antagonist, MK 886, during heterologous rat nephrotoxic serum (NTS) nephritis totally prevents proteinuria in the autologous phase. J. Am. Soc. Nephrol. 1:628.

54. Serhan, C.N. 1989. On the relationship between leukotriene and lipoxin production by human neutrophils: evidence for differential metabolism of 15-HETE and 5-HETE. Biochem. Biophys. Acta. 1004: 158-168.

55. Brady, H.R., U Persson, B. M. Brenner, and C. N. Serhan. 1990. Leukotrienes and lipoxins modulate neutrophil-mesangial cell adhesion: role of CD18/CD11 complex. Clin. Res. 38:275A.

56. Nicolau, K.C., Marron, B.E., Veale, C.A., Webber, S.E., Dahlen, S-E, Samuelsson, B., and Serhan, C. N. 1989.Identification of a vovel 7-cis-11-trans-lipxin A4 generated by human neutrophils total synthesis, spasmogenic activities, and comparison with other geometric isomers of lipoxin A4 and B4. Biochem, Biophys. Acta. 1003:44-53.

57. Marron, B.E., R.A. Spanecello, M.E. Elisseou, C.N. Serhan, and K.C. Nicolau. 1989. Synthesis of 19,19,20,20,20,-pentadeuterlipoxin A4 methyl ester and 19,19,20,20 20,-pentadeuterioarachidonic acid. Agents for use in the quantitative detection of naturally occurring eicosanoids, J. Org. Chem. 54:5522-5534.

NEUTROPHIL ADHESION TO GLOMERULAR MESANGIAL CELLS:

REGULATION BY LIPOXYGENASE-DERIVED EICOSANOIDS

Hugh R. Brady#, Mark D. Denton#, Barry M. Brenner#,
Charles N. Serhan*

Renal Division# and Hematology Division*, Department of
Medicine, Brigham & Women's Hospital, Harvard Medical
School, Boston, MA 02115, USA

NEUTROPHIL-GLOMERULAR CELL INTERACTION IN ACUTE GLOMERULONEPHRITIS

Glomerulonephritis (GN) is the leading cause of end-stage renal failure (1). While the pathophysiology of this condition is incompletely understood, polymorphonuclear leukocyte (PMN) and monocyte infiltration of the glomerulus is a characteristic early pathologic finding in many forms of human and experimental GN (reviewed in refs.1-4). For example, an intense PMN infiltrate is a prominant feature of several primary glomerular diseases in man, such as acute diffuse proliferative and crescentic GN (1-5, e.g. Fig.1), and also characterizes proliferative GN complicating systemic diseases, such as systemic lupus erythematosus (9). In addition, PMN infiltration is a common finding during the heterologous phase of experimental nephrotoxic serum nephritis, and glomerular hypercellularity and proteinuria can be abrogated in this model by prior depletion of phagocytes (6). Monocytic infiltration of the glomerulus has also been identified by morphologic and immunohistochemical techniques in clinical forms of postinfectious and crescentic GN (1-4), and the accelerated model of experimental nephrotoxic serum nephritis (1-4,7,8). In the latter condition, glomerular monocyte infiltration parallels the development of proteinuria, and, here also, prior depletion of monocytes by irradiation abrogates the development of proteinuria and glomerular hypercellularity (7). The mechanism(s) involved in leukocyte entrapment within the glomerulus in acute GN have not been fully established.

Leukocyte adhesion to endothelial cells through specific leukocyte adhesion molecules is a critical early event in leukocyte trafficking to sites of inflammation (reviewed in refs. 10,30,31). The glomerular capillary circulation is unique, however, in that circulating inflammatory cells also have the potential to adhere to mesangial cells (glomerular pericytes with smooth muscle and phagocytic properties) as the latter can be found interposed between endothelial cells, or may be in direct contact with circulating blood elements when the endothelium is denuded during the injury process (1-9). In addition, mesangial cells may contribute to the entrapment process by interaction with PMN which have diapedesed through

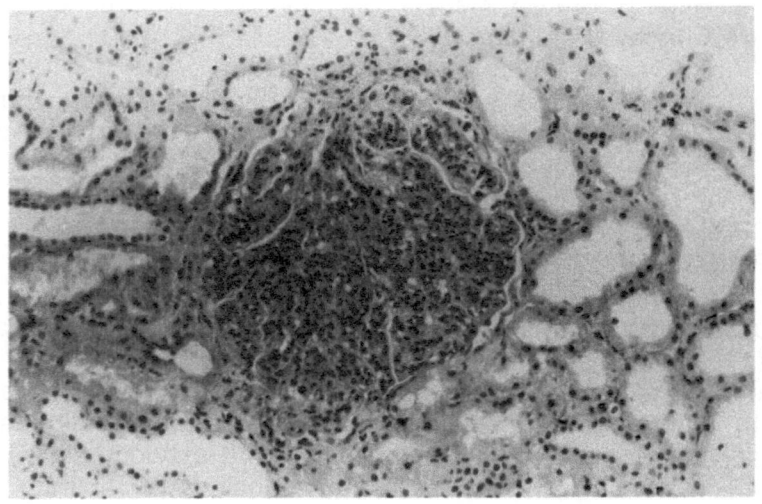

Figure 1. Photomicrograph of a renal biopsy specimen from a patient with poststreptococcal acute diffuse proliferative glomerulonephritis showing glomerular hypercellularity and intense PMN infiltration.

endothelium to the extravascular space. We studied PMN adhesion to mesangial cells using a quantitative monolayer adhesion assay as an in vitro model of glomerular inflammation (3). In particular, we assessed the influence of lipoxygenase products on this process, since these compounds have been implicated as important inflammatory mediators during the early stages of GN (vide infra, reviewed in ref. 12).

PMN-MESANGIAL CELL ADHESION: ACTIONS OF LEUKOTRIENES.

Enhanced leukotriene B4 (LTB4) production has been demonstrated in several models of GN, including the heterologous phase of experimental anti-glomerular basement membrane disease, and passive Heymann nephritis. While a causal relationship between LTB4 production and glomerular cell injury has not been demonstrated, several lines of evidence suggest that LTB4 may be important in mediating leukocyte activation under these circumstances. LTB4 causes PMN margination, diapedesis and migration to the extravascular space <u>in vivo</u> (reviewed in ref 21). LTB4 is chemotactic for phagocytes, and increases phagocyte adhesiveness for a variety of cells and artificial substrates <u>in vitro</u> (16,17). In addition, LTB4 is a potent stimulus for superoxide anion generation and lysosomal enzyme release by phagocytes (21), both of which have been implicated as mediators of tissue injury in GN (27). While increased production of peptidoleukotrienes (LTC4, LTD4, LTE4) has not been reported in GN, there is indirect evidence to suggest that these compounds also play a role in the pathophysiology of this disease. LTC4 and LTD4 cause a dramatic reduction in renal blood flow and glomerular filtration rate (GFR) <u>in vivo</u>, probably due to contraction of mesangial and vascular smooth muscle cells. LTD4 receptor antagonists abrogate the fall in GFR which occurs in experimental nephrotoxic serum nephritis (15). In addition, several investigators have reported that LTC4 and LTD4 may also increase PMN adhesiveness to cultured endothelial cells and artificial substrates <u>in vitro</u> (18,19). Given these observations, it was of interest to determine if leukotrienes provoke PMN adhesion to glomerular mesangial cells.

METHODS

PMN and Mesangial Cells

PMN and mesangial cells were isolated as previously described (11). Briefly, PMN were isolated from heparinized venous blood drawn from healthy volunteers by the standard procedure of Ficoll-Hypaque gradient centrifugation followed by dextran sedimentation. The isolated cells were suspended in Dulbecco's phosphate-buffered saline (PBS, pH 7.4) or RPMI 1640 medium with glutamine (GIBCO, NY), containing 5% bovine calf serum. These suspensions contained $98 \pm 1\%$ PMN as determined by light microscopy. Mesangial cells were cultured from glomeruli isolated from human cadaver kidney which could not be used for transplantation, or from bovine calf kidney, and identified by their characteristic morphology, positive immunofluorescent staining for myosin and negative staining for cytokeratin, and with factor VIII. Adhesion experiments were performed using monolayers between passage 5 and 15, except in the case of LTD4 when responsiveness decreased significantly after passage 10.

PMN-Mesangial Cell Adhesion

PMN adhesion to mesangial cells was assessed by an in vitro quantitative monolayer adhesion assay using [111]indium-labeled PMN and mesangial cell grown to confluence on 24-well plastic tissue culture plates (11). After coincubations, the wells were washed to remove non-adherent PMN, the contents solubilized, and radioactivity measured in a gamma counter (LKB Clinigamma). The number of PMN adherent per mm^2 of mesangial cell monolayer was calculated from the specific activity of the PMN preparation and the surface area of the mesangial cell monolayer. Agonist-induced adhesion was calculated by subtraction of adhesion in vehicle-treated controls.

PMN Aggregation

PMN aggregation (homotypic PMN adhesion) was monitored as changes in light transmittance using a four channel aggregation profiler (model PAP-4, Biodata)(11). PMN (5×10^6 cells/ml) were suspended in PBS, pH 7.45, at 37°C in siliconized cuvettes. Cytochalasin B (5 ug/ml) was added ~3 min before addition of stimuli. PMN aggregation was measured as a change in light transmittance over 10 min and was expressed as percentage of aggregation observed with equimolar concentrations of the synthetic peptide N-formyl-L-methionyl-L-leucyl-L-phenylalanine (FMLP).

Analysis of Leukotrienes

Leukotrienes were extracted from PMN-mesangial cell coincubations by utilization of a combination of techniques (11,13). For analysis of LTB4 metabolism, samples were injected into a gradient LKB dual-pump HPLC (Bromma, Sweden) equipped with an Altex Ultraspere-ODS (4.6 mm x 25 cm) column, injector, solvent controller and rapid spectral detector. The column was eluted with a gradient consisting of MeOH:H2O:acetic acid (65:35:0.01) as phase 1 (to -20 min), a linear gradient with MeOH:acetic acid (99.99:0.01) as phase 2 (30-50 min), and a flow rate of 1 ml/min. Post-HPLC analyses were performed with a Wavescan 2140-202 program and a Nelson Analytical 3000 series chromatography data system (5,11). For analysis of LTD4 metabolism, samples were injected into a system consisting of a Beckman gradient HPLC, an Altex Ultrasphere-ODS column, injector, and a Perkin-Elmer LC-75 spectrophotometric detector (11). The column was eluted with MeOH:H2O:acetic acid (65:35:0.01, pH 5.7, flow rate 1ml/min).

Reagents

Monoclonal antibodies against CD18 (TS1/18), CD11a (TS1/22),

CD11b (LM2/1) (14) were gifts from Drs. R. Rothlein and S.D. Marlin, Boehringer Ingelheim Pharmaceuticals (Ridgefield, CT). Monoclonal antibody against CD11c (CBRp150) was a gift from Drs. S. Stacker and T.A. Springer (Harv. Med. Sch.). Leukotrienes and lipoxins were obtained from Biomol Research Laboratories (Philadelphia, PA). The UV and chromatographic properties of each compound was checked before coincubation, and eicosanoid concentrations determined from their extinction coefficients.

Statistics

Data are expressed as means $\pm$ SE of at least three experiments and statistically significant differences between means ($p < 0.05$) were determined by use of Student's t test or ANOVA.

RESULTS

PMN and mesangial cells were coincubated for 30 minutes in the presence of 10^{-7}M LTB4 or LTD4 to determine if these compounds influence PMN-mesangial cell adhesion. Both LTB4 and LTD4 provoked adhesion in a concentration- and time-dependent manner (Table 1). Adhesion was rapid in onset, being observed within minutes of exposure, and was sustained over the 30 minute coincubation period.

The time course of LTB4-induced adhesion, specifically the sustained nature of the response, was intriguing as LTB4 is believed to be rapidly metabolized by PMN to inactive omega-oxidation products (21). Indeed, previous studies of LTB4-induced PMN-endothelial cell adhesion demonstrated a rapid, but short-lived process (8,9). LTB4 and LTD4 levels were monitored during PMN-mesangial cell coincubations to determine the fate of these compounds under these conditions. RP-HPLC analysis of materials recovered after 30 min coincubation revealed no decrement in LTB4 (Figure 2) or LTD4 levels (data not shown), and suggested that little leukotriene metabolism to inactive products had ocurred. The absence of LTB4 metabolism to inactive omega-oxidation products may relate to the presence of serum in the coincubation medium, as there is evidence that omega-oxidation of LTB4 by PMN is inhibited in the presence of serum or plasma proteins (12).

LTB4 appeared to provoke PMN-mesangial cell adhesion via an action with PMN. PMN in suspension underwent rapid homotypic adhesion upon exposure to LTB4, while increased PMN-mesangial cell adhesion was not observed when mesangial cells were exposed to LTB4 (10^{-7}M) for 30 minutes, followed by washing, prior to coincubation with PMN. In contrast, LTD4 appeared to provoke adhesion via an action on mesangial cells. LTD4 did not provoke homotypic adhesion of PMN in

Table 1. LTB4 and LTD4 provoke PMN-mesangial cell adhesion: concentration and time dependence.

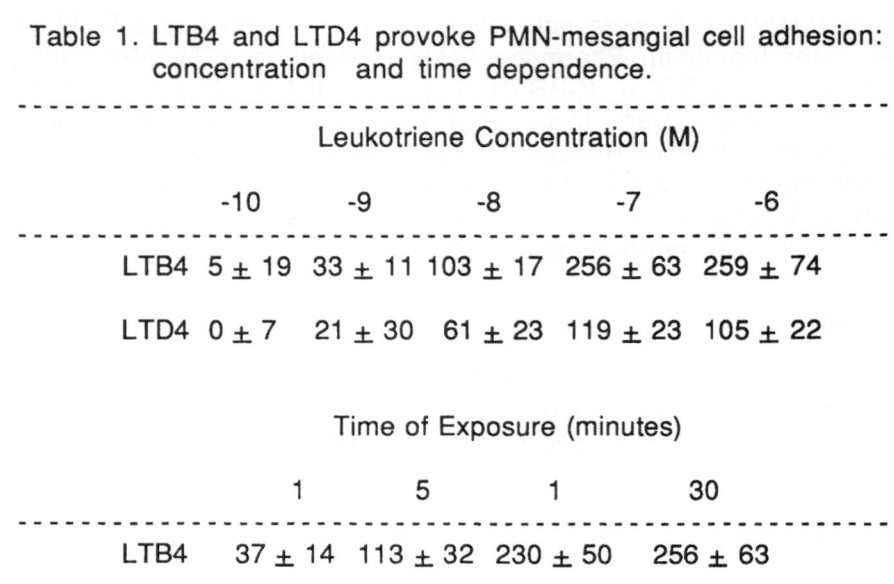

	Leukotriene Concentration (M)				
	-10	-9	-8	-7	-6
LTB4	5 ± 19	33 ± 11	103 ± 17	256 ± 63	259 ± 74
LTD4	0 ± 7	21 ± 30	61 ± 23	119 ± 23	105 ± 22

	Time of Exposure (minutes)			
	1	5	1	30
LTB4	37 ± 14	113 ± 32	230 ± 50	256 ± 63
LTD4	8 ± 27	44 ± 23	71 ± 30	119 ± 23

Data are PMN adherent per mm^2 of mesangial cells. In studies of dose-dependence, PMN and mesangial cells were coincubated for 30 minutes in the presence different concentrations of leukotrienes. In studies time dependence PMN and mesangial cells were coincubated for different periods of time in the presence of 10^{-7}M leukotrienes. Leukotriene-induced adhesion was calculated by subtraction of adhesion in vehicle-treated controls. Data are mean ± SE of 3-9 experiments. For comparison, adhesion provoke by 30 minute exposure to the synthetic chemotactic peptide FMLP (10^{-7}M) was 487 ± 126 PMN/mm^2.

suspension, and PMN did not show increased adhesiveness for mesangial cells when they alone were exposed to LTD4 (10^{-7}M for 30 min), prior to coincubation with mesangial cells. In support of these data, LTB4 has been previously demonstrated to be a potent chemoattractant and activator of PMN (21), without known actions on mesangial cells (22), while LTD4 has a well-defined receptor and stimulates mesangial cell contraction (22-24), but does not appear to directly affect PMN function (21).

The rapid onset of adhesion following exposure to leukotrienes (Table 1) suggested the involvement of pre-formed leukocyte adhesion molecules, rather than induction of adhesion molecule biosynthesis and surface expression. In keeping with this interpretation, LTB4- and LTD4-induced adhesion were not affected by inhibitors of protein (cycloheximide) or RNA (actinomycin D)

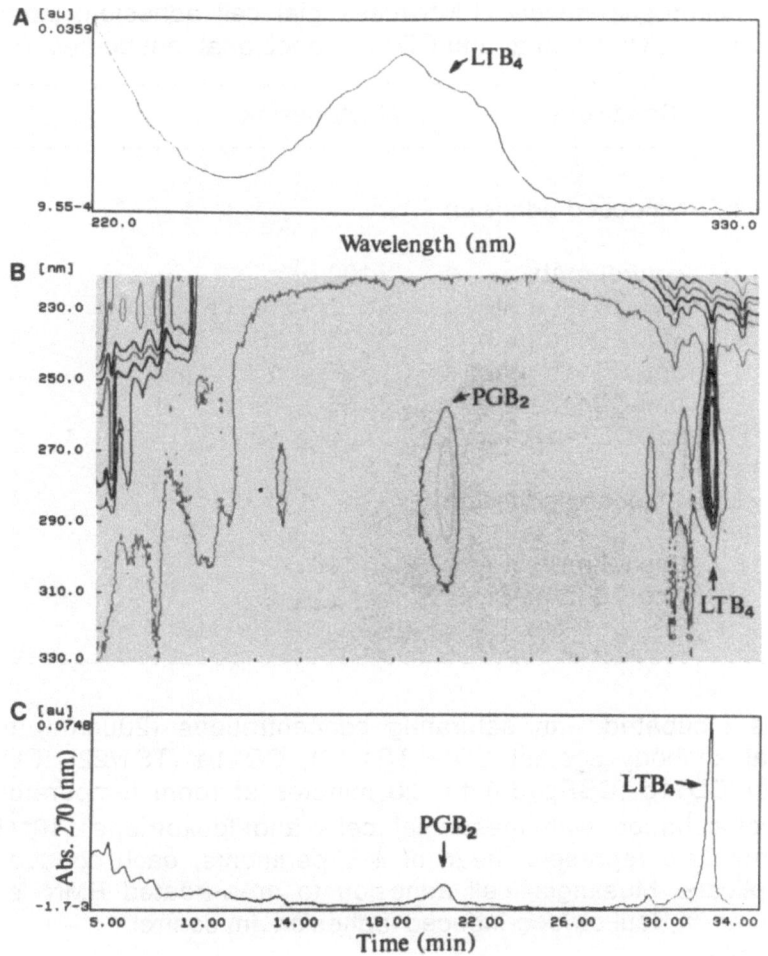

Figure 2. Results of RP-HPLC of material recovered after addition of LTB4 (1.5ug) to a PMN-mesangial cell coincubation. Materials were extracted as described under Methods. Post-HPLC analyses were performed with a Wavescan 2140-202 program and a Nelson Analytical 3000 series chromatography data system. A: UV spectrum of LTB4 peak. B: isogram recalled from the spectral data collected by scanning between 220 and 330nm. C: a single wavelength chromatogram (270nm) recalled from the spectral data. LTB4 recovery was calculated from it's extinction coefficient, with PGB2 as an internal standard for the extraction procedure. (Reproduced, with permission, from Am. J. Physiol. 259: F809-815, 1990).

synthesis (data not shown). LTB4-induced PMN adhesion to endothelial cells has been shown to be mediated through the CD11/CD18 family of leukocyte adhesion molecules (10,30). A blocking monoclonal antibody (mAb) against the common CD18 chain of these heterodimers was used to define the contribution of these

Table 2. Leukotriene-induced PMN-mesangial cell adhesion: Inhibition by anti-CD18 and anti-CD11 monoclonal antibodies (mAb).

Conditions	PMN Adhesion

LTB4-induced adhesion

control mAb	100
anti-CD18 mAb	5 ± 10
anti-CD11a mAb	75 ± 15
anti-CD11b mAb	25 ± 7
anti-CD11c mAb	55 ± 10
anti-CD11a+b+c	3 ± 10

LTD4-induced adhesion

control mAb	100
anti-CD18 mAb	100 ± 0

PMN were incubated with saturating concentrations (20ug/ml) of monoclonal antibody against CD18(TS1/18), CD11a (TS1/22), CD11b (LM2/1), or CD11c (CBRp150) for 20 minutes at room temperature prior to coincubation with mesangial cells and leukotriene (10^{-7}M for 30 min). Data represent mean of 4 experiments, each conducted in quadruplicate. Mesangial cell adhesion to mAb-treated PMN is expressed as % leukotriene-induced adhesion in control coincubations.

molecules to LTB4- and LTD4-induced PMN-mesangial cell adhesion (Table 2). LTB4-induced adhesion was completely inhibited in the presence of this mAb. In contrast, LTD4 adhesion was not affected suggesting that this compound provokes adhesion through other epitopes. Monoclonal antibodies against the individual CD11 subunits were used to probe the relative contributions of the individual CD11/CD18 leukocyte adhesion molecules to the LTB4-induced response. LTB4-induced adhesion was markedly attenuated by mAb against CD11b and, to a lesser extent, CD11c (Table 2). In contrast, mAb against CD11a afforded little inhibition of this response. LTB4 has been previously shown to increase CD11b/CD18 and CD11c/CD18 surface expression on PMN and monocytes within minutes of exposure by mobilization of molecules from intracellular vesicular stores, and anti-CD11b mAb attenuates LTB4-induced PMN-endothelial cell adhesion in vitro (10,30). Our data suggest that CD11b/CD18 (Mac-1, Mo1) and CD11c/CD18 (p150,95, LeuM5) support LTB4-induced adhesion of PMN to mesangial cells in a similar manner.

LEUKOTRIENE-INDUCED PMN-MESANGIAL CELL ADHESION: MODULATION BY LIPOXINS

Lipoxins are a series of lipoxygenase products, structurally and functionally distinct from leukotrienes, which are generated by the sequential actions of either 5- and 15 lipoxygenase in granulocytes, or 5- and 12-lipoxygenase during PMN-platelet interaction (17,21,28). There appears to be an inverse relationship between the amounts of leukotrienes and lipoxins formed in vitro, which suggests the potential for counterregulatory interactions between these two series of lipoxygenase products. Indeed, lipoxins have been shown to inhibit several actions of leukotrienes in vivo and in vitro (23,25,26,29). Lipoxin A4 (LXA4) abrogates the fall in renal plasma flow and glomerular filtration rate caused by infusion of LTD4 into rats, and reduces LTD4 binding to it's mesangial cell receptor in vitro (23). In addition, prior exposure of bronchial smooth muscle to LXA4 attenuates subsequent contractile responses to LTC4 (29). LXA4 can also inhibit LTB4-induced PMN chemotaxis in vitro and in vivo, and LTB4-induced phosphoinositide hydrolysis in PMN (25,26). It has not been determined whether lipoxin biosynthesis is increased during glomerular inflammation. However, glomerular mesangial cells can generate LXA4 from exogenous LTA4 in vitro (32), possibly via the action of mesangial cell 12-lipoxygenase, thus providing the potential for lipoxin biosynthesis in GN by infiltrating PMN alone, or through PMN-mesangial cell interaction. Given these observations, it was of interest to determine if lipoxins influence basal PMN-mesangial cell adhesion, or modulate leukotriene-induced responses.

PMN and mesangial cells were coincubated in the presence of LXA4 or LXB4 to determine if these compounds influenced basal adhesion. Neither LXA4 nor LXB4 (10^{-7}M - 10^{-10}M for 30 min) provoked statistically significant PMN adhesion to mesangial cells (data not shown). In contrast, exposure of PMN-mesangial cell coincubations to lipoxins (10^{-7}M) prior to addition of LTD4 resulted in significant inhibition of LTD4-induced adhesion (Figure 3). LXA4 appeared to be more potent than LXB4 in this regard. Neither LXA4 nor LXB4 inhibited LTB4-induced adhesion. These data suggest that lipoxins can modulate leukotriene-induced PMN-mesangial cell adhesion and provide further evidence that lipoxins and leukotrienes may have counterregulatory actions during inflammation.

SUMMARY

Our results suggest that leukotrienes promote rapid PMN adhesion to glomerular mesangial cells via actions on PMN (LTB4) and mesangial cells (LTD4). Leukotriene-induced adhesion appeared to be mediated by a CD11/CD18-dependent (LTB4) and -independent (LTD4) mechanisms. The specific epitopes mediating LTD4-induced adhesion remain to be defined. Lipoxins did not influence basal adhesion. In contrast, lipoxins markedly inhibited LTD4-, but not LTB4-induced responses. Further elucidation of the components of these adhesion processes, of the pathways for leukotriene and lipoxin biosynthesis in the inflammed glomerulus, and of the counterregulatory actions of lipoxins and leukotrienes may reveal sites for therapeutic intervention in GN.

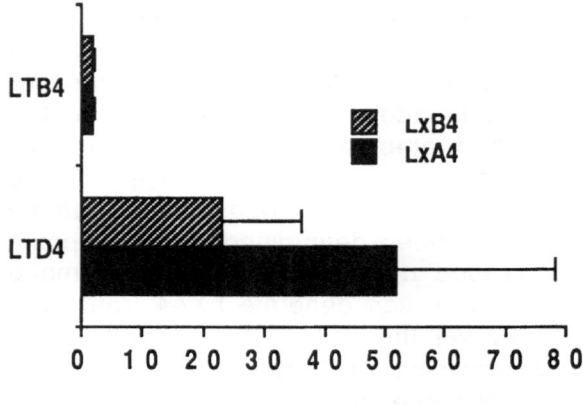

Figure 3. Lipoxins inhibit LTD4, but not LTB4-induced PMN-mesangial cell adhesion.

Data are mean ± SE of 3 experiments, each conducted in quadruplicate. Leukotriene-induced adhesion was not inhibited by vehicle alone.

ACKNOWLEDGEMENTS

These studies were supported in part by National Institutes of Health Grants DK-40839 (BMB), AI-26714, GM-38765, and PO1-HL-36028 (CNS). C.N. Serhan is a recipient of an Established Investigator Award from the American Heart Association and is a 1988 Pew Scholar in the Biomedical Sciences. Original studies referred to in this chapter are reported in ref. 3, and were presented at the 22nd Annual Meeting of the American Society of Nephrology, Washington, DC, December 1989 (abstract, Kidney Int 37: 346, 1990).

REFERENCES

1. C.B. Wilson and F.J. Dixon, The renal response to immunological injury, in "The Kidney," B.M. Brenner and F.C. Rector, eds., Saunders, Philadelphia, PA (1991).

2. R.J. Glassock, A.H. Cohen, S. Adler, and H. Ward, Primary glomerular diseases, in "The Kidney," B.M. Brenner and F.C. Rector, eds., Saunders, Philadelphia, PA (1991).

3. A.A. Eddy and A.F. Michael, Immunopathogenetic mechanisms of glomerular injury, in "Renal Pathology," C.C. Tisher and B.M. Brenner, eds., J.B. Lippincott, Philadelphia, PA, 111-155 (1990).

4. E.J. Lewis, T. Cavallo, J.T. Harrington, and R.S. Cotran, An immunopathologic study of rapidly progressive glomerulonephritis in the adult, Human Pathol. 2: 185 (1971).

5. A.J. Fish, R.C. Herdman, A:F. Michael, R.J. Pickering, and R.A. Good, Epidemic acute glomerulonephritis associated with type 49 streptococcal pyoderma. II. Correlative study of light, immunofluorescent, and electron microscopic findings, Am.J. Med. 48: 28 (1970).

6. C.G. Cochrane, E.R. Unanue, and F.J. Dixon. A role of polymorphonuclear leukocytes and complement in nephrotoxic nephritis, J. Exp. Med. 122: 99 (1978).

7. G.F. Schreiner, R.S. Cotran, V. Pardo, and E.R. Unanue. A mononuclear cell component in experimental immunological glomerulonephritis. J. Exp. Med. 147: 369 (1978).

8. H.M. Fillit and J.B. Zabriskie, Cellular immunology in glomerulonephritis. Am. J. Pathol. 109: 369 (1978).

9. M. Kashgarian and J.P. Hayslett. Renal involvement in systemic lupus erythematosus, in: "Renal Pathology," C.C. Tisher and B.M. Brenner eds., J.B. Lippincott, Philadelphia, PA, 380 (1990).

10. M. Patarroya and M.W. Makgoba, Leukocyte adhesion to cells: molecular basis, physiological relevance, and abnormalities, Scand. J. Immunol, 30: 129 (1989).

11. H.R. Brady, U. Persson, B.J. Ballermann, Barry M. Brenner, and Charles N. Serhan, Leukotrienes stimulate neutrophil adhesion to

mesangial cells: modulation with lipoxins, Am. J. Physiol. 259: F809 (1989).

12. E.A. Lianos, Eicosanoids and the modulation of glomerular immune injury, Kidney Int. 35: 985 (1989).

13. C.N. Serhan, HPLC separation and determination of the lipoxins, in "Methods in Enzymology: Arachidonate-Related Lipid Mediators," R.C. Murphy and F.A. Fitzpatrick, eds., academic, Orlando FL (in press).

14. C.W. Smith, S.D. Marlin, R. Rothlein, C. Toman, and C. Anderson, Cooperative interactions of LFA-1 and Mac-1 with intercellular adhesion molecule-1 in facilitating adherence and transendothelial migration of human neutrophils in vitro, J. Clin. Invest. 83: 637 (1989).

15. K. F. Badr, G.F. Schreiner, M. Wasserman, and I. Ishikawa, Preservation of glomerular capillary ultrafiltration coefficient during rat nephrotoxic serum nephritis by a specific leukotriene D4 receptor antagonist, J. Clin. Invest. 81: 1702 (1988).

16. M.A. Gimbrone, A.F. Brock, and A.I. Schafer, Leukotriene B4 stimulates polymorphonuclear leukocyte adhesion to cultured vascular endothelial cells. J. Clin. Invest. 74: 1552 (1984).

17. R.L. Hoover, M.J. Karnovsky, K.F. Austen, E.J. Corey, and R.A. Lewis. Leukotriene B4 action on endothelium mediates augmented neutrophil/endothelial adhesion. Proc. Natl. Acad. Sci. USA 81: 2191 (1984).

18. T.M. McIntyre, G.A. Zimmerman, and S.M. Prescott. Leukotriene C4 and D4 stimulate human endothelial cells to synthesize platelet-activating factor and bind neutrophils, Proc. Natl. Acad. Sci. USA 83: 2204 (1986).

19. E.J. Goetzl, L.L. Brindley, and D.W. Goldman, Enhancement of human neutrophil adherence by synthetic leukotriene constituents of the slow reacting substance of anaphylaxis, Immunology 50: 35 (1983).

20. R.M. McMillan, S.J. Foster, and P.A. Dieppe, Leukotriene B4 metabolism in human leukocytes: fact or artifact? Agents Actions 21: 355 (1987).

21. B. Samuelsson, S-E. Dahlen, J.A. Lindgren, C.A. Rouzer, and C.N. Serhan, Leukotrienes and lipoxins: structures, biosynthesis, and biological effects, Science Wash. DC 237: 1171 (1987).

22. M.S. Simonson, and M.J. Dunn, Leukotriene C4 and D4 contract rat glomerular mesangial cells. Kidney Int. 30: 524 (1986).

23. K.F. Badr, D.K. DeBoer, M. Schwartzberg, and C.N. Serhan, Lipoxin A4 antagonizes cellular and in vivo actions of leukotriene D4 in rat glomerular mesangial cells: evidence for competition at a common receptor, Proc. Natl. Acad. Sci. USA 86: 38 (1989).

24. R. Barnett, P. Golwasser, L.A. Scharschmidt, and D. Schlondorff, Effect of leukotrienes on isolated rat glomeruli and cultured mesangial cells. Am. J. Physiol. 250: F838 (1986).

25. T.H. Lee, C.E. Horton, U. Kyan-Aung, D. Haskard, A.E.J. Crea, and W. Spur, Lipoxin A4 and lipoxin B4 inhibit chemotactic responses of human neutrophils stimulated ny LTB4 and N-formyl-L-Methionyl-L-leucyl-L-phenylalanine, Clin. Sci. Lond. 77: 195 (1989).

26. P. Hedqqvist, J. Raud, U. Palmertz, J. Haeggstrom, K.C. Nicolau, S-E. Dahlen, Lipoxin A4 inhibits leukotriene B4-induced inflammation in the hamster cheek pouch, Acta Physiol. Scand. 137: 571 (1989).

27. W.G. Couser, Mediation of immune glomerular injury, J.A.S.N. 1: 13 (1990).

28. C.N. Serhan, and K-A. Sheppard, Lipoxin formation during human neutrophil-platelet interactions: evidence for the transformation of leukotriene A4 by platelet 12-lipoxygenase in vitro, J. Clin. Invest. 85: 772 (1990).

29. S-E. Dahlen, L. Franzen, J. Raud et al, Actions of lipoxin A4 and related compounds on smooth muscle preparations and on the microcirculation in vivo, Adv. Exper. Med. Biol. 229: 107 (1987).

30. M.A. Arnaout, Structure and function of the leukocyte adhesion molecules CD11/CD18, Blood 75: 1037 (1990).

31. T.A. Springer, Adhesion receptors of the immune system. Nature 346: 425 (1990).

32. R. Garrick, S.Y. Shen, S. Ogunc, P.Y-K. Wong, Transformation of leukotriene A4 to lipoxins by rat kidney mesangial cells, Biochem. Biophys. Res. Commun. 162: 626 (1989).

ENZYMATIC FORMATION AND REGULATORY FUNCTION OF LIPOXINS AND

LEUKOTRIENE B₄ IN RAT KIDNEY MESANGIAL CELLS

Renee Garrick and Patrick Y-K Wong

Departments of Physiology and Medicine
New York Medical College
Valhalla, N.Y. 10595

INTRODUCTION

Many metabolites of arachidonic acid (AA) have been demonstrated to act as second messengers for cellular functions and signal transduction. Following liberation from membrane phospholipids by the actions of phospholipases, arachidonic acid (AA) can be further metabolized by cyclooxygenase (CO), lipoxygenase (LO), and epoxygenase enzyme systems to generate oxygenated metabolites of AA that participate in the regulation of cell function.   There are at least three lipoxygenases responsible for the generation of biologically active metabolites from AA. These are the 5-, 12- and 15-lipoxygenases, which give rise to the predominately pro-inflammatory leukotrienes (LTs) and the mono and di-hydroxy eicosatetraenoic acids (HETEs). Recently lipoxins, a class of structurally related trihydroxytetraenes derived from AA and eicosapentaenoic acid (EPA) have been described (1,2). These compounds are formed by the sequential oxidation of these polyunsaturated fatty acids by 5-lipoxygenase and 15-Lipoxygenase or 12-Lipoxygenase (3,4). Lipoxins possess potent biological activities which are distinct from those of other lipoxygenase and cyclooxygenase products (5) and in many respects lipoxin have an apparent anti-inflammatory function. For example, prior exposure of neutrophils to LXA_4, inhibits LTB_4 induced phosphoinositol hydrolysis and impairs the chemotic response of polymorphonuclear cells to both LTB_4 and formyl-methionyl-leucyl-phenylalanine (fMLP) (6).

Recent investigations of glomerular microcirculation and glomerular mesangial cell functions suggest that in this tissue, lipoxins may modulate the effects of leukotrienes. For example, lipoxin A_4 has been shown to bind to the glomerular mesangial cell LTD_4 receptor, and to impair LTD_4-induced inositol phosphate generation (7). Along these same lines, preinfusion of LXA_4 antagonizes the LTD_4 induced fall in glomerular filtration rate, an effect that appears to be mediated by alterations in the contractile mechanism of the mesangial cell (8). Other studies indicate that LXA_4 modulates LTB_4-induced adherence of polymorphonuclear cells to mesangial cells (9).

Cell-Cell Interactions in the Release of Inflammatory Mediatiors
Edit ed by P. Y-K Wong and C.N. Serhan, Plenum Press, New York, 1991

In regards to the formation of lipoxins and other 5-lipoxygenase products, Serhan, et. al. (10) have demonstrated a dose dependent inverse relationship between the production of leukotriene B_4 and lipoxins from 15-HETE in polymorphonuclear cells stimulated with Ca^{+2} ionophore. We have previously reported that cultured rat glomerular mesangial cells contain lipoxygenases which convert exogenous LTA_4 to LXA_4 (11). In this communication we present evidence which suggests that cultured rat kidney mesangial cells possess two distinct enzyme systems which convert LTA_4 to LTB_4 and LXA_4. The conversion of the common intermediate, LTA_4, to LTB_4 and LXA_4 was found to be dose dependent and to vary in a reciprocal manner such that at low concentrations of LTA_4, the major product was LTB_4, whereas at high concentrations of LTA_4 the major product was LXA_4. The distinct biological activities of LXA_4 and LTB_4 and the finding that LTA_4 can be preferentially converted by the renal mesangial cells to either LTB_4 or LXA_4 suggests that a feed back regulatory system of self-defense may exist within the renal mesangium. Thus, in the presence of infiltrating inflammatory cells, as the concentration of LTB_4 and LTA_4 (derived from infiltrating cells) increases the transformation of LTA_4 to LTB_4 will decrease and the production of LXA_4 will increase. The LXA_4 can, in turn, modulate the proinflammatory effects of neutrophil and macrophage derived leukotrienes.

MATERIALS AND METHODS

LTA_4 methylester was converted to the Na salt by alkaline hydrolysis in methanol and NaOH (9:1) for four hours at 4°C. The LTA_4 substrate was either stored in hydrolysis solvent (stable for 72 hours as judged by UV analysis) or lyophilized and reconstituted with solvent before use. Authentic LTB_4 and LXA_4 standards were gifts from Dr. C. Pickett of Merck Frosst Canada. LTB_4 was estimated by radioimmunoassay kits obtained from commercial sources.

MESANGIAL CELLS PREPARATIONS AND INCUBATIONS

Using sterile techniques glomeruli were isolated as previously described from cortical section of male Sprague Dowley rats kidneys (11). Glomeruli were treated with collagenase, washed, then plated in RPMI media containing: 15% bovine calf serum, 25µg/ml fungizone, 100 Units/ml penicillin, 100 mg/ml streptomycin, 10mM L-glutamine, 15mM Hepes, 1mM sodium pyruvate and 0.66 Units/ml insulin. Mesangial cell outgrowth began after 2-4 days, and cells were passed 1:2 after 7 days in culture. Following the second pass, cells were identified as mesangial cells by morphologic, immunochemical, and biosynthetic criteria (11).

At the time of the experiment, cells were released from the culture plates with Trypsin-EDTA, and followed by neutralization of the trypsin with bovine serum. The cells were washed 3 times and resuspended in Krebs phosphate buffer. Viability of the cell was found to be greater than 97% as determined by trypan blue exclusion. Cells were incubated with LTA_4 in the presence of albumin (1µg/µl). To investigate the activity of 5-lipoxygenase, cells were incubated with 15-HPETE (25-150 µg/ml). Controls were run simultaneously with boiled cells or cells with vehicle alone. Sub-cellular fractions of lysed mesangial cells were prepared by differential ultracentrifugation at 10,000 and 105,000xg for 60 minutes.

PURIFICATION AND IDENTIFICATION OF PRODUCTS

Incubations were stopped with ethanol and the cells and media were then extracted with 10 volumes of ethanol. The ethanolic fraction was rotary evaporated to dryness and the residue was dissolved in methanol, and separated by RP-HPLC on a Water's dual pump system equipped with a reverse phase ultrasphere ODS column, (5μ, 10mm x 25cm). The products were eluted with a linear gradient from methanol/water/acetic acid (50:50:0.05, v/v) to methanol for 40 min at a flow rate of 1 ml/min. The wavelength detector was set at 301 nm (0-14 min), 270 nm (15-30 min) and 237 nm (31-40 min). Sample elution times were compared with authentic standards. For further purification, the lipoxin-like materials were methylated with diazomethane and rechromatographed in a second RP-HPLC equipped with the same column according to the method described by Serhan et al. (3,12). For characterization of the HETE productions, sample elution times were compared with authentic standards and the appropriate samples collected, dried with N_2 and resuspended in ethanol for analysis by radioimmunoassay.

UV SPECTROSCOPY AND GAS CHROMATOGRAPHY/MASS SPECTROMETRY

Samples eluted from the HPLC were evaporated to dryness under vacuum, resuspended in absolute ETOH (1ml), sonicated and examined with a Hewlett-Packard 8450 A-Diode Array ultraviolet/visible spectrophotometer. Lipoxin and leukotrienes concentrations were calculated using molar extinction coefficients of 50,000 and 40,000 respectively.

DATA ANALYSIS

Dose response of LXA_4 from LTA_4 substrate was determined by direct analysis of UV and HPLC peaks. Peak analysis was performed with Sigma Scan (Jandel Scientific Corp., Sausalito, CA) and digitilizer attached to IBM-AT computer.

RESULTS

Reverse phase HPLC analysis of the incubation products of mesangial cells with LTA_4 yielded 2 peaks (Figure 1), having a UV absorption at λmax 302nm with shoulders at 287 and 316nm, consistent with a conjugated tetraene structure of lipoxins (Fig 1, insert). Peak A, the major component from the HPLC co-eluted with synthetic LXA_4. The methyl ester of peak A, had identical HPLC retention times of LXA_4Me and trans LXA_4Me (12). Gas chromatography and mass spectroscopy analysis of peak A yield a spectrum identical to that of authentic LXA_4. Figure 2 demonstrates that when LTA_4 concentration was increased from 78 to 312 μM, LXA_4 production significantly increased. Incubations of LTA_4 alone, of boiled cells with LTA_4, and of cells with vehicle alone failed to produce any lipoxin or LTB_4-like

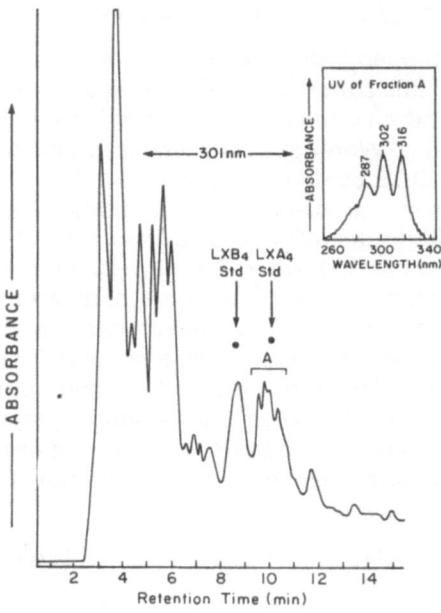

Figure 1. R.P. HPLC of mesangial cells (20x10⁶) incubated with LTA₄ (156µM) in Krebs Phosphate buffer. Elution of authentic Lipoxin standards shown by arrow. Insert shows U.V. spectroscopy of Peak A in methanol with characteristic " Tetraene" absorbance.

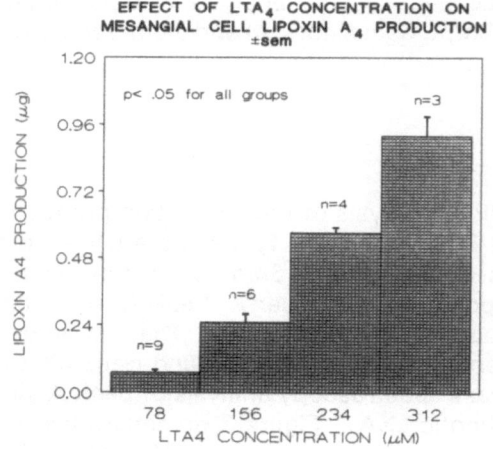

Figure 2. Effect of increasing concentrations of LTA₄ on the production of LXA₄ by mesangial cells.

substances. To further explore the enzymatic pathways responsible for the production of lipoxin from mesangial cell, cells were incubated with 15-HPETE (75μM to 415μM). These incubations failed to produce any lipoxin-like products. This finding suggests under these conditions 5-lipoxygenase activity in mesangial cells is not expressed, and that the formation of lipoxins depends on the actions of either 15- or 12-lipoxygenases. In agreement with this, incubation of mesangial cells with calcium ionophore A-23187 generated predominately 12-HETE and 15-HETE as detected by RP-HPLC and by RIA.

In addition, incubation of mesangial cells (2×10^7) with AA (30μM) and A23187 (3μM) failed to generate detectable levels of 5-HETE or LTB_4 as assessed by RP-HPLC and RIA. This further suggests that mesangial cells either lack 5-Lipoxygenase activity or that 5-lipoxygenase activity is suppressed in these cells. There is, however, strong evidence for the presence of LTA_4 hydrolase activity in mesangial cells. As shown in Figure 3, reverse phase HPLC analysis of incubations of mesangial cells with low concentrations of LTA_4 revealed three major peaks. One component, at a retention time of 20-21 minutes, co-eluted with LTB_4 standard, and displayed a UV spectrum characteristic of LTB4 with a λmax at 270nm and shoulders at 260nm and 280nm. GC/MS of the trimethylsilyether derivative revealed a characteristic mass ion of 492 and a fragmentation pattern identical to that of authentic LTB_4. As shown in Figure 4, LTB_4 formation varied directly with cell number, and its production was increased from 0.12 to 0.39μg as cell number increased from 5×10^6 to 40×10^6, indicating that LTB_4 production is dependent on cellular LTA_4 hydrolase activity. LTB_4 was not detected in incubations of boiled cells or vehicle controls incubated with LTA_4. Similarly, incubations of the 100,000 x g cytosolic fraction of mesangial cell with varying concentrations of LTA_4 produced LTB_4, whereas incubations of boiled supernatants with LTA_4 did not. The conversion of LTA_4 to LTB_4 suggests that mesangial cells have LTA_4 hydrolase activity. As shown in Figure 5, at low concentrations of the substrate LTA_4, LTB_4 is the major product, whereas at higher LTA_4 concentrations, LTB_4 production is significantly decreased and the production of LXA_4 is increased.

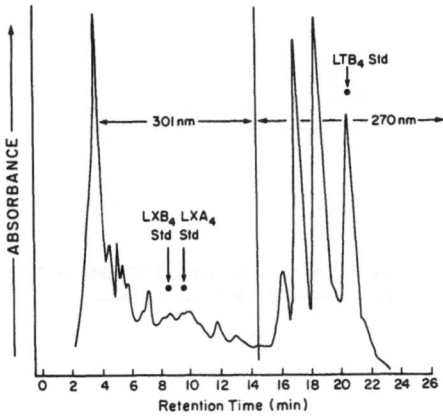

Figure 3. R.P. HPLC of mesangial cells (20×10^6) incubated with LTA_4
(78μM) under similar conditions. Elution of authentic
LTB_4 standard shown by arrow.

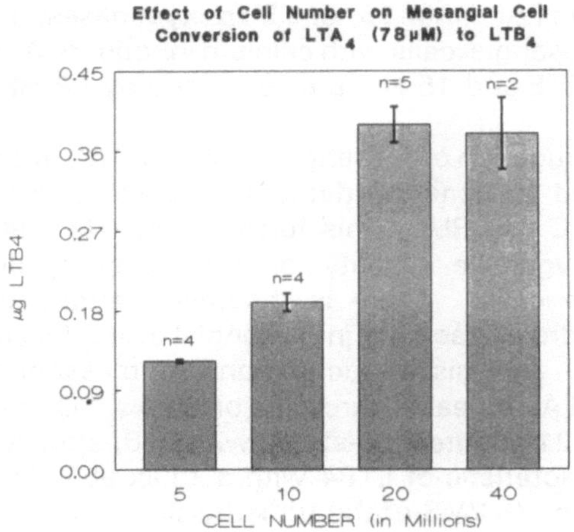

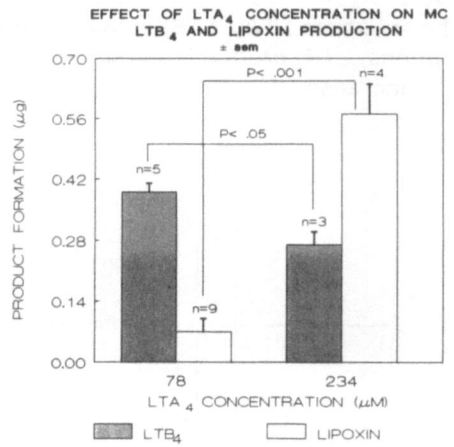

Figure 4. Effect of increasing cell numbers of mesangial cell on the conversion of LTA$_4$ (78μM) of LTB$_4$.

EFFECT OF LTA$_4$ CONCENTRATION ON MC
LTB$_4$ AND LIPOXIN PRODUCTION
± sem

Figure 5. Effect of LTA$_4$ concentration (78-234μM) on LTB$_4$ and LXA$_4$ production in mesangial cells.

DISCUSSION

The present studies demonstrated that cultured rat renal glomerular mesangial cells are capable of producing both LXA_4 and LTB_4 from a common intermediate, LTA_4. As shown in Figure 2, the production of LXA_4 from LTA_4 was dependent on substrate concentration. The amounts of LTA_4 are comparable to those employed in investigations of lipoxygenase products formation during neutrophil, platelet, and endothelial cell interactions (12-14). This concentration of LTA_4 had no effect on cell viability, as measured by trypan blue dye exclusion.

The conversion of LTA_4 to LXA_4 by mesangial cells suggests that these cells possess either 12- or 15-lipoxygenase activity. The recent detection of mRNA for 12-lipoxygenase in cultured mesangial cells further supports the presence of these lipoxygenase activities in this cell population (15). Mesangial cell 12-lipoxygenase activity could initiate LXA_4 synthesis from synthetic LTA_4 by the abstraction of hydrogen at C-13 of LTA_4 followed by omega-6 oxygenation. This has been demonstrated in studies with porcine leukocytes which revealed that 12-lipoxygenase possess omega -9 or omega-6 activity depending on the substrate (16). Further evidence for the presence of 12-lipoxygenase activity in mesangial cells comes from the finding that LTA_4 can be metabolized to both lipoxin A_4 and small amounts of trans LXA_4 (11). The latter compound may arise non-enzymatically from the formation of a delocalized cation intermediate following 12-lipoxygenation of LTA_4. An alternative mechanism for the transformation of LTA_4 to LXA_4 in mesangial cells may involve lipoxygenation of LTA_4 at C-15 by the action of 15-lipoxygenase followed by reduction to generate a 5(6) "epoxytetraene" intermediate (12). The latter would then undergo further hydrolysis to yield LXA_4 and its structural isomers. Further studies are needed to determine the detailed mechanism responsible for the conversion of LTA_4 to LXA_4 in this tissue.

The fact that mesangial cells were unable to convert 15-HPETE to LXA_4, or arachidonic acid to LTB_4 further supports the notion that 5-lipoxygenase is not expressed in this tissue. In contrast, the conversion of LTA_4 to LTB_4 by intact mesangial cells and by the cytosolic fraction of mesangial cells, suggests that mesangial cells possess LTA_4 hydrolase activity, and that in the rat this enzyme is distinct from 5-lipoxygenase. The presence of active LTA_4 hydrolase in these cells is further supported by the recent demonstration of mRNA for LTA_4 hydrolase in cultured rat mesangial cells (15). As shown in Figure 5, at low concentrations of the substrate LTA_4, LTB_4 is the major product, whereas at higher concentrations of LTA_4, the production of LTB_4 is significantly decreased while the production of LXA_4 is significantly increased. Thus, these data suggest that mesangial cells can convert LTA_4 to either LTB_4 or to lipoxins, and that the product generated is, in part, dependent upon the concentration of the substrate LTA_4. The finding that low concentrations of LTA_4 favors the production of LTB_4, suggests that the affinity of LTA_4 hydrolase for LTA_4 substrate is greater than that of either 12- or 15-Lipoxygenase. A likely source of LTA_4 is the activated leukocyte, and as demonstrated in non-renal systems (12,13,14), it is possible that via cell-cell interaction, the LTA_4 generated in leukocytes can be transferred to mesangial cells and further metabolized by LTA$_4$ hydrolase, or by 12- or 15-lipoxygenases to form either LTB$_4$ or lipoxins. The inverse dose-dependent formation of LTB$_4$ and LXA$_4$ by the mesangial cell may be an important local defense mechanism of the mesangium. Thus, as the numbers of infiltrating leukocytes (and therefore possibly the concentration of PMN derived LTA$_4$ and LTB$_4$) increases, the synthesis of LXA$_4$ by mesangial cell will also increase. The sudden increase in the concentration of LXA$_4$ in the micro-environment of the mesangium could act to suppress (or otherwise influence) the pro-inflammatory actions of LTB$_4$, presumably by inhibiting the further migration of leukocytes and other inflammatory cells into the mesangium, and modifying or counteracting the cellular effects of LTB$_4$.

REFERENCES

1. Serhan, C.N., M. Hamberg, B. Samuelsson. 1984. Lipoxins: Novel series biologically active compounds formed from arachidonic acid in human leukocytes. Proc Natl Acad Sci USA 81:5335-5339.
2. Wong, P.Y.K., R. Hughes, B. Lam. 1985 Lipoxene: A new group of tri-hydroxypentaenes of EPA from porcine leukocytes. Biochem Biophys Res Comm. 126:763-772.
3. Serhan, C.N., K.G. Nicolaou, S.E. Webber, C.A. Veale, S.E. Dahlen, T.J. Punstinen and B. Samuelsson. 1986, Lipoxin A: Stero chemistry and biosynthesis. J. Biol Chem, 261: 16340-16345.
4. Ueda, N., C. Yokoyana, S. Yamamuto, B.J. Fitzsimmons, J. Rokach, J.A. Oates, and A.R. Brash. 1987, Lipoxin synthesis by arachidonate 12 lipoxygenase purified from porcine leukocytes. Biochem Biophys Res Commun 149:1063-1069.
5. Samuelsson, B., S.E. Dahlen, J.A. Lindgren, C.A. Rouzer, C.N. Serhan. 1987. Leukotrienes and Lipoxins: Structures, Biosynthesis and Biological Effects. Sciences 237, 1171-1176.
6. Grandordy, B.M., H. Lacroix, E. Mavoungou, S. Krilis, A.E.G. Crea, B.W. Spur and T.H. Lee, 1990. Lipoxin A$_4$ inhibits phosphoinsitol hydrolysis in human neutrophils. Biochem. Biophys. Res. Commun. 167,1022-1029.
7. Badr, K.F., D.K. DeBoer, M. Schwartzberg, C.N. Serhan 1989. Lipoxin A$_4$ antagonizes cellular and in vivo actions of Leukotriene D$_4$ in rat glomerular mesangial cells: Evidence for competition of a common receptor. Proc. Natl. Acad. Sci. USA. 86, 3438-3442.
8. Badr, K.F., C.N. Serhan, K.C. Nicolaou, and B. Samuuelsson, 1987. The action of Lipoxin A in glomerular microciculatory dynamics in the rat. Biochem Biophys Res Commun. 145:408-914.
9. Brady, H.R., U.Perssan, B. Ballermann, B.Brenner, and C. N. Serhan, 1990. Leukotrienes stimulate neutrophil adhension to mesangial cells: modulation with lipoxins, Amer. J. Physiol. 259. F 809-815.
10. Serhan,C.N. 1989. The generation of Lipoxins and novel related compounds by human neutrophils: relationship to leukotriene production. in Adv. in Prost., Throm.,and Leukotriene Research, Vol. 19, Ed. by Samuelsson, B. Wong,P Y-K, and Sun, F.F., Raven Press, N.Y. P. 116-119.

11. Garrick, R., Shen S.Y., Ogunc, S., and P.Y.K. Wong. 1989. Transformation of leukotriene A_4 to lipoxins by rat kidney mesangial cells. Biochem Biophys Res Comm 162:626-633.
12. Serhan, C.N., and K. Sheppard, 1990. Lipoxin formation during human neutrophil-platelets interactions : evidence for the transformation of leukotriene A_4 by platelet 12-lipoxygenase in vitro. J. Clin. Invest. 85, 772-780.
13. Feinmark, S.J. and P.J. Cannon. 1986. Endothelial cell leukotriene C_4 synthesis results from intercellular transfer of leukotriene A_4 synthesis by polymorphonuclear leukocytes. J. Biol. Chem. 261, 16466-16472.
14. Edenius, C., J. Haeggstrom, and J.A. Lindgren. 1988. Transcellular conversion of endogenous arachidonic acid to lipoxins in mixed human platelet-granulocyte suspensions. Biochem Biophys Res Commun 157: 801-807.
15. Imai,E., R. Hoover, N. Makita, C. Funk and K.F. Badr, 1990. Abundance of 5-lipoxygenase, 15-lipoxygenase, 12-lipoxygenase and leukotriene A_4 hydrolase gene expression in cultured glomerular cells, Amer.J. Soc. Nephrol., 1: 443a.
16. Yokoyama, C., F. Shinijo, T. Yoshimoto, S. Yamamoto, J.A. Oates, and A.R. Brash, 1986. Arachidonate 12-lipoxygenase purified from porcine leukocytes by immunoaffinity chromatography and its activity with hyperoxyeicosatetraenoic acids, J. Biol. Chem. 261, 16714-16721.

CONTRIBUTORS

Dr. Kamal F. Badr
Division of Nephrology
Department of Medicine
Vanderbilt University
Nashville, TN 37232

Dr. John A. Badwey
Department of Cell Physiology
Boston Biomedical Research Institute
Department of Biological Chemistry
and Molecular Pharmacology
Harvard Medical School
Boston, MA 02115

Dr. Timothy D. Bigby
Department of Medicine
University of California and
Veterans Administration Medical
Center
San Diego, CA 92161

Kimberly A. Birch
Department of Pathology
Brigham and Women's Hospital and
Harvard Medical School
Boston, MA 02115

Dr. Kerry L. Blacker
Department of Medicine
Veterans Administration Medical
Center and University of California
San Francisco, CA 94121

Dr. Pierre Borgeat
Unité de Recherche Inflammation,
Immunologie et Rhumatologie
Centre de recherche du CHUL
Québec, G1V 4G2
Canada

Dr. Hugh Brady
Renal Division
Department of Medicine
Brigham and Women's Hospital and
Harvard Medical School
Boston, MA 02115

Dr. Barry M. Brenner
Renal Division
Department of Medicine
Brigham and Women's Hospital and
Harvard Medical School
Boston, MA 02115

Dr. Mark E. Brezinski
Hematology Division
Brigham and Women's Hospital and
Harvard Medical School
Boston, MA 02115

Dr. Johanne Carrier
Department of Pharmacology
Faculty of Medicine
University of Sherbrooke
Sherbrooke (P.Q.) Canada J1H 5N4

Dr. Hans-Erik Claesson
Department of Physiological
Chemistry II
Karolinska Institutet
S-104 01 Stockholm
Sweden

Dr. Sven-Erik Dahlén
Department of Physiology I and
Institute of Environmental Medicine
Karolinska Institutet
Stockholm, Sweden

Mark D. Denton
Renal Division
Department of Medicine
Brigham and Women's Hospital and
Harvard Medical School
Boston, MA 02115

Dr. Jiabing Ding
Department of Cell Physiology
Boston Biomedical Research Institute
Boston, MA 02114

Dr. Charlotte Edenius
Department of Physiological
Chemistry
Karolinska Institutet
Box 60 400
S-104 01 Stockholm, Sweden

Dr. Bruce M. Ewenstein
Division of Hematology
Brigham and Women's Hospital and
Harvard Medical School
Boston, MA 02115

Dr. Giora Feuerstein
Department of Cardiovascular
Pharmacology
Smith Kline Beecham Pharmaceuticals
King of Prussia, PA 19406

Dr. Stefano Fiore
Hematology Division
Brigham and Women's Hospital and
Harvard Medical School
Boston, MA 02115

Dr. Inger Forsberg
Department of Physiological
Chemistry
Karolinska Institutet
Box 60 400
S-104 01 Stockholm, Sweden

Dr. Renee Garrick
Department of Medicine
New York Medical College
Valhalla, NY 10595

Dr. Marc E. Goldyne
Department of Dermatology
Veterans Administration Medical
Center and University of California
San Francisco, CA 94121

Dr. Julian Gomez-Cambronero
Department of Physiology
University of Connecticut
Health Center
Farmington, CT 06030

Dr. Jesper Z. Haeggström
Department of Physiological
Chemistry II
Karolinska Institutet
S-104 01 Stockholm, Sweden

Dr. Joseph R. Hageman
Evanston Hospital and Northwestern
University Medical School
2650 Ridge Avenue
Evanston, IL 60201

Dr. David P. Hajjar
Departments of Pathology and
Biochemistry
National Institutes of Health
Specialized Center of Research in
Thrombosis
Cornell University Medical College
New York, NY 10021

Dr. Annie Hallée
Department of Pharmacology
Faculty of Medicine
University of Sherbrooke
Sherbrooke (P.Q.) Canada J1H 5N4

Dr. Per Hedqvist
Department of Physiology I
Institute of Environmental Medicine
Karolinska Institutet
Stockholm, Sweden

Dr. Katarina Heidvall
Department of Physiological
Chemistry
Karolinska Institutet
Box 60 400
S-104 01 Stockholm, Sweden

Dr. Peter M. Henson
National Jewish Center for
Immunology and Respiratory Medicine
1400 Jackson Street
Denver, CO 80206

Dr. Paul G. Heyworth
Department of Molecular and
Experimental Medicine
Research Institute of Scripps Clinic
La Jolla, CA

Dr. Ronald B. Holtzman
Evanston Hospital and Northwestern
University Medical School
2650 Ridge Avenue
Evanston, IL 60201

Brian C. Jacobson
Division of Hematology
Brigham and Women's Hospital
Boston, MA 02115

Der. Per-Johan Jakobsson
Department of Physiological
Chemistry II
Karolinska Institutet
S-104 01 Stockholm, Sweden

Dr. Stephen J. Lane
Department of Allergy and Allied
Respiratory Disorders
U.M.D.S.
Guy's Hospital, London SE1 9RT, U.K.

Dr. Johanne Laporte
Department of Pharmacology
Faculty of Medicine
University of Sherbrooke
Sherbrooke (P.Q.) Canada J1H 5N4

Dr. Tak H. Lee
Department of Allergy and Allied
Respiratory Disorders
U.M.D.S.
Guy's Hospital
London SE1 9RT, U.K.

Dr. Lennart Lindbom
Departments of Immunology and
Physiology
Karolinska Institutet, Stockholm and
Inflammation Research, Pharmacia,
Uppsala, Sweden

Dr. Jan Åke Lindgren
Department of Physiological
Chemistry
Karolinska Institutet
Box 60 400
S-104 01 Stockholm, Sweden

Dr. Claes Lundberg
Departments of Immunology and
Physiology
Karolinska Institutet, Stockholm and
Inflammation Research, Pharmacia,
Uppsala, Sweden

Dr. Jacques Maclouf
INSERM Unit 150, Hopital
Lariboisiere
75475 Paris Cedex 10, France

Dr. James L. Madara
Division of Gastrointestinal
Pathology
Department of Pathology
Brigham and Women's Hospital and
Harvard Medical School
Boston, MA 02115

Dr. Karim Maghni
Department of Pharmacology
Faculty of Medicine
University of Sherbrooke
Sherbrooke (P.Q.) Canada J1H 5N4

Dr. Juan F. Medina
Department of Physiological
Chemistry II
Karolinska Institutet
S-104 01 Stockholm, Sweden

Dr. Robert C. Murphy
National Jewish Center for
Immunology
1400 Jackson Street
Denver, CO 80206

Dr. Shirin Nash
Division of Gastrointestinal
Pathology
Department of Pathology
Brigham and Women's Hospital and
Harvard Medical School
Boston, MA 02115

Dr. Barbro Näsman-Glaser
Department of Physiological
Chemistry
Karolinska Institutet
Box 60 400
S-104 01 Stockholm, Sweden

Dr. Santosh Nigam
Eicosanoid Research, Department of
Gynaecological Endocrinology
Free University of Berlin
Germany

Dr. Björn Odlander
Department of Physiological
Chemistry II
Karolinska Institutet
S-104 01 Stockholm, Sweden

Dr. Cecil R. Pace-Asciak
Research Institute
Hospital for Sick Children
555 University Avenue
Toronto, Canada M5G 1X8

Dr. Remi Palmantier
Unité de Recherche Inflammation,
Immunologie et Rhumatologie
Centre de recherche du CHUL
Québec, G1V 4G2, Canada

Dr. Ulla Palmertz
Department of Physiology I and
Institute of Environmental Medicine
Karolinska Institutet
Stockholm, Sweden

Dr. Charles Parkos
Division of Gastrointestinal
Pathology
Department of Pathology
Brigham and Women's Hospital and
Harvard Medical School
Boston, MA 02115

Dr. Manuel Patarroyo
Departments of Immunology and
Physiology
Karolinska Institutet, Stockholm and
Inflammation Research, Pharmacia,
Uppsala, Sweden

Dr. Kenneth B. Pomerantz
Department of Medicine
National Institutes of Health
Specialized Center of Research in
Thrombosis
Cornell University Medical College
New York, NY 10021

Dr. Reuven Rabinovici
Department of Surgery
Jefferson Medical College
Philadelphia, PA 19107

Dr. Olof Rådmark
Department of Physiological
Chemistry II
Karolinska Institutet
S-104 01 Stockholm, Sweden

Dr. Johan Raud
Department of Physiology I and
Institute of Environmental Medicine
Karolinska Institutet
Stockholm, Sweden

Dr. Chantal Robidoux
Department of Pharmacology
Faculty of Medicine
University of Sherbrooke
Sherbrooke (P.Q.) Canada J1H 5N4

Dr. John M. Robinson
Department of Cell Biology,
Neurobiology and Anatomy
Program in Molecular, Cellular and
Developmental Biology
Ohio State University
Columbus, OH

Dr. Marek Rola-Pleszczynski
Immunology Division
Department of Pediatrics
Faculty of Medicine
University of Sherbrooke
Sherbrooke QC, Canada, J1H 5N4

Dr. Charles N. Serhan
Hematology Division
Brigham and Women's Hospital and
Harvard Medical School
Boston, MA 02115

Dr. Ramadan I. Sha'afi
Department of Physiology
University of Connecticut
Health Center
Farmington, CT 06030

Kelly-Ann Sheppard
Hematology Division
Brigham and Women's Hospital
Boston, MA 02115

Dr. Pierre Sirois
Department of Pharmacology
Faculty of Medicine
University of Sherbrooke
Sherbrooke (P.Q.) Canada J1H 5N4

Dr. Leif Stenke
Department of Physiological
Chemistry
Karolinska Institutet
Box 60 400
S-104 01 Stockholm, Sweden

Dr. Arnold Stern
New York University Medical Center
Department of Pharmacology
550 First Avenue
New York, NY 10016

Dr. Susanne Tornhamre
Department of Physiological
Chemistry
Karolinska Institutet
Box 60 400
S-104 01 Stockholm, Sweden

Dr. Jerome Vernick
Department of Surgery
Jefferson Medical College
Philadelphia, PA 19107

Dr. Anders Wetterholm
Department of Physiological
Chemistry II
Karolinska Institutet
S-104 01 Stockholm, Sweden

Dr. James R. Wilkinson
Department of Allergy and Allied
Respiratory Disorders
U.M.D.S.
Guy's Hospital, London SE1 9RT, U.K.

Dr. Mary L. Williams
Department of Pediatrics
Veterans Administration Medical
Center and University of California
San Francisco, CA 94121

Dr. Patrick Y-K Wong
Departments of Physiology and
Medicine
New York Medical College
Valhalla, NY 10595

Dr. Tian Li Yue
Department of Cardiovascular
Pharmacology
Smith Kline Beecham Pharmaceuticals
King of Prussia, PA 19406

INDEX

Electron microscopy (continued)
 Clara cell, 295
 Foà-Kurloff cells, 295
 lung cells, 252
 platelet-neutrophil adherence, 92
 transmigration of neutrophils, 332
 type II pneumocytes, 295
 Weibel-Palade body, 145
Endothelial cells, 6, 8, 9, 47, 123, 141-143, 145, 146-148, 159, 163, 164, 171, 186, 189, 213, 221, 252, 291, 300, 308, 339
 adhesion of leukocytes, 6, 186
 and biosynthesis of LTC$_4$, 308
 de-endothelialized area, 159
 effects of lipoxins, 123
 eicosanoids, 164
 endothelial leukocyte adhesion molecule 1 (ELAM-1), 8, 145
 injury, 9, 171
 monocyte-endothelial cell cultures, 308
 proliferation, 164
 protein kinase C, 146, 171
 secretion from, 141
 stimulus-response coupling, 146-148
 vWF, 141-143
Endothelial cell-derived relaxing factor (EDRF), 159
Endothelial-leukocyte adhesion molecule 1 (ELAM-1), see Endothelial cells
Endothelins, 20
Endotoxemia, 193, 194, 197, 198, 200
 endotoxin, 194, 197, 198, 200
 biological responses, 194
Enzyme immunoassay (EIA), 292
Eosinophil, 41, 227, 230, 269, 291, 292, 294, 299
 alveolar, 291
 margination of, 230
Epithelial cells, 184, 238, 239, 251, 260, 291, 329, 333, 340
 airway, 238, 239
 alveolar, 251, 260, 291
  type I, 251
  type II, 260, 291
 culture of, 239
 human intestinal, 184, 329
 human tracheal, 238
 neutrophil interactions, 333, 329

Epoxide hydrolase, 245, 297, 301, 307, 308, 310, 312, 314, 341, 367
 different functions, 314
 distribution, 312
 LTA$_4$ hydrolase, 245, 297, 301, 307, 310, 341, 367
 purified, 308
 zinc binding properties, 314
Epstein-Barr virus (EBV), 5
 EBV immortalized B cells, 5
Erythrocytes, 103, 105, 150, 244, 245
 neutrophil interactions, 103, 105
 sickle cell, 150
Erythropoietin, 41

Fibroblasts, 227, 312, 317, 319, 320
 3T3M, 317
 dermal, 320
 eicosanoid generation, 319
Foà-Kurloff cells, 291
FPL 55712, 259
 leukotriene receptor antagonist, 259

G proteins, 50, 125-126, 209, 225
 GTP binding regulatory proteins, 50, 209
 in lipoxin induced responses, 125-126
Gastrointestinal tract, 228, 229, 335
 effects of PAF on, 229
GC/MS, 241, 363 (see also Negative ion chemical ionization GC/MS)
 analysis of trimethylsilyl ester methyl ester derivatives, 241
Glomerular filtration rate (GFR), 337
Glomeruli, 228, 335, 343, 347
 injury to, 335, 343
 nephritis, 347
Glutathione-S-transferase, 122, 134, 235, 243, 298, 340
 in hepoxilin biosynthesis, 134
 in regulating lipoxin formation, 122
 activity in platelets, 122
Glycopolypeptides, 4
 GP90, 4
 GP130, 4
 GP155, 4
 GP160, 4
Granulocyte, 269, 272, 280

Granulocyte macrophage colony
          stimulating factor (GM-
          CSF), 20, 35-41, 44, 50,
          74, 79, 114, 214, 272
    biological actions of, 41
    definition of, 35
    effects on chemotaxis, 44
    gene, 37
    leukotriene generation, 74, 79
    in lipoxin formation, 114
    origins of, 36
    and PAF, 214
    protein synthesis, 50
    receptor, 38-40
Growth factors, 161, 162, 172, 339
    signal transduction, 172
GTP-ase, 209
N-(2-guanidinoethyl)-5-
          isoquinoline sulfonamide
          (HA1004), 25

Hamster cheek pouch, 186, 189
    effects of lipoxins, 189
    LTB$_4$ induced inflammation, 186
Heart, 228, 230
    coronary occlusion, 230
Hemorrhagic shock, 13
Hepoxilins, 133-138
    metabolism of, 134
HPLC, 76, 78, 79, 94, 104, 240,
          284, 292, 298, 364, 365
    analysis of platelets and
          neutrophils, 78, 94
    extracts from 9/HTEo⁻ cells
          incubated with LTA$_4$, 240
    lipoxin formation by nasal polyp
          tissue, 284
    photodiode array UV detector,
          79, 104
    reverse phase
      conversion of LTA$_4$ by
          mesangial cells, 365
      metabolites released by guinea
          pig alveolar eosinophils,
          298
Hydrocortisone, 273
Hydrogen peroxide, see Active
          oxygen species
Hydroxy-eicosatetraenoic acid
          (HETE), 23, 73, 76, 79,
          112, 115, 117, 119, 227,
          237, 283, 292, 302, 335,
          341
    5-HETE, 23, 227, 237, 302, 335
      priming by, 23
      role in superoxide generation,
          23
    12-HETE, 73, 237, 283, 335, 341
      ω-oxidation, 79
    15-HETE, 76, 283, 335
      acylation-deacylation, 115

Hydroxy-eicosatetraenoic acid
          (continued)
    15-HETE (continued)
      anti-inflammatory actions, 117
      formation, 112
      impact on aggregation, 119
      priming with, 117
Hydroperoxyeicosatetraenoic acids
          (HPETE), 205
    12-HPETE, 205
Hyperchlorous acid (HOCl), 103
Hyperoxia, 257, 263
    newborn, 263

Immunoglobulin (Ig), 227, 269, 310
    IgE, 269
    IgG, 227
    production in B cells, 310
Incubation procedures, 75
Indomethacin, 213, 255, 283
Inflammatory reactions, 14, 109,
          160, 209, 213, 226, 269
    chalones, 109
    inhibition by monoclonal
          antibodies to leukocyte
          adhesion molecules, 14
    milieu, 160
    origins of PAF, 226
    proinflammatory, 269
Integrin, 4
    cytoadhesion, 4
    leu CAM's, 4
    VLA proteins, 4
Interferon-gamma (INFγ), 10, 20,
          207, 339
Internal standards, 75, 298
    19-hydroxy-prostaglandin B$_2$, 75,
          298
    prostaglandin B$_2$, 75, 298
Ionophore, 21, 117, 227, 260, 273,
          289, 335
    A$_{23187}$, 21, 117, 227, 260, 273,
          289, 335
Ischemia, 230

Keratinocytes, 317, 319, 320
    basal, 320
    eicosanoid generation, 319
Kidney, 228, 330-338
    allograft, 228

Leu-CAM's, 2, 4, 5, 9, 50
    CD11/CD18, 2, 4, 5, 50
      monoclonal antibody to, 2
    regulation of, 9
Leukemia, 42, 212, 283
    chronic myelocytic, 283
    leukemia cell line (HL60), 212
    myeloblastic, 42
Leukocyte, 4, 9, 11, 19, 43, 45,
          255

Lipoxins (continued)
  role of G proteins in
      activation, 125-126
  role of 12-lipoxygenase, 121
  transcellular formation, 285
Lipoxygenase (LO)
  5-LO, 73, 76, 84, 104, 211, 235,
      237, 261, 301, 307, 313,
      335, 339, 361, 367
    activity, 84
    analysis of products, 76
    distribution of, 313
    effects of red blood cells,
        104
    5-LO activating protein
        (FLAP), 301
    monocyte, 237
    products in BAL fluids, 261
    transcellular metabolism, 307
        (see also Cell-cell
        interaction)
  12-LO, 74, 121, 133, 292, 342,
      361
    inhibition of, 74
    products, 292
    role in hepoxilin formation,
        133
    role in lipoxin formation, 121
    vasoactive, 342
  15-LO, 339, 361
Lipoxygenase inhibitors, 206, 209,
    229, 256-257
  corticosteroids, 229
  effects on cytokine production,
      206
  5-LO inhibitor, 256-257
  AA861, 256-257
Lung, 199, 228, 235, 251-253, 261,
    265, 272, 284, 294
  cell population, 294
  injury, 199, 251, 261
  LPS induced lung injury, 199
  metabolism of arachidonic acid,
      284
  microscopy, 252-253
  oxygen toxicity, 265
Lymphocyte, 1, 7-9, 206, 207, 213,
    310
  aggregation, 1
  differentiation, 310
  homing, 8
  LTA$_4$ hydrolase, 310
  lymphoblast antigen (LB-2), 7
  proliferation, 9

Macrophage, 159, 167, 193, 197,
    209, 211, 227, 256, 261,
    269, 272, 291, 292, 294,
    299, 302, 339, 362
  alveolar, 209, 211, 256, 269,
      272, 292, 302

Macrophage (continued)
  eicosanoid biosynthesis, 167
  electron micrograph, 261
Mass spectrometry, 111, 112 (see
    also GC/MS)
Mast cells, 269, 271
Megakaryocyte, 41
Mesangial cells, 336, 337, 340,
    347, 361
  binding to LTD$_4$, 337
Mezerein, 21, 22, 26
Mitomycin C, 324
Monocytes, 5, 12, 171, 197, 227,
    236, 271, 308, 310, 311
  adhesion of, 5, 171
  monocyte-endothelial cell
      cultures, 308
  released LTA$_4$, 310
Muramyl dipeptide (MDP), 211
Mycobacterium tuberculosis, 13
Myeloperoxidase, 103
Myeloproliferative disorders, 283

Na$^+$/H$^+$ exchanger, see
    sodium$^+$/hydrogen$^+$
    exchanger
Natural killer cells (NK), 300
Negative ion chemical ionization
    (NICI) GC-MS, 111, 112
Nephritis, 228, 338, 347
  glomerular, 347
Neutrophils, 19, 21, 44, 47, 53,
    91, 103, 105, 113, 121-
    125, 135-137, 200, 205,
    227, 271, 313, 329, 330,
    333, 335, 342, 347, 350,
    362
  chemotaxis, 44
  depletion, 342
  derived secretagogue (NDS), 333
  epithelial cell interactions,
      329, 333
  glomerular cell interaction, 347
  GM-CSF treated, 47, 53
  [111]indium-labeled, 350
  oxidase system, 19
  neutrophil mediated injury, 200
  platelet-neutrophil interaction,
      91, 113
  red blood cell interactions,
      103, 105
  responses to hepoxilins, 135-137
  responses to lipoxins, 121-125
  sodium$^+$/hydrogen$^+$ exchanger, 47
  synergistic stimulation, 21
  transepithelial resistance, 330
  transmigration, 330

Opsonin, 270
Oxygen, see Active oxygen species

382